Digital Course Materials

for

Understanding Human Communication

Fourteenth Edition

RONALD B. ADLER
GEORGE RODMAN
ATHENA DU PRÉ

Carefully scratch off the silver coating (e.g., with a coin) to see your personal redemption code.

This code can be used only once and cannot be shared.

Once the code has been revealed, this access card cannot be returned to the publisher. Access can also be purchased online during the registration process.

The code on this card is valid for two years from the date of first purchase. Complete terms and conditions are available at **oup-arc.com**

Access Length: 6 months from redemption of the code.

Your OUP digital course materials can be delivered several different ways, depending on how your instructor has elected to incorporate them into his or her course.

BEFORE REGISTERING FOR ACCESS, be sure to check with your instructor to ensure that you register using the proper method.

VIA YOUR SCHOOL'S LEARNING MANAGEMENT SYSTEM

Use this method if your instructor has integrated these resources into your school's Learning Management System (LMS)—Blackboard, Canvas, Brightspace, Moodle, or other

> Log in to your instructor's course within your school's LMS.

> When you click a link to a resource that is access-protected, you will be prompted to register for access.

> Follow the on-screen instructions.

> Enter your personal redemption code (or purchase access) when prompted on the checkout screen.

VIA OUP DASHBOARD

Use this method only if your instructor has specifically instructed you to enroll in a Dashboard course. **NOTE**: *If your instructor is using these resources within your school's LMS, use the Learning Management System instructions.*

> Visit **register.dashboard.oup.com** and select your textbook.

> Follow the on-screen instructions to identify your specific course section.

> Enter your personal redemption code (or purchase access) when prompted on the checkout screen.

> Once you complete your registration, you are automatically enrolled in your Dashboard course.

*For assistance with code redemption, Dashboard registration, or if you redeemed your code using the wrong method for your course, please contact our customer support team at **dashboard.support@oup.com** or 855-281-8749.*

OXFORD
UNIVERSITY PRESS

understanding
human
communication

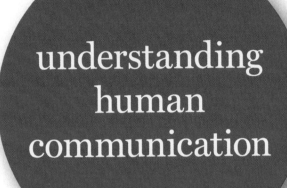

understanding human communication

FOURTEENTH EDITION

Ronald B. Adler   SANTA BARBARA CITY COLLEGE

George Rodman   BROOKLYN COLLEGE, CITY UNIVERSITY OF NEW YORK

Athena du Pré   UNIVERSITY OF WEST FLORIDA

OXFORD   NEW YORK
OXFORD UNIVERSICE PRESS

Oxford University Press is a department of the University of Oxford.
It furthers the University's objective of excellence in research, scholarship,
and education by publishing worldwide. Oxford is a registered trademark of
Oxford University Press in the UK and certain other countries.

Published in the United States of America by Oxford University Press
198 Madison Avenue, New York, NY 10016, United States of America.

© 2020, 2017, 2014, 2012, 2009, 2006, 2003 by Oxford University Press

Library of Congress Cataloging-in-Publication Data

Names: Adler, Ronald B. (Ronald Brian), 1946- author. | Rodman, George R.,
    1948- author. | du Pré, Athena, author.
Title: Understanding human communication / Ronald B. Adler, George Rodman,
    Athena du Pré.
Description: Fourteenth edition. | Oxford ; New York : Oxford University
    Press, 2019. | Includes bibliographical references and index.
Identifiers: LCCN 2019034688 (print) | LCCN 2019034689 (ebook) | ISBN
    9780190925697 (paperback) | ISBN 9780190925703 (epub)
Subjects: LCSH: Communication. | Interpersonal communication.
Classification: LCC P90 .A32 2019 (print) | LCC P90 (ebook) | DDC
    302.2—dc23
LC record available at https://lccn.loc.gov/2019034688
LC ebook record available at https://lccn.loc.gov/2019034689

9 8 7 6 5 4 3 2 1
Printed by LSC Communications, United States of America

Brief Contents

Preface *xv*
About the Authors *xxiv*

PART ONE FUNDAMENTALS OF HUMAN COMMUNICATION

1. Communication: What and Why   3
2. Communicating with Social Media   25
3. The Self, Perception, and Communication   51
4. Communication and Culture   77

PART TWO COMMUNICATION ELEMENTS

5. Language   101
6. Listening   127
7. Nonverbal Communication   155

PART THREE INTERPERSONAL COMMUNICATION

8. Understanding Interpersonal Communication   181
9. Managing Conflict   211

PART FOUR COMMUNICATION IN GROUPS, TEAMS, AND ORGANIZATIONS

10. Communicating for Career Success   239
11. Leadership and Teamwork   263

PART FIVE PUBLIC COMMUNICATION

12. Preparing and Presenting Your Speech   293
13. Speech Organization and Support   319
14. Informative Speaking   347
15. Persuasive Speaking   371

Notes *N-1*
Glossary *G-1*
Credits *C-1*
Index *I-1*

Contents

Preface *xv*
About the Authors *xxiv*

PART ONE FUNDAMENTALS OF HUMAN COMMUNICATION

1 Communication: What and Why 3

1.1 Characteristics of Communication 5
  Definition of Communication 5
  Models of Communication 8

1.2 Types of Communication 10
  Intrapersonal Communication 10
  Dyadic and Interpersonal Communication 11
  Small-Group Communication 11
  Organizational Communication 11
  Public Communication 12
  Mass Communication 12
  Social Media Communication 12

1.3 Communication in a Changing World 13
  Changing Technology 13
  Changing Discipline 15

1.4 Communication Competence 15
  There Is No "Ideal" Way to Communicate 16
  Competence Is Situational 16
  Competence Is Relational 16
  Competent Communicators Are Empathic 16
  Competence Can Be Learned 17
  Competence Requires Hard Work 17
  Competent Communicators Self-Monitor 17
  Competent Communicators Are Committed 18

1.5 Misconceptions About Communication 19
  Myth: Communication Requires Complete
    Understanding 19
  Myth: Communication Can Solve All Problems 19
  Myth: Communication Is a Good Thing 19
  Myth: Meanings Are in Words 20
  Myth: Communication Is Simple 20
  Myth: More Communication Is Always Better 20

Making the Grade 21
Key Terms 22
Activities 22

FEATURES

Understanding Diversity *The Limits of Language When
  Describing People* 7

@Work *Communication Skills and Career Success* 12

Understanding Communication Technology *Dear Social
  Media, I Need a Little Space* 14

Understanding Your Communication *What Type
  of Communicator Are You?* 18

Ethical Challenge *To Communicate or Not to
  Communicate?* 20

2 Communicating with Social Media 25

2.1 The Roles of Social
    and Mass Media 26
  Characteristics of Social Media 27
  Social Media Uses and Gratifications 27
  Masspersonal Communication 28

2.2 Mediated Versus Face-to-Face
    Communication 29
  Message Richness 29
  Synchronicity 32
  Permanence 32

2.3 Benefits and Drawbacks of Social Media 33
  Benefits of Social Media 33
  Drawbacks of Social Media 36

2.4 Influences on Mediated Communication 40
  Gender 40
  Age 41

2.5 Communicating Competently with Social
    Media 43

Maintaining Positive Relationships **43**
Protecting Yourself **45**

Making the Grade **47**

Key Terms **48**

Activities **49**

FEATURES

■ Understanding Your Communication *What Type of Social Media Communicator Are You?* **31**

▨ Ethical Challenge *How Honest Should You Be on Social Media?* **39**

■ @Work *Using LinkedIn for Career Success* **42**

⊘ Checklist *Use Social Media Courteously and Wisely* **43**

⊘ Checklist *Evaluating Online (Mis)information* **46**

3 The Self, Perception, and Communication **51**

🔎 3.1 Communication and the Self **52**
Self-Concept Defined **52**
Biology, Personality, and the Self **53**
External Influence on the Self-Concept **54**
Culture and the Self-Concept **55**
Self-Fulfilling Prophecies **56**

🔎 3.2 Perceiving Others **57**
Steps in the Perception Process **57**
Influences on Perception **59**
The Power of Narratives **61**
Common Perceptual Tendencies **62**
Empathy **64**

🔎 3.3 Communication and Identity Management **67**
Public and Private Selves **67**
Identity Management and Social Media **68**
Characteristics of Identity Management **69**
Identity Management in the Workplace **72**
Why Manage Identities? **72**

Identity Management and Honesty **73**

Making the Grade **74**

Key Terms **75**

Activities **75**

FEATURES

▨ Ethical Challenge *Looking Beyond Stereotypes* **55**

■ Understanding Diversity *Sexist Assumptions in Everyday Language* **60**

⊘ Checklist *Avoiding Stereotypes* **63**

⊘ Checklist *Check Your Perceptions Before Responding* **65**

■ Understanding Your Communication *How Emotionally Intelligent Are You?* **66**

▨ Understanding Communication Technology *Tweet from the Past: Forgivable or Inexcusable?* **69**

■ @WORK *Humblebragging in Job Interviews* **73**

4 Communication and Culture **77**

🔎 4.1 Understanding Cultures and Cocultures **78**
Differences and Similarities **79**
Salience of Differences **79**

🔎 4.2 How Cultural Values and Norms Shape Communication **80**
Individualism and Collectivism **81**
High and Low Context **83**
Uncertainty Avoidance **84**
Power Distance **85**
Talk and Silence **86**
Competition and Cooperation **86**

🔎 4.3 Cocultures and Communication **86**
Race and Ethnicity **87**
Regional Differences **87**

Gender Identity **88**
Sexual Orientation **88**
Religion **89**
Political Viewpoints **89**
Ability/Disability **90**
Age/Generation **90**
Socioeconomic Status **92**

🔎 4.4 Developing Intercultural Communication Competence **93**
Increased Contact **94**
Tolerance for Ambiguity **94**
Open-Mindedness **94**
Knowledge and Skill **97**
Patience and Perseverance **97**

Making the Grade 98

Key Terms 99

Activity 99

FEATURES

■ @Work *Power Distance and Culture in the Workplace* 85

⊘ Checklist *Discussing Politics Responsibly on Social Media* 90

■ Understanding Diversity *Communicating with People Who Have Disabilities* 91

■ @Work *Organizations Are Cultures, Too* 93

■ Ethical Challenge *Civility When Values Clash* 95

■ Understanding Your Communication *How Much Do You Know About Other Cultures?* 96

⊘ Checklist *Coping with Culture Shock* 98

PART TWO COMMUNICATION ELEMENTS

5 Language 101

🔵 5.1 The Nature of Language 102
　　Language Is Symbolic **103**
　　Meanings Are in People, Not in Words **103**
　　Language Is Rule Governed **104**

🔵 5.2 The Power of Language 107
　　Language *Shapes* Values, Attitudes, and Beliefs **107**
　　Language *Reflects* Values, Attitudes, and Beliefs **111**

🔵 5.3 Troublesome Language 114
　　The Language of Misunderstandings **114**

🔵 5.4 Disruptive Language 117
　　Confusing Facts and Inferences **118**
　　Presenting Opinions as Facts **119**
　　Personal Attacks **120**

🔵 5.5 Gender and Language 120

Making the Grade 124

Key Terms 125

Activities 125

FEATURES

■ Understanding Diversity *Language and Worldview* 108

■ @Work *What's in a Name?* 110

■ Ethical Challenge *Freedom of Speech* 111

■ Ethical Challenge *"Telling It Like It Is"* 116

⊘ Checklist *Avoiding Misunderstandings* 117

⊘ Checklist *Distinguishing Between Facts and Opinions* 120

⊘ Checklist *Choose Your Words Carefully* 121

■ Understanding Your Communication *How Do You Use Language?* 123

6 Listening 127

🔵 6.1 The Value of Listening 128

🔵 6.2 Misconceptions About Listening 129
　　Myth: Listening and Hearing Are the Same Thing **130**
　　Myth: Listening Is a Natural Process **132**
　　Myth: All Listeners Receive the Same Message **132**

🔵 6.3 Overcoming Challenges to Listening 133
　　Message Overload **133**
　　Noise **133**
　　Cultural Differences **134**

🔵 6.4 Faulty Listening Habits 135

🔵 6.5 Listening to Connect and Support 137
　　Be Sensitive to Personal and Situational Factors **137**
　　Allow Enough Time **138**
　　Ask Questions **138**
　　Listen for Unexpressed Thoughts and Feelings **138**
　　Encourage Further Comments **139**
　　Reflect Back the Speaker's Thoughts **140**
　　Consider the Other Person's Needs When Analyzing **141**
　　Reserve Judgment, Except in Rare Cases **141**

Think Twice Before Offering Advice **142**
Offer Comfort, If Appropriate **142**

🏢 6.6 Gender and Supportive Listening 143

🏢 6.7 Listening to Accomplish, Analyze,
  or Critique 144

Task-Oriented Listening **144**
Analytical Listening **147**
Critical Listening **148**

Making the Grade 151

Key Terms 152

Activities 152

FEATURES

 Understanding Communication Technology *Who Is Listening to You Online?* 131

✓ Checklist *Tips for Mindful Listening* 133

■ @Work *Multitasking Can Make You Less Productive* 134

✓ Checklist *Ways to Limit Social Media Distractions* 135

✓ Checklist *Techniques for Listening Nondefensively* 136

✓ Checklist *Temptations to Avoid When Asking Questions* 139

✓ Checklist *What to Consider Before Offering a Judgment* 142

✓ Checklist *Factors to Consider Before Offering Advice* 143

■ Ethical Challenge *The Good and Bad of Online Anonymity* 144

✓ Checklist *Ways to Offer Comfort* 144

✓ Checklist *How and When to Paraphrase* 146

✓ Checklist *Evaluating a Speaker's Message* 149

■ Understanding Your Communication *What Are Your Listening Strengths?* 150

7 Nonverbal Communication 155

🏢 7.1 Characteristics of Nonverbal
  Communication 157

Nonverbal Communication Is Unavoidable **158**
Nonverbal Communication Is Ambiguous **158**
Nonverbal Cues Convey Emotion **160**
Nonverbal Cues Influence Identities and
  Relationships **160**

🏢 7.2 Functions of Nonverbal Communication 161

Repeating **161**
Substituting **161**
Complementing **161**
Accenting **163**
Regulating **163**
Contradicting **163**
Deceiving **163**

🏢 7.3 Types of Nonverbal Communication 166

Body Movements **166**
Voice **168**
Appearance **168**
Touch **171**
Space **171**

🏢 7.4 Influences on Nonverbal Communication 174

Culture **175**
Gender **176**

Making the Grade 177

Key Terms 178

Activities 178

FEATURES

✓ Checklist *Three Ways to Convey Nonverbal Cues More Mindfully* 159

■ Understanding Your Communication *How Worldly Are Your Nonverbal Communication Skills?* 162

■ Understanding Communication Technology *Nonverbal Expressiveness Online* 164

■ @Work *Vocal Cues and Career Success* 169

■ Ethical Challenge *Appearance and Impression Management* 170

■ @Work *Touch and Career Success* 172

PART THREE INTERPERSONAL COMMUNICATION

8 Understanding Interpersonal Communication 181

8.1 Characteristics of Interpersonal Communication 182

What Makes Communication Interpersonal? **183**
How People Choose Relational Partners **183**
Content and Relational Messages **185**
Metacommunication **186**
Self-Disclosure **187**
Online Interpersonal Communication **189**

8.2 Communicating with Friends and Family 191

Unique Qualities of Friendship **191**
Friendship Development **191**
Gender and Friendship **193**
Family Relationships **195**

8.3 Communicating with Romantic Partners 197

Stages of Romantic Relationships **197**
Love Languages **201**
Male and Female Intimacy Styles **201**

8.4 Relational Dialectics 203

Connection Versus Autonomy **203**
Openness Versus Privacy **204**
Predictability Versus Novelty **204**

8.5 Lies and Evasions 205

Altruistic Lies **205**
Evasions **206**
Self-Serving Lies **206**

Making the Grade 207

Key Terms 208

Activities 208

FEATURES

☑ Checklist *Questions to Ask Yourself Before Self-Disclosing* 189
▪ Ethical Challenge *Is It Cheating?* 190
☑ Checklist *How to Be a Good Friend* 193
▪ Understanding Your Communication *What Kind of Friendship Do You Have?* 194
☑ Checklist *Strengthening Family Ties* 196
☑ Checklist *Meeting an Online Date for the First Time* 199
▪ Understanding Communication Technology *To End This Romance, Just Press "Send"* 200
▪ Understanding Your Communication *What Is Your Love Language?* 202

9 Managing Conflict 211

9.1 Understanding Interpersonal Conflict 213

Expressed Struggle **213**
Interdependence **213**
Perceived Incompatible Goals **213**
Perceived Scarce Resources **214**

9.2 Communication Climates 214

Confirming and Disconfirming Messages **215**
How Communication Climates Develop **217**

9.3 Conflict Communication Styles 220

Nonassertiveness **220**
Indirect Communication **221**
Passive Aggression **222**
Direct Aggression **222**
Assertiveness **223**

9.4 Negotiation Strategies 225

Win–Lose **225**
Lose–Lose **228**
Compromise **228**
Win–Win **228**

9.5 Social Influences on Conflict Communication 232

Gender and Conflict Style **232**
Cultural Approaches to Conflict **234**

Making the Grade 236

Key Terms 237

Activities 237

FEATURES

✓ Checklist *Rules for Fighting Fair* 216

▪ Understanding Communication Technology *Can You Hear Me Now?* 218

▪ Understanding Your Communication *What's the Forecast for Your Communication Climate?* 219

▪ Communication Technology *You Can't Take It Back* 220

▪ Ethical Challenge *"It's Nothing!"* 221

▪ @Work *Dealing with Sexual Harassment* 223

✓ Checklist *Protecting Yourself from an Abusive Partner* 223

▪ Understanding Your Communication *How Assertive Are You?* 226

▪ Ethical Challenge *Negotiating with a Bully* 227

▪ Understanding Diversity *They Seem to Be Arguing* 235

PART FOUR COMMUNICATION IN GROUPS, TEAMS, AND ORGANIZATIONS

⑩ Communicating for Career Success 239

🔹 10.1 Setting the Stage for Career Success 241

Developing a Good Reputation **241**
Managing Your Online Identity **241**
Cultivating a Professional Network **243**

🔹 10.2 Pursuing the Job You Want 244

Preparing Application Materials **244**
Planning for a Job Interview **247**
Participating in a Job Interview **249**

🔹 10.3 Organizational Communication Factors 252

Culture in the Workplace **253**
Patterns of Interaction **253**
Power in the Workplace **254**

🔹 10.4 Communicating in a Professional Environment 256

Avoiding Common Communication Mistakes **256**
Communicating Well as a Follower **257**
Communicating in a Professional Manner Online **259**

Making the Grade 260

Key Terms 261

Activities 261

FEATURES

✓ Checklist *Strategies to Meet Networking Prospects* 243

✓ Checklist *What to Include in a Cover Letter* 247

✓ Checklist *"What Is Your Greatest Weakness?"* 248

▪ Ethical Challenge *Responding to Illegal Interview Questions* 249

▪ Understanding Communication Technology *Interviewing by Phone or Video* 252

▪ Understanding Your Communication *How Good a Follower Are You?* 258

⑪ Leadership and Teamwork 263

🔹 11.1 Communication Strategies for Leaders 265

Characteristics of Effective Leaders **265**
Leadership Approaches **265**
Trait Theories of Leadership **266**
Situational Leadership **266**
Transformational Leadership **269**

🔹 11.2 Communicating in Groups and Teams 271

What Makes a Group a Team? **271**
Motivational Factors **272**

Rules in Small Groups **273**
Individual Roles **273**

🔹 11.3 Making the Most of Group Interaction 275

Enhance Cohesiveness **276**
Managing Meetings Effectively **276**
Using Discussion Formats Strategically **279**

🔹 11.4 Group Problem Solving 280

Advantages of Group Problem Solving **281**
Stages of Team Development **283**

A Structured Problem-Solving Approach **283**
Problem Solving in Virtual Groups **287**

Making the Grade **288**

Key Terms **290**

Activities **290**

FEATURES

☑ Checklist *Demonstrating Your Leadership Potential* 265

■ @Work *"I'll Do It Myself"—Or Should I?* 268

☑ Checklist *Working with a Difficult Boss* 269

■ Understanding Your Communication *What's Your Leadership Style?* 270

☑ Checklist *Getting Slackers to Do Their Share* 273

☑ Checklist *Dealing with Difficult Team Members* 275

☑ Checklist *Enhancing Group Productivity* 276

■ Ethical Challenge *Balancing Overly Talkative and Quiet Group Members* 277

☑ Checklist *Coping with Information Overload* 278

☑ Checklist *Making the Most of a Brainstorming Session* 279

■ @Work *The Power of Constructive Dialogue* 280

■ Understanding Diversity *Maximizing the Effectiveness of Multicultural Teams* 282

☑ Checklist *Stages in Structured Problem Solving* 285

■ Understanding Communication Technology *Developing Trust Long Distance* 288

PART FIVE PUBLIC COMMUNICATION

Preparing and Presenting Your Speech 293

12.1 Getting Started 295

Choosing Your Topic **295**
Defining Your Purpose **295**
Writing a Purpose Statement **295**
Stating Your Thesis **296**

12.2 Analyzing the Speaking Situation 297

The Listeners **297**
The Occasion **301**

12.3 Gathering Information 301

Online Research **302**
Library Research **302**
Interviewing **302**
Survey Research **303**

12.4 Managing Communication Apprehension 304

Facilitative and Debilitative Communication Apprehension **304**
Sources of Debilitative Communication Apprehension **304**
Overcoming Debilitative Communication Apprehension **306**

12.5 Presenting Your Speech 307

Choosing an Effective Type of Delivery **307**
Practicing Your Speech **308**

12.6 Guidelines for Delivery 308

Visual Aspects of Delivery **308**
Auditory Aspects of Delivery **310**

12.7 Sample Speech 312

Making the Grade **316**

Key Terms **316**

Activities **317**

FEATURES

■ Ethical Challenge *If I Adapt, Do I Lose My Integrity?* 299

☑ Checklist *Evaluating Websites* 302

■ Understanding Your Communication *Speech Anxiety Symptoms* 307

☑ Checklist *Practicing Your Presentation* 309

■ Ethical Challenge *Speaking Sincerely to Distasteful Audiences* 310

13 Speech Organization and Support 319

13.1 Structuring Your Speech 320
Your Working Outline **321**
Your Formal Outline **322**
Your Speaking Notes **324**

13.2 Principles of Outlining 324
Standard symbol **324**
Standard Format **325**
The Rule of Division **325**
The Rule of Parallel Wording **325**

13.3 Organizing Your Outline into a Logical Pattern 326
Time Patterns **326**
Space Patterns **327**
Topic Patterns **327**
Problem-Solution Patterns **328**
Cause-Effect Patterns **328**
Monroe's Motivated Sequence **329**

13.4 Beginnings, Endings, and Transitions 330
The Introduction **330**
The Conclusion **332**
Transitions **333**

13.5 Supporting Material 334
Functions of Supporting Material **334**
Types of Supporting Material **336**
Styles of Support: Narration and Citation **339**

13.6 Sample Speech 340

Making the Grade 344

Key Terms 345

Activities 345

FEATURES

Understanding Your Communication *Main Points and Subpoints* 326

Understanding Diversity *Nontraditional Patterns of Organization* 329

Checklist *Effective Conclusions* 333

@Work *Organizing Business Presentations* 335

Understanding Communication Technology *Plagiarism in a Digital Age* 337

14 Informative Speaking 347

14.1 Types of Informative Speaking 349
By Content **350**
By Purpose **350**

14.2 Informative Versus Persuasive Topics 351
Type of Topic **351**
Speech Purpose **351**

14.3 Techniques of Informative Speaking 351
Define a Specific Informative Purpose **351**
Create Information Hunger **353**
Make It Easy to Listen **353**
Use Clear, Simple Language **353**
Use a Clear Organization and Structure **354**

14.4 Using Supporting Material Effectively 355
Emphasizing Important Points **356**
Generating Audience Involvement **356**
Using Visual Aids **359**

Using Presentation Software **360**
Alternative Media for Presenting Graphics **361**
Rules for Using Visual Aids **362**

14.5 Sample Speech 363

Making the Grade 367

Key Terms 368

Activities 368

FEATURES

Understanding Your Communication *Are You Overloaded?* 349

Understanding Diversity *How Culture Affects Information* 352

Checklist *Techniques of Informative Speaking* 353

@Work *The Pros and Cons of Presentation Software* 363

15 Persuasive Speaking 371

🛈 **15.1 Characteristics of Persuasion 373**
Persuasion Is Not Coercive **373**
Persuasion Is Usually Incremental **373**
Persuasion Is Interactive **374**
Persuasion Can Be Ethical **374**

🛈 **15.2 Categorizing Persuasive Attempts 376**
By Type of Proposition **376**
By Desired Outcome **377**
By Directness of Approach **377**
By Type of Appeal: Aristotle's Ethos, Pathos, and Logos **378**

🛈 **15.3 Creating a Persuasive Message 379**
Set a Clear, Persuasive Purpose **379**
Structure the Message Carefully **380**
Use Solid Evidence **382**
Avoid Fallacies **382**

🛈 **15.4 Adapting to the Audience 385**
Establish Common Ground **385**
Organize According to the Expected Response **386**

Neutralize Potential Hostility **386**

🛈 **15.5 Building Credibility as a Speaker 386**
Competence **387**
Character **388**
Charisma **389**

🛈 **15.6 Sample Speech 389**

Making the Grade 394

Key Terms 395

Activities 395

FEATURES
Ethical Challenge *You Versus the Experts* 375
Understanding Diversity *Cultural Differences in Persuasion* 383
Understanding Your Communication *Persuasive Speech* 387
@Work *Persuasion in the World of Sales* 388

Notes *N-1*
Glossary *G-1*
Credits *C-1*
Index *I-1*

Preface

The case for learning about human communication is compelling. Consider the practical benefits: Effective communicators are more likely than others to be popular among their peers,[1] land the jobs they want,[2] succeed in their careers,[3] and be considered appealing friends[4] and romantic partners.[5]

Communication is more than a collection of techniques. *Understanding Human Communication* introduces readers to the scholarship that underlies everyday skills. We invite you and your students to look at any page of this book and ask: *Is the content important, clearly explained, and useful?*

Approach

This new edition builds on the successful approach that has served more than one million students. Rather than focusing solely on either skills or scholarship, *Understanding Human Communication* embraces the idea that each enhances the other. Reader-friendly content is up to date and clear without being simplistic. Real-life examples and engaging images make concepts interesting, clear, and relevant to students' lives.

New to This Edition

Updates in this edition reflect the changing world that *Understanding Human Communication* seeks to explain.

- **Expanded Coverage of Social Media.** Chapter 2 (new in this edition) is dedicated to social media, including the associated advantages and dangers, the masspersonal and hyperpersonal nature of social media, and the challenges of managing communication across a multitude of platforms. In addition, readers will find information about online communication throughout the book. Discussions include identity management through social media (Chapter 3), the influence of bots and trolls (Chapter 4), when to put technology aside (Chapter 6), the lack of inhibition online (Chapter 9), establishing a professional identity online (Chapter 10), and avoiding plagiarism in online research (Chapter 12).

- **Additional Strategies for Career Success.** Chapters 10 and 11 have been revised to extend general coverage of teamwork and leadership to organizational and workplace communication. Discussions include strategies for landing a desirable job, adapting to organizational cultures, and succeeding as a team member and leader. In addition, *@Work* boxes throughout the text offer tips on networking (Chapter 2), developing communication skills that pay off at work (Chapter 6), using vocal cues and touch in professional environments (Chapter 7), and dealing with sexual harassment in the workplace (Chapter 9).

- **Updated Discussions of Communication, Gender, and Culture.** Chapter 4 focuses exclusively on culture and communication, with updated explorations of race and ethnicity, regional differences, sexual orientation, gender identity, religion, physical abilities, age/generation, and socioeconomic status, as well as a new section on political viewpoints. The discussion of culture extends throughout the book, with topics such as traditional patterns in the way men and women communicate (Chapter 2), generational differences in social media use (Chapter 2), sexist assumptions in everyday language (Chapter 3), nongendered pronouns (Chapter 5), cultural listening styles (Chapter 6), nonverbal communication differences around the world (Chapter 7), organizational culture in the workplace (Chapter 10), the challenges and advantages of multicultural teamwork (Chapter 11), and examples of phenomena such as "mansplaining" in public speaking (Chapter 14).

- **"On Your Feet" speech activities.** New prompts incorporate public-speaking activities and integrate with **GoReact**, an interactive platform for video assignments.

- **Expanded Focus on Ethics.** "Ethical Challenge" sidebars throughout the book have been updated and expanded to explore honesty on social media (Chapter 2), the hidden influence of stereotypes (Chapter 3), the line between innocent online communication and digital infidelity (Chapter 8), negotiating with people who don't fight fair (Chapter 9), responding to illegal job interview questions (Chapter 10), and audience adaptation in a time of political polarity (Chapter 15).

- **Coverage of Media Literacy.** Given the importance of media literacy skills today, we have included new sections on evaluating the credibility, quality, and accuracy of online stories and information (e.g., Chapter 2 includes a section titled "Don't Believe Everything You See"; Chapter 4 presents a checklist on discussing politics responsibly on social media; and Chapter 5 includes a checklist on distinguishing between facts and opinions).

- **New and Expanded Coverage in Each Chapter.** Changes include the following:

  ○ Chapter 1 (**Communication: What and Why**) includes expanded coverage of the way communication functions in today's technological world and the sensitive nature of language when describing groups of people.

  ○ Chapter 2 (**Communicating with Social Media**) considers the unique qualities of mediated communication as well as strategies for using social media to enhance understanding and build relationships while avoiding social isolation, deception, and loss of privacy.

  ○ Chapter 3 (**The Self, Perception, and Communication**) includes new coverage of emotional intelligence, gender diversity, and identity management. A new "Ethical Challenge" feature invites readers to look below the surface of their fears and stereotypes.

  ○ Chapter 4 (**Communication and Culture**) includes expanded coverage of identity and intersectionality theory. The chapter features new strategies for engaging in respectful online discourse and coping with culture shock. It also introduces coverage of political viewpoints as a form of diversity.

  ○ Chapter 5 (**Language**) opens with a new segment on the use of nonbinary pronouns to reflect diverse gender identities. The chapter now includes a segment on the increased use of profanity and name calling in society, with insights from coordinated management of meaning and speech acts. Another new segment helps readers learn to distinguish between facts and opinions.

o Chapter 6 (**Listening**) features expanded coverage of how and why to listen respectfully to people who have different viewpoints. It presents updated real-life examples that incorporate a diverse array of listening contexts and challenges. A new description of Feynman's Technique offers a process to boost comprehension of material that is complex but important.

o Chapter 7 (**Nonverbal Communication**) now features a discussion of tattoos as a form of nonverbal communication. The chapter also introduces coverage of expectancy violation theory and tips for sending and interpreting nonverbal cues mindfully.

o Chapter 8 (**Understanding Interpersonal Communication**) includes tips for avoiding relationship-damaging communication patterns. It also presents new coverage of parental and sibling communication patterns.

o Chapter 9 (**Managing Conflict**) has been reconfigured to focus on productive ways to address differences at work and in interpersonal relationships. The chapter now includes a "Rules for Fighting Fair" checklist and a "You Can't Take It Back" feature about the pros and cons of using mediated communication to manage conflict.

o Chapter 10 (previously Chapter 9, Communicating in Groups and Teams) is now titled **Communicating for Career Success** and has been refocused to cover both organizational and workplace communication. It presents strategies for landing a desirable job, adapting to a new work environment, and being a highly valued follower. In addition, a new section on organizational communication covers power, patterns of interaction, and culture in the workplace.

o Chapter 11 (previously Chapter 10, Solving Problems in Groups and Teams) is now titled **Leadership and Teamwork**. Featuring updated research, the chapter introduces leadership concepts and foundational information about group and team work before segueing into coverage of collaborative problem solving. The section titled "Making the Most of Group Interaction" presents new tips for conducting effective meetings, an expanded ethical challenge box about dealing with quiet and talkative team members, updated coverage of brainstorming, and a new feature on round robin discussions.

o Chapter 12 (**Preparing and Presenting Your Speech**) has increased its emphasis on public speaking skill builders and opportunities for student practice.

o Chapter 13 (**Speech Organization and Support**) now has additional coverage and examples of the roles and formats of different types of speech outlines.

o Chapter 14 (**Informative Speaking**) now features a new sample speech on how different languages may shape the way people think.

o Chapter 15 (**Persuasive Speaking**) now includes more coverage of Aristotle's Rhetorical Triad.

Learning Tools

- **Checklists** throughout the book provide handy information and tips to help students build their communication skills. Topics include how to use social media courteously (Chapter 2), perception checking (Chapter 3), coping with culture shock (Chapter 4), avoiding misunderstandings (Chapter 5), listening mindfully (Chapter 6), managing dialectical tensions (Chapter 8), creating positive communication climates (Chapter 9), and getting slackers to do their share (Chapter 11).

- **Understanding Your Communication quizzes** invite students to evaluate and improve their communication skills. These include quizzes about communication style (Chapter 1), emotional intelligence (Chapter 3), intercultural sensitivity (Chapter 4), listening strengths (Chapter 6), friendship types (Chapter 8), interpersonal communication climates (Chapter 9), follower styles (Chapter 10), and leadership approaches (Chapter 11).

- **Learning Objectives** correspond to major headings in each chapter and coordinate with the end-of-chapter summary and review. They provide a clear map of what students need to learn and where to find that material.

- A **Making the Grade** section at the end of each chapter helps students test and deepen their mastery of the material. Organized by learning objective, this section summarizes key points from the text and presents questions and prompts to help students understand and apply the material.

- **Understanding Communication Technology** boxes highlight the increasingly important role of technology in human communication.

- **Understanding Diversity** boxes provide in-depth treatment of intercultural communication topics.

- **@Work** boxes show students how key concepts from the text operate in the workplace.

- **Ethical Challenge** boxes engage students in considering whether honesty is always the best policy, the acceptability of presenting multiple identities, how to deal effectively with difficult group members, and more.

- **Key Terms** are boldfaced on first use and listed at the end of each chapter.

- A **running glossary** in the margins helps students learn and review new terms.

- **Activities** at the end of each chapter help students apply the material to their everyday lives. Additional activities are available in the Instructor's Manual (*The Complete Guide to Teaching Communication*) at www.oup.com/he/adler-uhc14e.

- **Ask Yourself** prompts in the margins invite students to apply the material to their own lives.

- **On Your Feet** prompts provide confidence-building opportunities to get students speaking in class before undertaking formal presentations.

- An **enhanced support package** for every chapter (described in detail below) includes video links, pre- and post-reading quizzes, activities, discussion topics, examples, **GoReact**-powered tools for recording and uploading student speeches for assessment, an online gradebook, and more.

Teaching and Learning Support

The 14th edition of *Understanding Human Communication* contains a robust package of digital materials that make teaching more efficient and learning more effective. Instructors and students alike will be pleased to find a complete suite of resources.

The course package works with a **variety of learning management systems**, including Blackboard Learn, Canvas, Moodle, D2L, and Angel. Course cartridges allow instructors to create course websites that integrate resources available on the Ancillary Resource Center. Contact your Oxford University Press representative for access or for more information about these supplements or customized options.

Enhanced Ebook

An accessible, multi-device enabled enhanced ebook version integrates a rich assortment of digital resources, including interactive "Understanding Your Communication" diagnostics, self-quizzes, and videos. The enhanced ebook is available via your preferred textbook eBook vendor. After purchase, it can be accessed through your LMS.

For Instructors

The **Ancillary Resource Center (ARC)** at www.oup.com/he/adler-uhc14e is a convenient, instructor-focused website that provides access to all of the up-to-date teaching resources for this text, while guaranteeing the security of grade-significant resources. In addition, it allows OUP to keep instructors informed when new content becomes available. The following items are available on the ARC:

- The **Enhanced Ebook**, an accessible, multi-device enabled version, integrates a rich assortment of digital resources, including interactive "Understanding Your Communication" diagnostics, self-quizzes, and videos.

- *The Complete Guide to Teaching Communication*, written by coauthor Athena du Pré, provides a complete syllabus, teaching tips, preparation checklists, grab-and-go lesson plans, high-impact learning activities, handouts, links to relevant video clips, and coordinating PowerPoint lecture slides.

- A comprehensive **Computerized Test Bank** includes 60 exam questions per chapter in multiple-choice, short-answer, and essay formats. The questions have been revised for this edition, are labeled according to difficulty, and include the page reference and chapter section where the answers may be found.

- **PowerPoint slides** include key concepts, video clips, discussion questions, and other elements to engage students. They correspond to content in the lesson plans, making them ready to use and fully editable so that preparing for class is faster and easier than ever.

- *Now Playing*, **Instructor's Edition**, includes an introduction on incorporating film and television segments in class, as well as video clips, viewing guides and assignments, sample responses to discussion questions in the student edition, and an index by subject. A companion website to *Now Playing* is available at www.oup.com/us/nowplaying. It features video clips from previous editions.

- **TED Talks** on key topics show students how studying communication can enrich their own lives. In addition, quizzes following each talk help students internalize what they've learned.

- **Integration with GoReact** allows for uploading speech videos in response to both suggested prompts and instructor-created activities.

For Students

- *The Digital Study Guide* offers videos, activities, tutorials, chapter outlines, review questions, worksheets, practice quizzes, flashcards, and other study tools. The site is ideal for students who are looking for extra study material online. Students can access the full suite of resources via a code included with each new print or ebook purchase. Several of these resources are also available without a code at www.oup.com/he/adler-uhc14e.

Code Required (Comes with New Book Purchase):

- TED Talk Videos with Quizzes
- Chapter Outline Quizzes
- Pre- and Post-Reading Quizzes
- "Understanding Your Communication" Activities
- Matching Quizzes
- Speech Activities
- Selected Supporting Videos and Concept Animations with Quizzes
- Self-Quizzes (Expanded)

Open Access:

- Selected Supporting Videos
- Concept Animations
- Self-Quizzes (Basic)
- Flashcards
- Exam Prep Questions

Acknowledgments

Anyone involved with creating a textbook knows that success isn't possible without the contributions of many people.

We owe a debt to our colleagues. Thanks yet again to Russ Proctor, University of Northern Kentucky, for sharing his work and insights. We thank the following educators whose reviews helped shape this edition:

Marcee Andersen	*Tidewater Community College*
Manuel G. Avilés-Santiago	*Arizona State University*
Adam Burke	*Hawaii Pacific University*
Sherry L. Dean	*Richland College–Dallas County Community College District*
Andrew Herrmann	*East Tennessee State University*
Tricia Hylton	*Seneca College*
Elaine Jansky	*Northwest Vista College*
Angela King	*Cape Cod Community College*
Brett Maddex	*St. Petersburg College and Harrisburg Area Community College*
Anne McIntosh	*Central Piedmont Community College*
Denise Menchaca	*Northeast Lakeview College*
Jennifer Millspaugh	*Richland College/Grayson College*
Emily Normand	*Lewis University*
Leslie Ramos Salazar	*West Texas A&M University*
Sara Shippey	*Austin Community College*
Karin Wilking	*Northwest Vista College*
Archie Wortham	*Northeast Lakeview College*
Yingfan Zhang	*Suffolk County Community College*

We also continue to be grateful to the many educators whose reviews of previous editions continue to bring value to this book: **Theresa Albury**, Miami Dade College; **Deanna Armentrout**, West Virginia University; **Miki Bacino-Thiessen**, Rock Valley College; **Marie Baker-Ohler**, Northern Arizona University; **Kimberly Batty-Herbert**, South Florida Community College; **Mark Bergmooser**, Monroe County Community College; **Pete Bicak**, SUNY Rockland; **Brett N. Billman**, Bowling Green State University; **Shepherd Bliss**, Sonoma State University; **Jaime Bochantin**, University of North Carolina, Charlotte; **Beth Bryant**, Northern Virginia Community College, Loudoun; **Jo-Anne Bryant**, Troy State University–Montgomery; **Ironda Joyce Campbell**, Pierpont Community and Technical College; **Patricia Carr Connell**, Gadsden State Community College; **Cheryl Chambers**, Mississippi State University; **Kelly Crue**, Saint Cloud Technical & Community College; **Dee Ann Curry**, McMurry University; **Amber Davies-Sloan**,

Yavapai College; **Heather Dorsey**, University of Minnesota; **Rebecca A. Ellison**, Jefferson College; **Gary G. Fallon**, Broward Community College and Miami International University of Art and Design; **Amber N. Finn**, Texas Christian University; **Lisa Fitzgerald**, Austin Community College; **David Flatley**, Central Carolina Community College; **Sarah Fogle**, Embry-Riddle Aeronautical University; **Cole Franklin**, East Texas Baptist University; **Mikako Garard**, Santa Barbara City College; **Karley Goen**, Tarleton State University; **Samantha Gonzalez**, University of Hartford; **Betsy Gordon**, McKendree University; **Sharon Grice**, Kirkwood Community College–Cedar Rapids; **Donna L. Halper**, Lesley University; **Lysia Hand**, Phoenix College; **Deborah Hill**, Sauk Valley Community College; **Lisa Katrina Hill**, Harrisburg Area Community College–Gettysburg Campus; **Brittany Hochstaetter**, Wake Technical Community College; **Emily Holler**, Kennesaw State University; **Milton Hunt**, Austin Community College; **Maria Jaskot-Inclan**, Wilbur Wright College; **Kimberly Kline**, University of Texas at San Antonio; **Carol Knudson**, Gateway Tech College–Kenosha; **Kara Laskowski**, Shippensburg University of Pennsylvania; **Jennifer Lehtinen**, State University of New York at Orange; **Amy K. Lenoce**, Naugatuck Valley Community College; **Kurt Lindemann**, San Diego State University; **Judy Litterst**, St. Cloud State College; **Natashia Lopez-Gomez**, Notre Dame De Namur University; **Allyn Lueders**, East Texas Baptist University; **Jennifer McCullough**, Kent State University; **Bruce C. McKinney**, University of North Carolina–Wilmington; **Brenda Meyer**, Anoka Ramsey Community College–Cambridge; **Jim Mignerey**, St. Petersburg College; **Randy Mueller**, Gateway Technical College, Kenosha; **Kimberly M. Myers**, Manchester College and Indiana University–Purdue University Fort Wayne; **Gregg Nelson**, Chippewa Valley Technical College, River Falls; **Kim P. Nyman**, Collin College; **Catriona O'Curry**, Bellevue Community College; **Emily Osbun-Bermes**, Indiana University–Purdue University at Fort Wayne; **Christopher Palmi**, Lewis University; **Doug Parry**, University of Alaska at Anchorage; **Daniel M. Paulnock**, Saint Paul College; **Cheryl Pawlowski**, University of Northern Colorado; **Stacey A. Peterson**, Notre Dame of Maryland University; **Kelly Aikin Petkus**, Austin Community College–Cypress Creek; **Evelyn Plummer**, Seton Hall University; **Russell F. Proctor**, Northern Kentucky University; **Shannon Proctor**, Highline Community College; **Robert Pucci**, SUNY Ulster; **Terry Quinn**, Gateway Technical College, Kenosha; **Elizabeth Ribarsky**, University of Illinois at Springfield; **Delwin E. Richey**, Tarleton State University; **Charles Roberts**, East Tennessee State University; **Dan Robinette**, Eastern Kentucky University; **B. Hannah Rockwell**, Loyola University Chicago; **Dan Rogers**, Cedar Valley College; **Theresa Rogers**, Baltimore City Community College, Liberty; **Michele Russell**, Northern Virginia Community College; **John H. Saunders**, University of Central Arkansas; **Gerald Gregory Scanlon**, Colorado Mountain College; **David Schneider**, Saginaw Valley State University; **Cady Short-Thompson**, Northern Kentucky University; **Kim G. Smith**, Bishop State Community College; **Karen Solliday**, Gateway Technical College; **Patricia Spence**, Richland Community College; **Sarah Stout**, Kellogg Community College; **Linda H. Straubel**, Embry-Riddle University; **Don Taylor**, Blue Ridge Community College; **Raymond D. Taylor**, Blue Ridge Community College; **Cornelius Tyson**, Central Connecticut State University; **Curt VanGeison**, St. Charles Community College; **Lori E. Vela**, Austin Community College; **Robert W. Wawee**, The University of Houston–Downtown; **Kathy Wenell-Nesbit**, Chippewa Valley Technical College; **Shawnalee Whitney**, University of Alaska, Anchorage; **Princess Williams**, Suffolk County Community College; **Rebecca Wolniewicz**, Southwestern College; and **Jason Ziebart**, Central Carolina Community College.

Many thanks are due to colleagues who developed and refined elements of the ancillary package:

Christie Kleinmann, *Lee University*	Pre- and Posttests
Jennifer James, *Volunteer State Community College*	Test Bank
Ellen Bremen, *Highline College*	Matching and Self-Quizzes
Birgit Meadows, *Lee University*	Outline Matching Quizzes
Karen Anderson-Lain, *University of North Texas*	Exam Prep Questions
Athena du Pré (author)	PowerPoints/Instructor's Manual

The enhanced package that is the result of their efforts will help instructors teach more effectively and students succeed in mastering the material in this text.

In an age when publishing is becoming increasingly corporate, impersonal, and sales driven, we continue to be grateful for the privilege and pleasure of working with the professionals at the venerable Oxford University Press. They blend the best old-school practices with cutting-edge thinking.

Executive Editor Keith Chasse was a heroic advocate and mentor in the creation of this book, going so far as to travel the country personally to meet with students and educators. Interim Executive Communication Editor Karon Bowers stepped in to deftly manage final details. Senior Development Editor Lauren Mine is an author's best friend. She embodies the best of all the communication skills we describe here—a great listener, writer, friend, and coach. Assistant Editor Katlin E. Kocher was helpful in every way, always with a great attitude and never-ending knowledge of publishing. Assistant Editor Alyssa Quinones was a valuable member of the *UHC* team, coordinating the digital resource package, image updates, and countless other details. Senior Media Editor Mike Quilligan saw to the execution of a greatly enhanced digital media package for this edition. Senior Production Editor Bill Murray's steady hand and Art Director Michele Laseau's design talents have transformed this project from a plain manuscript into the handsome book you are now reading. Marketing Manager Sheryl Adams and the entire OUP sales team have gone the extra mile in bringing this book to users and supporting their teaching efforts. We are grateful to Sherri Adler and Sandy Cooke for their resourcefulness and the artistic sense they applied in choosing photos in these pages.

Finally, as always, we thank our partners Sherri, Linda, and Grant for their good-natured understanding and support while we've worked on this edition for more than a year. When it comes to communication, they continue to be the best judges of whether we practice what we preach.

Ron Adler
George Rodman
Athena du Pré

About the Authors

Ronald B. Adler is Professor of Communication, Emeritus, at Santa Barbara City College. He is coauthor of *Interplay: The Process of Interpersonal Communication*; *Essential Communication*; *Looking Out, Looking In*; and *Communicating at Work: Principles and Practices for Business and the Professions*.

George Rodman is Professor in the Department of Television and Radio at Brooklyn College, City University of New York, where he founded the graduate media studies and undergraduate TV writing programs. He is the author of *Mass Media in a Changing World*, *Making Sense of Media*, and several books on public speaking, as well as the coauthor of *Essential Communication*.

Athena du Pré is Distinguished University Professor of Communication at the University of West Florida. She is the author of *Communicating About Health: Current Issues and Perspectives* and coauthor of *Essential Communication*, as well as other books, journal articles, and chapters on communicating effectively.

understanding
human
communication

Communication: What and Why

CHAPTER OUTLINE

1.1 Characteristics of Communication 5
Definition of Communication
Models of Communication

1.2 Types of Communication 10
Intrapersonal Communication
Dyadic and Interpersonal Communication
Small-Group Communication
Organizational Communication
Public Communication
Mass Communication
Social Media Communication

1.3 Communication in a Changing World 13
Changing Technology
Changing Discipline

1.4 Communication Competence 15
There Is No "Ideal" Way to Communicate
Competence Is Situational
Competence Is Relational
Competent Communicators Are Empathic
Competence Can Be Learned
Competence Requires Hard Work
Competent Communicators Self-Monitor
Competent Communicators Are Committed

1.5 Misconceptions About Communication 19
Myth: Communication Requires Complete Understanding
Myth: Communication Can Solve All Problems
Myth: Communication Is a Good Thing
Myth: Meanings Are in Words
Myth: Communication Is Simple
Myth: More Communication Is Always Better

MAKING THE GRADE **21**
KEY TERMS **22**
ACTIVITIES **22**

LEARNING OBJECTIVES

1.1

Compare and contrast how linear and transactional models illustrate key characteristics of communication.

1.2

Distinguish between communication in a variety of contexts.

1.3

Describe how changing technology affects communication.

1.4

Analyze elements of effective and ineffective communication.

1.5

Replace common misconceptions about communication with more accurate information.

Communication has a profound impact on the relationships we form with others and how well we are able to reach our goals. Nathan Vass uses communication to build relationships and express himself artistically.

What role does communication play in the quality of your relationships at school, at work, and in your personal life?

Describe the communication style of someone you admire. In what ways does that person display respect for others and encourage mutual understanding?

What communication skills will be most essential in your career? Why?

AS A BUS DRIVER IN SEATTLE, Washington, Nathan Vass meets a lot of people. But what sets him apart is his ability to connect with them. He knows the names of hundreds of regular passengers. They look forward to riding his bus and even bring him cookies now and then. Some riders purposefully sit near Vass to talk about hard times, loss, accomplishments, and everyday happenings. They appreciate that he's an attentive listener and an engaging storyteller.[1] "I think we've all had those moments where we have a brief encounter with a stranger on the street that is profoundly meaningful because it emphasizes the commonalities we all share," he says. "Driving the bus is an opportunity to have those special moments all day." Vass has not only received hundreds of customer commendations but has been featured in numerous news stories and podcasts. He has even been honored as one of Seattle's "Most Influential People."[2,3]

A University of Washington graduate with a degree in fine arts, Vass is also a writer, photographer, and filmmaker. He says he enjoys these occupations nearly as much as driving a bus, which "fuels the art."[4] His blog, *The View from Nathan's Bus,* is popular around the country, and his recent book, *The Lines That Make Us: Stories from Nathan's Bus,* has been reprinted multiple times to keep up with demand.

Vass's success makes a strong case for the importance of communication. No matter what you do, communication is sure to play a central role in your life. The average person spends 7 out of 10 waking hours communicating with family members, friends, coworkers, teachers, and even strangers.[5] With computers, phones, tablets, and all the rest, it's possible to carry on several conversations at once. Of course, the quality of communication doesn't always match its quantity. In this chapter, we begin to explore what happens when people exchange messages and how to make wise choices when interacting with others.

Why do people spend so much time communicating? There's good reason: Communication satisfies many needs. Here are just a few:

- **Identity management.** How you dress, act, and speak can help you create and display your identity so that others understand you better.

- **Social connection.** Communication provides a means of fitting in socially and, at the same time, learning about people who are different from you. Social connection is so important that lonely people typically experience abnormally high levels of pain, depression, fatigue, and illness.[6,7] Evidence suggests not only that loneliness affects people at a psychological level, but also that the stress can cause changes in body chemistry and compromise immune systems.[8]

- **Relationship management.** Communication allows you to establish and manage relationships. People are typically happiest when they are with friends they consider to be good communicators[9] and with romantic partners who are good listeners and who share personal feelings and information.[10]

- **Goal accomplishment.** Language helps you share goals and work with others—everything from asking your hairstylist for a new look to launching a successful career.

- **Education and persuasion.** Public speaking skills in particular can help you educate and influence others.

Communication skills are a vital element in any successful career. It's probably no surprise that several of the top qualities employers look for in job candidates involve communication skills—including the ability to work well with team members, write and speak well, and influence others.[11] Communication can also help you fit in and get up to speed when you join a new organization.[12] And this skill set is just as important outside of work. It's a means of getting things done, learning, having fun, giving and receiving affection, feeling included, and managing conflict.[13]

1.1 Characteristics of Communication

What does it mean to communicate? We begin by examining three defining characteristics, then consider ways of modeling the process.

Definition of Communication

Defining *communication* isn't as simple as it might seem. People use the word in a variety of ways that are only vaguely related:

- A dog scratches at the back door to be let out.
- Data flows from one computer database to another.
- A satellite transmits a signal.
- Strangers who live thousands of miles apart build a relationship via social media.
- Locals offer directions to a group of confused-looking people from out of town.
- A civic leader encourages citizens to get more involved in helping homeless populations.

For starters, we'll be exploring *human* communication. Animal behavior, computer networks, and satellite communications aren't our focus, so that rules out the first three examples in the list above. In fact, there is a difference between *communications* (with an "s") and communication (without an "s"). (With an "s" generally refers to the technologies that enable exchange of information.) The last

Communication (without an "s") refers to the study of how people share messages. That's the primary focus of this book. *Communications* (with an "s") usually refers to technologies that enable the exchange of messages.

Can you give examples of "communication" versus "communications"?

communication The process of creating meaning through symbolic interaction.

symbol An arbitrary sign used to represent a thing, person, idea, or event in ways that make communication possible.

three examples do meet our definition of **communication**: *the process of creating meaning through symbolic interaction.* Let's unpack this definition to explore three fundamental qualities of communication.

Communication Is a Process   Most people think about communication as if it occurs in discrete, individual acts: telling a joke, asking for a raise, and so on. But communication is actually a continuous, ongoing process. Consider, for example, a conversation about politics in which you disagree with others. Your response will most likely depend on a host of considerations: your relationship to them, your opinion of their analytical skills, how open- (or closed-) minded you think they are, whether past disagreements have been handled respectfully or contentiously, whether you think the conversation is serious or casual . . . the list goes on. The response you choose will shape the reaction you get, both in the moment and going forward. Metaphorically speaking, communication isn't a series of episodes pasted together like photographs in a scrapbook. Instead, it's more like a movie in which meaning emerges as interrelated experiences unfold.

Communication Is Relational   Communication isn't something you do *to* others. Rather, it is something you do *with* them. Like many types of dancing, communication depends on the involvement of a partner. Great dancers who don't consider and adapt to the skill level of their partner can make both dancer and partner look bad. In communication and in dancing, even two highly skilled partners must work at adapting and coordinating their efforts.

Psychologist Kenneth Gergen captures the relational nature of communication well when he points out that our success depends on our interaction with others. As he says, "one cannot be 'attractive' without others who are attracted, a 'leader' without others willing to follow, or a 'loving person' without others to affirm with appreciation."[14]

Because communication is relational, it's often a mistake to suggest that just one person is responsible for a relationship. Rather than blaming each other for a disappointing outcome, it's usually better to ask, "How did we handle this situation poorly, and what can we do to make it better?"

Communication Is Symbolic   Words are **symbols** in that they represent people, things, ideas, and events, making it possible for us to communicate about them.

One feature of symbols is their arbitrary nature. For example, there's no logical reason why the letters in the word *book* should stand for what you're reading now. Speakers of Spanish call it a *libro*, and Germans call it a *buch*. Even in English, another term would work just as well as long as everyone agreed to use it in the same way. Symbolic communication allows people to think or talk about the past, explain the present, and speculate about the future.

People are most likely to understand each other if they interpret symbols in the same way. For example, bus driver/artist Nathan Vass recognized it as a compliment when a regular passenger called him a "beast." When Vass countered, "No, you're the beast man," the passenger said, "No, no no, I'm not the friggin' beast, dude . . . YOU'RE the Beast. Don't be modest man."[15]

On the other hand, conflicts can arise when people attach different meanings to a symbol. Decision makers in several states and countries have advocated for replacing the International Symbol of Access—an image of a person in a wheelchair—with a new version. Some prefer the new symbol for its more active implication.[16] Others feel that it marginalizes wheelchair users who aren't as active as the image suggests. One critic charges that the new symbol looks "like a person in a wheelchair race."[17] Others object to both symbols because they imply that everyone with a

Symbols—whether linguistic or graphic—can be powerful but ambiguous. Some feel the older International Symbol of Access (at left) depicts wheelchair users as passive and incapable, while the one on the right conveys more energy and control. Others feel that the newer symbol implies that everyone in a wheelchair is, or should be, athletic.

Think of a verbal or visual symbol that could be interpreted in more than one way. How might communication be affected depending on the interpretation?

UNDERSTANDING DIVERSITY

The Limits of Language When Describing People

Language is a useful tool, but it often falls short or exacerbates problems. Consider how we describe racial and ethnic groups. Terms such as *black*, *white*, and *brown* are inadequate to describe the rich diversity within and between groups. Moreover, putting labels on people can be misleading and divisive.

What connotations are attached to color words? White is often associated with purity and black with negativity. "I'm troubled when clients or colleagues use the word *black* to refer to negative or unwanted traits, in phrases such as *black sheep* and *black mark*," says a therapist who identifies as black.[18] Using terms such as *black* and *white* also implies that the people so described are opposites.

Some people identify as brown, particularly if they are from certain regions of North Africa, Asia, or Latin America. However, they may have a range of skin tones from dark to pale. As observed on the podcast *Code Switch*, race is about much more than skin color.[19]

Terms such as *Asian American* and *Latinx American* highlight diversity in the American population. Non-Latinx white people, however, are usually referred to as Americans without a qualifier. The implication is that they are somehow more authentically American than other racial groups.[20]

How should you decide what racial or ethnic terms to use and when? First, consider whether referring to someone's race or ethnicity is relevant to a situation. For example, when telling a story about the nurse who cared for you, is the nurse's race an important detail? On the other hand, pretending not to see race—to be "color blind"—is also problematic. Talani Shoneye describes an uncomfortable moment in her workplace when a colleague struggled to point her out without mentioning her race. Every description (white shirt, brown hair, and so on) described more than one person in the group, but Shoneye was the only black person in sight. "So I helped her out," she says, "Waving my hands I cheerfully replied, 'It's me, the black woman over here.'"[21] Others, such as Donya Momenian, an American with Iranian heritage, wish there were more terms in everyday use to describe membership in diverse groups. Momenian says that people view her as brown sometimes and white at others, but standardized forms in the United States seldom give her the opportunity to identify as Middle Eastern, which she feels more accurately describes her identity.[22]

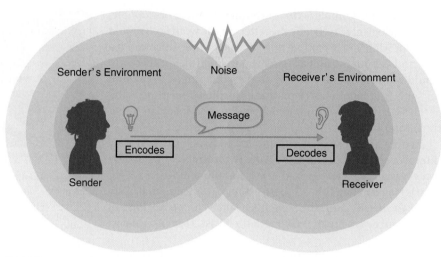

FIGURE 1.1 Linear Communication Model

ASK YOURSELF

What terms do you prefer
when people describe
you? Which, if any, make
you cringe or feel misun-
derstood? How might you
consider changing your language to avoid bias?

physical disability uses a wheelchair. "When you have a disability but you're not in a wheelchair, using an accessible parking space or restroom often makes you the target of all kinds of catty and hostile behavior," observes one analyst.[23] The strong feelings surrounding this issue underscore the importance of symbols in our lives.

You might wish that communication didn't rely so heavily on words, behaviors, and images that can be interpreted in many different ways. But that's the nature of symbols. The "Limits of Language" feature on page 7 explores the nuances of using words to describe racial and ethnic groups, and in Chapter 5 we'll discuss the nature of symbols in more detail.

Models of Communication

So far we have introduced a basic definition of communication and considered its characteristics. This information is useful, but it only begins to describe the process we will examine throughout this book. One way to deepen your understanding is to look at some models that describe what happens when two or more people interact. Over the years, scholars have developed an increasingly accurate and sophisticated view of this process.

linear communication model A characterization of communication as a one-way event in which a message flows from sender to receiver.

sender The originator of a message.

encode To put thoughts into symbols, most commonly words.

message A sender's planned and un-planned words and nonverbal behaviors.

receiver One who notices and attends to a message.

decode To attach meaning to a message.

noise External, physiological, and psychological distractions that interfere with the accurate transmission and re-ception of a message.

Linear Communication Model   Researchers once viewed messages as something that one person sends in some form to another.[24] In the **linear communication model** shown in Figure 1.1, communication is like tossing a ball: A **sender encodes** ideas and feelings into a **message** conveyed to a **receiver**, who **decodes** (attaches meaning to) it. The idea is that people take turns being either a sender or a receiver.

At first glance, the linear model suggests that communication is a straightfor-ward matter: If you choose your words correctly, your message should get through without distortion. But in reality, misunderstandings are common, even in close relationships.[25]

Why are misunderstandings so common? One factor is what scholars call **noise**, a broad category that includes any force that interferes with the accurate reception of a message. Three types of noise can disrupt communication—external, physi-ological, and psychological.

- *External noise* (also called *physical noise*) includes factors outside of a person that are distracting or make hearing difficult. For example, a text message that catches your eye might divert your attention during a lecture, or a lawnmower might make it difficult to hear the professor's remarks clearly.

- *Physiological noise* involves biological factors in the receiver or sender that interfere with accurate reception. For example, you might find it difficult to listen well if you are tired, sick, or hungry.

- *Psychological noise* refers to thoughts and feelings that interfere with the ability to express or understand a message accurately. If you come home annoyed or preoccupied, it will probably be harder to give the people you live with the attention they deserve.

A communication **channel** is the method by which a message is conveyed between people. Channels include face-to-face contact and various **media** (intervening mechanisms) such as phones and other screens. The channel you use can make a big difference in the effect of a message. For example, if you wanted to say "I love you," a generic ecard probably wouldn't have the same effect as a handwritten note. Likewise, saying "I love you" for the first time on the phone might be a different experience from saying the words in person. Media can facilitate personal messages and public ones.

The linear model also shows that communicators occupy different **environments**—fields of experience that influence how they interpret others' behavior. In communication terminology, *environment* refers not only to a physical location but also to the personal experiences and cultural backgrounds that participants bring to a conversation. It's easy to imagine how your position on economic issues might differ depending on whether you are struggling financially or are well off, and how your thoughts on immigration reform might depend on how long your family has lived in this country.

The model in Figure 1.1 shows that the environments of two people overlap. This area represents the background that the communicators have in common. The overlap may be quite large if they belong to the same groups, have had similar life experiences, and share many of the same opinions. By contrast, communication may be challenging if participants' shared environment is small. That doesn't mean you should avoid it. Communicating with people who are different from you— perhaps of different ages, socioeconomic status, abilities, or ethnicities—can be rewarding. In Chapter 4, we discuss the advantages and skills involved in intercultural communication, and information throughout this book offers ways to bridge the gaps that separate each of us to a greater or lesser degree.

The linear model provides a good start to understanding communication, but it oversimplifies the factors involved. For example, if you recall the definition of communication we discussed earlier, you'll note that the linear model doesn't depict communication as an ongoing process. The transactional communication model, which we will cover next, presents a more accurate picture in several respects.

Transactional Communication Model   The **transactional communication model** shows that people usually send and receive messages simultaneously. Consider what happens when a friend yawns as you complain about your family or when a business associate blushes at one of your jokes. Even though you are a "sender" in these episodes, you're also a "receiver" who is aware of the other person's reactions. While it's true that some types of mass communication flow in a one-way manner, most personal communication involves a simultaneous, two-way exchange.[26] As depicted in Figure 1.2, the roles of "sender" and "receiver" that seemed separate in the linear model are redefined as "communicators" to reflect the fact that we receive, decode, and respond to other people's behavior at the same time that they receive and respond to ours.

Nonverbal reactions such as yawning and blushing, as well as verbal and written responses, are forms of **feedback**—a communicator's response to another's

channel The medium through which a message passes from sender to receiver.

media Communication mechanisms such as phones and computers used to convey messages between people.

environment Both the physical setting in which communication occurs and the personal perspectives of the parties involved.

transactional communication model A characterization of communication as the simultaneous sending and receiving of messages in an ongoing process that involves feedback and includes unintentional (often ambiguous) messages.

feedback A receiver's response to a sender's message.

ASK YOURSELF

Think of a recent conversation. Write down not what the other person said but what you were thinking. Did psychological noise interfere with your ability to listen well? If so, how?

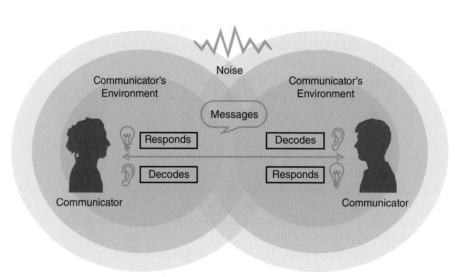

FIGURE 1.2 Transactional Communication Model

message. In person, you might field questions and comments following a presentation. Online, the comments viewers append to a status update, tweet, photo, or video give you an idea of how they feel about it. Even advertisements and news stories, which were once considered one-way communication, are now the subject of influential feedback. A few years ago, Nivea skin care discontinued an advertising campaign with the tagline "White Is Purity" after white supremacists praised it on social media and others condemned the slogan as racist.[27]

Another difference between the linear and transactional model is that the transactional includes unintentional messages. Your facial expressions, gestures, postures, and vocal tones may offer information to others even when you aren't aware of them.[28] As you have certainly experienced, these messages can be powerful but confusing. For example, if you look at the ceiling during an important conversation, the other person may assume (rightly or wrongly) that you disagree or aren't paying attention.

intrapersonal communication
Communication that occurs within a single person.

1.2 Types of Communication

Nathan Vass says he loves face-to-face communication because it allows him to make a personal connection with others. Via his blog, book, photography exhibits, and film projects, he also enjoys sharing stories with a wider range of people than he may ever meet in person. He's got a point. Each method of communicating has unique qualities. In this section, we consider seven communication contexts, each with its own characteristics, advantages, and challenges.

Intrapersonal Communication

Intrapersonal communication involves "communicating with oneself."[29] One way that each of us communicates internally is by engaging with our inner voice. Take a moment and listen to your own self-talk before reading on. It may have said something like, *"What inner voice? I don't have any inner voice!"* That's the "sound" of your thinking.

Intrapersonal communication affects almost every type of interaction. A conversation with a new acquaintance is shaped by your self-talk (*"I'm making a fool out of myself"* or *"She likes me!"*). Take this idea further by imagining your thoughts in each of the following situations:

- You're planning to approach a stranger you would like to get to know better.
- You pause a minute and look at the audience before beginning a 10-minute speech.
- The boss yawns while you are asking for a raise.
- A friend seems irritated lately, and you're not sure whether you are responsible.

You'll read more about self-talk later. Much of Chapter 3 deals with how what you think shapes how you relate to others, and part of Chapter 14 explains how the right kind of intrapersonal communication can minimize anxiety when delivering a speech.

Dyadic and Interpersonal Communication

Social scientists use the term **dyadic communication** to describe two people interacting, either in person or via mediated channels. Dyadic conversations are the most common type of personal communication. Even communication within larger groups (such as classrooms, parties, and work environments) often consists of multiple, shifting dyadic encounters.

It's a common misconception that dyadic communication is the same as interpersonal communication. As Chapter 8 will explain, not all dyadic communication is interpersonal (think of idle banter about sports or the weather). And not all interpersonal communication is dyadic (consider a deep conversation between several close friends).

Small-Group Communication

In **small-group communication**, each person can participate actively with the other members. Small groups are a fixture of everyday life. Your family is an example. So are athletic teams, colleagues at work, and groups of students collaborating on a class project.

Small groups exhibit characteristics that aren't present in a dyad. With two communicators, there's no chance of being outvoted. In a group, however, the majority of members can overrule or put pressure on those in the minority. And for better and worse, groups take risks that members wouldn't take if they were alone or in a dyad. On the plus side, groups may be more creative than dyads, if only because there are more people from whom to draw ideas. Finally, group communication is affected strongly by the leader. Leadership and teamwork are so important to career success that Chapter 11 focuses extensively on them.

Organizational Communication

Larger, more permanent collections of people engage in **organizational communication** when they work collectively to achieve goals. Organizations operate for a variety of reasons: commercial (such as a corporation), nonprofit (such as the Humane Society or Habitat for Humanity), political (a government or political action group), health-related (a hospital or doctor's office), and even recreational (a YMCA or sports league).

Organizational communication involves specific roles (e.g., sales associate, general manager, coach) that shape what people communicate about and how they relate to one another. As you'll read in Chapter 4, each organization develops its own culture and traditions that influence how members behave. The "Communication Skills and Career Success" feature on page 12 shows the advantages that effective communicators enjoy in the career world.

dyadic communication Two-person communication.

small-group communication Communication within a group of a size such that every member can participate actively with the other members.

organizational communication Interaction among members of a relatively large, permanent structure (such as a nonprofit agency or business) in order to pursue shared goals.

ASK YOURSELF (?)

In your daily life, how much of an overlap is there between dyadic and interpersonal communication?

Communication Skills and Career Success

When more than 3,000 U.S. adults were asked what skills are most important "to get ahead in the world today," communication skills topped the list—ahead of math, writing, logic, and scientific ability.[30]

Employers agree. Representatives from a wide range of industries rank the "ability to verbally communicate with persons inside and outside the organization" as the most essential skill for career success. In fact, they rate communication skill as more important than "technical knowledge related to the job."[31] Other research reinforces the value of communication. An analysis of almost 15 million job advertisements from across all occupations revealed that the ability to speak and write effectively was the most requested skill, identified twice as often as any other quality.[32]

The reasons are clear. Regardless of the job, people spend most of their working lives communicating.[33] The average American office worker receives nearly 12,000 emails every year and spends the equivalent of 111 work days responding to them.[34] When you also consider telephone and face-to-face conversations, instant messaging, team meetings, videoconferences, presentations, and many other types of interaction, it's undeniable that communication is at the heart of the workplace.[35]

Evidence such as this shows that communication skills can make the difference between a successful and a disappointing career. For more detail on increasing your efficiency and productivity at work, see Chapters 10 and 11, as well as *Communicating for Career Success* at www.oup.com/he/adler-uhc14e.

Public Communication

public communication Communication that occurs when a group is too large for everyone to contribute. It is characterized by an unequal amount of speaking and by limited verbal feedback.

Public communication occurs when a group becomes too large for all members to contribute. It's generally characterized by an unequal amount of speaking among the members. One or more people are likely to deliver their remarks to others, who act as an audience. Even when audience members have the chance to ask questions and post comments (in person or online), the speakers are still mostly in control and do most of the talking. This underscores the leadership potential of being a great public speaker. The last four chapters of this book describe the steps you can take to prepare and deliver an effective speech.

Mass Communication

mass communication The transmission of messages to large, usually widespread audiences via TV, Internet, movies, magazines, and other forms of mass media.

Mass communication consists of messages that are transmitted to large, widespread audiences via electronic and print media such as websites, magazines, television, and radio, to name just a few. Mass communication is often less personal and more of a product than the other types of communication we have examined. That's because most mass messages are aimed at a large audience without personal contact between senders and receivers.

Social Media Communication

Social media use doesn't fit neatly within the groupings we've just described. A tweet or post might reach your closest friends *and* thousands of people you don't know. This can be inspiring, as when a grocery clerk's kindness to an autistic teenager became a viral video.[36] The clip was viewed by tens of thousands of people and became the focus of stories carried by CNN, ABC News, *The Washington Post*, and other news outlets. More attention isn't always better, however. UC Berkeley graduate student Connor Riley learned this lesson the hard way when she tweeted "Cisco just offered me a job! Now I have to weigh the utility of a fatty paycheck against the daily commute to San Jose and hating the work." Needless to say, the company withdrew its offer.[37]

Star Wars actor Kelly Tran faced months of racist, sexist trolling on social media.

In your experience, what are the benefits and drawbacks of using various social media platforms?

Social media posts can be devastating. Following the release of *Star Wars: The Last Jedi*, actor Kelly Tran faced racist, sexist trolling that continued for more than a year. The harassment led her to delete her social media accounts. As she wrote in an essay for *The New York Times*, "Their words seemed to confirm what growing up as a woman and a person of color already taught me: that I belonged in margins and spaces, valid only as a minor character in their lives and stories."[38]

Social media play such an important role in today's communication environment that Chapter 2 is devoted entirely to the subject. Throughout the book, you'll also find guidance on discussing politics respectfully on social media (Chapter 4), distinguishing between facts and inflammatory bot posts (Chapter 5), avoiding social media distractions to be a better listener (Chapter 6), using your online identity to help with a job search (Chapter 10), and more.

As useful as social media are, they can be overwhelming. The "Understanding Communication Technology" feature considers the reasons some people declare social media holidays now and then.

1.3 Communication in a Changing World

Today's communicators are equipped with a range of technologies that, even two decades ago, would have been the stuff of fantasy and science fiction. As you read the following, ask yourself: *How have communication tools made your life richer? In what ways can they be overwhelming or distracting?*

Changing Technology

Communication technology is changing more rapidly than ever before. Until about 5,000 years ago, written language didn't exist and face-to-face speech was the primary form of communication. For thousands of years after that, most people were still unable to read and write. Books were scarce, and the amount of written information available was small. Speaking and listening were the predominant communication "technologies."

UNDERSTANDING COMMUNICATION TECHNOLOGY

Dear Social Media, I Need a Little Space

Something interesting happened when tech writer Eric Griffith left social media: nothing. "I didn't miss out on anything world-shaking, the globe kept on spinning," Griffith says. "Eventually, it felt utterly normal to not be on Twitter all the time."[39] He did, however, experience a personal rise in productivity and uninterrupted time with friends. No one is suggesting a total break from the Twitter feed. Well, some people are. But they usually announce that via—you guessed it—social media.

Here are the most common reasons people give for limiting time on social media:

- **To Reclaim Time and Focus**
  Taken to extremes, social media drain effort that could be spent on other activities. On average, college students spend about 3 times more of their day on cell phones than they do on school and 4 times more than they spend on work-related activities.[40,41]

- **To Live in the Moment**
  The lure of online activities can detract from the here and now. "How many of you have sat down for dinner with friends or family at a restaurant, you take out your phone to check Facebook or Twitter?" asks reformed social media junkie Chris Mullen.[42] To participate in the online chatter, "you take a pic of everyone at dinner. Then you tweet 'at din w/ fam,'" Mullen predicts. All the while, the people closest to you (physically speaking) are distanced from interacting with you.

- **To Give Your Ego a Break**
  "One of the main problems with social media is you are often bombarded by others' accomplishments," observes a writer for *Elite Daily*.[43] It's a distorted reality in that few people post unflattering photos of themselves or broadcast their failures. A steady diet of Facebook, Twitter, and Instagram can make life seem dull and inadequate by comparison.

- **To Break the Approval Craving**
  Social media posts are often designed to get a validating response from others. That can foster an unhealthy dependency. "If you are waiting for people to like your status or your photos, you might be seeking and spending a lot of time waiting on others to approve of the online version of yourself," warns pop culture writer Christen Grumstrip.[44]

All in all, our relationship with social media is best described as "it's complicated." One online story urging people to cut back on use of social media garnered more than 2 million Facebook likes,[45] which suggests more than a little ambivalence about that advice. And no wonder. A little space doesn't mean a permanent breakup. Eric Griffith, who had quit Twitter, is now back on it. But the break gave him a healthy perspective. As with any relationship, he says, "when [social media] takes over your life, it's not a good thing."[46]

By the mid-1700s, literacy grew in industrial societies, giving ordinary people access to ideas that had previously been available only to the most privileged. By the end of the 1800s, the telegraph made it possible for everyday people to send messages over vast distances.

The first half of the 1900s introduced a burst of communication technology. The invention of the telephone, radio, and then television extended the reach of personal and business relationships and made mass communication possible.

By the dawn of the 21st century, cellular technologies and the Internet broadened people's ability to communicate beyond the dreams of earlier generations. Pocket-sized telephones have made it affordable and easy to talk, send data, and exchange images with people around the globe. Now, new fiber-optic technology allows for tens of millions of phone calls every second.[47,48] Videoconferencing now allows us to see one another's facial cues, body movements, and gestures almost as if we were face to face.

The accelerating pace of innovations in communication technology is astonishing:[49]

- It took 38 years for radio to reach 50 million listeners.
- It took television only 13 years to capture the same number of viewers.
- It took less than 4 years for the Internet to attract 50 million users.
- Facebook added 100 million users in less than 9 months.

Changing Discipline

The study of communication has evolved to reflect the changing world. The first systematic analysis of how to communicate effectively focused on public address. Aristotle's *Rhetoric*, written about 2,500 years ago,[50] laid out specific criteria for effective speaking (called the "Canons of Rhetoric") which are still used today. (You'll read about them in Chapter 15.)

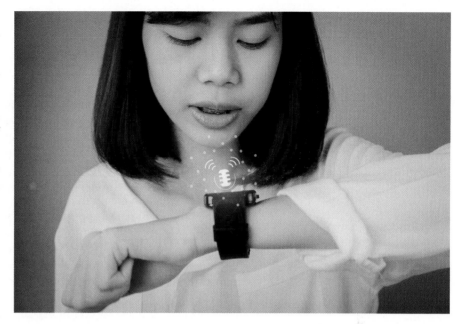

Communicating via wristwatch seemed like science fiction to earlier generations, but even more sophisticated technologies are commonplace today.

Think about the communication technologies you use. How would your life be different without them?

In the early 1900s, the study of communication expanded into a social science with a more diverse focus. As messages began to reach large numbers of people via print, film, and broadcasting, scholars began to study their effects on people's attitudes and behaviors. Propaganda became an important focus of research during and after World War II.[51]

In the 1950s and 1960s, researchers began asking questions about human relationships in family and work settings,[52] leading to early research about small-group communication.[53] Decision making and other small-group communication processes emerged as a major area of study and continue to fascinate researchers today.

Communication scholars now study a wide range of phenomena, including how relationships develop, the nature of social support, the role of emotions, how honesty and deception operate, and how new technologies affect interpersonal relationships. Other branches of the discipline examine how health care providers and clients interact, the influence of gender, and how people from different backgrounds communicate with one another, to name just a few. As you can see, the scope of the field has expanded far beyond its roots in public speaking.

1.4 Communication Competence

Most scholars agree that **communication competence** involves achieving one's goals in a manner that, ideally, maintains or enhances the relationship in which it occurs.[54,55] As you read about the elements of competence, think about how you can apply them.

communication competence The ability to achieve one's goals through communication and, ideally, maintain healthy relationships.

There Is No "Ideal" Way to Communicate

Some successful communicators are serious, while others are lighthearted. Some are talkative, while others are quiet. And some are straightforward, while others are subtle. Just as beautiful art takes many forms, there are many kinds of competent communicators. Others may inspire you to improve your communication skill, but ultimately it's important to find approaches that suit you.

Many struggling communicators are easy to spot by their limited approaches. Some are chronic jokers. Others seem always to be argumentative. Still others are quiet in almost every situation. Like a piano player who knows only one tune or a chef who can prepare only a few dishes, these people rely on a small range of communication strategies again and again, whether or not they are successful. By contrast, competent communicators have a wide repertoire from which to draw, and they choose the most appropriate behavior for a given situation.

Competence Is Situational

Because competent behavior varies from one situation and person to another, it's a mistake to think people are either great communicators or they aren't. It's more accurate to talk about *degrees* or *areas* of competence. You might be skillful socializing at a party but less successful making small talk with professors during office hours. This means it's an overgeneralization to say, in a moment of distress, "I'm a terrible communicator!" It would be more accurate to say, "I didn't handle this situation very well, even though I'm better in others."

Competence Is Relational

Because communication is transactional—something we do *with* others rather than *to* them—behavior that is competent in one relationship or culture won't necessarily succeed in others.

An important study on relational satisfaction illustrated that what constitutes satisfying communication varies from one relationship to another.[56] Researchers found that friendships were most satisfying when partners possessed harmonious communication styles. The same principle holds true in the case of jealousy. People deal with insecurity in a variety of ways, including keeping close tabs on their partner, acting indifferent, decreasing affection, talking the matter over, and acting angry. Researchers have found that no type of behavior is effective or ineffective in every relationship. Approaches that work well with some people are hurtful to others. Findings like these demonstrate that competence arises out of developing ways of interacting that work for you and for the other people involved.

Competent Communicators Are Empathic

You have the best chance of developing an effective message when you understand the other person's point of view. And because people don't always express their thoughts and feelings clearly, the ability to *imagine* how an issue might look from someone else's viewpoint is an important skill. **Cognitive complexity** is the ability to understand issues from a variety of perspectives. For instance, imagine that your longtime friend seems angry with you. Rather than jumping to a conclusion, you might consider a range of possibilities: Is your friend offended by something you have done? Did something upsetting happen earlier in the day? Or perhaps nothing is wrong, and you're just being overly sensitive. That's why listening is so important. It helps you understand others and judge how they are responding to your messages. Because empathy is such an important element of communicative competence, part of Chapter 3 and all of Chapter 6 are devoted to the topic.

cognitive complexity The ability to understand issues from a variety of perspectives.

Competence Can Be Learned

Communication is, at least in part, learnable. Systematic education (such as the class in which you are now enrolled) and a little training can produce dramatic results.[57,58] Communication skills also increase through trial and error and observation. Think of your own experience: You've learned from your own successes and failures and by seeing what works well, and not so well, for the people around you.

Competence Requires Hard Work

Knowing *how* to communicate effectively is necessary, but it isn't sufficient to be competent. Hard work is also an ingredient. For example, one study revealed that, even when college students were capable of paying attention to important conversational cues, they often didn't put in the effort to do so, perhaps because they were preoccupied or distracted.[59]

Developing competence takes time. Simply reading about communication skills in the following chapters won't guarantee that you can use them flawlessly right off the bat. As with any other skill—playing a musical instrument or learning a sport, for example—the road to competence in communication can be a long one. You can expect that your first efforts at communicating differently will be awkward. After some practice you should become more skillful, even though you will still have to think about the new way of speaking or listening. Finally, after repeating the new skill again and again, you will find you can perform it without much conscious thought.

ASK YOURSELF

Who are the most competent communicators you know? How does their behavior reflect the elements of competence described here?

Competent Communicators Self-Monitor

Psychologists use the term **self-monitoring** to describe the process of paying close attention to one's own behavior and using these observations to make effective choices. High self-monitors are good at observing themselves from a detached viewpoint, which helps them understand the impressions they make on others. A high self-monitor might think, *"I'm making a fool out of myself," "I'd better speak up now,"* or *"This approach is working well. I'll keep it up."*

High self-monitors are more likely than low self-monitors to develop relationships with many types of people[60] and to recognize ways they might improve as communicators.[61] Self-monitoring is especially helpful in real time. Consider the mother of a 3-year-old who reflected: "I disciplined my son and he threw a tantrum that I thought was so funny that I disciplined him again just so I could video it. . . . After uploading it on Instagram I thought, 'What did I just do?'"[62] Awareness of her actions came too late to change that situation, but perhaps it will help her be more of a self-monitor in the future.

> **self-monitoring** Paying close attention to one's own behavior and using these observations to make effective choices.

What Type of Communicator Are You?

 Answer the questions below for insight about your approach as a communicator.

1. You are puzzled when a friend says, "The complex houses married and single soldiers and their families."[63] What are you most likely to do?

   a. Tune out and hope your friend changes topics soon.

   b. Declare, "You're not making any sense."

   c. Ask questions to be sure you understand what your friend means.

   d. Nod as if you understand, even if you don't.

2. You are working frantically to meet a project deadline when your phone rings. It's your roommate, who immediately launches into a long, involved story. What are you most likely to do?

   a. Pretend to listen while you continue to work on your project.

   b. Interrupt to say, "I don't have time for this now."

   c. Listen for a few minutes and then say, "I'd like to hear more about this, but can I call you back later?"

   d. Listen and ask questions so you don't hurt your friend's feelings.

3. You are assigned to a task force to consider the parking problem on campus. Which of the following are you most likely to do during task force meetings?

   a. Talk in a quiet voice to the person next to you.

   b. Express frustration if meetings aren't productive.

   c. Ask questions and take notes.

   d. Spend most of your time listening quietly.

4. Your family is celebrating your brother's high school graduation at dinner. What are you most likely to do at the table?

   a. Ask for dessert in a take-out container so you can leave early.

   b. Keep your cell phone handy so you won't miss anything your friends post.

   c. Give your undivided attention as your brother talks about his big day.

   d. Paste a smile on your face and make the best of the situation, even if you feel bored.

INTERPRETING YOUR RESPONSES

For insight about your communication style, see which of the following best describes your answers. (More than one may apply.)

Distracted Communicator

If you answered "a" to two or more questions, you have a tendency to disengage. It may be that you are shy, introverted, or easily distracted. You needn't change your personality, but you are likely to build stronger relationships if you strive to be more attentive and proactive. Chapter 6 offers tips that may help you become a more active listener.

Impatient Communicator

If you answered "b" more than once, you tend to be a straight-talker who doesn't like delays or ambiguity. Honesty can be a virtue, but be careful not to overdo it. Your tendency to "tell it like it is" may come off as bossy or domineering at times. The perception-checking technique in Chapter 3 offers a good way to balance your desire for the truth with concern for other people's feelings.

Tactful Communicator

If you answered "c" multiple times, you are able to balance assertiveness with good listening skills. Your

willingness to actively engage with people is an asset. Use tips throughout the book to enhance your already strong communication skills.

Accommodator

If you answered "d" two or more times, you tend to put others' needs ahead of your own. People probably appreciate your listening skills but wish you would speak up more. Saying what you feel and sharing your ideas can be an asset both personally and professionally. Tips in Chapters 11 and 15 may help you become more assertive and confident without losing your thoughtful consideration for others.

Competent Communicators Are Committed

One feature that distinguishes effective communication in almost any context is commitment. Nathan Vass, the bus driver profiled at the beginning of the chapter, demonstrates again and again his commitment to the people he meets. He once helped a passenger load furniture onto the bus. He has encouraged people on the street to jump aboard the bus to escape assailants, and Vass's listening skills are so legendary that people talk about riding with him as "bus therapy."[64] As a result, Vass is often able to stop fights by appealing to people by name and asking them to calm down as a favor to him. And riders who have come to know him lend assistance when there is trouble on the bus.

The same principles work in other relationships. People who are emotionally committed to a relationship are more likely than others to talk about difficult subjects and to share personal information about themselves. This commitment can strengthen relationships and contribute to a heightened sense of well-being.[65] Take the "Understanding Your Communication" quiz "What Type of Communicator Are You?" for insight about your communication strengths and challenges.

ASK YOURSELF ?

Describe the most skillful communicator you know. What makes this person so effective? What can you learn from them to be a more accomplished communicator yourself?

1.5 Misconceptions About Communication

Having spent time talking about what communication is, we should identify some things it is not. Understanding these myths can boost your effectiveness and prevent misunderstandings.

Myth: Communication Requires Complete Understanding

Most people operate on the flawed assumption that the goal of all communication is to maximize understanding between communicators. In fact, there are times when complete comprehension isn't the primary goal.[66] For example, when people in the United States ask social questions such as *"How's it going?"* they typically aren't interested in the details of how the other person is doing. Beyond formalities, competent communicators may also deliberately create ambiguous messages.[67] For example, consider what you might say if someone you care about asks a personal question that you don't want to answer such as *"Do you think I'm attractive?"* or *"Is anything bothering you?"*

Myth: Communication Can Solve All Problems

"If I could just communicate better . . ." is the sad refrain of many unhappy people who believe that if they could express themselves more effectively, their relationships would improve. This is sometimes true, but it's an exaggeration to say that communicating—even communicating clearly—is a guaranteed cure-all.

Myth: Communication Is a Good Thing

In truth, communication is neither good nor bad in itself. Rather, its value comes from the way it is used. Communication can be a tool for expressing warm feelings and useful facts, but under different circumstances words and actions can cause both physical and emotional pain.

"My wife understands me."

Source: Cartoonbank.com

Myth: Meanings Are in Words

As communication scholars put it: *Meanings are in people, not in words.*

What they mean is that it's a mistake to think that, just because you use and understand a word in one way, others will too. Think about arguments over concepts such as privilege, freedom, and honor. It's easy to see how people can view complex ideas like these quite differently. On a more granular level, think of specific words that have multiple possible meanings. For example, the word *wicked* is common slang in New England and can mean either "really" or "excellent," but if you said "She's wicked" in the South, it might be an insult. Chapter 5 addresses the problems that come from mistakenly assuming that meanings rest in words.

Myth: Communication Is Simple

Most people assume that communication is an aptitude that people develop without the need for training—rather like breathing. After all, you've been swapping ideas with one another since early childhood, and there are lots of people who communicate pretty well without ever having taken a class on the subject. However, regarding communication as a natural ability is a gross oversimplification. Communication skills are a lot like athletic ability: With training and practice, even the most inept of us can learn to be more effective, and those who are talented can always become better.

Myth: More Communication Is Always Better

Although it's true that not communicating enough is a mistake, there are situations in which *too much* communication is ill advised. Sometimes we "talk a problem to death," going over the same ground again and again without making any headway. And there are times when communicating too much can actually aggravate a problem. You've probably had the experience of "talking yourself into a hole"—making a bad situation worse by pursuing it too far. There are even times when *no communication* is the best course. Any good salesperson will tell you that it's often best to stop talking and let the customer think about the product. And when two people are angry and hurt, they may say things they don't mean and will later regret. One key to successful communication, then, is to share an adequate amount of information in a skillful manner when the time is right. The *Ethical Challenge* feature "To Communicate or Not to Communicate" gives you a chance to consider how you would respond in a range of situations.

ON YOUR FEET

Describe a time when a misunderstanding led to a mistake or to a comical situation. How did interpretations of the initial message(s) differ?

ETHICAL CHALLENGE To Communicate or Not to Communicate?

Read the scenarios below and consider in each case whether it would be best to (a) say what you are thinking, (b) listen, or (c) remove yourself from the situation. Explain your reasons.

- A close relational partner of yours was hurt by something you said. You have apologized, but your partner still wants to talk about the issue.

- A classmate is considering a job offer that would require leaving school and moving to another city. You think it's a bad idea, but your classmate seems excited about it.

- Your ex calls at all hours of the night wanting to "talk," which usually means asking you over and over why the relationship ended.

- A coworker confides in you about a family issue that has the coworker worried and upset.

DISCUSSION: What moral obligation, if any, do you have to express your thoughts and feelings in each of these situations?

Throughout the chapter we have followed Nathan Vass—artist, bus driver, and communicator extraordinaire. Let's close with one more story about him. One night soon after midnight, Vass had completed his shift when he realized that a young man who looked to be 18 or 19 was still on the bus. Vass listened to the teen, whose name was Leroy, and learned that he had just arrived in Seattle that day after leaving a bad family situation in Philadelphia. Vass suggested some places Leroy might find shelter and gave him $20. Within the week, the two ran into each other in town. "This was only four days later," Vass remembers, noting that "Leroy had not even found a change of clothes" yet.[68] However, Leroy had already lined up four job interviews, one of which led to full-time employment, and, eventually, a home of his own. Fifteen years later, the two are still friends, and Leroy says that Vass's kindness that night saved his life. "Nathan gave me hope," he says.[69]

The bus is life on a small stage, Vass says. People come together. They make a difference in each other's lives. And very often, "the journeys [are] more interesting than the destinations."[70]

MAKING THE GRADE

At www.oup.com/he/adler-uhc14e, you will find a variety of resources to enhance your understanding, including video clips, animations, self-quizzes, additional activities, audio and video summaries, and interactive self-assessments.

OBJECTIVE 1.1 Compare and contrast linear and transactional models in terms of how they illustrate key characteristics of communication.

- Communication is the process of creating meaning through symbolic interaction.
- Communication is influenced by many factors, including environment, noise, and the channel through which messages are shared.
- Because it acknowledges that people send and receive messages simultaneously, the transactional model of communication is more accurate than the linear model that preceded it.

  > Describe a real-life example that illustrates how communication is process-oriented, relational, and symbolic.

  > Have you ever overreacted to something someone said, only to realize later that you misinterpreted that person's intent? How can the knowledge that communication is an ongoing process help you put things in perspective?

  > Pause for a moment to consider what is on your mind right now. How might these thoughts create physiological noise that could interfere with your ability to listen closely? Brainstorm some techniques people might use to listen carefully rather than being distracted by their own thoughts.

OBJECTIVE 1.2 Distinguish between communication in a variety of contexts.

- Communication operates in many contexts, including intrapersonal, dyadic and interpersonal, small group, organizational, public, mass media, and social media.

  > Provide examples of communication in three different contexts.

  > In what way would you most like to improve your communication relevant to interpersonal communication? To small-group communication? To public speaking?

  > Overall, what goals would you most like to set for yourself as a communicator? Which contexts will your goals involve?

OBJECTIVE 1.3 Describe how changing communication technology affects communication.

- After millennia with little change, the pace of innovation in communication technologies has accelerated in the last 150 years.

- The study of communication has also evolved, taking a more expansive approach to human interaction.
  > How do the communication technologies you use differ from those of your parents' generation? Your grandparents'?
  > How is your life enriched by using today's communication technologies? Do these technologies diminish the quality of your relationships in any ways? If so, how?

OBJECTIVE **1.4**   **Analyze the difference between effective and ineffective communication.**

- Communication competence is situational and relational, and it can be learned.
- Competent communicators are able to choose and perform appropriately from a wide range of behaviors. They are also flexible, cognitively complex self-monitors who can take the perspective of others and who are committed to important relationships.
  > Give examples from your own life to illustrate the situational and relational nature of communication competence.
  > Based on the criteria in the "Communication Competence" section, how would you rate your own communicative skills?
  > Imagine that someone you know often tells jokes that you find offensive. Reflect on which of the following you are most likely to do: (a) say nothing, (b) ask a third party to talk to the jokester, (c) hint at your discomfort, (d) use humor to comment on your friend's insensitivity, or (e) express your discomfort in a straightforward way and ask your friend to stop. Do you feel the strategies you use are effective?

OBJECTIVE **1.5**   **Replace common misconceptions about communication with more accurate information.**

- Communication doesn't always require complete understanding.
- Communication is not always a positive factor that will solve every problem.
- Meanings are in people, not in words.
- Communication is neither simple nor easy.
- More communication is not always better.
  > Give examples of how each misconception described in this chapter can lead to problems.
  > Which misconceptions have been most problematic in your life? Give examples.
  > How can you avoid succumbing to the misconceptions listed in this chapter? Give examples of how you could communicate more effectively.

KEY TERMS

channel p. 9
cognitive complexity p. 16
communication p. 6
communication competence p. 15
decode p. 8
dyadic communication p. 11
encode p. 8
environment p. 9
feedback p. 9
intrapersonal communication p. 10
linear communication model p. 8
mass communication p. 12
media p. 9
message p. 8
noise p. 8
organizational communication p. 11
public communication p. 12
receiver p. 8
self-monitoring p. 17
sender p. 8
small-group communication p. 11
symbol p. 6
transactional communication model p. 9

ACTIVITIES

1. Test for yourself whether communication is frequent and important by observing your interactions for one day. Record every occasion in which you are involved in some sort of human communication as it is defined in this chapter. Based on your findings, answer the following questions:
   - What percentage of your waking day is involved in communication?
   - What percentage of time do you spend communicating in the following contexts: intrapersonal, dyadic, small group, and public?
   - What percentage of your communication is devoted to satisfying each of the following types of needs: physical, identity, social, and practical? (Note that you might try to satisfy more than one type at a time.)

   Based on your analysis, describe 5 to 10 ways you would like to communicate more effectively. For each item on your list of goals, describe who is involved (e.g., "my boss," "people I meet at parties") and how you would like to communicate differently (e.g., "act less defensively when criticized," "speak up more instead of waiting for them to approach me"). Use this list to focus your studies as you read the remainder of this book.

2. Construct a diary of the ways you use social media in a three-day period. For each instance when you use social media (email, a social networking website, phone, Twitter, etc.) describe:

- The kind(s) of social media you use
- The nature of the communication (e.g., "Wrote on friend's Facebook wall," "Reminded roommate to pick up dinner on the way home.")
- The type of need you are trying to satisfy (information, relational, identity, entertainment)

Based on your observations, describe the types of media you use most often and the importance of social media in satisfying your communication needs.

Communicating with Social Media

2

CHAPTER OUTLINE

2.1 The Roles of Social and Mass Media 26

Characteristics of Social Media
Social Media Uses and Gratifications
Masspersonal Communication

2.2 Mediated Versus Face-To-Face Communication 29

Message Richness
Synchronicity
Permanence

2.3 Benefits and Drawbacks of Social Media 33

Benefits of Social Media
Drawbacks of Social Media

2.4 Influences on Mediated Communication 40

Gender
Age

2.5 Communicating Competently with Social Media 43

Maintaining Positive Relationships
Protecting Yourself

MAKING THE GRADE **47**

KEY TERMS **48**

ACTIVITIES **49**

LEARNING OBJECTIVES

2.1

Identify the distinguishing characteristics of social media, which combine elements of mass and personal communication.

2.2

Outline cases for using social media versus face-to-face communication, noting important differences.

2.3

Identify the benefits and drawbacks of mediated communication, and use this knowledge to maximize effectiveness and minimize the potential for harm.

2.4

Describe how factors such as gender and age can influence social media use.

2.5

Understand how to use social media competently to minimize misunderstandings, maintain positive relationships, and protect yourself.

Chloe Rose Stuart-Ulin's unusual job raises some important questions about the use of social media.

How do individual users interact with wide audiences via social media?

How does your use of social media platforms affect how others perceive you?

What ethical obligations do communicators have to represent themselves accurately on social media?

LOOKING AT HER PHOTO, you probably wouldn't mistake Chloe Rose Stuart-Ulin for a middle-aged man. But online, many people have.

In addition to being a journalist, Stuart-Ulin has served as a paid virtual dating assistant. Part of her job was to monitor the dating app accounts of heterosexual male clients to help identify appealing partners for them. But that's not all: her job also entailed communicating online as if she were the man profiled.

"Oh, you like Pink Floyd?" she might write, "Cool. I saw them in concert in '77." If you guessed that Stuart-Ulin wasn't part of the rock scene in the 1970s, you're correct. "This technically isn't a fib," she says. "My client did see Pink Floyd in 1977—though I wasn't born until 1992."[1]

In an article she wrote for the news site *Quartz*, Stuart-Ulin says that being an online dating assistant gave her insight into the hectic lives of hopeful singles. They simply don't have time to keep up with the demands of online dating, she says. Her job also led her through the "moral gray area" of impersonating someone else online. As an employee of ViDA, a matchmaking service that aims to save clients time, she was expected to use stereotypical assumptions in her communication. These included *"Rule 1: Don't make her think too hard," "Women want to date the alpha male,"* and *"Alpha males don't apologize."*

Stuart-Ulin ultimately resigned, disturbed by both the deceit and the sexist attitudes the work embraced, but ViDA still employs about 80 other client impersonators. She has a question for you: "Have you unknowingly flirted with a professional . . . ? Me, even?"

2.1 The Roles of Social and Mass Media

What do you think of when you hear *media*? As you read in Chapter 1, the term refers to channels through which messages flow. (The singular form is *medium*.) Print media (like newspapers and magazines) have been around for centuries. Electronic media (telephone, telegraph, and television) are more recent. And digital media began to grow only one generation ago.

Chapter 1 defined *mass communication* as messages transmitted to large, widespread audiences. Until the early 21st century, creators of mass media were almost always professional **gatekeepers** who controlled what information the public

gatekeeper In mass media, professionals who control the content of public messages.

received. Audiences were passive consumers who had little direct influence on the content of public messages.

Corporate mass media giants like Fox News and CNBC still exert a powerful influence, framing information and reinforcing beliefs. In the entertainment world, content producers such as Netflix and HBO shape and reflect public tastes. But today, not all media are aimed at mass audiences or controlled by corporate gatekeepers. **Social media** are dynamic websites and applications that enable individual users to create and share content or to participate in social networking. If you blog, tweet, or post or follow content on platforms such as Snapchat, Twitch, Instagram, Tumblr, YouTube, Meetup, or Facebook, you're using social media. If you use LinkedIn, BranchOut, or other professional networking sites, you're harnessing social media for career success. With the growth of social media, the number of channels to reach large audiences has increased. As a result, the agenda-setting power of mass media has diminished.[2]

> **social media** Dynamic websites and applications that enable individual users to create and share content or to participate in personal networking.

Characteristics of Social Media

Social media are similar to other channels of communication in some respects and different in others. Here we consider some distinguishing qualities. As you read, think about how these characteristics align with your communication practices.

User-Generated Content   Social media provide a channel for individual users—not just big organizations—to generate content. Take a look at the YouTube home page and you'll see an overwhelming array of content created by nonprofessionals, most of which would have been impossible to share so widely before social media. Some of these videos are silly (e.g., "Chimpanzee Riding on a Segway"). Others can be helpful: "How to Build a Gaming PC" or "Warning Signs of Alcoholism." Before social media, material like this was difficult or impossible to share.

Variable Audience Size   Whereas the mass media are aimed at large audiences, the intended audience in social media can vary. You typically address emails, texts, and other messages to a single receiver or maybe a few. In fact, you'd probably be embarrassed to have some of your personal messages circulate more widely. By contrast, blog updates, tweets, and other social media posts are often aimed at much larger groups of receivers. A quick look illustrates the range of podcast topics published by amateurs: Gridiorn Gals, Ready, Set, Real Estate!, Cook with Me, Homemade Camera Podcast, Inside Winemaking, Buddhist Geeks, Podcast for the Urban Survivalist . . . the list goes on almost endlessly.

Interactivity   The recipients of your social media messages can—and often do—talk back. For example, Snapchat allows users to spontaneously send in-the-moment experiences with friends and followers, who can easily respond with photos of their own.[3] This sort of back-and-forth sharing reflects the difference between traditional print media, in which communication is essentially one way, and far more interactive web-based social media.

ON YOUR FEET

What are the most common ways you use social media? How well do these satisfy your personal and practical needs?

Social Media Uses and Gratifications

In the mid-20th century, researchers began to study the question, "What do media do to people?" They sought answers by measuring the effects of print and broadcast media on users. Did programming influence viewers' use of physical violence? Did it affect academic success? What about family communication patterns?

In the following decades, researchers began to explore a different question: "What do people do *with* media?"[4] This branch of study became known as *uses and gratifications* theory. In the digital age, researchers continue to explore how we use both social media and face-to-face communication. The uses fall into four broad categories:[5]

Lisa Quinones-Fontanez felt lost when her son Norrin was diagnosed with autism. To share her story, she created the blog *Atypical Familia* to provide other parents with reassurance and hope.

How personal is your social media presence? Do you want to reach an audience beyond your circle of friends and family? If so, why?

TABLE 2.1

Common Types and Purposes of Social Media Content

CONTENT TYPE	PURPOSE	EXAMPLES
General social sharing	Connecting with individuals and larger audiences	Facebook, Twitter
Media sharing	Posting primarily photos, videos, and other media	Instagram, Snapchat, YouTube
User-posted news	Sharing news, providing personal updates, and exchanging ideas	reddit, Digg, CaringBridge
Crowdsourcing	Finding answers to questions	Quora, Innocentive
Discover, save, and share content	Discovering, saving, and sharing content	Pinterest, Flipboard, Scoop.it
Advocacy	Promoting causes	Human Rights Watch, Sierra Club
Career advancement	Exploring career options, networking	LinkedIn, Indeed, Glassdoor
Blogging and publishing	Publishing content online	WordPress, Tumblr, Medium
Consumer reviewing	Finding and evaluating goods and services	Yelp, Zomato, TripAdvisor
Interest-based discussion	Sharing interests and hobbies	Goodreads, MusoCity, Houzz
Social shopping	Spotting trends, following brands, and making purchases	Polyvore, Etsy, Fancy
Goods and services trading	Trading goods and services	Airbnb, Lyft, Taskrabbit
Dating	Finding romantic partners	Bumble, OkCupid, Match

1. **Information.** Asking questions such as: What do people think of a new film or musical group? Can anybody trade work hours this weekend? Is there a good Honda mechanic nearby? Can your network provide leads on getting your dream job?

2. **Personal relationships.** Seeing what your friends are up to, tracking down old classmates, announcing changes in your life to the people in your personal networks, and finding a romantic partner.

3. **Personal identity.** Observing others as models to help you become more effective, getting insights about yourself from trusted others, and asserting your personal values and getting feedback from others.

4. **Entertainment.** Gaming online with a friend, sharing your music playlists with others, joining the fan base of your favorite star, and finding interest or activity groups to join.

Thinking of social media specifically, we can identify some general categories of content types. Acknowledging that there is some overlap between these categories, Table 2.1 outlines some primary types and uses of content.

Masspersonal Communication

With the advent of social media, the boundaries between interpersonal and mass communication have blurred. **Masspersonal communication** is a term some scholars use to describe the overlap between personal and public communication.[6] Figure 2.1 illustrates the fuzzy boundaries where some communication channels meet.

Consider, for example, video sharing services like YouTube, Vimeo, and Twitch. They provide a way for individuals to publish their own content (e.g., a graduation or a baby's first birthday party) for a limited number of interested viewers. Some videos go viral, receiving thousands or even millions of hits. For example, over 1 million viewers have watched a series of lengthy, mostly wordless YouTube videos showing medical student Jamie Lee studying at her desk.[7] Go figure.

Whereas YouTube is a source for prerecorded video, Twitch makes it possible to stream personal content to a wide audience in real time. It's a tool for sharing used

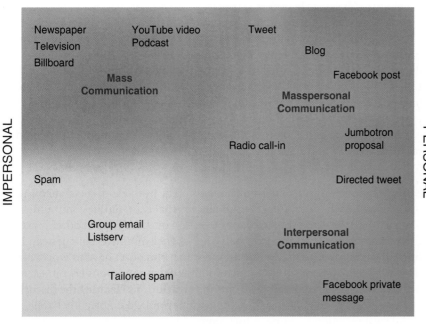

PUBLIC

Newspaper   YouTube video   Tweet
Television   Podcast
Billboard   Blog

        Mass   Facebook post
        Communication
                                Masspersonal
                                Communication

IMPERSONAL   Jumbotron   PERSONAL
                        Radio call-in   proposal

Spam   Directed tweet

        Group email
        Listserv   Interpersonal
                                        Communication

        Tailored spam
                                        Facebook private
                                        message

PRIVATE

FIGURE 2.1 Examples of Mass, Interpersonal, and Masspersonal Communication
Adapted from O'Sullivan & Carr, 2017.

by roughly 2.2 million monthly broadcasters, including would-be social media influencers, artists, comedians, hobbyists, musicians, athletes, sports commentators, and chefs.[8] Some streamers use Twitch as a source of income, while others do so purely for the exposure.

Twitter is another example of the fuzzy boundary between personal and mass media. You may broadcast updates to a small group of followers ("I'm at the concert— Great seats!"). Once it's been sent, though, followers can retweet it, share it, and use it to form opinions about you. In fact, you may have a wider audience in mind as you create your message; otherwise, you could have sent a private text or email.

2.2 Mediated Versus Face-To-Face Communication

As Figure 2.2 on page 30 shows, both face-to-face and mediated channels are important ways to communicate. What does in-person communication have in common with mediated options? How is it different?

In some ways, mediated and face-to-face communication are quite similar. Both include the elements described on pages 8–9: senders, receivers, channels, feedback, and so on. Both are used to satisfy physical, identity, social, and practical needs, as described in Chapter 1. Despite these similarities, the two forms of communication differ in some important ways.

Message Richness

Social scientists use the term **richness** to describe the degree to which nonverbal cues can clarify a message.[9] As Chapter 7 explains in detail, face-to-face communication is

richness The degree to which nonverbal cues can clarify a verbal message.

leanness The lack of nonverbal cues to clarify a message.

hyperpersonal communication The phenomenon in which digital interaction creates deeper relationships than arise through face-to-face communication.

rich because it abounds with nonverbal cues that hint at what others mean and feel.[10] By comparison, message **leanness** is the lack of nonverbal cues; mediated channels are much leaner than in-person communication.

To appreciate how message richness varies by medium, imagine you haven't heard from a friend in several weeks and decide to ask, "Are you okay?" Your friend replies, "I'm fine." The descriptiveness of that response would depend on whether you received it via text message, a phone call, or in person. You could probably tell a great deal more from a face-to-face response because it would contain a richer array of cues, such as facial expressions and vocal tone. By contrast, a text message is lean because it contains only words and possibly emoji. A voice message—containing vocal cues but no visual ones—would probably fall somewhere in between.

Identity Management via Mediated Channels   The leanness of social media messages presents another challenge. Without nonverbal cues, online communicators can create idealized—and sometimes unrealistic—images of one another. As you'll read in Chapters [self] and [nonverbal], the absence of nonverbal cues can help communicators manage their identities. Online, you don't have to worry about bad breath, blemishes, or stammering responses. Such conditions encourage what Joseph Walther calls **hyperpersonal communication**, accelerating the discussion of personal topics and relational development beyond what normally happens in face-to-face interaction.[11] Young online communicators self-disclose at higher rates and share more emotions than they would in person, often leading to a hastened (and perhaps premature) sense of relational intimacy.[12] This accelerated disclosure may explain why communicators who meet online sometimes have difficulty shifting to face-to-face relationships.[13]

Richer doesn't always mean better. When you want people to focus on what you're saying rather than on your appearance, leaner communication can be the way to go. Emails and texts don't reveal the quiver in your voice, the sweat on your

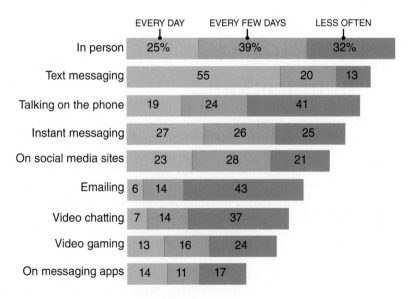

% of all teens who spend time with friends

	EVERY DAY	EVERY FEW DAYS	LESS OFTEN
In person	25%	39%	32%
Text messaging	55	20	13
Talking on the phone	19	24	41
Instant messaging	27	26	25
On social media sites	23	28	21
Emailing	6	14	43
Video chatting	7	14	37
Video gaming	13	16	24
On messaging apps	14	11	17

FIGURE 2.2 Media Used to Keep in Touch with Friends

UNDERSTANDING YOUR COMMUNICATION

What Type of Social Media Communicator Are You?

Answer the following questions for insight about your online communication style.

1. **When you and your friends go on a deep-sea fishing trip, you are nauseated the entire time. What are you most likely to do?**

   a. Post selfies in which you pretend to be having a great time.

   b. Tweet about how much you detest the ocean and warn others never to take a similar trip.

   c. Post pictures of your friends boozing it up on deck. So funny!

   d. Set social media aside and try to be fully present with your friends, even if the circumstances aren't ideal.

   e. Concentrate on making it back to shore. You'll think about social media later. Maybe.

2. **Imagine that you work for an outdoor furniture company that encourages you to promote products via your personal social media accounts. What are you most likely to do?**

   a. Post pictures of yourself using the furniture around the pool, at the beach, at the park, and so on.

   b. Use social media to tell all your friends and followers about the benefits of your products compared to others.

   c. Post a message such as, "I need to boost my sales numbers. Place an order, please!"

   d. Post promotional messages only to people you think will be interested in the products.

   e. Decline to use your personal social media in this way, reasoning that friendship and business are best kept separate.

3. **Which of the following best describes your social media presence?**

   a. I like to post pictures that show my activities and hobbies—my morning workout, the delicious dish I had for lunch, cute pictures of my pet, and so on.

   b. I mostly use social media to spread the word about issues I think are important, such as elections and social justice.

   c. My posts reflect how I'm feeling that day—whether it's happy, sad, frustrated, or excited.

   d. I strive for a balance between how much I "talk" and how much I "listen" to others on social media.

   e. I don't have much of an online presence.

4. **Which of the following best describes you?**

   a. I have more friends and followers on social media than most people do.

   b. I gravitate to people on social media whose opinions are similar to mine.

   c. I sometimes cringe when I look back at things I have shared on social media.

   d. I use social media mostly to stay in touch with people I care about.

   e. I wonder how people find the time to keep up with social media.

INTERPRETING YOUR RESPONSES

For insight about your communication style, see which of the following best describes you. More than one may apply.

Publicist

If you answered "a" more than once, you realize the image-enhancing potential of social media. This can help you present yourself and others in a favorable light, as a good publicist would. Just be sure that you are genuine in what you post and that you don't get too caught up in pleasing others. Publicists sometimes struggle to balance superficial relationships online (page 36) with close relationships in person.

Advocate

If you answered "b" multiple times, you often use social media to garner support for ideas and issues. At their best,

advocates inspire others (page 35); at their worst, they can come off as bossy or domineering. The trick is to stand up for what you believe, but also listen to others and keep an open mind.

Emoter

If you answered "c" two or more times, you tend to post emotional messages. This can help you connect with others and build strong relationships. It can also get you in trouble. Emoters are sometimes hyperpersonal online (page 32) in ways they may regret later. Avoid this pitfall by using the asynchronous (page 32) nature of social media to your advantage. Speak from the heart, but pause before posting.

Empath

If "d" answers best describe you, you use social media to foster

connections with others (page 35). You are likely to be an empathic listener who considers how your messages will be received, both in person and online. The segment on mindful listening in Chapter 6 will probably resonate with you.

Abstainer

If you saw yourself reflected mostly in "e" options, you aren't likely to annoy your online friends with minutia or self-serving messages. Don't abstain from social media entirely, however. There are advantages to being on board. For example, Chapter 10 presents strategies for building an online presence that will impress potential employers and others in the career world.

Because of their high potential for ambiguity, mediated messages are often subject to misinterpretation.

When have your mediated messages been misunderstood? What were the consequences?

polymediation The range of communication channel options available to communicators.

synchronous communication Communication that occurs in real time.

asynchronous communication that occurs when there's a lag between the creation and reception of a message.

forehead, or the clothing you're wearing. One study found that the text-only format of most online messages can bring people closer by minimizing the perception of differences due to gender, social class, race or ethnicity, and age.[14]

Ambiguity in Mediated Messages Because most mediated messages are leaner than the face-to-face variety, they can be harder to interpret with confidence. For example, irony and attempts at humor can be misunderstood. As a receiver, it's important to clarify your interpretations before jumping to conclusions. Adding phrases such as "just kidding" or an emoji such as ☺ can help your lean messages become richer, but your sincerity could still be interpreted as sarcasm. As a sender, think about how to send unambiguous messages (as much as that is possible) so that you aren't misunderstood.

Choosing a Communication Channel Social scientists use the term **polymediation** to address the range of options (e.g., Facebook, Instagram, Whatsapp, phone calls, texts) available to communicators when they are choosing how to engage with others.[15] Knowing which channel to choose can make a big difference in how a message will be received and understood. Design students seeking employment are wise to present a portfolio of photographs and drawings to demonstrate their work, realizing that the words on a resume aren't the best way to showcase their talents. Sometimes it can be best to avoid using words at all. *Transmediation* is a term used to describe recasting a message from one medium (e.g., written or spoken language) into other media (e.g., music, art).[16] If you've ever expressed regrets or sympathy by sending flowers rather than speaking or writing, you've practiced transmediation.

Synchronicity

Communication that occurs in real time, such as through in-person or phone conversations, is **synchronous**. By contrast, **asynchronous communication** occurs when there's a lag between receiving and responding to messages. Voice mail messages are asynchronous. So are "snail mail" letters, emails, and Tweets. When you respond to asynchronous messages, you have more time to carefully consider your wording or to ask others for advice about what to say. You might even choose not to respond at all. You can ignore most problematic text messages without much fallout. But that isn't a good option if the person who wants an answer gets you on the phone or confronts you in person.

Even if you want to respond, asynchronous media give you the chance to edit your reply. You can mull over different wording or even ask others for advice about what to say. On the other hand, delaying a response to an asynchronous message can send a message of its own, intentionally or not ("I wonder why she hasn't texted me back?").

Permanence

What happens in a face-to-face conversation is transitory. By contrast, the text and video you send via mediated channels can be stored indefinitely and forwarded to others. Sometimes permanence is useful.[17] You might want a record documenting your boss's permission for time off work or confirmation for the vacation rental reservation you've made online.

In other cases, though, permanence can work against you. It's bad enough to blurt out a private thought or lash out in person, but at least there's no visible record of your indiscretion. By contrast, a regrettable text message, email, or web posting can be archived virtually forever. Even worse, it can be retrieved and forwarded in nightmarish ways. In some parts of the world, laws have enshrined a partial "right to be forgotten." This allows users to seek removal of some obsolete and/or potentially damaging, private information.[18] Venture capitalist Hunter Walk proposed an even more drastic remedy: a legally mandated "start over" button would allow users of social networks to delete all their data, clear out their feeds and friend lists, and begin with a fresh account.[19] Obviously, no such magical device now exists.

A single indiscretion could come back to haunt you years later.

What can you do to avoid online embarrassments?

Some mediated platforms are designed to prevent message permanence. Snapchat, on which content typically disappears within 10 seconds, is the most popular of these time-limited instant messaging services.[20] The ephemeral nature of this app encourages less inhibited communication than on more permanent channels. In-depth interviews revealed that Snapchatters feel free to express themselves without worrying about how their messages look and sound because they know those messages will disappear.[21] As a result, the respondents said that Snapchat is more enjoyable and puts them in a more positive mood than other digital platforms. But because it encourages flirting (there's no enduring evidence of it!), Snapchat generates more partner jealousy than more permanent sites like Facebook.[22] About half the college-age respondents surveyed reported sending drunk photos via Snapchat—and between 13% and 20% admitted sending snaps involving sexting or "legally questionable activities." The ability to capture screen shots makes it a risky bet that images you want to disappear will truly vanish forever, even on Snapchat.) Given risks like these, the best advice is to take the same approach with mediated messages that you do in person: Think twice before saying something you might later regret.

2.3 Benefits and Drawbacks of Social Media

By now it's clear that social media can be a boon for connecting with others. You can also see that using these technologies carries some risks. In the following section, we take a closer look at the benefits and risks of mediated communication.

Benefits of Social Media

Steve Jobs, the legendary cofounder of Apple Computer, once suggested that personal computers should be renamed "*inter*personal computers."[23] He had a point: Social media have great potential to bring people together and enhance the quality of their relationships. In one survey, over 80 percent of U.S. teens say social media help them feel more connected to their friends.[24]

Research suggests that many couples who meet via online dating services go on to enjoy long-term relationships.

How might beginning a relationship online differ from beginning one in person?

Opportunities to Connect In the jargon of social science, mediated channels offer "low-friction opportunities" to create and maintain close relationships.[25] About two-thirds of teens say they have made new friends online, and 9 in 10 say they keep in touch with established friends via technology.[26]

The world of courtship and dating has changed with advances in social media. Online dating services were originally viewed as last-ditch options for the romantically challenged. Skeptics questioned how well a computer could match people together and whether relationships started online could be successful in person. Research has put many of those concerns to rest.[27] In one survey, more than one-third of the 19,000 married respondents said their marital relationship began online.[28] Couples who meet online stay together about as much as those who met in person, and those who stay together transition to marriage more quickly and, on average, have happier marriages.[29]

Online dating has many advantages.[30] Trying to find a compatible partner can be challenging, and online dating services help identify prospective partners with similar backgrounds and interests. Online dating sites also streamline the dating process: You can review prospective dates before you invest time and energy in meeting in person. And online dating can reduce some of the initial awkwardness that may come with seeking romantic partners.

Along with romance, social media can play a role in the startup of other close relationships. Discussion boards, blog sites, and online forums can create a sense of "virtual community" between strangers.[31] Whether you're a follower of Premiere League soccer, an avid environmentalist, or a devotee of punk rock, you can find like-minded people online. These virtual community members often provide social support for each other.

Communicating online can be especially helpful for people who are shy.[32] One study found a positive connection between shyness, Facebook use, and friendship quality.[33] The researchers concluded that social networking services provide "a comfortable environment within which shy individuals can interact with others."

Social media can be especially useful for those who find it difficult to get out and about.[34] Mediated friendships can help alleviate feelings of loneliness.[35] Electronic communication isn't a replacement for the face-to-face variety, but it expands the world for people who seek connection beyond the people they already know.

Sustaining and Enriching Relationships Social media offer a powerful way to both maintain and rekindle or deepen relationships.[36] College students cite "keeping in touch" as one of the main reasons for using social media.[37] The asynchronous capabilities of social networking tools help friends stay in touch without having to connect in real time.[38]

The masspersonal nature of blogging—addressing a large audience with personal information—makes it a logical choice for keeping in touch. Imagine that you want to share the news about landing a good job. You can inform everyone via a single blog entry. In addition, the interactivity of blogs allows readers to add their own responses (such as congratulations), which creates an environment for further maintenance of communication. Toward this end, blogs and other social networking sites provide more maintenance "bang" for the message-sending "buck."[39]

Mediated channels are especially important for sustaining the growing number of long-distance romantic relationships. Some 3 million Americans live apart from their spouses for reasons other than divorce or discord,[40] and between 25% and 50% of college students are currently in long-distance relationships.[41] One study demonstrated the value of video chats in maintaining such relationships.[42] For partners who used technologies such as Skype and FaceTime, the number of daily interactions was lower than those who lived together, but their exchanges were longer and included more personal disclosures. One researcher explained why: "Seeing someone's face and having those facial expressions really makes a big difference. Sometimes when we're on the telephone, we can be distracted, but if you're sitting down for a video chat, then you're really focused on each other."[43] Some scholars suggest that interaction via social media can actually be *more* effective than face-to-face interaction at improving the quality of a relationship.[44]

Social Support   Before social media, finding support for personal problems usually meant reaching out to friends, family members, and perhaps trusted members of one's local community. Those personal contacts are still important, but today social media provide an alternative source of support for matters ranging from marital problems[45] to substance abuse[46] to suicide prevention[47] to coping with senseless acts of violence.[48] The words of one recovering alcoholic demonstrate the value of online support. After reaching out for help

> I began to get emails, phone calls, text messages, tweets and other digital notes from people around the world. Some offered kind words. Some offered support. Many people shared their own stories of addiction. In my darkest times, these notes would come. And always, without question, they pulled me back from the brink.[49]

About 20% of Internet users have gone online to find others with similar health problems.[50] When asked why, a common response is that they feel more comfortable talking with like-minded people with whom they have few formal ties—particularly when the health issues are embarrassing or stigma-laden. For example, researchers in one study explored how blogs offer social support for people who are morbidly obese.[51] These sites often give rise to interactive communities in which people with similar conditions share their struggles and offer each other affirming feedback. One blogger in the study put it this way:

> When I have a bad week on the scale or a problem I don't know how to handle, all I have to do is write up an entry and post it on the blog. My readers are always full of good advice, comments and support.

Because online support groups and blogs are relatively anonymous and the participants are similar, they can offer help in ways that make strangers seem like close friends.

Advocacy and Fundraising   The "mass" dimension of social media has dramatically increased the power of individuals and informal groups to change society. Consider the effects of the following examples of movements that spread via social media: MeToo (victims of sexual offenses), Black Lives Matter (protesting abuses of authority), OscarsSoWhite (diversity in film), NeverAgain (gun violence). Spurred by social media, roughly half of Americans reported having been civically active in the past 12 months.[52] The same power has resonated around the world: Je Suis Charlie in France showed solidarity in the face of terrorist attacks. In India, DelhiGangRape shone a spotlight on the culture of violence against women. In 2014, the Twitter campaign #BringBackOurGirls helped to alert the world to the plight of 276 Nigerian schoolgirls who were kidnapped by the terrorist group Boko Haram.[53] Beyond advocacy, social media also provide a way to raise funds quickly in support of a candidate or cause. The crowdfunding platform Gofundme has enabled raising over $5 billion for a wide range of causes—personal, professional, and political.

Drawbacks of Social Media

Despite all its advantages, mediated communication isn't a replacement for face-to-face interaction. One study of college students who frequently use text-based messaging concluded that "nothing appears to compare to face-to-face communication in terms of satisfying individuals' communication, information, and social needs."[54] Furthermore, there's a synergistic relationship between text-based messages, phone contact, and in-person communication. If you regularly communicate with friends and family online, it's likely that you will also call them and try to see them more often.[55] In other words, few close relationships use mediated channels to the exclusion of in-person communication.

Along with the potential benefits, mediated relationships can have a dark side.[56] Understanding the potential drawbacks can help you guard against them.

Superficial Relationships   Social scientists have concluded that most people can only sustain about 150 relationships.[57] (That figure has been termed "Dunbar's number" in recognition of Oxford University anthropologist Robin Dunbar, who established it.) If you're lucky, you have an inner circle of five "core" people and an additional layer of 10 or 15 close friends and family members.[58] Beyond that lies a circle of roughly 35 reasonably strong contacts.[59] That leaves about 100 more people to round out your group of meaningful connections. You almost certainly don't have the time or energy to sustain relationships with many more people.

Dunbar's number is much smaller than the array of "friends" that many people claim on social networking sites. For example, some Facebook users seem proud to have hundreds, or even thousands, of social media friends. Dunbar explored the discrepancy between "true" and mediated friends by comparing the online exchanges of people with thousands of friends to those who identified smaller numbers of online relationships.[60] He discovered that, regardless of how many online friends users claimed, they only maintained relationships with the same number of people—roughly 15. As Dunbar put it, "'People obviously like the kudos of having hundreds of friends but the reality is that they're unlikely to be bigger than anyone else's."[61]

Besides being superficial, a large number of online "friendships" can yield diminishing returns. You may impress others if you list 150 friends in your profile, but research shows that as that number doubles or triples, you're likely to be perceived in less flattering terms, such as superficial or desperate.[62] Some scholars have suggested that seeking an unrealistically large number of social media friends might be compensation for low self-esteem.[63]

Even in close relationships, communication technologies can lead to more superficial communication. MIT professor Sherry Turkle explains:

> As we ramp up the volume and velocity of online connections, we start to expect faster answers. To get these, we ask one another simpler questions; we dumb down our communications, even on the most important matters. It is as though we have all put ourselves on cable news.[64]

"It says no one really knows who he is, but that he's got 400,000 followers on Twitter."

Source: Cartoonist Group

Social Isolation   There's a correlation between loneliness and what social scientists call a *preference for online social interaction*.[65] Heavy users prefer to interact with others online, which can detract from in-person relationships and result in feelings of loneliness.[66]

Two factors help explain how and why, for some people, online communication crowds out face-to-face interaction. The first involves social skills—or more accurately, a lack of those skills. People who typically struggle to communicate successfully in person because of nervousness or anxiety can communicate online without facing many challenges.[67] They can edit thoughts and transmit them when and how they want, and even construct identities that are more attractive than their in-person presence.

As online interaction proves successful, lonely and socially anxious people who struggle with social interaction offline receive positive feedback from others online. As you read earlier, these people may begin to feel respected and important online but disconfirmed offline.[68] This leads to increasing dependence on and desire for online interpersonal interaction.

People who spend excessive time on the Internet may begin to experience problems at school or work and withdraw further from their offline relationships.[69] If they already had trouble interacting with people face to face, that problem may get even worse. It's hard to say where the cycle begins and ends. Are people socially awkward because they play interactive online games all day—or do people play interactive online games all day because they are socially awkward and can escape a not-so-kind reality? It's almost impossible to say, but the effect is the same in terms of isolation from in-person networks.

Relational Deterioration   Although social media can enhance relationships in healthy doses, heavy social media usage can lead to relational problems. Researchers in one study found a negative relationship between interpersonal intimacy and excessive involvement in online social networking.[70] Other studies have revealed that the mere presence of mobile devices can have a negative effect on closeness, connection, and conversation quality during face-to-face discussions of personal topics.[71] You can probably think of times when you've heard or said, "Put away that phone and talk to me!" Some even blame social media for relational cheating and breakups.[72] Although this position may be extreme, it's important to recognize that online affairs can be as serious as the in-person variety.[73]

After studying U.S. demographic data, researchers found a correlation between social network use, marital dissatisfaction, and divorce.[74] Facebook usage in particular emerged as "a significant predictor of divorce rate and spousal troubles." The authors make clear, however, that social media may be as much a symptom of relational problems as a cause in that "men and women troubled by their marriage may turn to social media for emotional support."

Mental Health Issues   Along with relational deterioration, excessive use of social media can take a toll on users. At the most basic level, too much time online can lead to sleep deprivation.[75] There is a link between heavy reliance on mediated communication and conditions including depression, loneliness, and social anxiety.[76] People who spend excessive time online may begin to experience problems at school or work and withdraw further from their offline relationships. Many people who pursue exclusively online social contacts do so because they have social anxiety or low social skills to begin with. For these people, retreating further from offline relationships may diminish their already low social skills.

Use of social media can also lead to diminished self-esteem. Most user profiles paint an unrealistically positive portrait, so it's easy to understand why viewers who take them at face value feel inferior by comparison.[77] Problems are especially great for *passive* consumers of social media.[78] By contrast, more active social media users feel more connected and better about themselves.

Deception   Nev Schulman, a 20-something New York photographer, was flattered and intrigued when a little girl named Abby from Michigan began sending him fan mail and paintings based on his work. Nev and Abby struck up an online friendship, and soon he was exchanging romantic messages with Abby's older sister, Megan. Nev was intrigued by the soulful songs Megan claimed to have written and by the beautiful photos of herself she posted online. When Nev and his buddies visited Michigan to meet Megan and her family, they discovered he had been duped. "Megan" was actually Abby's mother and had done the paintings herself. The photos she had posted of "Megan" were actually of a stranger. These real-life events are depicted in the documentary *Catfish*. Nev parlayed the lessons he learned into *Catfish, a* TV show in which he tries to help online communicators connect in person. These meetings often reveal deceit.

Although the *Catfish* film and show capture extreme cases of deceit, online misrepresentation is common. That's not surprising since it's possible to craft deceptive messages online in ways that wouldn't be possible in face-to-face communication where nonverbal cues might give a person away. For instance, dating profiles can use dated or edited photos as well as outright lies, underreporting weight and overreporting height.[79] Some rationalize their decision, claiming it's not really deception since they intend to lose a few pounds in the future. Others explain that identity misrepresentation is a social norm—"Everyone else is doing it so I need to as well." In other cases, online representations are outright lies. People declare they are single when they're actually involved in a romantic relationship, and others salt their LinkedIn profiles with jobs they never held. Given the unreliable nature of online self-characterizations, it's probably a good idea to view them with at least a little skepticism.

Along with the moral dimension of deception, research suggests that seriously misrepresenting yourself can damage your reputation, especially with strangers and new acquaintances. In one study, experimenters asked college students to identify items they thought were misleading on the Facebook profiles of a close friend and an acquaintance.[80] The results indicated that people were more willing to give their close friends "a pass" for their misrepresentations, but they were likely to think that acquaintances were hypocritical or untrustworthy for doing the same.

The advantages of smartphones are obvious. But overuse can get in the way of healthy relationships. For a tragicomic example, see the YouTube video "I Forgot My Phone."

How would being without your phone for the day affect both your own experience and your perception of others?

ETHICAL CHALLENGE How Honest Should You Be on Social Media?

Less than a third of social media users report their self-characterizations are always honest.[81] Why is this figure so high—and what are the effects of lying online? What does it mean to be honest on social media? Several types of behavior may be at play:

- **Fabrication**—deliberately telling a false story or misrepresenting facts (e.g., lying about your age, appearance, experience, or relationship status).

- **Omission**—leaving out significant information (e.g., talking about an accomplishment while failing to acknowledge others involved).

- **Distortion**—giving a misleading account or impression (e.g., posting photos of your life as nothing but one fun experience after another).

- **Exaggeration**—close cousin of distortion: representing some aspect of yourself as better than it is (e.g., overstating your leadership experience in your LinkedIn profile).

- **Denial**—refusing to acknowledge a truth (e.g., categorically denying past actions or statements).

What is your moral obligation to represent yourself faithfully? Before the dawn of social media, philosopher Immanuel Kant suggested that in some cases omission might be appropriate, but in other circumstances we are obliged to tell the complete truth.[82] Not all ethicists have shared Kant's rigid standards of truth telling, however. Some deem altruistic lies acceptable in certain situations. Utilitarian philosophers claim that the morality of a mistruth can be justified if it leads to the greatest happiness for the greatest number of people. Philosopher Sissela Bok offers some circumstances in which deception may be justified: doing good, avoiding harm, and protecting a larger truth.[83]

DISCUSSION: In what ways have you potentially misrepresented yourself on social media, and why? How intentional were these choices?

ACTIVITY: Draft a set of guidelines for using social media honestly. For example, ask whether it is ever acceptable to fabricate, omit, distort, exaggerate, or deny information on social media if you would not do so in person. Why or why not?

Stalking and Harassment   You probably have searched the Internet to find out more about someone you find interesting, perhaps an intriguing stranger, a former friend, or a romantic partner. **Online surveillance** is a discreet way of monitoring the social media presence of unknowing targets.

Although occasional online surveillance is relatively harmless, it's a problem if it escalates into an unhealthy obsession or even full-blown **cyberstalking**.[84,85] One team of researchers found that cyberstalkers are most often males monitoring their female ex-partners.[86] People who discover that they're being cyberstalked can suffer the same types of mental and emotional trauma experienced in offline stalking. If you believe you're under unwanted surveillance by someone you know, you can alert legal authorities and victim assistance professionals. You also might want to consider getting off social media until you feel safe again.[87]

Cyberstalking can feel creepy and threatening. But, even more hateful, it isn't as painful as **cyberbullying**—a malicious act in which one or more parties aggressively harass a victim online, often in public forums.[88] Cyberbullies can create hateful messages on social networking sites and circulate disparaging texts, emails, and photos about their victims. Cyberbullying has become a widespread phenomenon with some dire consequences.[89] Although middle school is the peak period for cyberbullying, it can start as early as grade school and continue into the college years and beyond.[90] Almost 6 in 10 teens report being the target of online harassment.[91] Cyberbullying has been linked to a variety of negative consequences, including poor academic performance, depression, withdrawal, psychosomatic pain, drug and alcohol abuse, and even suicide.[92] News like this is especially discouraging when you consider that 81% of cyberbullies admit that they bully simply because "it's funny."[93]

Because cyberbullying is a relatively recent phenomenon, researchers are busy compiling data about the process and its outcomes. Here are a few of their findings:

A key to stopping cyberbullying is blowing the whistle on perpetrators. Unfortunately, many adolescents are reluctant to report bullying for fear that the bully will

online surveillance Discreet monitoring the social media presence of unknowing targets.

cyberstalking Ongoing monitoring of the social presence of a person.

cyberbullying A malicious act in which one or more parties aggressively harass a victim online, often in public forums.

retaliate against them or that parents who realize they are being bullied will revoke their social media privileges. Teens are more likely to tell their friends about online harassment, so many school programs encourage peer-led support and intervention.

Authorities agree that cyberbullying will remain a problem as long as it stays a secret. If you're being bullied online, keep copies of the harassing messages—and then contact an appropriate teacher, administrator, or supervisor. Most schools and companies have policies to help protect you. And if you know someone who is being victimized—especially if it's a young person—be receptive and help arrange professional intervention. Open communication is vital to bringing cyberbullying out of the shadows.

2.4 Influences on Mediated Communication

Who we are affects the way we use social media and other forms of mediated communication. Two of the strongest influences on personal use are gender and age.

Gender

As in face-to-face interaction, men and women tend to communicate differently online.[94,95] Here are a few patterns revealed in the research:

- In general, men use more large words, nouns, and profanity than women. On the other hand, females use more personal pronouns, verbs, and hedge phrases ("I think," "I feel").

- Whereas males and females use the word "we" about equally as often, they do so in different ways. Women are more likely to use what's known as the "warm we" ("We have so much fun together"), while men are more inclined toward the "distant we" ("We need to do something about this).

- Females use more emotion words and first-person singular pronouns on Facebook than men do, as revealed by an extensive study of more than 15 million Facebook status updates involving about 75,000 volunteers.[96] For their part, men were more likely to make object references (talking about things rather than people) and to use swear words.

Gender differences are partly a matter of perception. In one study, participants were given randomly selected gendered avatars—some matching their biological sex, some not. Communicators who were assigned feminine avatars expressed more emotion, made more apologies, and used more tentative language than did those with masculine avatars. In other words, participants adapted their language to match linguistic gender stereotypes.[97]

Stereotypes don't always match reality, however. Computers programmed to analyze the content of online posts correctly identify the sex of an author about 72% of the time (50% is chance). This suggests that communicators don't always reflect traditional patterns, especially among adults.

Online language differences between the sexes are more pronounced among adolescents. When researchers examined the word choices of teenage boys and girls in chat rooms, they found that teen males were more active and assertive. They initiated online interactions more often than the girls and made more flirtatious proposals such as "Any hotties wanna chat?". In general, the females were more reactive in their online posts, using terms such as *wow*, *omg*, and *lmao* in response to others' comments.[98] The researchers noted that these accentuated differences were probably due to the age of the participants and that some of the distinctions would likely recede in adulthood.

Age

If you were born after the early 1990s, mediated communication probably feels as natural as breathing. It's a different story for people who grew up in a world without the technology that most take for granted today.[99] People who have firsthand experience with telegrams, floppy disks, and dial-up modems have experienced life without social media.

Age isn't the only factor determining digital natives. Socioeconomic status and country of origin also play roles. Nevertheless, there are some clear trends in the preferred communication modes for different generations. It's probably no surprise that texting, emailing, and telephoning form an age continuum: older communicators are more likely to prefer phone conversations, and teens are more likely to text and to consider phone calls as annoying and even intrusive.[100,101,102] Contemporary parent–child arguments may include questions such as "Why don't you just call?" followed by "Why don't you just text me back?" It's also not surprising that younger communicators use social networking sites more than older communicators do, although the gap isn't as large as it used to be.[103]

These age distinctions may not hold true in the future. Today's texting teens won't necessarily become prolific emailers or phone callers in middle age, and older communicators are joining the digital revolution at rapid rates.[104] But for now, you might want to consider the age of your message recipient when deciding whether to text, email, or call. Your choice may have a bearing on when—or even if—you get a response.

Along with channel preferences, age also shapes what topics people discuss when using mediated communication. The researchers who produced the word cloud in the "Gender" section also analyzed age differences in more than 15 million Facebook posts.[105] Here are some of the findings:

- Not surprisingly, school was a major topic for 13- to 18-year-olds. Typical terms in teens' messages include "homework," "math," and "prom." Abbreviations such as "lol," "jk," and "<3" were also common.

- 19- to 22-year-olds more often post about college. Typical terms include "semester," "studying," and "campus." Other lifestyle choices were also prominent in their use of "drunk," "tattoo," and a host of swear words.

- By their mid-20s, people tend to shift toward topics that involve words such as "office," "pay/paying," and "wedding." But communication at this stage isn't all serious: "beer" is still a common term.

Source: Ron Barrett http://ronbarrettart.com/Editorial.html

- 30- to 65-year-olds post often about *family*. Typical terms include "daughter/son," "pray/prayer," "friends," and "country."
- Use of the word "we" in Facebook posts tends to increase with age, while use of "I" decreases. This suggests the increasing importance of friendships and relationships as people age.

Age differences even apply to the use of punctuation. Older communicators use the same rules they learned in school and have a distinct "digital accent." One example involves the use, or nonuse, of periods in online messages. Whereas older communicators use periods in the way they were taught in school, lifelong digital communicators often interpret them as a sign of negative feelings.[106] Other forms of punctuation can also signal age, as one author describes:

> . . . unless you want to be viewed as straight-up geriatric, it's, best to avoid the comma at all costs.
>
> The only person who still does this is my dad, who also signs his texts "ILY [I love you], Daddy," as if I didn't know who was texting me in the first place. Trust me when I say that no human under 50 is using this particular shortcut.[107]

@WORK

Using LinkedIn for Career Success

With over 500 million members, LinkedIn is the go-to networking tool to advance your career. You can use it as a billboard to build your personal brand, presenting yourself the way you want to be seen. LinkedIn is especially useful for expanding your professional network, connecting you with people who can help you succeed. If you are actively searching for a job, you can make yourself visible to recruiters searching for someone with your skill set.

Here are some tips for using LinkedIn effectively.

☐ **Write a compelling summary** In 2,000 characters or less, make viewers want to read about your educational and professional history, skills, and accomplishments. This summary is likely to be the first section viewers see, so it's essential to identify your qualifications and what sets you apart.

☐ **Use key words in your profile** Your LinkedIn profile acts as an online resume. Think about what skills, experience, and personal traits you want to highlight, using key words that viewers will be searching for. Look at job descriptions and other profiles as points of reference.

☐ **Think about your audience** Who is your target audience? What do you want them to see or not see? Consider changing your visibility settings to optimize the reach of your message.

☐ **Keep it professional** This isn't the place to brag about your spring break. Be sure to make your profile public so anyone can view it. Include a headshot that looks professional.

☐ **Seek (and provide) positive recommendations** Display endorsements from professors, colleagues, supervisors, or clients. It's fine to ask them to highlight qualities you want to promote: your work, attitude, skills, achievements, professionalism, and ethics. Consider paying it forward by recommending others, especially when viewers would find your perspective valuable.

☐ **Build your online portfolio** Your profile can serve as a portfolio of your work. When appropriate, incorporate multimedia samples to showcase your skills and accomplishments.

☐ **Proofread everything you write** A single error can demolish your credibility.

☐ **Expand your professional network** Use LinkedIn to connect with people who can enhance your career success. Personalized invitations to connect are often more successful than the default.

🎯 2.5 Communicating Competently with Social Media

Perhaps you've found yourself in situations like these:

- You want to bring up a delicate issue with a friend, family member, or colleague.
- You aren't sure whether to do so in person, on the phone, or via a mediated channel such as texting or email.
- You're enjoying a film at the theater—until another moviegoer starts a cell phone conversation.
- A friend posts a picture of you online that you would rather others not see.
- Someone you care about is spending too much time online, crowding out face-to-face interactions.
- Reading a text message while you walk to an appointment, you bump into another pedestrian.
- Even though you know it's a bad idea, you read a text message while driving and nearly hit a cyclist.
- You are copied on an email or text message that was obviously not meant for your eyes.
- You receive so many messages that you have trouble reading them all and meeting school and work deadlines.

None of these situations would have existed a generation ago. They highlight the need for a set of social agreements that go beyond the general rules of communicative competence outlined in this chapter. The following pages offer some guidelines that have evolved in recent years. Although they won't cover every situation involving mediated communication, they can help you avoid some problems and deal most effectively with others that are bound to arise.

Maintaining Positive Relationships

Gaining goodwill and respect are just as important in mediated relationships as in face-to-face communication. Here are some guidelines. Mediated communication calls for its own rules of etiquette. Here are a few. (The checklist "Use Social Media Courteously and Wisely" provides a handy reminder of these rules.)

Respect Others' Need for Undivided Attention   It might be hard to realize that some people are insulted when you divide your attention between your in-person conversational partner and distant contacts. As one observer put it, "While a quick log-on may seem, to the user, a harmless break, others in the room perceive it as a silent dismissal. It announces: 'I'm not interested.'"[108]

As Figure 2.3 shows, most people understand that using mobile devices in social settings is distracting and annoying. Chapter 6 has plenty to say about the challenges of listening effectively when you're multitasking. Even if you think you can understand others while dealing with communication media, it's important to realize that they may perceive you as being rude.

Keep Your Tone Civil   If you've ever shot back a nasty reply to a text or instant message, posted a mean comment on a blog, or forwarded an

CHECKLIST ✓

Use Social Media Courteously and Wisely

Here are a few reminders to keep you from having regrets about your use of social media.

- ☐ Choose the best medium for your message. (It may not be social media.)
- ☐ Be careful what you publish, keeping in mind that it may reach a larger audience than you intended.
- ☐ Reserve time to give people you care about undivided attention, without the distraction of devices.
- ☐ Keep your tone online civil. Don't post a message you wouldn't deliver in person when you are calm.
- ☐ Be mindful that cell phone use can be annoying to people around you.
- ☐ Avoid posting information or images that might embarrass you (or others) in the future.
- ☐ Don't use devices while you are behind the wheel.

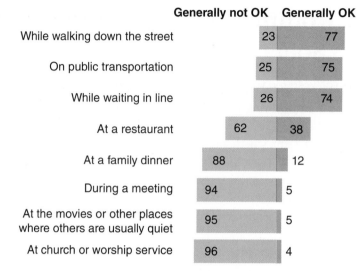

	Generally not OK	Generally OK
While walking down the street	23	77
On public transportation	25	75
While waiting in line	26	74
At a restaurant	62	38
At a family dinner	88	12
During a meeting	94	5
At the movies or other places where others are usually quiet	95	5
At church or worship service	96	4

FIGURE 2.3 "When Is It Acceptable to Use a Cell Phone?

trolling Attacking others via online channels.

disinhibition The tendency to transmit messages without considering their consequences.

embarrassing email, you know that it's easier to behave badly when the recipient of your message isn't right in front of you.

Trolling is attacking others through online channels. The academic term for transmitting messages without considering their consequences is **disinhibition**. Whatever you call it, this sort of abuse is more likely in mediated channels than in face-to-face contact.[109] [Could use specific examples here of how people are meaner online than in person]

Hostile, abusive language, usually sent anonymously, is a sad fact of online life. When you encounter these sorts of messages, the best strategy is to not respond.

How would your civility online rate using the checklist on page 43?

Respect Privacy Boundaries   Sooner or later you're bound to run across information about others that you suspect they would find embarrassing. If your relationship is close enough, you might consider sending out a for-your-information alert. In other cases, you may intentionally run a search about someone.

Be Mindful of Bystanders   If you spend even a little time in most public spaces, you're likely to encounter communicators whose use of technology interferes with others: restaurant patrons whose phone alerts intrude on your conversation, pedestrians who are more focused on their handheld device than on avoiding others, or people in line who are trying to pay the cashier and use their cell phone at the same time. If you aren't bothered by this sort of behavior, it can be hard to feel sympathetic with others who are offended by it. Nonetheless, this is another situation in which the platinum rule applies: Consider treating others the way *they* would like to be treated.[110]

1. **Don't assume you understand.** Before jumping to conclusions about the meaning of a message or the intentions of the sender, consider alternate interpretations. Ask the other person for clarification. The skill of perception checking (Chapter 3, page 65) provides a template: "When you posted a barfing emoji after my message, I thought you were mocking what I said. But maybe you meant it as a joke. I'm confused—what *did* you mean?"
2. **Seek common ground.** Start discussions by looking for beliefs you share instead of focusing on differences. You may be surprised to find that disagreements aren't as stark as they first seem. If you're debating politics, for example, try to find areas where you agree with others—making the world better for the next generation, seeking fairness, reducing economic inequality, and so on.
3. **Assume you have something to learn.** As Bill Nye noted, "Everyone you will ever meet knows something you don't."[111] Even if you disagree, you can learn by exploring the thinking behind positions that differ from yours.
4. **Try to be charitable.** Instead of interpreting ambiguity in the worst possible way, start by taking other people's words at face value, assuming that what they *actually* say is sincere.

Protecting Yourself

Respecting others is only part of competent online communication. You also need to keep yourself safe. The following pages offer tips on how to do so.

Be Careful What You Post   You may be forever haunted by mistakes in the form of text and photos posted online. As a cautionary tale, consider the case of 10 students whose acceptance offers were rescinded after university officials discovered their posts in a private Facebook group for incoming students.[112] The content included images mocking the deaths of children, the Holocaust, and sexual assault. Harvard's policy reserves the right to withdraw acceptance offers if an "admitted student engages in behavior that brings into question his or her honesty, maturity, or moral character."

Don't Believe Everything You See   Some stories offer clues that they're untrustworthy: excessive exclamation points, capital letters, and misspellings ("THIS IS NOT A HOAXE!"). But not all misinformation is this easy to spot. The checklist "Evaluating Online (Mis)information" offers tips to help you evaluate the quality and accuracy of information you read online.

Some incautious posts can go beyond misguided attempts at humor. Consider the phenomenon of "sexting." One survey revealed that 10% of young adults

Evaluating Online (Mis)information

In an environment where anybody can publish claims, no matter how bogus, it's vitally important to separate solid information from information that is either intentionally or unintentionally inaccurate or misleading. This checklist can help you think critically about content you come across online.

☐ **Consider the source.** Only the most gullible reader would swallow a story from the news parody *The Onion* (Sample headline: "Archaeological Dig Uncovers Ancient Race of Skeleton People.") Most false news stories are more dangerous, often coming from sites that disguise themselves as legitimate. Some examples: a site branded abcnews.com.co (not the actual URL for ABC News), and the *Boston Tribune* (whose "contact us" page lists only a Gmail address).

☐ **Check the author.** Another tell-tale sign of a fake story is often the byline. Sometimes a search of the writer's name will reveal that he or she has dubious or nonexistent credentials. Another tipoff is seeing the same writer's name on most or all of the stories on a particular site.

☐ **Check your biases.** Recognizing your own lack of objectivity can be difficult. What's known as *confirmation bias* leads people to put more stock in information that reinforces their existing beliefs while discounting information that doesn't. Ask yourself how someone less disposed to your belief system would react to information you accept uncritically.

☐ **Consult fact checkers.** Plenty of services can help validate or debunk stories that appear in your news feed. Among the best are FactCheck.org, Snopes.com, the *Washington Post* Fact Checker, and PolitiFact.com.

between the ages of 14 and 24 have texted or emailed a nude or partially nude image of themselves to someone else, and 15% have received such pictures or videos of someone else they know.[113] Perhaps even more disturbing, 8% reported that they had been forwarded nude or partially nude images of someone they knew.[114]

When minors are involved, authorities can make arrests for manufacturing, disseminating, or even possessing child pornography. Far worse, some teens have committed suicide when explicit photos were posted online.[115] Even without such dire consequences, it's not hard to imagine the unpleasant consequences of a private photo or text going public.

Balance Mediated and Face Time   It's easy to make a case that many relationships are better because of social media. And as you've already read in this chapter, some research supports this position. But even with all the benefits of communication technology, your own experience probably supports research saying that too much time online is unhealthy. One way to reveal the impact of social media on your life is to try a "detox" in which you stay off your devices for a period of time. Participants in one study who were instructed to not use social media for one week exhibited withdrawal symptoms, such as craving and boredom.[116]

Although experts disagree about whether Internet addiction disorder (IAD) is a certifiable addiction or just a symptom of another issue, they suggest several strategies for reining in excessive use of digital media. Unlike other addictions, such as those to drugs and alcohol, treatment for Internet addiction focuses on moderation and controlled use of the Internet rather than abstinence. If you are worried about your Internet use:

- Keep track of the amount of time you spend online so you can accurately assess whether it's too much.

- Plan a limited amount of online time in your daily schedule and see if you can stick to your plan.

- Make a list of problems in your life that may have occurred because of your time spent online.

- If you do not feel able to change your behavior on your own, seek the help of a counselor or therapist.

Be Safe   Many people fail to realize the hazards of sharing certain information in public forums, and other people don't realize that what they are posting is public. You may post your "on vacation" status online, assuming that only your friends can see your message. But if a reader uses a public computer or lets another friend see your page, unintended recipients are viewing your information.[117]

As a rule, don't disclose information in a public-access medium that you would not tell a stranger on the street. Even personal emails present a problem: They can be forwarded, and accounts can be hacked. The safest bet is to assume that mediated messages can be seen by unintended recipients, some of whom you may not know or trust.

Careless use of social media can damage more than your reputation. Using a cell phone while driving is just as dangerous as driving

"I need a more interactive you."

Source: Mick Stevens The New Yorker Collection/The Cartoon Bank.

under the influence of alcohol or drugs. Cell phone use while driving (handheld or hands-free) lengthens a driver's reaction time as much as having a blood alcohol concentration at the legal limit of .08%.[118] In one year alone, United States drivers distracted by cell phones cause almost 3,500 deaths and almost 400,000 injuries.[119] Even a hands-free device doesn't eliminate the risks. Drivers carrying on phone conversations are 18% slower to react to brake lights. They also take 17% longer to regain the speed they lost when they braked.[120]

MAKING THE GRADE

At www.oup.com/he/adler-uhc14e, you will find a variety of resources to enhance your understanding, including video clips, animations, self-quizzes, additional activities, audio and video summaries, interactive self-assessments, and more.

OBJECTIVE 2.1   Identify the distinguishing characteristics of social media, which combine elements of mass and personal communication.

- Social media are distinguished from other forms of communication by user-generated content, variable audience size, and interactivity.

- Uses and gratifications theory explains that social media serve the following functions: providing information, facilitating personal relationships, defining personal identity, and entertainment.

- Social media can be "masspersonal," combining elements of both mass and interpersonal communication.

    > How does your use of social media reflect the characteristics of user-generated content, variable audience size, and interactivity?

    > For a 1-week period, describe how your use of social media reflects the uses and gratifications described on pages 27–28.

    > To what extent are your social media "masspersonal" in nature according to the description on pages 28–29?

OBJECTIVE **2.2** **Outline cases for using social media versus face-to-face communication, noting important differences.**

- Mediated communication is less rich than face-to-face interaction. For that reason, it is more prone to ambiguity. When asynchronous, mediated communication is easier to use for managing one's identity. The permanent nature of mediated content means that regrettable behavior can have long-lasting repercussions.

- Paradoxically, mediated communication can be hyperpersonal, accelerating the discussion of personal topics and relational development beyond what normally happens in face-to-face interaction.

- Keeping the above characteristics in mind can help communicators decide which communication channel is most appropriate and effective in a given situation.

    > Apply the information in this section to a representative sample of your communication, explaining in each case what communication channel is most effective.

    > Describe a situation in which choosing a different communication channel might have led to better results.

OBJECTIVE **2.3** **Identify the benefits and drawbacks of mediated communication, and use this knowledge to maximize effectiveness and minimize the potential for harm.**

- Use of social media has the potential for great benefit, including opportunities to connect with others you might not see face to face. In addition, using social media can help sustain and enrich personal relationships, provide social support, and promote social advocacy.

- The drawbacks of social media use include potential harm to offline relationships and increased risk of depression, loneliness, and social anxiety. It's more difficult to detect deception in mediated relationships, and social media can provide an avenue for abusive behavior including stalking, harassment, and cyberbullying.

    > How is your life richer from using social media? How much do the downsides described in this section affect your personal and relational well-being?

OBJECTIVE **2.4** **Describe how factors such as gender and age can influence social media use.**

- Social media content and word choice vary by gender. To some extent, however, the perception of gender influences beliefs in a sender's profile more than biological sex.

- People raised with the Internet differ significantly in their orientation toward social media than do those who

learned online communication as adults. Even for "digital natives," the content of mediated communication changes according to age.

    > Interview a digital immigrant who lived in the world before social media. How different has the digital immigrant's life been in terms of the benefits and risks outlined in this section?

    > Use the material in this section to compare your use of social media with the use of classmates or others with different gender orientations. Based on your findings and the information in this section, how can you adapt your communication when interacting with people from different demographic groups?

OBJECTIVE **2.5** **Understand how to use social media competently to minimize misunderstandings, maintain positive relationships, and protect yourself.**

- Maintaining positive relationships online comes from respecting others' needs for undivided attention, using a civil tone, preserving privacy boundaries, and being mindful of bystanders.

- Self-protection includes being careful about what you post, being a critical consumer of online information, balancing mediated and face time, and using mobile devices safely.

    > Ask several social media contacts to evaluate how well you follow the guidelines on pages 43–45 to maintain positive relationships. Based on the feedback you receive, how can you become a more competent user of social media?

    > Use the guidelines on pages 45–47 to evaluate how well you protect yourself when using social media. What are your strengths? In what areas do you need to improve?

KEY TERMS

asynchronous communication p. 32

cyberbullying p. 39

cyberstalking p. 39

disinhibition p. 44

gatekeeper p. 26

hyperpersonal communication p. 30

leanness p. 30

masspersonal communication p. 28

online surveillance p. 39

polymediation p. 32

richness p. 29

social media p. 27

synchronous communication p. 32

trolling p. 44

ACTIVITIES

1. Come up with a list of "do's and don'ts" for communicating on social media.

    a. What social media practices would you list as rude or ineffective? Why?

    b. What social media practices would you encourage? Why?

    c. What do you think would happen if everyone followed your advice?

2. Study the LinkedIn profile of someone you consider to be highly successful. Based on what you see:

    a. How would you describe that person's brand?

    b. What are the person's main accomplishments?

    c. What about the profile is most impressive to you?

    d. What additions or revisions might make the profile even better?

The Self, Perception, and Communication

3

CHAPTER OUTLINE

3.1 Communication and the Self 52
Self-Concept Defined
Biology, Personality, and the Self
External Influence on the Self-Concept
Culture and the Self-Concept
Self-Fulfilling Prophecies

3.2 Perceiving Others 57
Steps in the Perception Process
Influences on Perception
The Power of Narratives
Common Perceptual Tendencies
Empathy

3.3 Communication and Identity Management 67
Public and Private Selves
Identity Management and Social Media
Characteristics of Identity Management
Identity Management in the Workplace
Why Manage Identities?
Identity Management and Honesty

MAKING THE GRADE **74**
KEY TERMS **75**
ACTIVITIES **75**

LEARNING OBJECTIVES

3.1

Identify the communicative influences that shape the self-concept.

3.2

Explain perceptual tendencies and situational factors that influence perceptions of others.

3.3

Describe how identity management operates in both face-to-face and online communication.

Kayla's

low self-esteem is fueled primarily by the messages she receives from others. As you read about how communication shapes the self-concept, ask yourself these questions:

How do you view yourself? Would an objective observer agree with your self-perception?

What messages from the media and personal interactions have shaped your self-concept?

How does your communication shape the way others regard themselves?

self-concept A set of largely stable perceptions about oneself.

"**THE TOPIC OF** today's video is being yourself," proclaims Kayla Day (Elsie Fisher) in the movie *Eighth Grade*. An awkward 13-year-old, Kayla spends time creating positive-sounding videos for her YouTube channel. But following her own advice proves tricky, and her view count stays in the low single digits.

Kayla eats alone at lunch and is voted Most Quiet. Other girls snub her. In the glow of her screen, Kayla can't resist comparing herself to classmates and celebrities. Online, everyone seems happy and confident. Even looking at her crush's Instagram feed sends her into a tailspin.

Kayla's biggest fan is her single father (Josh Hamilton), whose clumsy attempts to help ("I think you're sooo cool!") leave Kayla rolling her eyes. She puts on her headphones to tune out the world. Toward the end of the film, the kind encouragement of a high school senior hints that Kayla's life will get better. But for now, she is struggling.

3.1 Commmunication and the Self

Who are you? How did you come to view yourself this way? How does the way you see yourself shape your communication with others? In this section, we consider these types of questions as we explore the idea of self-concept.

Self-Concept Defined

The **self-concept** is a set of largely stable perceptions individuals have of themselves. You might imagine the self-concept as a mental mirror that reflects how you view yourself. It shows what is unique about you and what makes you both similar to and different from others. The picture may involve your gender identity, age, religion, and occupation. It's also likely to include your physical features, emotional states, talents, likes and dislikes, values, and roles.

Take a few minutes to list as many ways as you can to identify who you are. You'll need this list later in the chapter. Try to include all the major characteristics that describe you, such as:

- Groups with whom you identify (e.g., Southerners, musicians, business majors)
- Your common moods or feelings
- Your appearance and physical condition
- Your career goals
- Your social traits
- Talents you possess or lack
- Your race and ethnicity
- Your intellectual capacity
- Your gender identity

- Your sexual orientation
- Your belief systems (religion, philosophy)
- Your social roles

Of course, to make this self-portrait even close to complete, your list would have to be hundreds—or even thousands—of words long. And not all items on your list are equally important to you. For example, you might define yourself primarily by your social roles (parent, veteran), culture (Mexican American, Chinese), or beliefs (libertarian, feminist). Others might define themselves more in terms of physical qualities (tall, Deaf), or accomplishments and skills (athletic, scholarly).

An important element of the self-concept is **self-esteem**, which involves evaluations of self-worth. For example, if your *self-concept* includes being athletic or tall, your *self-esteem* indicates how you feel about these qualities: "I'm glad that I am athletic" or "I am worried about being so tall." There's a powerful link between communication and self-esteem. It's probably no surprise that people who have close, supportive interactions with others are more likely to have high self-esteem.[1,2] The same principle works in reverse. People with high self-esteem are more likely than others to take a chance on starting new relationships[3] and showing affection to others,[4] which can enhance their self-esteem even more.

Research also shows that people who feel good about themselves are more likely than others to believe and enjoy compliments.[5] They are also more resilient in the face of criticism or even cyberbullying. For example, individuals with healthy self-esteem are more likely than those with low self-esteem to report bullying and to see bullies as immature and eager to prove their own status.[6] That's not to say that bullying is okay or can always be shrugged off. It does suggest, however, that being silent or self-critical can make unkind comments feel even worse.

Despite its obvious benefits, self-esteem doesn't guarantee success in personal and professional relationships.[7] People with an exaggerated sense of self-worth may mistakenly *think* they make a great impression, even though the reactions of others don't always match this belief. It's easy to see how people with an inflated sense of self-worth could irritate others by coming across as condescending know-it-alls.[8]

Biology, Personality, and the Self

Take another look at the list of terms you used to describe yourself. You'll almost certainly find some that reflect your **personality**—characteristic ways you think and behave across a variety of situations. Personality tends to be stable throughout life: Research suggests that temperament at age 3 is highly predictive of how a person will behave as an adult.[9]

To some extent, genes determine personality. Researchers estimate that people inherit about 40% of their personality traits.[10] If your parents or grandparents are shy, for example, you may have a genetic tendency to be reserved around strangers. Or you might have inherited a tendency to be novelty seeking, emotionally expressive,[11] or assertive,[12] or to exhibit many other traits.

self-esteem The part of the self-concept that involves evaluations of self-worth.

personality The set of enduring characteristics that define a person's temperament, thought processes, and social behavior.

High or low self-esteem can make all the difference in how you communicate with others.
How would you describe your level of self-esteem? How is it affected by the messages you receive in face-to-face exchanges and via social media?

That's far from the whole story, though. You may have a disposition toward some personality traits, but you can do a great deal to control how you actually communicate.[13] Even shy people can learn how to reach out to others, and those with aggressive tendencies can learn to communicate in more sociable ways. One author put it this way: "Experiences can silence genes or activate them. Even shyness is like Silly Putty once life gets hold of it."[14] Throughout this book you will learn about communication skills that, with practice, you can build into your repertoire.

So far we have defined the self-concept, but you may be asking what it has to do with human communication. We can begin to answer this question by exploring factors that have helped shape your self-concept.

External Influence on the Self-Concept

reflected appraisal The influence of others on one's self-concept.

social comparison Evaluating oneself in comparison to others.

Identity is shaped not only by genetics but by communication with others. The term **reflected appraisal** describes the influence of others on one's self-concept. Imagine crying uncontrollably during a business meeting, yelling in a library, or wearing an unfashionable outfit or hairstyle. Odds are you would be mortified if you thought others (even strangers) disapproved of you. And if their opinions of you were consistent over time, you might start to think of yourself as they did— perhaps as highly emotional, the class clown, or a geek.

As you go through life, you are likely to internalize many of the messages you receive. A teacher, a special friend or relative, or even a barely known acquaintance can leave an imprint on how you view yourself.

The messages we receive from others can be positive, but sometimes they are profoundly negative. People who defy social expectations are at particular risk of being judged, rejected, and even physically harmed. For example, transgender and nonbinary people in the United States face high rates of attack.[15] The Ethical Challenge feature "Looking Beyond Stereotypes" considers how stereotypes can affect the messages we send to others.

Along with messages from others, your own assessments shape your self-concept. A major part of this process is **social comparison**: evaluating yourself in comparison to others.

For her work on *Orange Is the New Black*, Laverne Cox became the first openly transgender actor nominated for a primetime Emmy. Cox has opened up about the staying power of harmful messages: "It's a struggle every day . . . not to become that eight-year-old who was bullied and chased home from school," she says.

In what ways do the people and media images around you affect your sense of belonging?

Are you attractive? Successful? Intelligent? The answer depends on who you measure yourself against. Researchers in one study found that people who spend more-than-average time viewing photos on social media often report reduced self-esteem afterward.[16] Mass media messages play a role as well. In the United States, higher exposure to media images increases the likelihood that young women will see themselves as overweight and will develop eating disorders.[17] Adolescent American boys with high exposure to media images are likely to feel they are not as slender or as muscular as society expects them to be.[18] These patterns tend to persist even when people know the images have been artificially enhanced, as they often are.[19]

As people grow older, they tend to develop a more enduring sense of who they are and consequently are slightly less affected by other people's opinions of them.[20] Over time, you might find that constructive criticism no longer damages your self-confidence very much. Yet, if you have negative feelings about yourself,

ETHICAL CHALLENGE Looking Beyond Stereotypes

"The thought of being around a transgender person is extremely uncomfortable to me and I don't know exactly why.... Is there any way for me to come to grips with this?"[21]

An online reader wrote these words to Matt Kailey, who is a transgender individual. Kailey's response was encouraging: "It takes guts to do some self-reflection, realize you have an issue, and take steps to resolve it." Kailey then posed the following questions (which we have paraphrased). Ask yourself the same questions as you consider any group of people you tend to stereotype.

- *Is it possible your response is a reflection of cultural rules and values?* Even if you don't agree with the messages you've encountered, you may have absorbed them to some degree.

- *Is your reaction based on fear of the unknown?* It's a human tendency to feel threatened by what is unfamiliar. Take an honest account of what scares you about a stereotyped group. Then consider: Are your fears rational?

- *Why not reach out?* One of the best ways to dispel a stereotype is firsthand contact. Rather than shying away, strike up a conversation with someone you might otherwise avoid. If the person is transgender, "you don't have to talk about 'trans stuff,'" says Kailey. Instead, "talk about sports, the weather, taxes, television shows, movies, music—you're likely to find something you have in common. Once you do this, you might find that your fear or disgust starts to dissipate."[22]

you may still cling to those feelings. If you see yourself as a terrible student, for example, you might respond to a high grade by thinking, "I was just lucky" or "The professor must be an easy grader."

You might argue that not every part of one's self-concept is shaped by others. Some traits are objectively recognizable through self-observation. Nobody needs to tell you that you are tall, speak with an accent, can run quickly, and so on. However, the *significance* people attach to such attributes depends greatly on their social environment and culture, a topic we consider next.

Culture and the Self-Concept

Cultural assumptions influence what people think about themselves and others. For example, someone who looks attractively thin and healthy to an American might look stressed and unwell to a Jamaican.[23] And as you'll read in Chapter 4,

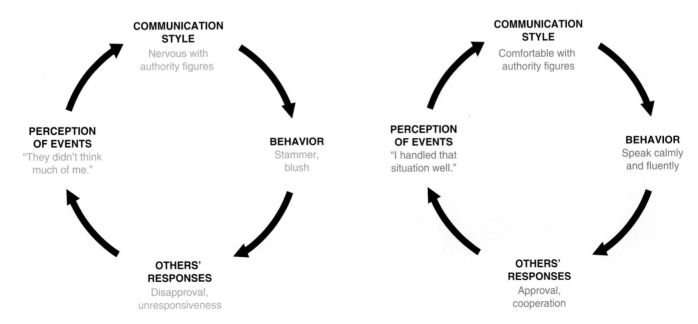

FIGURE 3.1 The Relationship Between the Self-Concept and Behavior

members of most Western cultures are highly *individualistic*, meaning they value individuality and independence. By contrast, members of some cultures—many Asian ones, for example—are traditionally more *collectivistic*, meaning they value group identity more than autonomy.

Cultural dynamics have powerful implications for communication. In job interviews, for example, people from the Americas, Canada, Australia, and Europe are likely to describe and even exaggerate their accomplishments. However, people in many Asian cultures value modesty instead. Job candidates raised to honor traditional Chinese values may downplay their accomplishments, assuming the interviewer will consider them both highly accomplished and admirably humble.[24]

Self-Fulfilling Prophecies

The self-concept both shapes and is shaped by communication. If, for example, your self-concept involves feeling "nervous with authority figures," you'll probably behave in nervous ways during a job interview or an interaction with a professor. That nervous behavior is likely to influence how others view you, which in turn will shape how they respond to you. The response may reinforce that aspect of your self-concept, thus affecting your future behavior. This cycle illustrates the chicken-and-egg nature of communication and the self-concept, as shown in Figure 3-1.

The self-concept is such a powerful force that it can influence future behavior—both your own and that of others. A **self-fulfilling prophecy** occurs when a person's expectation of an outcome and subsequent behavior make the outcome more likely to occur. This happens all the time. For example, think of the following experiences:

self-fulfilling prophecy A prediction or expectation of an event that makes the outcome more likely to occur than would otherwise have been the case.

- You expect to be nervous during a job interview, and that causes you to answer questions poorly.

- You anticipate having a good (or terrible) time at a party, and you act in ways that fit your prediction.

- A teacher or boss explains a new task to you, saying you probably won't do well. You take these comments to heart, and as a result you don't try very hard.

- A friend describes someone you are about to meet, saying you won't like the person. You then look for—and find—reasons to dislike the new acquaintance.

In each of these cases, the outcome happened at least in part because of the expectation that it *would* happen.

There are two types of self-fulfilling prophecies: those that influence your own behavior and those affect the behavior of others. As an example of the first type, think about sports experiences in which you psyched yourself into playing either better or worse than usual. The same principle operates for public speakers: Self-confidence can improve performance, while anxiety can hurt it.[25] (Chapter 12 offers advice on overcoming communication apprehension.)

In one study, communicators who predicted intense conflict episodes were likely to be highly emotional during them and to engage in personal attacks.[26] On the bright side, people who expect friendliness and acceptance usually act friendly and outgoing themselves.[27] As you might expect, they find a warmer reception than people who are fearful of rejection. Self-fulfilling prophecies can involve a feedback loop between how you feel and how you behave. Researchers have found that smiling, even if you're not in a good mood, may give you a more positive outlook.[28]

"*I don't sing because I am happy. I am happy because I sing.*"

Source: Edward Frascino The New Yorker Collection/The Cartoon Bank

In the second type of self-fulfilling prophecy, expectations influence others' behavior.[29] This principle was demonstrated in a classic experiment.[30] Researchers told teachers that 20% of the children in a certain elementary school showed unusual potential for intellectual growth. The names of the students were actually drawn randomly, but 8 months later, the children who had been identified as "unusually gifted" showed significantly greater gains on tests than the remaining children. Directly or indirectly, the teachers had communicated the message "I think you're bright" to the selected students. This message had affected the students' self-concepts, which ultimately affected their performance.

This type of self-fulfilling prophecy has been demonstrated in a wide range of settings. In medicine, for example, patients who unknowingly receive placebos—substances that have no curative value—may respond favorably, as if they had received an effective drug. The self-fulfilling prophecy operates in families as well. If, on the one hand, parents belittle their children, the children's self-concepts may soon incorporate this idea, and they may fail at many of the tasks they attempt. On the other hand, if children are told they are capable, or lovable, or kind, there is a much greater chance that they will behave in those ways.[31]

The self-fulfilling prophecy is an important force in communication, but it doesn't explain all behavior. There are certainly times when expecting a particular outcome won't bring it about. For example, believing you'll do well in a job interview when you're not qualified for the position is unrealistic. Similarly, there will be people you don't like and occasions you won't enjoy, no matter what your attitude.

As we keep these qualifications in mind, it's important to recognize the tremendous influence that self-fulfilling prophecies play in our lives. To a great extent we are what we believe we are. In this sense, we and those around us constantly create and refine our self-concepts, and thus ourselves.

ASK YOURSELF

How might your self-concept affect both your academic performance and your communication patterns?

🧍 3.2 Perceiving Others

perception A process in which people use sensory data to reach conclusions about others and the world around them.

Online blogger Tiffany Tan has a good friend she describes as unique, creative, and talented. Her friend has autism. "That might make it seem like she's different," says Tiffany, "but let's be real, we're all different."[32] Tan's attitude illustrates the importance of **perception**, the way people regard others and the world around them. The beliefs and attitudes we bring to interactions help shape those interactions.

One of the most powerful functions of communication is to find common ground. Perceptions can either help toward that goal or impede it. This section focuses on how perceptions influence communication.

Steps in the Perception Process

Psychologist William James described an infant's world as "one great blooming, buzzing confusion."[33] Babies are bombarded by unfamiliar stimuli nearly all the time. You probably feel the same way when you move to a new city, take on a new position at work, or enter a room

We tend to focus on features that stand out.

What might others fail to recognize about you if they noticed only your most obvious features?

FIGURE 3.2 Which do you see first: the faces or the vase? It depends how you organize the information from the image.

full of people you don't know. How people deal with stimuli in their environment shapes the way they communicate. We sort out and make sense of stimuli in three steps: selection, organization, and interpretation.

Selection   Because you are exposed to more input than you can possibly manage, the first step in perception is **selection**, paying attention to some stimuli while ignoring others. Some external factors help shape what you notice about others. For example, stimuli that are *intense* often attract attention. Something that is louder, larger, or brighter than its surroundings stands out.

You are also likely to pay attention to *contrasts* or *changes* in stimulation, whereas unchanging people or things tend to be less noticeable. This principle explains why you may take consistently wonderful people for granted. It's only when they are no longer there that you appreciate their admirable qualities.

Personal goals also shape how people make sense of others. For example, someone on the lookout for romance will be especially aware of attractive potential partners, whereas someone who is new in town might focus more on potential friends.

Your *emotional state* also shapes what you select. To some extent, your mood reflects your self-talk—what you tell yourself about a situation. People whose internal dialogue focuses on a bright future typically experience better moods and more happiness than people who focus on negative past events.[34] Focusing on either the positive or the negative can create a spiral. For example, if you're happy about your relationship, you're more likely to interpret your partner's behavior in a charitable way. This outlook, in turn, can lead to greater happiness. Of course, the same process can work in the opposite direction. In one study, spouses who felt uncertain about the status of their marriage were likely to perceive their partners' conversations with strangers as relational threats.[35]

selection The perceptual act of attending to some stimuli in the environment and ignoring others.

organization The perceptual process of organizing stimuli into patterns.

Organization   After selecting information from the environment, the mind performs a process called **organization**, arranging this information to make meaning from it. Psychologists call this the *figure–ground* principle of perception. A classic illustration of this principle is the drawing in Figure 3.2. Which do you see first: a vase or faces in profile? Your answer will depend on how you mentally organized the image. In a similar way, how you organize information about others' appearance and behavior affects how you think about them. Are your perceptual filters attuned to clothing? Age? Skin color?

TABLE 3.1

Factors that Affect Interpretation of Behavior

FACTOR	EXAMPLE
Degree of familiarity	You may perceive people who are familiar to you as more extraverted, stable, friendly, and conscientious than people you regard as strangers.
Relational satisfaction	If a friend or romantic partner hurts your feelings, you are more likely to forgive and forget if you are generally happy in the relationship than if you are dissatisfied with it.
Personal experience	If landlords have unfairly charged you in the past, you might be skeptical about an apartment manager's assurances of refunding your cleaning deposit.
Assumptions about human behavior	A boss who assumes that people are lazy and avoid responsibility may misjudge an employee's contributions.
Expectations	If you go into a conversation expecting a hostile attitude, you're likely to hear a negative tone in the other person's voice—even if that tone isn't intended by the speaker.
Knowledge of others	If you know a friend has just been rejected by a lover or fired from a job, you'll interpret their aloof behavior differently than if you were unaware of what happened.

Online environments often provide less nonverbal information than is available in person, so what is available can be highly influential. In one study, experimenters asked college students to form impressions of fictional characters based solely on their online names—such as "packerfan4" and "stinkybug." The respondents assigned attributes, including biological sex, ethnicity, and age, to the supposed owners of these names.[36]

Interpretation   Once you have selected and organized your perceptions, the next step is interpreting them in a way that makes sense to you. **Interpretation** plays a role in virtually every type of communication. How you make sense of things may lead you to conclude that a person who smiles at you across a crowded room is interested in romance—or, conversely, is simply being polite. You might interpret a friend's kidding as a sign of affection or irritation. And you might conclude that someone who says "drop by anytime" does (or doesn't) really mean it. Table 3.1 summarizes several factors that cause people to interpret others' actions in one way or another.

> **interpretation** The perceptual process of attaching meaning to stimuli that have previously been selected and organized.

Although we have talked about selection, organization, and interpretation separately, the three phases of perception can occur in differing sequences. For example, a babysitter's past interpretation (such as "Jason is a troublemaker") can influence future selection (Jason's behavior becomes especially noticeable) and organization (when there's a fight, the babysitter may assume that Jason started it). As with all communication, perception is an ongoing process in which it's hard to pin down beginnings and endings.

Influences on Perception

A variety of factors influence how people select, organize, and interpret data about themselves and others. We touch on a few of those factors here, including physiology, stereotypes, and relational roles. (In Chapter 4, we will discuss social and cultural influences in more depth.)

Physiological Factors   Numerous physiological (functional biological) factors can affect perception and communication.[37] One is a person's *developmental stage*. For example, it may be developmentally normal for a young child to behave in ways that would seem egocentric, selfish, and uncooperative in an adult. A 4-year-old may not notice that adults are weary of reading the same storybook over and over.

Health and nutrition also influence communication. When you are ill or have been working long hours, the world can seem quite different than when you are well rested. People who are sleep deprived perceive time intervals as longer than they really are, so the 5 minutes they spend waiting for a friend may seem longer, leaving them feeling more impatient than they otherwise would.[38] Diet and nutrition can affect communication as well. One study found that teenagers who reported that their families did not have enough to eat were almost twice as likely as other teens to have difficulty getting along with others.[39]

Biological cycles also affect perception and communication. People have daily cycles in which all sorts of changes occur, including variations in body temperature, sex drive, alertness, tolerance to stress, and mood.[40] These cycles can affect the way they relate to others. Researchers in one study found that men had more negative views of their partners after a night in which they didn't sleep well. The same study showed that women's sleep patterns were disrupted when they perceived problems in the relationship. The researchers observed that sleep, mood, and interpersonal communication are mutually influential.[41]

"How is it gendered?"

Source: Edward Koren The New Yorker Collection/The Cartoon Bank

What captures your attention about makeup vlogger and Cover Girl cosmetics ambassador Nura Afia?

What beliefs, attitudes, and stereotypes shape the way you perceive people in everyday life?

Some differences in perception are rooted in *neurology,* the functioning of the nervous system. For instance, people with ADHD (attention-deficit/hyperactivity disorder) are easily distracted from tasks and have difficulty delaying gratification.[42] Those with ADHD might find a long lecture boring and tedious, while other audience members are fascinated by it.[43,44] People with bipolar disorder experience significant mood swings in which their perceptions of events, friends, and even family members shift dramatically. The National Institute of Mental Health estimates that between 5 million and 7 million Americans are affected by these two disorders[45]—and many other psychological conditions also influence people's perceptions.

Stereotypes   Social and cultural expectations—for example, those related to gender, race, and ethnicity—have a powerful influence on identity, perception, and communication. Part of this power lies in **stereotypes**—widely held but oversimplified or inaccurate ideas tied to social categorization. Once people apply a stereotype, they may focus on behaviors that seem to support it.

For most people, *gendered expectations* began the moment they were born—or even before, with the selection of clothing, blankets, toys, and nursery décor. Yet we often don't question the assumptions or stereotypes that may underlie these expectations. For example, a mother might say to her daughter, "Don't act that way. It isn't ladylike." But what exactly does that mean, and where does that idea come from?

It's important to note that *sex* and *gender* are not the same thing. **Sex** is a biological category (e.g., male, female, intersex), whereas **gender** is a socially constructed set of expectations about what it means to be "masculine," "feminine," or "transgender," for example. Gendered expectations vary culturally and are always evolving.

It's also important to debunk the idea that masculine and feminine behaviors occupy opposite poles on a single continuum. Instead, many theorists now embrace the idea of a **gender matrix** that recognizes gender as a multidimensional collection of qualities.[46,47] Here are the main ideas behind this perspective:

- Simple labels are not sufficient to capture the complexity of gender.[48]
- Social expectations about gender can feel needlessly confining.[49]

UNDERSTANDING DIVERSITY

Sexist Assumptions in Everyday Language

"You don't want kids? Aren't you worried you won't be fulfilled?"

"Nice car. Overcompensating for something?"

If you grew up in the United States, you can probably guess which statement above is more frequently directed toward women and which more toward men. Gendered assumptions are remarkably consistent.

The videos *48 Things Women Hear in a Lifetime (That Men Just Don't)* and *48 Things Men Hear in a Lifetime (That Are Bad for Everyone)* call attention to the powerful role of communication in perpetuating gender stereotypes. Among the comments directed toward women are "Don't be a slut," "You'd be much prettier if you smiled," and "You're going to let someone else raise your kids when you go back to work?"[50] Men are likely to hear remarks such as

"She has you whipped," "You're the *nurse*? I thought you were the doctor," and "Do you mind that she makes more money than you?"[51]

The wallop of such statements is what they imply about culturally approved ways to enact gender. Women, it is suggested, should be physically attractive but chaste and committed to motherhood. Men are admonished to avoid being feminine and to focus on achieving dominance in romance and the workplace. Underlying these statements is the assumption that everyone should conform to gender stereotypes.

"Alone, each remark may seem like no big deal," reflects one observer, "but listening to them all together shows just how pervasive and exhausting subtle sexism really is."[52]

- People exhibit different gender qualities depending on the context, their mood, and what they are trying to accomplish.[53]

Gender-related stereotypes can be slow to change, especially when emphasized by prevalent media images.[54,55] On a more encouraging note, children who are exposed to diverse media images of gender tend to feel less limited by stereotypes about gendered behavior, as indicated by toy selection and interaction with peers.[56] "My generation is poised to eliminate 'traditional' gender roles completely," declares 30-something Dan Schwabel.[57] Even if his prediction does not entirely come true, he makes a good point: Changing expectations can lead to broader communication options for people of all genders.

Race and ethnicity can also influence our identities and perceptions of others, in turn influencing communication. Even when communicators do not intend to be racist, stereotypes can contribute to **implicit bias**: unconsciously held associations about a social group.[58] Participants in one study initially rated white managers to be more competent, achievement-oriented, and manipulative than black managers, and black managers to be more interpersonally skillful and less polished than their white counterparts.[59] However, those judgments largely disappeared when the participants were informed that all of the managers were successful.

Categorizing people according to their appearance can be problematic in a number of ways. Casual remarks known as *microaggressions* can be hurtful, especially when they're directed at members of marginalized groups. Consider someone like Heather Greenwood, who is biracial. Strangers often ask how her children can be so fair skinned when she is dark. Some speculate that she is the nanny or joke that her kids must have been "switched in the hospital" with a white woman's. The implication, Greenwood says, is that a legitimate family is either one color or another. She hears comments such as these nearly every day. "Each time is like a little paper cut, and you think, 'Well, that's not a big deal.' But imagine a lifetime of that. It hurts," she says.[60]

Relational Roles　Think back to the "Who am I?" list you made earlier in this chapter. It's likely your list included roles you play in relation to others: daughter, roommate, spouse, friend, and so on. Roles like these don't just define who you are—they also affect your perceptions.

Take, for example, the role of parent. As most new mothers and fathers will attest, having a child alters the way they see the world. They might perceive their crying baby as a helpless soul in need of comfort, whereas nearby strangers are simply annoyed by the sounds. As the child grows, parents often pay increasingly close attention to the messages in the child's environment. One father we know said he never noticed how much football fans curse until he took his 6-year-old to a game with him. In other words, his role as father affected what he heard and how he interpreted it.

The roles involved in romantic love can also dramatically affect perception. You may see your sweetheart as more attractive than other people do, and you might overlook faults that others notice.[61] Your romantic role can also change the way you view others. One study found that when people are in love, they view other romantic candidates as less attractive than they normally would.[62]

The Power of Narratives

Narratives are stories people create to help make sense of the world.[63] One person's narrative is often quite different from someone else's. For example, if you were taking an unsatisfying class, you might cast the professor as boring. The professor, however, might characterize the students as unmotivated and blame the class environment on them.

stereotype　A widely held but oversimplified or inaccurate idea tied to social categorization.

sex　A biological category such as male, female, or intersex.

gender　Socially constructed roles, behaviors, activities, and attributes that a society considers appropriate.

gender matrix　A construct that recognizes gender as a multidimensional

implicit bias　Unconsciously held associations about a social group.

narratives　The stories people create and use to make sense of their personal worlds.

This question was probably not intended as harmful, but consider the potentially negative assumptions behind it.

Have you ever experienced or committed microaggressions?

After they take hold, narratives offer a framework for interpreting behavior and shaping communication. For example, in blended families, the stories children tell about their stepparents both shape and reflect how they feel about them.[64] On the one hand, children who view stepparents as sudden or unwanted are likely to be wary or even hostile toward them. On the other hand, regarding stepparents as welcome or even idealized additions leads to more positive behavior. Later in this chapter, we will look at a tool called *perception checking*, which is helpful in bridging the gap between different narratives.

A classic study of long-term happy marriages demonstrates that shared narratives don't have to be accurate to be powerful.[65] Couples who report being happily married after 50 or more years seem to collude in a relational narrative that doesn't always reflect the facts. They often report minimal conflict, even when objective analysis reveals frequent disagreements and challenges. They offer the most charitable interpretations of each other's behavior. They seem willing to forgive, or even forget, transgressions. The conclusion is not that happy couples have a poor grip on reality but that the stories we tell ourselves affect our relationships. Research such as this demonstrates that experiences are created subjectively with others through communication.[66]

Common Perceptual Tendencies

Some of the greatest obstacles to understanding arise from **attribution**—the process of attaching meaning to behavior. People attribute meaning to both their own actions and the actions of others, but they often use different yardsticks. Following are six perceptual tendencies that can lead to inaccurate attributions—and to communication problems.

1. Clinging to First Impressions   Labeling people according to first impressions is an inevitable part of the perception process. Given limited information, it's quick and easy to conclude that "She seems cheerful," "He seems sincere," or "They sound awfully conceited." Problems arise when these impressions are inaccurate but you continue to cling to them, ignoring conflicting information. Suppose, for instance, that you mention a new neighbor to a friend. "Oh, I know him," your friend replies. "He seems nice at first, but it's all an act." Perhaps this appraisal is off base, but it will probably influence the way you respond to the neighbor. Your response may in turn influence your neighbor's behavior, creating a negative self-fulfilling prophecy.

Snap judgments are particularly likely in mediated communication. In text-based channels, receivers get a sense of senders based largely on their writing style. This means that typos and grammatical mistakes can damage your image, especially in the workplace. Researchers in one study found that 74% of people notice spelling and grammar errors on corporate websites, and most of them avoid doing business with those companies as a result. It follows that employees with poor writing skills are a liability rather than an asset. One analyst compares sloppy writing to sloppy dressing at work.[67]

The power of first impressions is important in personal relationships as well. A study of college roommates found that those who had positive initial impressions of each other were likely to have positive subsequent interactions, manage their conflicts constructively, and continue living together.[68] The converse was also true: Roommates who got off to a bad start tended to spiral negatively—hence the old adage, "You never get a second chance to make a first impression." Given the almost unavoidable tendency to form first impressions quickly, the best advice is to keep an open mind. Be willing to change your opinion if your first impressions were mistaken.

2. Categorizing People   We not only form strong first impressions, but we tend to categorize people. This is where stereotypes may come into play: We may unfairly

attribution The process of attaching meaning.

In her TED talk, "The Danger of a Single Story," award-winning Nigerian novelist Chimamanda Ngozi Adichie warns that by limiting yourself to a single narrative, you may fail to achieve your full potential.

What primary narratives do you use to characterize a person or group? What might other narratives offer?

assume that others have particular attributes as members of a group. If you have ever been judged or rebuffed on the basis of race, class, gender, or sexual orientation, then you know the feeling of being stereotyped.

3. Judging Ourselves More Charitably Than We Judge Others Although we're quick to be critical of others, we tend to judge ourselves in generous terms. Social scientists call this tendency *fundamental attribution error*, or simply the **self-serving bias**.[69] When others suffer, we often blame the problem on their personal qualities. By contrast, when *we* suffer, we find explanations outside ourselves. Consider a few examples:

- When someone else botches a job, you might think they did not try hard enough. When you botch a job, you blame unclear directions or inadequate time.

- When someone else lashes out angrily, you may say that person is being moody or oversensitive. When you blow off steam, you point to the pressure you are under.

- When others don't reply to your text or email, you might assume they are inconsiderate, disrespectful, or unprofessional. When you don't reply, you may say you were too busy, you didn't see the message, or it didn't seem necessary to reply.[70]

As these examples show, uncharitable attitudes toward others affect communication. Your harsh opinions can lead to judgmental messages, which trigger defensive responses. At the same time, you may be defensive when people question your behavior—which can keep you from trying to improve.

4. Paying More Attention to Negative Impressions than to Positive Ones Imagine being on a first date with someone who says "My ex loved this restaurant" or "I hate my job." If these statements set off alarm bells for you, they might outweigh your consideration of the person's more appealing qualities. **Negativity bias** is the tendency to focus more on negative impressions than on positive ones.[71,72] Scientists speculate that it may have evolved as a survival advantage. Spotting the one threatening element in the environment may ultimately be more important than focusing on all the safe ones.[73] But this tendency can skew judgments, affecting relationships and careers. Potential employers may rule out otherwise qualified job candidates after finding a negative post on social media.[74] One lesson here is to pause and reflect on the big picture. Just as you would hate to be judged for one statement or mistake, focusing only on a negative quality in someone else might cause you to treat that person unfairly.

5. Overgeneralizing People tend to generalize based on a single positive or negative trait or experience. When someone has one positive quality, you might unduly assume other positive qualities—a bias that scholars call the **halo effect**.[75] For example, people often suppose that physically attractive people are more intelligent than others, even when they are not.[76] The converse of the halo effect is known as the **horns effect**: perceiving others in an unfairly negative light on the basis of a single negative trait or experience.[77]

6. Gravitating to the Familiar People tend to favor characteristics and ideas that are similar to their own. In one study involving social media,

CHECKLIST ✓

Avoiding Stereotypes

- ☐ Keep in mind that someone's most noticeable characteristics are not necessarily the most important.

- ☐ Appreciate others as unique individuals instead of assuming they are defined by membership in a particular group.

- ☐ Admit that stereotypes are faulty.

self-serving bias The tendency to judge others harshly but to cast oneself in a favorable light.
negativity bias The perceptual tendency to focus more on negative indicators than on positive ones.
halo/horns effect A form of bias that overgeneralizes positive or negative traits.

ON YOUR FEET

Think of a time when your initial impression changed once you got to know someone. Consider which perceptual tendencies were involved. Prepare to share your thoughts in a brief presentation.

empathy The ability to imagine another person's point of view.

sympathy Compassion for another's situation.

participants were more likely to rate a profiled person as likeable if they perceived common interests and shared-group membership, such as attending the same university.[78]

A preference for similarities often leads people to mistakenly project their own attitudes and ideas onto others. For example, you might assume an off-color joke won't offend a friend, but it does. Others don't always think or feel the way we do, and assuming similarities can lead to problems.

When you become aware of differences, resist the temptation to distance yourself or demean the other person. Competent communicators are able to talk respectfully about viewpoints that differ from their own. As one blogger puts it, "The overall objective of expressing your views is supposed to be to encourage conversation and gain/provide new perspectives. It is not to demean and disprove."[79]

Empathy

One solution to many communication challenges is to increase **empathy**, the ability to imagine another person's perspective.

Dimensions of Empathy Empathy involves three dimensions.[80] The first, *perspective taking*, involves temporarily adopting the viewpoint of another person—attempting to view the issue as that person does. This requires setting aside your own opinions and suspending judgment. The second dimension, *emotional experience*, requires understanding what the other person is feeling—knowing that person's fear, joy, sadness, and so on. Empathizing allows you to experience the other's perceptions, in effect, to become that person temporarily. The third dimension is *genuine concern*, caring for another person.

It's easy to confuse empathy with sympathy, but the concepts are different in two important ways. **Sympathy** means you feel compassion for another person's predicament, whereas empathy means you have a personal sense of what that predicament is like. Consider the difference between sympathizing with a homeless person and empathizing with that person—imagining what it is like to be in that person's position. A second difference is that you can empathize with a difficult relative, a rude stranger, or even a criminal without feeling much sympathy for that person. Empathizing allows you to understand other people's motives without requiring you to agree with them.

People vary in their ability to understand how others feel. This is partly because of the abilities they are born with. Some people seem to have a hereditary capacity for greater empathy than do others.[81] But most people can develop their empathic capacity by being aware and attentive. Parents and teachers can help by role-modeling empathic behaviors[82] and by pointing out to children how their actions affect others ("Look how sad Jessica is because you took her toy. Wouldn't you be sad if someone took away your toys?"). Children who are encouraged to understand how others feel and who are rewarded for being considerate tend to demonstrate more empathy and prosocial behavior than their peers do.[83]

There is no consistent evidence that the ability to empathize is greater for one sex or another. On average, teenage girls tend to develop adult levels of empathy sooner than boys do, but the boys usually catch up.[84] Women are commonly considered to be more empathic toward children than men are, in part because of traditional parenting roles. But the latest research shows that people of any sex who identify with certain characteristics, such as

"How would you feel if the mouse did that to you?"

Source: William Steig The New Yorker Collection/The Cartoon Bank

being affectionate, gentle, and nurturing, are typically more responsive to babies' nonverbal cues than other adults are.[85]

Perception Checking   Communication skills provide a way of enhancing both understanding and empathy. In short, the best way to understand another person's perspective is to ask about it.[86] Simply asking "How are you feeling?" or "What's your opinion?" beats assuming you know how that person feels or thinks.

Beyond simple questions, the skill of **perception checking** offers a structured way to boost your understanding and empathy as well as minimize defensiveness and show respect. It involves three steps.

The first is to *reference a specific behavior*, as in "I noticed you left the room during my presentation." It's important to avoid jumping to conclusions, making value judgments, or presuming that you know how the other person is feeling. For example, it's not fair to say, "You are so rude" or "I can tell you didn't like my ideas." Such statements are likely to cause defensiveness, and they signal that you have already made up your mind without trying to understand the other person's feelings.

The second step is to *offer two options* that may be true. This opens the way for conversation and makes it clear that you don't presume to know the other person's motives. For example, you might say, "When you left, I wasn't sure if something important came up or if you were disappointed in my presentation."

The final step is simple but important. Simply *ask the other person* to tell you what they were (or are) feeling. You might simply ask "Help me understand: What *was* going on?"

Here are some examples of perception checking that include all three steps:

> *"Hey, when you slammed the door* [behavior], *I wasn't sure whether you were mad at me* [first interpretation] *or just in a rush."* [second interpretation] *Are we good, or do you want to talk?"* [request for clarification]

> *"You haven't laughed much in the last couple of days* [behavior]. *I wonder whether something's bothering you* [first interpretation] *or whether you're just feeling quiet.* [second interpretation] *What's up?"* [request for clarification]

As you can see, perception checking takes a respectful approach that implies "I know I'm not qualified to understand your feelings without some help." Of course, it can succeed only if your nonverbal behavior reflects the open-mindedness of your words. An accusing tone of voice or a hostile glare will suggest you've already made up your mind about the other person's intentions.

Emotional Intelligence   Part of empathizing involves **emotional intelligence (EI)**—the ability to understand and manage your own emotions and deal effectively with the emotions of others. The idea was made famous by psychologist Daniel Goleman,[87] who proposes that EI has five dimensions:

- *Self-awareness* involves understanding how you feel. For example, when something great happens to your best friend, you might realize you feel happy but also a little jealous.

perception checking A three-part method for verifying the accuracy of interpretations, including an objective description of the behavior, two possible interpretations, and a request for more information.
emotional intelligence (EI) The ability to understand and manage one's own emotions and to deal effectively with the emotions of others.

Check Your Perceptions Before Responding

The next time you feel puzzled or upset about someone's behavior, don't jump to conclusions. Instead, follow these steps to find out more.

☐ **Describe the behavior.** *("You didn't say anything in response to my ideas.")*

☐ **Suggest at least two interpretations of the behavior.** *("Maybe you needed some time to think about it. Or perhaps my ideas weren't what you had in mind.")*

☐ **Request clarification about how to interpret the behavior.** *("I value your opinion. What are the pros and cons of my ideas, as you see them?")*

How Emotionally Intelligent Are You?

 Answer the questions below for insights about how EI influences you as a communicator.

1. A friend says something that hurts your feelings. What are you most likely to say?

   a. "That is so insensitive! I can't believe you just said that!"

   b. "I feel hurt by what you said."

   c. "That makes me feel bad. Tell me why you feel that way."

   d. Say nothing. You'll get over it.

2. It's Monday morning and you feel great. What are you most likely to do?

   a. Take the day off. This feeling is too good to waste at work.

   a. Announce to everyone at work, "I feel like a million bucks!"

   b. Channel your positive energy into being a great team member.

   c. Set your emotions aside and get to work. You'll enjoy yourself later.

3. Your usually talkative roommate is quiet today and seems to be looking out the window rather than focusing on the book he's trying to read for school. What are you most likely to do?

   a. Tell him, "Focus! That book's not going to read itself."

   b. Say that you understand because you've had a hard day too.

   c. Ask if anything is bothering him and then listen attentively to what he says.

   d. Give him some space. He's probably just tired.

4. The grade on your research paper is not as high as you had hoped. How are you most likely to respond?

   a. Fume about what an idiot the professor is.

   b. Post on social media that you are sad and discouraged today.

   c. Go over the paper carefully to learn what you might do better next time.

   d. Tell yourself, "What's done is done" and try to forget about it.

INTERPRETING YOUR RESPONSES

For insight about your emotional intelligence, see which of the following best describes your answers. (More than one may apply.)

Emotionally Spontaneous

If you answered "a" to two or more questions, you tend to display your emotions expressively and spontaneously. This can be a bonus, but be careful not to let feelings get the best of you. Your unfiltered declarations may sometimes offend or overwhelm others, and they may prevent you from focusing on what other people are thinking and feeling. Suggestions for perception checking (page 65) and self-monitoring (page 71) may help you strengthen the empathy and self-regulation components of EI.

Emotionally Self-Aware

If you answered "b" to two or more questions, you tend to be aware of your emotions and express them tactfully. You score relatively high in terms of emotional intelligence. Just be careful to pair your self-awareness with active interest in others. You may feel impatient with people who are not as emotionally aware as you are. Stay tuned for listening tips and strategies in Chapter 6.

EI Champion

If you answered "c" to two or more questions, you balance awareness of your own emotions with concern for how other people feel. Your willingness to be self-reflective and a good listener will take you far. Communication strategies throughout this book provide opportunities to build on your already strong EI.

Emotion-Avoidant

If you answered "d" to two or more questions, you tend to downplay emotions—yours and other people's. While this may prevent you from overreacting to situations, it may also make it difficult to build mutually satisfying relationships and to harness the benefits of well-managed emotions. You may sometimes feel that others are taking advantage of you, when they actually don't know how you feel. The tips for self-disclosure presented in Chapter 8 may be especially useful to you.

- *Self-regulation* requires managing emotions effectively. People who lack self-regulation may lose their temper, say things they wish they hadn't, or be plagued by remorse or envy. Conversely, good self-regulators use emotions in positive ways. For example, you might channel your nervousness about public speaking into being more dynamic before an audience.

- *Internal motivation* is finding the inner strength and determination to accomplish important goals. Consider success as a student. You may assume people who do well in school are smarter than others, but research shows that their success often depends more on confidence and perseverance than on intelligence.[88]

- *Empathy* involves being willing and able to imagine a situation from another person's point of view. This is a crucial factor in developing strong and trusting relationships.[89]

- *Social skills* allow a person to use all these factors to build strong relationships. For example, organizations led by people with high emotional intelligence are usually more successful than others. In part, that's because these leaders are self-aware, exercise emotional control, and are good at understanding how employees and customers feel.[90]

Take the "Understanding Your Communication" quiz on page 66 to consider your own emotional intelligence.

3.3 Communication and Identity Management

Like Kayla, the main character in the film *Eighth Grade*, most of us rely on **identity management**: communication strategies meant to influence how others view us. Whether or not identify management strategies are successful, they occupy an important place in the messages we construct.

Public and Private Selves

To understand why identity management exists, we must discuss the notion of self in more detail. So far we have referred to the "self" as if each of us had only one identity. In truth, every person has many selves, some private and others public. Often these selves are quite different.

The **perceived self** is a reflection of the self-concept. Your perceived self is the person you believe yourself to be in moments of honest self-examination. We call the perceived self "private" because you are unlikely to reveal all of it to another person. You can verify the private nature of the perceived self by reviewing the self-concept list you developed while reading page 52. You'll probably find some elements of yourself there that you would not disclose to many people, and some that you would not share with anyone. You might, for example, be reluctant to share some feelings about your appearance ("I think I'm rather unattractive"), your intelligence ("I'm not as smart as I wish I were"), your goals ("The most important thing to me is becoming rich"), or your motives ("I care more about myself than about others").

In contrast to the perceived self, the **presenting self** is a public image—the way you want to appear to others. In most cases, the presenting self a person seeks to create is a socially approved image: diligent student, loving partner, conscientious worker, loyal friend, and so on. Social norms often create a gap between the perceived and presenting selves. For example, you may present yourself as more confident than you feel. The "Understanding Communication Technology" feature in

identity management Strategies used by communicators to influence the way others view them.
perceived self The person we believe ourselves to be in moments of candor. It may be identical to or different from the presenting and ideal selves.
presenting self The image a person presents to others. It may be identical to or different from the perceived and ideal selves.

Both in person and online, you have a choice about how much information to reveal about yourself. The challenge is how to present an accurate persona without revealing information you might later regret.

How congruent are your public and private selves? Are you over- or undersharing?

face The socially approved identity that a communicator tries to present.

facework Verbal and nonverbal behavior designed to create and maintain a communicator's face and the face of others.

"On the Internet, nobody knows you're a dog."

Source: Peter Steiner The New Yorker Collection/The Cartoon Bank

the section explores the challenges of presenting a positive self online while still being genuine.

Sociologist Erving Goffman used the word **face** to describe the presenting self, and he coined the term **facework** to describe the verbal and nonverbal ways people maintain their own presenting image and the images of others.[91] He argued that each of us can be viewed as a kind of playwright who creates roles that we want others to believe, as well as a performer who acts out those roles.

Facework involves two tasks: managing our own identity and communicating in ways that reinforce the identities that others are trying to present.[92] You can see how these two goals operate by recalling a time when you have used self-deprecating humor to defuse a potentially unpleasant situation. Suppose, for example, that a friend gave you confusing directions to a party that caused you to be late. "Sorry I got lost," you might have said. "I'm a terrible navigator." This sort of mild self-putdown accomplishes two things at once. It preserves the other person's face by implicitly saying, "It's not your fault." At the same time, your mild self-debasement shows that you're a nice person who doesn't find faults in others or make a big issue out of small problems.[93]

Now that you have a sense of what identity management is, we can look at some characteristics of this process.

Identity Management and Social Media

Rachel Leonard had it made. Newly wed to the love of her life, she could relax on her front porch and enjoy a beautiful view of the Blue Ridge Mountains, all the while looking forward to the birth of their first child. Well, to be more accurate, *virtual* Rachel had all those things.

Rachel's social media posts included happy wedding pictures, gorgeous mountain scenes, and pregnancy updates. But real-life Rachel was grappling with a difficult pregnancy and a growing realization that she had married the wrong person. And the beautiful scenery? The mountain view straight ahead *was* gorgeous, "but if you looked to the left, you could see this huge factory," she admits, adding, "Of course, I didn't take [or post] pictures of the factory because why would you do that?"[94]

Rachel faced a common dilemma rooted in self-concept, communication, and perception: She wanted to present herself favorably to others. At the same time, she craved the genuine approval of people who understood and accepted her as she was. Concerns such as these are central to the communication choices people make. Here we consider the impact of social media in managing that delicate balance.

Conventional wisdom suggests that face-to-face communication is richer and more meaningful than mediated messages in terms of boosting self-esteem, but that isn't always true. Research with adults age 35 or younger suggests that text-based interactions—such as through emails, texts, and tweets—often contribute to self-esteem more than in-person and telephone conversations may. The reasons are twofold. First, people (at least in that age bracket) tend to disclose things about themselves in writing that they wouldn't share in person. And second, technology makes it possible to receive support, even from people who aren't available to offer it in person.[95] The caveat is that self-esteem is not boosted when people present an unrealistically positive image of themselves online.

UNDERSTANDING COMMUNICATION TECHNOLOGY

Tweet from the Past: Forgivable or Inexcusable?

"It was a stupid tweet and immature of me." When he was 21, a message that Ryan Rolison tweeted at age 15 came back to haunt him. Rolison had just been drafted by the Denver Rockies in 2018 when he joined the ranks of athletes and celebrities whose offensive social media posts have attracted national attention. Six years earlier, Rolison had tweeted in reference to then-newly reelected president Barack Obama: "Well we have one hope left . . . if someone shoots him during his speech."[96] Rolison deleted the tweet but not before images of it went viral.

How you judge Rolison's offense and apology are matters of perception. He has said, "I had no idea what I was talking about, and it was immature of me to post something like that. People know that's not who I am."[97]

Some are inclined to forgive him. Sports analyst and attorney Exavier Pope tweeted of Rolison: "He apologized and admitted he made a dumb mistake. It's a lesson he should speak to local kids to give back. Let's use that same standard of forgiveness for all kids in our country."[98]

Others make the point that, youthful indiscretion or not, advocating for the assassination of a U.S. president is a serious offense,[99] and the racial undertones of targeting the country's first black president are ugly and inflammatory.[100]

Taking a wide-angle view, sports publicist Lauren Walsh suggests that people should consider whether or not an offensive post from the past is part of an ongoing pattern:

> It comes down to, who are they today? If they are a different person . . . then you have to give them a little bit of the benefit of the doubt because we all were different people back then. However, if their actions today are still reflecting things that they put out back then, then of course they absolutely need to be held accountable to all of that.[101]

How might your current social media posts be perceived in the future? What steps can you take to make sure that what you post doesn't damage your reputation and offend others, even years into the future?

It probably won't surprise you that people tend to strategically post photos that make them appear attractive and socially engaged with others.[102,103] This is okay within limits. But ultimately, confidence arises from a sense of being accepted for the genuine you. College students who accept their own strengths and weaknesses are more likely to show their true selves on social media. Consequently, they enjoy the security of knowing that others like them for who they really are, imperfections and all.[104] As one social media analyst puts it, stop *trophy hunting*—trying to find that perfect picture or story that will play well on social media—and enjoy your life.[105] Share what happens naturally, not what you have manufactured to impress others.

Characteristics of Identity Management

Now that you have a sense of what identity management is, we can look at some characteristics of this process.

People Have Multiple Identities In the course of even a single day, most people take on a variety of roles: respectful student, joking friend, friendly neighbor, and helpful worker, to suggest just a few. In his Netflix special, comedian W. Kamau Bell humorously presents some of his different identities. As one observer describes it:

Sometimes [Bell is] speaking as a parent, who has to go camping because his kids enjoy camping. Sometimes he's speaking as an African-American, who, for ancestral reasons, doesn't see the appeal of camping ("sleeping outdoors *on purpose?*"). Sometimes—as in a story about having been asked his weight before boarding a small aircraft—he's speaking as "a man, a heterosexual, cisgender *Dad* man." (Hence: "I have no idea how much I weigh.")[106]

Like W. Kamau Bell, who jokes about how his identity changes in different contexts, none of us has a single self. We present ourselves differently, depending on the circumstances.

What multiple selves do you show the world? How authentic and appropriate are each of them?

ASK YOURSELF **?**

Have you ever felt that society imposes an identity on you that you don't want or can't live up to? If so, in what ways does your communication behavior support that identity? In what ways do you challenge people's expectations about you?

You may even play a variety of roles with the same person. With your parents, for instance, perhaps you acted sometimes as a responsible adult (*"You can trust me with the car!"*) and at other times as a helpless child (*"I can't find my socks!"*). At some times—perhaps on birthdays or holidays—you were a dedicated family member, and at other times you may have played the role of rebel.

The ability to construct multiple identities is one element of communication competence. For example, your the style of speaking and the language you use can reflect choices about the social identity you wish to construct. We recall a colleague who was also minister of a Southern Baptist congregation. On campus his manner of speaking was typically professorial, but a visit to hear him preach one Sunday revealed a speaker whose style was much more animated and theatrical, reflecting his identity in that context. Likewise, one scholar pointed out that bilingual Latinxs in the United States often choose whether to use English or Spanish depending on the kind of identity they seek in a given conversation.[107]

Identity Management Is Collaborative   As people perform like actors trying to create a persona (character), their "audience" is made up of other actors who are trying to create their own. Identity-related communication is a kind of theater in which people collaborate with other actors to improvise scenes.

You can appreciate the collaborative nature of identity management by thinking about how you might handle a minor complaint about a friend's behavior:

You: By the way, Jenny said she texted you last week to invite us to her party. If you let me know, I guess I missed it.

Friend: Oh, sorry. I meant to tell you in person, but our schedules have been so crazy this week that I haven't seen you.

You: *(in friendly tone of voice)* That's okay. Maybe next time you can just forward the text to me.

Friend: No problem.

In this upbeat conversation, both you and your friend accept each other's bids for identity as basically thoughtful people. As a result, the conversation runs smoothly. Imagine, though, how different the outcome would be if your friend didn't accept your role as "nice person":

You: By the way, Jenny said she texted you last week to invite us to her party. If you let me know, I guess I missed it.

Friend: *(defensively)* Okay, so I forgot. It's not that big a deal. You're not perfect either!

At this point, you might persist in trying to play the original role: "Hey, I'm not mad at you, and I know I'm not perfect!" Or, you might switch to the role of "unjustly accused person," responding with aggravation, "I never said I was perfect. But we're not talking about me here." The point here is that virtually all conversations provide an arena in which communicators construct their identities in response to the behavior of others. As you read in Chapter 1, communication isn't made up of discrete events that can be separated from one another. Instead, what happens at one moment is influenced by what each party brings to the interaction and by what has happened in their relationship up to that point.

Identity Management Can Be Conscious or Unconscious   At this point, you might object to the notion of strategic identity management, claiming that most of your communication is spontaneous and not a deliberate attempt to present yourself in a certain way. However, you might acknowledge that some of your communication involves a conscious attempt to manage impressions.

Sometimes, you may be highly aware of managing your identity, perhaps during a job interview or a first date. **Frame switching** involves adopting different perspectives based on the cultures and situations in which you find yourself.[108] You may know people who aren't very good at this. They tend to say things that are inappropriate for the situation and offend others by being more aggressive, casual, or standoffish than others expect them to be. Identity management can be especially hard work for people who operate in two or more cultures. A Filipino American man describes the duality of his work and family life this way: "At my first job I learned I had to be very competitive and fighting with my other coworkers for raises all the time, which I was not ready for—being brought up as nice and quiet in my family."[109]

In other cases, you may act largely out of habit or an unconscious sense of what is appropriate.[110] For example, people tend to smile and display sympathetic expressions more during in-person conversations than they do on the phone.[111] They probably don't consciously think, "Since the other person can't see me, I'll alter my nonverbal displays." Reactions like these are often instantaneous and outside of their conscious awareness. Another kind of unconscious face management involves so-called **scripts**—habitual behaviors people have developed over time. When you find yourself in familiar situations, such as greeting customers at work or interacting with friends or family members, you probably slip into these roles quite often. In many cases, scripts are effective and save us mental energy. They can have downsides, however, when we don't think clearly about the demands of the moment or how others may interpret our words and actions.

People Differ in Their Degree of Identity Management Some people are more aware of their identity management behavior than others. So-called **high self-monitors** pay close attention to their own behavior and to others' reactions, adjusting their communication to create the desired impression. By contrast, **low self-monitors** express what they are thinking and feeling without much attention to the impression their behavior creates.[112] There are advantages and disadvantages to both approaches.

High self-monitors are generally good actors and good "people readers" who can act interested when bored or friendly when they feel quite the opposite. This approach allows them to handle social situations smoothly, often putting others at ease and getting a desired reaction from them. For example, high self-monitors tend to post pictures and messages on Facebook that make them seem especially outgoing, which correlates to a higher than average number of "Likes."[113] The downside is that they are unlikely to experience events completely because a portion of their attention is always devoted to viewing the situation from a detached position.

People who score low on the self-monitoring scale behave differently from their more self-conscious counterparts. They have a simpler, more focused idea of who they are and who they want to be. Low self-monitors are likely to have a more limited repertoire of behaviors, so they act in more or less the same way regardless of the situation. This means that they are easy to read. "What you see is what you get" might be their motto. However, a lack of flexibility may sometimes cause them to seem awkward or tactless.

It's probably clear that neither extremely high nor low self-monitoring is the ideal. There are some situations in which paying attention to yourself and adapting your behavior can be useful, but sometimes, reacting without considering the effect on others is a better approach. The desirability of a range of behaviors demonstrates again the notion of communicative competence outlined in Chapter 1: Flexibility is the key to successful communication.

frame switching Adopting the perspectives of different cultures.

script Habitual, reflexive way of behaving.

high self-monitors People who pay close attention to their own behavior and to others' reactions, adjusting their communication to create the desired impression.

low self-monitors People who express what they are thinking and feeling without much attention to the impression their behavior creates.

Identity Management in the Workplace

Some advisors encourage workers to "just be yourself" on the job. But there are times when disclosing information about your personal life can damage your chances for success.[114] This is especially true for people with "invisible stigmas"—traits that, if revealed, may cause them to be viewed unfavorably.[115]

Many parts of a worker's identity have the potential to be invisible stigmas: religion (evangelical Christian, Muslim), sexual orientation (gay, lesbian, bisexual), and health (bipolar, HIV positive). What counts as a stigma to some people (e.g., politically progressive, conservative) might be favored in another organization.[116]

As you consider how to manage your identity at work, consider the following:

- *Proceed with caution.* In an ideal world, everyone would be free to reveal themselves without hesitation. But in real life, total candor can have consequences, so it is best to move slowly.

- *Assess the organization's culture.* If people in your workplace seem supportive of differences—and especially if they appear to welcome people like you—then revealing more of yourself may be safe.

- *Consider the consequences of not opening up.* Keeping an important part of your identity secret can also take an emotional toll.[117] If keeping quiet is truly necessary, you may be better off finding a more welcoming place to work.

- *Test the waters.* If you have a trusted colleague or manager, consider revealing yourself to that person and asking advice about whether and how to go further. But realize that even close secrets can leak, so be sure the person you approach can keep confidences.

Another aspect of identity management at work involves letting other people know about your achievements. See the "@Work" box on humblebragging for more about striking a balance between self-promoting and bragging.

Why Manage Identities?

Why bother trying to shape others' opinions? Sometimes people create and maintain a front to follow social rules. As a child, you probably learned to act polite, even when bored. Part of growing up consists of developing a set of manners for various occasions: meeting strangers, attending school, going to religious services, and so on. Young children who haven't learned all the do's and don'ts of polite society often embarrass their parents by behaving inappropriately ("Mommy, why is that man so fat?"). But by the time they enter school, behavior that might have been excusable or even amusing isn't acceptable. Good manners are often aimed at making others more comfortable.

Social rules govern behavior in a variety of settings. It would be impossible to keep a job, for example, without meeting certain expectations. Salespeople are obliged to treat customers with courtesy. Employees should appear reasonably respectful when talking to the boss. Some forms of clothing would be considered outrageous at work. By agreeing to take on a job, you sign an unwritten contract that you will present a certain face at work, whether or not that face reflects the way you feel at a particular moment.

Even when social roles don't dictate the proper way to behave, people often manage identities for a second reason: to accomplish personal goals. You might, for example, dress up for a visit to traffic court in the hope that your front (responsible citizen) will convince the judge to treat you sympathetically. You might act friendlier and more lively than you feel on meeting a new person so that you will appear

@WORK

Humblebragging in Job Interviews

Life is hard for humblebraggers. Every golden moment inspires an ostensibly modest complaint about a dazzling accomplishment. Can't you just feel the envy—and anguish—of their social media followers?

> Why does the Mercedes dealership always have fresh baked hot cookies?! Don't they understand how mean that is?

> Why can't I look cool when I meet @TomHanks & he hands me his Emmy? Instead I get so excited & look like a goober.

> I'm wearing a ponytail, rolled out of bed from a nap, at the bar w/my guy and guys r still hitting on me. Like really?

After reading even a few humblebrags, you probably agree that they are . . . well, annoying. A Harvard study confirms that trying to pass off a brag as a complaint usually doesn't fool anyone.[118] After reading humblebraggers' online posts, participants in the study gave them low marks in terms of likability, sincerity, and competence.

Despite the blatant faux modesty of humblebrags, sometimes it's necessary to self-promote, especially in the world of work. People were humblebragging in job interviews long before the term was invented: "My greatest weakness? It's probably that I'm a perfectionist." Self-serving comments like this sound phony. But are they worse than honest confessions such as, "I'm not very organized" or "I'm not a team player"? How can you share your accomplishments without being self-defeating on one hand or boastful on the other?

In the second part of their humblebragging study, the Harvard team put those questions to the test. The researchers asked college students to respond in writing to the classic job interview question: "What is your biggest weakness?" Trained evaluators then considered how likely they would be to hire the students based on their responses. In the end, humblebraggers ("I find myself doing a lot of favors for others") were less likely to be hired than those who revealed honest weaknesses ("I sometimes tend to procrastinate").

Honesty carries the day, say the researchers: "While people do not love braggers or complainers, they at least see them as more sincere than humblebraggers."[119]

likable. You might smile and preen to show the attractive stranger at a party that you would like to get better acquainted. In situations like these. you aren't being deceptive as much as putting your best foot forward.

All these examples show that it is difficult—even impossible—not to create impressions. After all, you have to send some sort of message. If you don't act friendly when meeting a stranger, you have to act aloof, indifferent, hostile, or in some other manner. If you don't act businesslike, you have to behave in an alternative way: casual, goofy, or whatever. Often the question isn't whether or not to present a face to others; the question is only which face to present.

Identity Management and Honesty

Managing identities doesn't necessarily make you a liar. In fact, it's almost impossible to imagine how to communicate effectively without selecting a front to present in one situation or another. You would never act the same way with strangers as you do with close friends, for example, or talk to a 2-year-old the same way you talk to a peer.

Each of us has a repertoire of faces—a cast of characters—and part of being a competent communicator is choosing the best role for the situation. In countless situations every day, you have a choice about how to act. When meeting a new roommate, you face the challenge of deciding how much to reveal about yourself and how soon: *Is it appropriate to say right off the bat that you snore? That you're overcoming a painful breakup? That you really hope the two of you will become good friends?*

It's an oversimplification to say there is only one honest way to behave in a particular circumstance. Instead, impression management involves deciding which face—which part of your identity—to reveal and when. In Chapter 8 we'll talk more about the rewards and risks of disclosing personal information.

It's also worth noting that not all misrepresentations are intentional. Researchers have used the term *foggy mirror* to describe the gap between participants' self-perceptions and a more objective assessment. Overweight online daters who describe themselves as being "average" in weight might be engaging in wishful thinking rather than telling an outright lie.[120]

Bo Burnham, who wrote and directed the movie *Eighth Grade*, says he wanted to depict the ups and downs of adolescence, and teen actor Elsie Fisher was the perfect choice to play Kayla.[121] Together, they tried to show the messy work of establishing a social identity at age 13—acne, awkward conversations, social media bravado, insecurities, and all.

MAKING THE GRADE

At www.oup.com/he/adler-uhc14e, you will find a variety of resources to enhance your understanding, including video clips, animations, self-quizzes, additional activities, audio and video summaries, interactive self-assessments, and more.

OBJECTIVE 3.1   Identify the communicative influences that shape the self-concept.

- The self-concept is a set of largely stable perceptions about oneself.
- Although some personality characteristics are innate, communication and cultural/social factors also help shape the self-concept.
- The self-concept can feed into self-fulfilling prophecies about communication behavior.
  - > Name at least four factors that influence a person's self-concept. Explain the role of communication relevant to each of these factors.
  - > Describe a relationship or event that had a powerful impact on your self-concept. What role did communication play?
  - > Present a scenario in which a self-fulfilling prophecy might help someone achieve an important goal.

OBJECTIVE 3.2   Explain perceptual tendencies and situational factors that influence perceptions of others.

- Perception is a multistage process that includes selection, organization, and interpretation of information.

- Perceptions are influenced by a range of factors, including physiology, stereotypes, and relational roles.
- We often incorporate our perceptions into personal narratives that not only tell a story but suggest a particular interpretation that others may accept or challenge.
- Perceptual tendencies and errors can affect the way we view and communicate with others.
- Empathy and emotional intelligence are valuable tools for increasing understanding of others and hence communicating more effectively with them, both in person and online.
- Perception checking is one tool for increasing the accuracy of perceptions and for increasing empathy.
  - > Have you ever been judged on the basis of your sex or gender? In what ways have you judged other people? What perceptual tendencies contribute to such judgments? What advice do you have for avoiding hurtful and unfair judgments?
  - > List and explain at least five perceptual tendencies that shape how individuals make judgments about other people.
  - > Describe an experience in which you met someone new. How did your impressions of that person evolve in the context of the perceptual phases of selection, organization, and interpretation?
  - > How could greater empathy have changed an interpersonal conflict you have experienced?

OBJECTIVE 3.3   Describe how identity management operates in both face-to-face and online communication.

- Identity management consists of strategic communication designed to influence others' perceptions of an individual.

- Identity management is usually collaborative.
- Identity management occurs for two reasons: (1) to follow social rules and conventions and (2) to achieve a variety of content and relational goals.
- Communicators engage in creating impressions by managing their manner, appearance, online posts, and the settings in which they interact with others.
- Although identity management might seem manipulative, it can be an authentic form of communication. Because each person has a variety of faces that they can present, choosing which one to present is not necessarily being dishonest.
  - > Explain how people use facework and frame switching to manage their private and public identities.
  - > Describe at least three different identities you present. Compare and contrast the communication strategies you use to support each.

KEY TERMS

attribution p. 62

emotional intelligence (EI) p. 65

empathy p. 64

face p. 68

facework p. 68

frame switching p. 71

gender p. 61

gender matrix p. 61

halo/horns effect p. 63

high self-monitors p. 71

identity management p. 67

implicit bias p. 61

interpretation p. 59

low self-monitors p. 71

narratives p. 61

negativity bias p. 63

organization p. 58

perceived self p. 67

perception p. 57

perception checking p. 65

personality p. 53

presenting self p. 67

reflected appraisal p. 54

script p. 71

selection p. 58

self-concept p. 52

self-esteem p. 53

self-fulfilling prophecy p. 56

self-serving bias p. 63

sex p. 61

social comparison p. 54

stereotype p. 60

sympathy p. 64

ACTIVITIES

1. Think of a situation in which you and others have achieved relational harmony by sharing a narrative. Then think of a situation in which you and another person used different narratives to describe a set of circumstances. What were the consequences of having different narratives?

2. Keep a 1-day log listing the identities you create in different situations: at school and at work, and with strangers, various family members, and different friends. For each identity:

   a. Describe the persona you are trying to project (e.g., "responsible son or daughter," "laid-back friend," "attentive student").

   b. Explain how you communicate to promote this identity. What kinds of things do you say (or not say)? How do you act?

Communication and Culture

4

CHAPTER OUTLINE

4.1 Understanding Cultures and Cocultures 78
Differences and Similarities
Salience of Differences

4.2 How Cultural Values and Norms Shape Communication 80
Individualism and Collectivism
High and Low Context
Uncertainty Avoidance
Power Distance
Talk and Silence
Competition and Cooperation

4.3 Cocultures and Communication 86
Race and Ethnicity
Regional Differences
Gender Identity
Sexual Orientation
Religion
Political Viewpoints
Ability/Disability
Age/Generation
Socioeconomic Status

4.4 Developing Intercultural Communication Competence 93
Increased Contact
Tolerance for Ambiguity
Open-Mindedness
Knowledge and Skill
Patience and Perseverance

MAKING THE GRADE **98**

KEY TERMS **99**

ACTIVITY **99**

LEARNING OBJECTIVES

4.1

Analyze how salient cultural and cocultural differences can affect communication both within and across groups of people.

4.2

Explain the communicative influences of individualism and collectivism, high and low context, uncertainty avoidance, power distance, talk and silence, competition, and cooperation.

4.3

Evaluate the influence of factors such as race and ethnicity, gender identity, religion, age, and political viewpoints on communication.

4.4

Implement strategies to increase intercultural and cocultural communication competence.

As you read about communication and culture, ask yourself these questions:

How does your own cultural identity affect the way you communicate?

How does bias affect intercultural communication?

What could you do to communicate more frequently and more successfully with people from backgrounds different from your own?

culture The language, values, beliefs, traditions, and customs people share and learn.

coculture A group that is part of an encompassing culture.

NEW JERSEY HIGH SCHOOL grads Priya Vulchi and Winona Guo could have gone straight to college. Instead, they spent a gap year traveling to all 50 states, listening to ordinary Americans describe how race, privilege, and gender shape their lives. Guo and Vulchi's book *Tell Me Who You Are: Sharing Our Stories of Race, Culture, and Identity* is a record of those conversations.

In one of their TED talks, "Lessons of Cultural Intimacy," Vulchi and Guo describe the challenges of communicating across cultures. They say that "race is not only one of the things that affect us the most, but also one of the most difficult to talk about." They urge readers to follow in their footsteps by reaching out and listening to people from diverse backgrounds.

By talking with more than 500 strangers across the United States, Guo and Vulchi put their intercultural communication skills to good use. They write, "Millions of Americans—both white people and people of color—still don't come close to understanding people outside their mostly homogenous immediate communities."[1] The challenge is to keep listening and keep learning.

You don't have to travel to another country or state or even another town to encounter people from different backgrounds and cultures. Society is not only more interconnected but more diverse than ever—in terms of culture, race and ethnicity, physical ability, personal identity, family background, and more. As you'll see in this chapter, this diversity presents communication benefits as well as challenges. We will explore both. The main themes in this chapter include:

- When culture does—and when it doesn't—affect communication
- The values and norms that can shape interaction between people from different backgrounds
- How diversity within a culture shapes communication
- How to communicate effectively with people from different backgrounds

4.1 Understanding Cultures and Cocultures

As we use the term, **culture** refers to "the language, values, beliefs, traditions, and customs people share and learn."[2] Intercultural communication—communication between people from different cultures—happens both across and within societies. A **coculture** is a group that is part of an encompassing culture. The children of immigrants, for example, might be immersed in mainstream American culture while still identifying with the customs of their parents' homeland. Features of other cocultures include:

- Age (e.g., teen, senior citizen)
- Race/ethnicity (e.g., African American, Latinx, White)
- Sexual orientation (e.g., LGBTQ)
- Physical challenges (e.g., wheelchair users, Deaf persons)
- Language (e.g., native and non-native speakers)
- Religion (e.g., Mormon, Muslim)
- Activity (e.g., biker, gamer)

You may identify with one or more of these groups and many others—each of which may have its own communication style and vocabulary.

Differences and Similarities

Sometimes there are obvious differences in communication across cocultures. For example, it's no surprise that Millennials (born in the 1980s through early 1990s) and members of Generation Z (born between the mid-1990s and early 2010s) rely more on social media than their parents do.[3] At other times, people overestimate differences. Based on her appearance and name, people often assume that Rupa Shenoy is not American. If she says "I'm from Iowa," they are likely to ask, "No, where are you *from*?" It can be discouraging to feel "othered" in that way. At the same time, Shenoy says she has learned that most people are genuinely curious and just want to know her better.[4]

Membership in cocultures can be a source of identity, enrichment, and pride. However, when negative stereotypes (Chapter 3) are involved, **intergroup communication** between members of different cocultures can be limited or hurtful. For example, in a poll of U.S. high school students, more than a third said they would be reluctant to spend time outside of school with classmates who had physical disabilities. The reasons they gave were awkwardness and the assumption that they wouldn't have much in common.[5]

intergroup communication Interaction between members of different cultures or cocultures.

Stereotypes may seem insurmountable. But the news isn't all bad. Those who interact daily with people from different backgrounds are less likely to be prejudiced than those who don't.[6] (We'll talk more about this later in the chapter.) Positive media images also have an influence. *Sesame Street* and *Power Rangers* have introduced characters with autism,[7] and ABC's *The Good Doctor* portrays a physician who has autism. In one study, teenagers who watched a video featuring teens with autism were subsequently less likely than their peers to view people with the condition negatively.[8]

Salience of Differences

Depending on the situation, differences may seem more or less noticeable or relevant. Intercultural and intergroup communication—at least as we'll use the terms here—don't always occur when people from different backgrounds interact. Those backgrounds must have a significant impact on the exchange before we can say that culture has made a difference. Social

In the award-winning comedy series *Jane the Virgin*, the title character Jane Villanueva (Gina Rodriguez) is a smart, middle-class, religious Latinx mom who struggles to find the right path in the face of personal and romantic dilemmas.

What cocultures are most fundamental to your identity? How do your roots shape your communication?

The TV show *The Good Doctor* features Shaun Murphy (Freddie Highmore), a surgeon who has autism. Murphy is socially awkward at times but also compassionate, perceptive, and an excellent practitioner.

If you were challenging stereotypes about a particular group or coculture, what would you like people to know?

salience   How much weight people attach to a particular phenomenon or characteristic.

in-groups   Groups with which one identifies.

out-groups   Groups one views as different from oneself.

scientists use the term **salience** to describe how much weight people attach to characteristics in a particular situation. Consider a few examples in which culture has little or no salience:

- A group of preschool children is playing together in a park. It does not matter to them that their parents come from different countries.

- Members of a school athletic team are intent on winning the league championship. During a game, cultural distinctions aren't salient.

- A husband and wife were raised in different religious traditions. Most of the time their religious heritage makes little difference and the partners view themselves as a unified couple.

These examples show that in order to view ourselves as members of a culture or coculture, there has to be some distinction between "us" and "them." Social scientists use the label **in-groups** to describe groups with which a person identifies and is emotionally connected, and **out-groups** to describe those they view as different and with whom they have no sense of affiliation.[9]

Although it's eye-opening to look at cultural differences, it's important not to overstate their influence on communication. There are sometimes greater differences *within* cultures than *between* them. Within every culture, members display a wide range of communication styles. For example, Jaclyn Samson, whose parents are from the Philippines, is frustrated when people typecast her in ways that don't reflect who she really is. In school "people saw me as the quiet Asian girl in class who was good at math and kept to herself," she says, when actually "I was outgoing and talkative . . . I had a ton of friends, and to be quite honest, I was terrible at math."[10] She makes a good point: Cultural differences are generalizations—broad patterns that do not apply to every member of a group.

🔘 4.2 How Cultural Values and Norms Shape Communication

Growing up in the Netherlands, Daniëlle didn't anticipate that she would fall in love with a Sudanese man. However, she and her-now husband Hussam connected right away. "Even though we were from different continents, we had an insane amount of things in common," she says. "We both loved to read the same books and liked playing around with graphic design. We understood each other."[11] Even so, Daniëlle was initially nervous, based on stereotypes she had heard about Arab men. Over time, she learned a valuable lesson. "We are not the stereotypes people have about us," she says. "We are all just people, with differences and similarities, strengths and weaknesses, habits and customs."

Better understanding diverse norms and values is one way to reduce the uncertainty of communicating with people from different cultures. Here is a look at six patterns that help distinguish cultures around the world. Unless communicators are aware of these differences, they may consider people from other cultures to be unusual—or even offensive—without realizing that their apparently odd behavior comes from following a different set of beliefs and (often unwritten) rules about the socially accepted way to communicate.

Individualism and Collectivism

Chapter 3 introduced the concepts of individualism and collectivism and described how they affect identity. Members of **individualistic cultures**—including the United States, Canada, and Great Britain—tend to view their primary allegiance to themselves as individuals. Conversely, communicators in **collectivistic cultures**—such as China, Korea, and Japan—place higher value on loyalty and obligations to an in-group such as one's extended family, community, or employer.[12] Individualistic and collectivistic cultures also have different approaches to communication, as the following patterns illustrate.

individualistic culture A culture in which members focus on the value and welfare of individual members more than on the group as a whole.

collectivistic culture A culture in which members focus more on the welfare of the group as a whole than on individual identity.

Names　When asked to identify themselves, individualistic Americans, Canadians, Australians, and Europeans usually respond with their given name and then their surname. But collectivistic Asians do it the other way around. They are likely to begin with their family name and provide their given name second.[13] Their primary emphasis is on group membership rather than individual identity. It's easy to imagine the confusion created when Western forms ask for a person's "first name" and "last name," assuming that everyone orders them the same way.

Pronouns　When American-born Ann Babe moved to South Korea to teach English, she was confused when a colleague said, "Our husband is also a teacher." Then Babe came to understand that South Koreans' use of pronouns reflects a cultural emphasis on collectivism. They use *our* and *my* (also *we* and *I*) interchangeably but prefer the collective terms because they seem less egocentric. That helped her understand why South Korean students were initially put off by English, which sounded selfish to them. As one student put it, everything in English is "my, my, my" and "me, me, me."[14]

Direct or Indirect Speech　Individualistic cultures are relatively tolerant of conflicts and tend to use a direct, solution-oriented approach. In Hungary, where individualism is highly valued, the locals pride themselves on being straight talkers who don't shy away from sensitive topics. "Hungarians are self-expressed and to-the-point," says a Californian who married a woman from Hungary and moved there, adding, "If someone has the slightest problem with something, they're going to let you know. . . . That's just the way it is here. Don't take it personally—tempers flare, decibels rise."[15]

By contrast, members of collectivistic cultures are less direct, often placing greater emphasis on harmony and group interests.[16] "Everyone is soooo polite," observes Spike Dailey of his time in Japan, saying, "Two of the most common phrases you'll here in Japanese are 'sumimasen' and 'gomenasai.' They mean 'excuse me' and 'I'm sorry,' respectively." The emphasis there is on maintaining good relations and not causing offense.

Competition or Teamwork　Individualistic cultures are characterized by self-reliance and competition. A website for international students explains, "Americans thrive on competition. From a young age, children are encouraged to work hard and try their best. . . . Even Girl Scouts vie to sell the most cookies."[17]

As portrayed in the film *Blindspotting*, Collin (Daveed Diggs) and Miles (Rafael Casal) have been best friends since childhood. In most cases, their long and loyal relationship has much more salience than their ethnicities. At times, though, race moves to the forefront.

What cultural factors are most salient in your important relationships? Do those factors change from time to time?

By contrast, collectivistic cultures are more focused on collaboration and group effort.[18] As a result, members are often adept at seeing others' point of view. In one study, Chinese and American players were paired together in a game that required them to take on the perspective of their partners.[19] The collectivist Chinese were markedly more successful at this than the Americans. Of course, it's important to recall that cultural differences are generalizations. There are elements of collectivism in Western cultures and of individualism in Asian cultures.[20]

Humility or Self-Promotion   In high-context cultures, it's considered polite to downplay one's accomplishments. To investigate this issue, researchers asked female college students in Iran to answer questions about the quality of instruction at the school.[21] However, the real purpose of the interviews was to see how the women would respond when the researchers complimented them. When a research assistant told a student "Your shoes are really beautiful," the student replied, "This is not so stylish." Another time, after a researcher complimented a participant's phone, the student said, "You can have it." The lead researcher reflects that, in the collectivistic Iranian culture, these responses are courteous ways to demonstrate modesty and to avoid implying that you are better than anyone else. (To members of the Iranian culture, it's clear that "you can have it" is most often a polite response to a compliment, not an actual invitation to assume ownership.)

By contrast, individualistic Americans are relatively comfortable talking about their personal accomplishments or possessions and even saying they are "the best" at something. This doesn't mean they are more self-absorbed than people in other cultures. From an American's perspective, sharing news of an accomplishment can be a means of including others in a happy life event. For example, if an American student says "I made an A on the test!" a friend may say, "That's great! Let's celebrate." In other circumstances, "boasting" can be a form of verbal play, as when one person tells another "My team will thrash yours in the big game" even if that is very unlikely. These behaviors are not usually considered rude. In fact, self-promotion is expected, and even respected, in an individualistic culture. An employee or job candidate who downplays accomplishments may be seen as lacking confidence or being ill prepared.

Communication Apprehension   Cultural differences can also affect the level of anxiety people feel when communicating. In societies in which the desire to conform is great, a higher degree of communication apprehension is evident. For example, South Koreans exhibit more conflict avoidance and more apprehension about speaking in front of people than do members of individualistic cultures such as the United States.[22] It's important to realize that different levels of communication apprehension don't mean that shyness is a "problem" in some cultures. In fact, the opposite is true: In these cultures, reticence is valued. When the goal is to avoid being the nail that sticks out, it's logical to feel nervous when you make yourself appear different by calling attention to yourself. A self-concept that includes being assertive might make a Westerner feel proud, but in much of Asia, people may consider it inappropriate to speak forthrightly with out-group members.

"I gotta be me" could be the motto of a Westerner, but "If I hurt you, I hurt myself" is closer to the collective way of thinking. Table 4.1 illustrates some of the differences between individualistic and collectivistic cultures.

High and Low Context

Cultures also differ in how much they rely on words versus situational cues.[23,24] Members of **low-context cultures** use *language* mostly to express thoughts, feelings, and ideas. They prefer specific, straightforward communication. Mainstream cultures in the United States, Canada, northern Europe, and Israel fall toward the low-context end of the scale. By contrast, members of **high-context cultures** rely heavily on unspoken and situational cues. As the term suggests, they interpret meaning based more on the *context* in which a message is delivered—the nonverbal behaviors of the speaker, the history of the relationship, the general social rules that govern interaction between people, and so on. They place a higher value on social harmony than on direct communication. People in most Asian, Latin, and Middle Eastern cultures fit this pattern. Here we explore three communication activities that illustrate the difference between high and low context.

low-context culture A culture that uses language primarily to express thoughts, feelings, and ideas as directly as possible.

high-context culture A culture that relies heavily on subtle, often nonverbal cues to maintain social harmony.

Doing Business　Imagine a meeting that involves professionals from around the world. Members of low-context cultures will probably want to "get down to business" quickly by talking about the details of a deal or idea. They tend to be verbal, direct, and results-oriented. Talking for very long about other matters, such as personal information, can seem to them like a waste of time.[25]

By contrast, "in countries such as Mexico, conversations are first and foremost an opportunity to enhance the relationship," observe international business consultants Melissa Hahn and Andy Molinsky.[26] From a high-context perspective, focusing on business without first developing a relationship can feel dismissive and disrespectful. It implies that the other person has no value. And because members of high-context cultures focus on nonverbal cues and on what *isn't* said, they may wonder how they are supposed to trust or understand the other person with getting to know them first.

TABLE 4.1

The Self in Individualistic Versus Collectivistic Cultures

ELEMENT	INDIVIDUALISTIC CULTURES	COLLECTIVISTIC CULTURES
Priorities	Personal independence and self-reliance	Harmony and the needs of others
Expectations	Caring for oneself and one's immediate family	Caring for extended families and not putting oneself first
Response to conflict	Tendency to address conflict directly	Emphasis on harmony; indirect approach to conflict
Values	Autonomy, competition, self-promotion	Duty, humility, loyalty, fitting in

Sources: Cai, D. A., & Fink, E. L. (2002). Conflict style differences between individualists and collectivists. *Communication Monographs, 69*, 67–87.
Croucher, S. M., Galy-Badenas, F., Jäntti, P., Carlson, E., & Cheng, Z. (2016). A test of the relationship between argumentativeness and individualism/collectivism in the United States and Finland. *Communication Research Reports, 33*(2), 128–136.
Merkin, R. (2015). The relationship between individualism/collectivism. *Journal of Intercultural Communication, 39*, 4.

Managing Conflict   Members of low-context cultures are likely to state their concerns or complaints up front, whereas people raised in high-context cultures usually hint at them.[27,28] An exchange student gives an example:

> Suppose a guy feels bad about his roommate eating his snacks. If he is Chinese, he may try to hide his food secretly or choose a certain time to say, "My snacks run out so fast, I think I need to buy more next time." Before this, he also may think about whether his roommate would hate him if he says something wrong. But Americans may point out directly that someone has been eating their food.

The roommate from China may feel it is obvious, based on the situation and his indirect statement, that he is upset about the roommate eating his food. But the American—who may expect his friend to say outright if he is upset—may miss the point entirely. Table 4.2 summarizes some key differences in how people from low- and high-context cultures use language. The point is to understand each other better by recognizing differences and, ideally, by finding ways to meet everyone's goals.

Uncertainty Avoidance

Uncertainty may be universal, but cultures have different ways of coping with it. **Uncertainty avoidance** is the degree to which members of a culture feel threatened by ambiguous situations and how much they try to avoid them.[29] As a group, residents of some countries (including Singapore, Great Britain, Denmark, Sweden, Hong Kong, and the United States) are relatively unthreatened by change, whereas others (such as natives of Belgium, Greece, Japan, and Portugal) are more likely to find new or ambiguous situations uncomfortable.

A culture's degree of uncertainty avoidance is reflected in the way its members communicate. In countries that avoid uncertainty, people value rules, regulations, and traditions that provide a sense of order and predictability. For example, college students from France (which scores high in uncertainty avoidance) tend to prefer professors they consider to be old, experienced, highly knowledgeable, and clear about classroom rules.[30] However, American college students, who are typically more comfortable with ambiguity, gravitate to professors who build strong relationships with them, are open-minded, and are easy to talk to.[31]

The misunderstandings that arise from these cultural differences can be striking. A person who stands out as different or who expresses ideas that challenge the status quo may be considered untrustworthy or even dangerous by people who avoid uncertainty, but as a visionary and an inspiration by those who are comfortable with change. The take-away lesson is that, even within a culture, people are likely to have varied reactions to ambiguous information and to communicate differently about it.

> **uncertainty avoidance** The cultural tendency to seek stability and to honor tradition instead of welcoming risk, uncertainty, and change.

TABLE 4.2

High- and Low-Context Communication Styles

ELEMENT	LOW CONTEXT	HIGH CONTEXT
How information is conveyed	The majority of information is conveyed in explicit verbal messages, with less focus on the situational context.	Information is conveyed in contextual cues such as time, place, relationship, and situation. There is less reliance on explicit verbal messages.
Listening habits	Listeners are likely to be distracted by other stimuli and their own thoughts.	Listeners are usually attentive to nonverbal cues and situational factors.
Approaches to conflict	Conflict is often addressed directly and verbally.	People are likely to manage conflict indirectly, through hints and nonverbal cues.

Power Distance

Power distance refers to the gap between social groups who possess a great amount of resources and influence and those who have less. Members of cultures with low power distance believe in minimizing the difference between various social classes. Rich and poor, educated and uneducated groups may still exist, but there's a pervasive belief in low-power-difference cultures that one person is as good as another regardless of their station in life. On the other hand, in cultures that observe high power distance, it's considered natural and respectful to treat some people as more important than others.

Austria, Denmark, Israel, and New Zealand are among the most egalitarian countries, meaning that residents are disposed to consider everyone to be equal. Less extreme but on the same end of the scale are the United States and Canada, where employees typically expect that upper-level leaders will ask for and listen to their opinions.[32] (See the *@Work* feature for more details on power distance in the workplace.)

Countries with a high degree of power distance include the Philippines, Mexico, Venezuela, India, Singapore, Japan, and Thailand.[33] Traditionalists in those countries are likely to expect obedience and to honor the opinions of authority figures without question. For example, medical patients in Pakistan, Japan, and Thailand seldom question their physicians' advice or ask them questions because they think that might seem disrespectful.[34]

In cultures that honor high power distance, it may seem rude to treat everyone the same way. In the Japanese workplace, for example, new acquaintances exchange business cards immediately, which helps establish everyone's relative status. The oldest or highest ranking person receives the deepest bows from others, the best seat, the most deferential treatment, and so on. This treatment is not regarded as elitist or disrespectful. Indeed, treating a high-status person the same as everyone else would seem rude.

> **power distance** The degree to which members of a group are willing to accept a difference in power and status.

@WORK

Power Distance and Culture in the Workplace

"Siew Tian, why don't you speak up? I know you have something to say, and you're not saying it." American Kate Sweetman remembers being bewildered when her Indonesian colleague resisted this invitation to share her ideas.[35] Tian's reluctance is easier to understand if you know that she grew up in a culture that observes high power distance.

In countries with high power distance, employees are not accustomed to having input about workplace matters. In fact, they may feel uncomfortable when given freedom to make decisions or when prompted or encouraged to speak up. The reverse is true in cultures with low power distance, where employees may feel unappreciated when they aren't consulted or when their opinions are ignored.

The potential for misunderstanding is great. The behavior demonstrated by a deferential employee may be regarded as admirably dutiful, submissive, and respectful by managers accustomed to high power distance. But leaders accustomed to low power distance may assume that quiet and nonassertive employees lack initiative and creativity—traits that help people gain promotions in their culture.

Given these differences, it's easy to understand why multinational companies are advised to consider fundamental differences in communication values and behavior when they launch operations in a new country.

Do you work with people whose beliefs about power distance differ from your own? If so, how do those differences show up when you work together? Can you think of ways to manage these differences that will improve both personal relationships and workplace effectiveness?

Jaclyn Samson, a student at West Washington University, blogs about college life and the challenges and joys of growing up in a multiethnic environment. She says that people often assume she will be quiet and shy, but she is actually outgoing and enjoys social activities and public speaking.

Do people ever make assumptions about you based on appearance or stereotypes? If so, how does that affect the way they communicate with you?

ON YOUR FEET

Prepare to share an example of a communication episode in which you exhibited one of the cultural patterns described in this chapter (e.g., individualism, collectivism, power distance, silence). In what ways is your identity shaped by groups to which you belong (e.g., your family, hometown, college, clubs, religion)?

Talk and Silence

Beliefs about the value of talk differ from one culture to another.[36] Members of Western cultures tend to view talk as desirable and use it for social purposes as well as to perform tasks. Silence can feel embarrassing and awkward in these cultures, since it's likely to be interpreted as lack of interest, unwillingness to communicate, hostility, anxiety, shyness, or a sign of interpersonal incompatibility.

Yet, members of many cultures tend to perceive talk quite differently. Unlike most Westerners, who find silence uncomfortable, Japanese and Chinese traditionalists more often believe that remaining quiet is proper when there is nothing to be said. To many Asians, a talkative person is often considered a show-off or a fake. Silence is also valued in some European cultures. For example, Swedes are more reluctant than Americans to engage in small talk with strangers.[37] And members of some Native American communities honor silence. Traditional members of Western Apache tribes maintain silence when others lose their temper. As one member explained, "When someone gets mad at you and starts yelling, then just don't do anything to make him get worse."[38]

Competition and Cooperation

Cultures are a bit like people in that they may be regarded as competitive, cooperative, or somewhere in the middle. Competitive cultures—including those in Italy, Nigeria, and Great Britain—embody qualities such as independence, competitiveness, and assertiveness.[39] In those cultures, women are often expected to take care of home and family life, whereas men are expected to shoulder most of the financial responsibilities.

Gender roles are less differentiated in cooperative cultures—which emphasize equality, relationships, cooperation, and consensus building.[40] In Iceland, the Netherlands, and Norway, both men and women tend to consider harmony and cooperation to be more important than competition. When doing business, members of cooperative cultures are likely to strive for mutually satisfying outcomes, whereas people from competitive cultures often negotiate to "win" and consider it a "bargain" when a deal is resolved to their benefit. This short-term gain may result in damaged relationships, however, if people on either side feel disrespected or treated unfairly.

4.3 Cocultures and Communication

Much of how individuals view themselves and how they relate to others grows from their cultural and cocultural identity—the groups with which they identify. Where do you come from? What's your ethnicity? Your religion? Your sexual orientation? Your age?

In the following pages we will look at some—though by no means all—of the factors that help shape cultural identity, and hence the way people perceive and communicate with others. As you read on, think about other cocultures that might be added to this list.

Race and Ethnicity

Race is a social, not a biological, construct to describe a group of people who share physical and cultural traits and potentially a common ancestry. It is not a reliable indicator of individual differences. As one analyst puts it, "There is less to race than meets the eye. . . . Knowing someone's skin color doesn't necessarily tell you anything else about him or her."[41] There is more genetic variation within races than between them. For example, some people with Asian ancestry are short, but others are tall. Some have sunny dispositions, while others are sterner. Some are terrific athletes, while others were born clumsy. The same applies to people from every background. Even within a physically recognizable population, personal experience plays a far greater role than superficial characteristics such as skin color.[42]

The election of Sadiq Khan as the mayor of London was a triumph of multiculturalism. Khan told a reporter that he was more than a Muslim: "I'm a Londoner, I'm a European, I'm British, of Asian origin, of Pakistani heritage.

To what cocultures do you belong? How do these memberships affect your communication?

Ethnicity, another social construct, refers to the degree to which a person identifies with a particular group, usually on the basis of nationality, culture, religion, or some other perspective.[43] This goes beyond physical indicators. For example, a person may have physical characteristics that appear Asian, but they may identify more strongly as a Presbyterian or a member of the working class.

People are multidimensional, and factors such as race and ethnicity intersect in profound and complex ways with gender and class. **Intersectionality theory** proposes that each person experiences life at the intersection of multiple factors, whose interplay gives rise to a unique perspective and collection of experiences all their own.[44,45] Each of our perspectives is unique. Philosopher Kwame Anthony Appiah reinforces the point:

> When I was a student at the University of Cambridge in the 1970s, gay men were *très chic*: You couldn't have a serious party without some of us scattered around like throw pillows. Do my experiences entitle me to speak for a queer farmworker who is coming of age in Emmett, Idaho? Nobody appointed me head gay.[46]

With intersectional theory in mind, treat the descriptions in this chapter not as a formula for understanding particular individuals, but as a way of exploring factors that can affect one another.

race A social construct to describe a group of people who share physical and cultural traits and potentially a common ancestry.

ethnicity A social construct that refers to the degree to which a person identifies with a particular group, usually on the basis of nationality, culture, religion, or some other unifying perspective.

intersectionality theory The idea that people are influenced in unique ways by the complex overlap and interactions of multiple identities and social factors.

Regional Differences

Regional stereotyping can lead to judging people based on their accents or where they grew up. Researchers in one experiment asked human resource professionals to rate the intelligence, initiative, and personality of job applicants after listening to a brief recording of their voices. The speakers with recognizable regional accents—from the southern United States or New Jersey, for example—were tagged for lower-level jobs, whereas those with less pronounced speech styles were recommended for higher-level jobs that involved more public contact.[47] Reactions such as these show how pervasive and powerful regional stereotypes can be.

"Anything you say with an accent may be used against you."

Source: Paul Noth The New Yorker Collection/The Cartoon Bank

For years, CNN host Anderson Cooper did not publicly disclose that he is gay. He considered it private information, feared it might put him in danger, and was concerned about career ramifications. After coming out, Cooper said, "The tide of history only advances when people make themselves fully visible."

What factors affect your decision to share personal information with others?

Gender Identity

As you read in Chapters 3 and 5, the classification of gender into two groups—the *gender binary*—has given way to a less constrictive, more inclusive notion of gender identity. *Masculine* and *feminine* are only two adjectives in a constellation of gender-related qualities. Other identities include nonbinary, gender independent, gender creative, gender expansive, gender diverse, and gender queer,[48] to name just a few. *They/them* have become the singular pronouns of choice for those who don't identify with *he/his* or *she/hers*.

Gender identity may change over time and is not strictly tied to physical features. *Cisgender* people have a gender identity that matches the sex they were assigned to at birth. *Transgender* individuals don't feel that their biological sex is a good description of who they are. Lewis Hancox, a filmmaker and comedian who is transgender, reflects that people who transition to a different gender identity encounter some of the same ambiguities. Having been through a public transition himself, he recommends that trans individuals love and accept themselves, connect with people who are supportive, have patience with those who don't immediately embrace the change, and keep in mind all the qualities that make them unique and special as individuals. "It took me a long time to realize that being transgender didn't make me any less of a guy, or more importantly, any less of a person," Hancox says. "We're all different in our own right and we should embrace those differences."[49]

Sexual Orientation

Another dimension of social identity is sexual orientation. It is described in many ways, such as gay, straight, lesbian, bisexual, and queer. The social climate has become more receptive to LGBTQ individuals than in the past, at least in most of the developed world. As one report put it,

> Most LGBTQ people who are now adults can recall feeling that they were "the only one" and fearing complete rejection by their families, friends, and associates should their identity become known. Yet today, in this era of GSAs [gay–straight alliances] . . . openly gay politicians and civil unions; debates about same-sex marriage, gay adoption, and gays in the military; and a plethora of websites aimed specifically at LGBTQ youth, it is hard to imagine many youths who would believe they are alone in their feelings.[50]

At ItGetsBetter.org, people can post messages to emphasize to LGBTQ youth that any harassment they may be experiencing is not their fault and that people care about making it better. Since the It Gets Better Project launched in 2010, people have posted more than 60,000 videos, which, all together, have been viewed more than 50 million times.[51] The site's creators say it shows what can happen when communication technology and good intentions come together.[52]

All the same, revealing a sexual orientation other than straight can be challenging. "Every time I meet someone new I must decide if, how, and when I will reveal my sexual orientation,"[53] reflects Jennifer Potter, a physician who is gay. She says it's often easy to "pass" as heterosexual, but then she experiences the awkwardness of people assuming she has a boyfriend or husband.

And whereas individuals in some cultures are receptive to diverse sexual orientations, others are more disapproving. People who are not straight may face ridicule, discrimination, and even violence. On average, nearly 1 in 5 hate crimes in the United States is motivated by a sexual orientation bias.[54]

Religion

In some cultures, religion is the defining factor in shaping in- and out-groups. As fears of terrorism have grown, peace-loving Muslims living in the West have often been singled out and vilified. Yasmin Hussein, who works at the Arab American Institute, reflects:

> Many Muslims and individuals of other faiths who were thought to be Muslim have been attacked physically and verbally. Young children have been bullied at schools, others told to go back home and social media has become at times (a lot of the time) an ugly place to be on.[55]

In an effort to dispel unrealistic stereotypes, tens of thousands of Muslims have joined the #NotInMyName social media movement in which they denounce terrorist groups such as ISIS and condemn violence in the name of their religion.

Religion shapes how and with whom many people communicate. For example, members of the Orthodox Jewish community consider it important to marry within the faith.[56] Members of evangelical churches are likely to view parents as family decision makers and to honor children for following their advice without question.[57] Religious activities such as reading scripture at home are most common among Jehovah's Witnesses and members of the Church of Jesus Christ of Latter-Day Saints (Mormons), evangelical churches, and historically black churches.[58] These congregations are also the most likely to believe there is only one true religion.

Political Viewpoints

Political discussions have become one of the most contentious examples of intercultural communication in the United States. As one analyst joked, "Liberals like to have fun. Conservatives like to have fun. Liberals and Conservatives like to have fun with each other . . . unless they're talking about politics."[59]

Conventional wisdom advises people to steer conversations away from politics and

In recent years, many people have hardened into opposing political camps that distrust and disrespect one another, preventing productive discussion.

What might you discover you have in common if you listened more respectfully to others with differing political positions?

social media bots Automated systems that generate and distribute social media posts.

social media trolls Individuals whose principal goal is to disrupt public discourse by posting false claims and prejudiced remarks, usually anonymously.

CHECKLIST ✓

Discussing Politics Responsibly on Social Media

Experts offer the following tips for engaging in responsible political discourse on social media:[68]

☐ **Don't assume that what you read or watch online is true.** Sometimes it's difficult to know which sources to trust. Check to see if multiple sources with different perspectives report the same information.

☐ **Don't flame or troll.** Resist the urge to post messages that are designed to anger or belittle others. As one article puts it, "Disagreement is perfectly normal," but using those differences to bash or inflame others is unacceptable.[69]

☐ **Be open-minded to differing opinions.** Resist the temptation to block responsible messages that differ from your own. You'll never learn anything new if you aren't willing to hear about diverse perspectives.

religion, recognizing the sensitive nature of talking about deeply held beliefs. Perhaps for that reason, many people use electronic means to express their views. In the United States, nearly 2 out of 3 people who use social media have posted messages encouraging others to join a political or social cause.[60] The results are not indicative of a respectful, open exchange of ideas, however. Among frequent social media users, 57 percent say that people post messages that are insensitive to diverse others, and 42 percent have blocked users whose views differ from their own.[61]

Even worse are so-called bots and trolls. **Social media bots** (short for robots) are automated systems that generate and distribute social media posts. **Social media trolls** are individuals whose principal goal is to disrupt public discourse by posting false claims and prejudiced remarks, usually behind a mask of anonymity. In a recent study by the Pew Research Center, only 1 in 5 people surveyed said they think that online harassment and fake news (defined for this purpose as untrue accounts spread through social media) will decrease in the coming years. All others think they will stay the same or worsen.[62,63] With this in mind, see the checklist in this section for tips about communicating respectfully about political views.

Ability/Disability

It's important to keep in mind that abilities or disabilities are just a part of a person's identity. Debates abound over whether to use what's known as "person-first language" versus "identity-first language" to describe people with disabilities. Describing someone as "a person who is blind" (person-first) can be more accurate and less constricting than calling her "a blind person." On the other hand, some people with disabilities prefer identity-first language (e.g., Deaf culture), which places the disability first, sometimes to differentiate among multiple disabilities but generally to avoid being treated as an "other."[64]

Sometimes people become part of certain communities as a result of disabilities or abilities. For example, the shared experiences of deafness can create strong bonds. Most notably, distinct languages build a shared worldview and solidarity. There are Deaf schools, Deaf competitions (e.g., Miss Deaf America), Deaf performing arts (including Deaf comedians), and other organizations that bring Deaf people together. One former airline pilot who lost his hearing described his trip to China: "Though we used different signed languages, these Chinese Deaf people and I could make ourselves understood; and though we came from different countries, our mutual Deaf culture held us together. . . . Who's disabled then?"[65]

Age/Generation

We tend to think of getting older as a purely biological process. But age-related communication reflects culture at least as much as biology. In many ways, people learn how to "do" various ages—how to dress, how to talk, and what not to say and do—in the same way they learn how to play other roles in their lives, such as student or employee.

Relationships between older and younger people are shaped by cultural assumptions that change over time. At some points in history, older adults have been regarded as wise, accomplished, and even magical.[66] At others, they have been treated as unwanted surplus and uncomfortable reminders of mortality and decline.[67]

Today, Western cultures mostly honor youth, and attitudes about aging are more negative than positive. On balance, people over age 40 are twice as likely as younger ones to be depicted in the media as unattractive,

Communicating with People Who Have Disabilities

Research has revealed some clear guidelines for interacting with people who have disabilities.[70,71] They include the following:

1. Speak directly to people with disabilities, rather than looking at and talking to their companions.

2. If you volunteer assistance, wait until your offer is accepted before proceeding. Then listen to or ask for instructions.

3. When meeting a person who is visually impaired, identify yourself and others who may be with you.

4. Remember that a wheelchair is part of the user's personal body space, so don't touch or lean on it without permission.

5. When you're talking with a person who has difficulty speaking, be patient and wait for the person to finish. Ask short questions that require a nod, shake of the head,

or brief answers. Never pretend to understand if you are having difficulty doing so.

6. With a person who is Deaf, speak clearly, slowly, and expressively to determine whether the person can read your lips. (Not all people who are Deaf can read lips.)

7. Don't be embarrassed if you happen to use common expressions such as "See you later" or "Did you hear about that?" that seem to relate to a person's disability.

8. Avoid stereotyping people with disabilities, and don't let disabilities distract you from appreciating their full range of qualities. In the ways that matter most, they are just like everyone else.

Notice that these recommendations are mostly similar to the ways we communicate with everyone else. People with disabilities are people first and should be treated as such.

bored, and in declining health.[72] And people over age 60 are still underrepresented in the media. Despite negative stereotypes, the data on personal satisfaction present a different story. Studies show that, overall, people in their 60s are just as happy as people in their 20s.[73]

Unfavorable attitudes about aging can show up in interpersonal relationships. Even though gray or thinning hair and wrinkles don't signify diminished cognitive capacity, they may be interpreted that way—with powerful consequences. People who believe older adults have trouble communicating are less likely to interact with them. When they do, they tend to use the mannerisms listed in Table 4.3.[74] Even when these speech styles are well intentioned, they can have harmful effects. Older adults who are treated as less capable than their peers tend to perceive *themselves* as older and less capable.[75,76] And challenging ageist treatment presents seniors with a dilemma: Speaking up can be taken as a sign of being cranky or bitter, reinforcing the stereotype that older adults are curmudgeons.[77]

That doesn't mean communication is perfectly easy for young people. Teens and young adults typically feel intense pressure, both internally and from people around them, to establish their identity and prove themselves.[78] At the same time, adolescents typically experience what psychologists call a **personal fable** (the sense that they are different from everybody else) and an **imaginary audience** (a heightened self-consciousness that makes it seem as if people are always observing and judging them).[79]

Communication challenges also can arise when members of different generations work together. For example, Millennials tend to have a much stronger need for affirming feedback than previous generations.[80] Because of their strong desire for achievement, they tend to want clear guidance on how to do a job correctly—but they do not want to be micromanaged when they do it. After finishing the task, they have an equally strong desire for praise. To a Baby Boomer boss, that type of guidance and feedback may feel more like a nuisance. In the boss's experience, "no news is good news," so the absence of negative feedback should be praise enough.

personal fable A sense common in adolescence that one is different from everybody else.

imaginary audience A heightened self-consciousness that makes it seem as if people are always observing and judging you.

TABLE 4.3

Patronizing Speech Directed at Older Adults

ELEMENT	DEFINITION AND EXAMPLE
Simplified grammar	Use of short sentences without multiple clauses. "Here's your food. You can eat it. It is good."
Simplified vocabulary	Use of short words rather than longer equivalents. Saying *dog* instead of *Dalmatian*, or *big* instead of *enormous*.
Endearing terms	Calling someone "sweetie" or "love."
Increased volume, reduced rate	Talking LOUDER and s-l-o-w-e-r!
High and variable pitch	Using a slightly squeaky voice style, and exaggerating the pitch variation in speech (a "singsong"-type speech style).
Use of repetition	Saying things over and over again. Repeating. Redundancy. Over and over again. The same thing. Repeated. Again. And again . . .
Use of baby-ish terms	Using words like *doggie* or *choo-choo* instead of *dog* or *train*: "Oh look at the cute little doggie, isn't he a coochie-coochie-coo!"

Source: J. Harwood. (2007). *Understanding communication and aging: Developing knowledge and awareness.* Newbury Park, CA: Sage, p. 76.

Neither perspective is wrong. But when members of these cocultures have different expectations, miscommunication can occur.

Socioeconomic Status

Social class can have a major impact on how people communicate. Individuals living in the United States typically identify themselves as belonging to the working class, middle class, or upper class, and feel a sense of solidarity with people in the same social strata.[81] This is especially true for working-class people, who tend to feel that they are united both by hardship and by their commitment to hard, physical work. One working-class college student put it this way:

> I know that when all is said and done, I'm a stronger and better person than they [members of the upper class] are. That's probably a horrible thing to say and it makes me sound very egotistical, but . . . it makes me more glad that I've been through what I've been through, because at the end of the day, I know I had to bust my a** to be where I want, and that makes me feel really good.[82]

The communication styles of people from different social classes can have consequences throughout life. College professors often find that students who are raised not to challenge authority can have a difficult time speaking up, thinking critically, and arguing persuasively.[83] This may factor into workplace communication as well. **Organizational culture** reflects a relatively stable, shared set of rules about how to behave and what is important. In everyday language, culture is the insiders' view of "the way things are around here." If the expectation is that employees should be assertive and persuasive to get ahead, new speech and language, clothing, and nonverbal patterns may be necessary to gain acceptance.[84] See the "@Work" box in this section for more about workplace cultures.

Even within the same family, education level can create intercultural challenges. First-generation college students often feel the intercultural strain of "trying to live simultaneously in two vastly different worlds" of school and home. Many say that they alter their communication patterns between these two worlds, censoring their speech with classmates and professors to avoid calling attention to their status, and with family members to avoid threatening and alienating them.[85]

organizational culture A relatively stable, shared set of rules about how to behave and a set of values about what is important.

Organizations Are Cultures, Too

Not all the rules and values of an organization are written down. And some that are written down aren't actually followed. Perhaps the workday officially ends at 5 p.m., but you quickly notice that most people stay until at least 6:30. That says something about the culture. Or even though it doesn't say so in the employee handbook, employees in some companies consider one another as extended family, taking personal interest in their coworkers' lives. That's culture, too.

Because you're likely to spend as much time at work as you do in personal relationships, selecting the right organization is as important as choosing a best friend. Research shows that people are likely to enjoy their jobs and do them well if they believe that the organization's values reflect their own and that its values are consistently and fairly applied.[86] For example, at Nordstrom department store team members are rewarded for offering great customer service without exception. On the other hand, a boss who talks about customer service but violates those principles cultivates a culture of cynicism and dissatisfaction.

Ask yourself these questions when considering whether a specific organization's culture is a good fit for you. (Notice how important communication is in each case.)

- How do people in the organization present themselves in person and on the telephone? Are they welcoming and inviting?

- Are customers/clients happy with the service and quality provided?
- Do members of the organization have the resources and authority to do a good job?
- Do employees have fun? Are they encouraged to be creative?
- Is there a spirit of cooperation or competition among team members?
- What criteria are used to evaluate employee performance?
- What happens during meetings? Is communication open or highly scripted?
- How often do people leave their jobs to work somewhere else?
- Do leaders make a point of listening, respecting, and collaborating with employees?
- Do people use their time productively, or are they bogged down with inefficient procedures or office politics?

Communication is the vehicle through which people both create and embody culture. At a personal and organizational level, effective, consistent, value-based communication is essential to success.

4.4 Developing Intercultural Communication Competence

Adapting to cultural differences isn't always easy. Americans may be surprised to hear that, despite their willingness to talk to strangers, the United States can be a hard place to make friends. Robin Luo, the exchange student from China we mentioned earlier, says it's hard to tell when Americans are genuinely interested in friendship and when they are just being polite.

"When I have talked to someone more than 15 minutes, I would think we can go further as friends," Luo says. "But Americans may consider it a casual conversation, and they may say, 'Let's get together' or 'I'll call you' but not follow through." He says he generally admires the optimistic attitude and forthrightness he has encountered in the United States, but he is puzzled by some of the communication patterns of his fellow students.

In this section, we consider what it takes to adapt to culturally diverse ways of communicating. To a great degree, interacting successfully with strangers calls for the same ingredients of general communicative competence outlined in Chapter 1.

It's important to have a wide range of behaviors and to be skillful at choosing and performing the most appropriate ones in a given situation. A genuine concern for others plays an important role. Cognitive complexity and the ability to empathize also help. Finally, self-monitoring is important because it is often necessary to make midcourse corrections in your approach when dealing with strangers.

But beyond these basic qualities, communication researchers have worked hard to identify qualities that are unique, or at least especially important, ingredients of intercultural communicative competence. We conclude the chapter with a look at five factors that are central to cultivating intercultural competence: increased contact, tolerance for ambiguity, open-mindedness, knowledge and skill, and patience and perseverance.

Increased Contact

More than a half century of research confirms that, under the right circumstances, spending time with people from different backgrounds leads to a host of positive outcomes: reduced prejudice, greater productivity, and better relationships.[87,88] The link between exposure and positive attitudes is called the *contact hypothesis*.[89]

It's encouraging to know that increased contact with people from stigmatized groups can transform hostile attitudes. For example, door-to-door canvassers in Los Angeles were able to dramatically change attitudes of area residents by engaging in 10-minute conversations aimed at breaking down stereotypes.[90] After canvassers engaged residents in thinking more deeply about the rights of transgender people, the residents reported feeling increased empathy.

Beyond face-to-face contact, the Internet offers a useful way to enhance contact with people from different backgrounds.[91] Online venues make it relatively easy to connect with people you might never meet in person. The asynchronous nature of most online contact reduces the potential for stress and confusion that can easily come in person. It also makes status differences less important: When you're online, gaps in material wealth or physical appearance are much less apparent.

Tolerance for Ambiguity

When you encounter communicators from different cultures, the level of uncertainty is especially high. Consider the basic challenge of communicating in an unfamiliar language. Native English speaker Lauren Collins reflects on the challenges and rewards of falling in love with someone who spoke a language (French) that she didn't know.[92] As she got to know her now-husband Oliver, she says, they relied heavily on nonverbal cues and sometimes misunderstood them. "I constantly thought Oliver looked irritated," Collins says, until she realized they simply had different cultural expectations about when and how much to smile. The language gap has sometimes been frustrating. "One day, Oliver told me that speaking to me in English felt like 'touching me with gloves,'" she remembers. At other times, it has offered comic relief, as when Collins inadvertently told her French mother-in-law that she "had given birth to a Nespresso machine." Over time, Collins says, she and Oliver have learned to speak each other's language pretty well. In a way she says, we are "an exaggerated version of every couple. We all have to learn how to talk" to each other.

Without tolerance for ambiguity, the mass of often confusing and sometimes downright incomprehensible messages that impact intercultural interactions would be impossible to manage. Some people seem to come equipped with this sort of tolerance, whereas others have to cultivate it. One way or the other, that ability to live with uncertainty is an essential ingredient of intercultural communication competence.

Open-Mindedness

Being comfortable with ambiguity is important, but without an open-minded attitude a communicator will have trouble interacting competently with people from

ON YOUR FEET

What examples of intercultural or cocultural incompetence have you witnessed, either in person or on social media? How could the communicator(s) involved have handled the situation more competently?

different backgrounds. To understand open-mindedness, it's helpful to consider three traits that are incompatible with it. **Ethnocentrism** is an attitude that one's own culture is superior to others. An ethnocentric person thinks—either privately or openly—that anyone who does not belong to their in-group is somehow strange, wrong, or even inferior. Travel writer Rick Steves describes how an ethnocentric point of view can interfere with respect for other cultural practices:

Source: © 2011 Malcolm Evans

> We [Americans] consider ourselves very clean and commonly criticize other cultures as dirty. In the bathtub we soak, clean, and rinse, all in the same water. (We would never wash our dishes that way.) A Japanese visitor, who uses clean water for each step, might find our way of bathing strange or even disgusting. Many cultures spit in public and blow their nose right onto the street. They couldn't imagine doing that into a small cloth, called a hanky, and storing that in their pocket to be used again and again.

Too often we think of the world in terms of a pyramid of "civilized" (us) on the top and "primitive" groups on the bottom. If we measured things differently (maybe according to stress, loneliness, heart attacks, hours spent in traffic jams, or family togetherness), things might stack up differently.[93]

Ethnocentrism leads to an attitude of **prejudice**—an unfairly biased and intolerant attitude toward others who belong to an out-group. (Note that the root term in *prejudice* is "pre-judge.") An important element of prejudice is stereotyping—exaggerated generalizations about a group. Familiarity can change attitudes, however, as American attitudes toward same-sex marriage have shown. As the practice has become legally sanctioned, there has been a dramatic increase in acceptance of gay and lesbian marriage. By 2018, 67% of Americans supported same-sex marriage, compared to 27% in 1996.[94]

Another barrier to diversity occurs in the form of **hegemony**, the dominance of one culture over another. A common example is the impact of Hollywood around the world. People who are consistently exposed to American images and cultural ideas can begin to regard them as desirable. For example, in South Korea, about 1 in 3 women has plastic surgery by the time she is 29, most often hoping that her eyelids, cheekbones, or noses will look more "Western" as a result.[95] The women typically say they will feel "prettier" and enjoy greater success in life if they look like the American women they see in the media.[96]

On a more encouraging note, whereas the mass media tend to perpetuate stereotypes, online communication can sometimes help people understand each other better. When students studying German at an American university connected

ethnocentrism The attitude that one's own culture is superior to other cultures.

prejudice An unfairly biased and intolerant attitude toward others who belong to an out-group.

hegemony The dominance of one culture over another.

ETHICAL CHALLENGE Civility When Values Clash

Most people acknowledge the importance of treating others from different cultural backgrounds with respect. But what communication obligations do you have when another person's cultural values differ from yours on fundamental matters, such as abortion or gender equity? How should you behave when confronted with views you find shocking or abhorrent?

How Much Do You Know About Other Cultures?

 Answer the following questions to test your knowledge about what is culturally appropriate around the world.

1. **Japanese visitors are in town. You've heard that Japanese custom involves gift-giving. What should you know about this practice?**

    a. It's important that gifts be expensive and of the finest quality.

    b. Avoid gifts that come in threes, as in three flowers or three candies.

    c. It's preferable to sign the accompanying card in green ink rather than black.

    d. It is not customary to wrap gifts in Japan.

2. **You are interacting with a person who is Deaf and who uses an interpreter. What should you do?**

    a. Address your comments to the interpreter, then look to see how the Deaf person reacts.

    b. Maintain eye contact with the Deaf person rather than the interpreter.

    c. Offer to communicate in written form so that the interpreter will be unnecessary.

    d. Speak very slowly and exaggerate the movements your mouth makes.

3. **While traveling in China, you should be aware of which rule of dining etiquette?**

    a. It's considered rude to leave food on your plate.

    b. You should put your drinking glass on your plate when you finish eating.

    c. Cloth napkins are just for show there. Use a paper napkin to wipe your mouth.

    d. Avoid sticking your chopsticks upright in your food when you are not using them.

4. **You are meeting with a group of Arab business associates for the first time. What should you know?**

    a. They favor greetings that involve shaking hands and kissing on each cheek.

    b. It's polite to say no if an Arab host offers you coffee or tea.

    c. Men tend to be touch avoidant and to stand at least 3 feet from one another during conversations

    d. They consider the left hand unsanitary, and they hold eating utensils only with their right hand.

INTERPRETING YOUR RESPONSES

Read the following explanations to see how many questions you got right.

Question 1

Gift-giving is an important ritual in Japan, but gifts needn't be extravagant or expensive. The number 3 is fine, but avoid gifts that involve 4 or 9, as these numbers rhyme with the Japanese words for "death" and "suffering," respectively, and are considered unlucky.[97] Black is associated with death or bad luck, so green ink, which symbolizes good luck, is preferred.[98] Gift wrapping is expected and is even considered an art form. The correct answer is c.

Question 2

The short answer is: Treat Deaf people with the same courtesy as you treat anyone else, which means maintaining eye contact and focusing on them. If the circumstances make it difficult for the interpreter to see clearly, make arrangements in advance so that it is not an issue.[99] The correct answer is b.

Question 3

Cultures vary in terms of whether it is rude to eat everything or rude not to. In China, leaving a little food on your plate lets your hosts know that they provided plentifully for you. However, sticking your chopsticks upright in your food evokes thoughts of funerals, where it's customary to place a stick of

lighted incense upright in a container of rice.[100] The correct answer is d.

Question 4

Members of Arab cultures may shake hands and kiss on each cheek, but usually only with people they already know well. A handshake is more appropriate for an introductory business meeting.[101] It's polite to accept a host's offer of coffee or tea. Men tend to speak at close distances (less than 3 feet) unless the conversation involves a woman, in which case it is typically considered rude to touch or crowd her.[102] It's considered unclean to eat with one's left hand (even if you are left handed),[103] harkening back to days when the left hand was used for personal hygiene. The correct answer is d.

online with students at a German university, most of them were successful in establishing ask-and-share conversations. This result was especially pronounced when they used instant messaging rather than asynchronous posts, which were typically less conversational and interactive.[104]

Knowledge and Skill

Attitude alone doesn't guarantee success in intercultural encounters. Communicators need enough knowledge of other cultures to have a clear sense of which approaches are appropriate. The rules and customs that work with one group might be quite different from those that succeed with another. The ability to shift gears and adapt one's style to the norms of another culture or coculture is an essential ingredient of communication competence.[105]

One school of thought holds that uncertainty can motivate relationship development—to a point. For example, you may be interested in a newcomer to your class because he's from another country. However, if attempting a conversation with him heightens your sense of uncertainty and discomfort, you may abandon the idea of making friends. The basic premise of anxiety uncertainty management theory is that, if uncertainty and anxiety are too low or too high, we are likely to avoid communicating.[106]

How can a communicator learn enough about other cultures to feel curious but not overwhelmed? Scholarship suggests three strategies.[107]

- *Passive observation* involves noticing what behaviors members of a different culture use and applying these insights to communicate in ways that are most effective.
- *Active strategies* include reading, watching films, and asking experts and members of the other culture how to behave, as well as taking academic courses related to intercultural communication and diversity.[108]
- *Self-disclosure* involves volunteering personal information to people from the culture with whom you want to communicate. One type of self-disclosure is to confess your cultural ignorance: "This is very new to me. What's the right thing to do in this situation?" This approach is the riskiest of the three described here because some cultures may not value candor and self-disclosure as much as others. Nevertheless, most people are pleased when strangers attempt to learn the practices of their culture, and they are usually quite willing to offer information and assistance.

Take the "Understanding Your Communication" quiz to gauge how much you know about communication norms in other cultures. Don't be discouraged if you don't make 100%. The important point is to be curious and eager to learn.

Patience and Perseverance

Becoming comfortable and competent in a new culture or coculture may be ultimately rewarding, but the process isn't easy. After a "honeymoon" phase, it's typical to feel confused, disenchanted, lonesome, and homesick.[109] To top it off, you may feel disappointed in yourself for not adapting as easily as you expected. This stage—which typically feels like a crisis—has acquired the labels *culture shock* or *adjustment shock*.[110]

You wouldn't be the first person to be blindsided by culture shock. Barbara Bruhwiler, who was born in Switzerland and has lived in South Africa for 5 years, says she loves her new home but still experiences moments of confusion and distress.[111] Likewise, when Lynn Chih-Ning Chang came to the United States from Taiwan for graduate school, she cried every day on the way home from class.[112] All her life, she had been taught that it was respectful and ladylike to sit quietly and listen,

Coping with Culture Shock

☐ **Don't be too hard on yourself.** It's normal to feel disoriented and frustrated after the initial excitement of being in a new culture wears off a bit.

☐ **Homesickness is normal.** You can expect to long for familiar environments and people sometimes. Accept these feelings and take comfort in knowing that homesickness usually diminishes over time.

☐ **Expect progress and setbacks.** Remember the "draw back and leap" pattern. People tend to take steps forward and backward as they adjust to a new culture. This is normal.

☐ **Reach out to others.** Your current friends were once strangers to you. Give yourself a chance to get to know others. It may be the beginning of lifelong friendships.

so she was shocked that American students spoke aloud without raising their hands, interrupted one another, addressed the teacher by his or her first name, and ate food in the classroom. What's more, Chang's classmates answered so quickly that, by the time she was ready to say something, they were already on a new topic. The same behavior that made her "a smart and patient lady in Taiwan," she says, made her seem like a "slow learner" in the United States.

Communication theorist Young Yum Kim has studied cultural adaptation extensively. She says it's natural to feel a sense of push and pull between the familiar and the novel.[113] Kim encourages sojourners to regard stress as a good sign. It means they have the potential to adapt and grow. With patience, the sense of crisis begins to wane, and once again, there's energy and enthusiasm to learn more.

The transition from culture shock to adaptation and growth is usually successful, but it isn't a smooth, linear process. Instead, people tend to take two steps forward and one step back, and to repeat that pattern many times. Kim calls this a "draw back and leap" pattern.[114] Above all, she says, if people are patient and they keep trying, the rewards are worth it. The checklist in this section offers some useful tips for coping when the process feels overwhelming.

Communication can be a challenge while you're learning how to operate in new cultures, but it can also be a solution.[115] Chang, the Taiwanese student adapting to life in America, learned this firsthand. At first, she says, she was reluctant to approach American students, and they were reluctant to approach her. Gradually, she got up the courage to initiate conversations, and she found that her classmates were friendly and receptive. Eventually, she made friends, began to fit in, and successfully completed her degree.

MAKING THE GRADE

At www.oup.com/he/adler-uhc14e, you will find a variety of resources to enhance your understanding, including video clips, animations, self-quizzes, additional activities, audio and video summaries, interactive self-assessments, and more.

OBJECTIVE 4.1   Analyze how salient cultural and cocultural differences can affect communication both within and across groups of people.

- Communicating with people from different backgrounds is more common today than ever before. Some encounters involve people from different cultures, whereas others involve communicating with people from different cocultures within a given society.

- Cultural differences are not a salient factor in every intergroup encounter.

- Although cultural characteristics are real and important, they are generalizations that do not apply equally to every member of a group.

> Provide two examples of intercultural communication from your own experience—one in which your affiliation with a particular group or culture was salient and one in which it was not.

> Think of a cultural group with which you identify. Now describe a conversation that would be interpreted one way by members of that group and another way by out-group members.

> List three common generalizations about college students. How well does each of those generalizations describe you? What can you learn about generalizations from this exercise?

OBJECTIVE 4.2   Explain the communicative influences of individualism and collectivism, high and low context, uncertainty avoidance, power distance, talk and silence, competition and cooperation.

- Some cultures value autonomy and individual expression, whereas others are more collectivistic.

- Some pay close attention to subtle, contextual cues (high context), whereas others pay more attention to the words people use (low context).

- Cultures vary in their acceptance of uncertainty.

- Authority figures are treated with more formality and often greater respect in cultures with high power distance than in those with low power distance.

- Depending on the cultural context, people may interpret silence as comforting and meaningful or, conversely, as an indication of awkwardness or disconnection.

- Some cultures value competition, independence, and assertiveness, whereas others emphasize equality, relationships, cooperation, and consensus building.

  > Compare and contrast the communication techniques that job applicants from high and low power distance cultures might use.

  > Analyze the assertion that "young people today have no respect for authority" by applying the notion of power distance to specific examples.

  > Do you identify more with individualism or collectivism? In what ways?

OBJECTIVE 4.3 **Evaluate the influence of factors such as race and ethnicity, gender identity, religion, age, and political viewpoints on communication.**

- People often make assumptions about others based on their apparent race, but appearance is an unreliable indicator of cultural and personal differences.

- People have multiple cultural identities that intersect in unique ways.

- Gender diversity goes far beyond the simplistic dichotomies of male or female and gay or straight.

- People from different regions of the same country may uphold different beliefs about what is valued and "normal" in terms of communication.

- People may identify with others of similar gender identity, religion, political beliefs, age, physical ability, and socioeconomic status in ways that define them as members of a culture or coculture.

  > Explain the difference between race and ethnicity and defend why one is a more reliable indicator of cultural identity than the other.

  > List three identities that apply to you. How does living at the intersection of those identities make your life experience different from that of a person who shares one or two of those identities but not all three?

OBJECTIVE 4.4 **Implement strategies to increase intercultural and cocultural communication competence.**

- Cultural competence is typically enhanced by contact with people who have different beliefs.

- It's useful to become comfortable with uncertainty and open to the idea of different, but equally valid, worldviews.

- It often takes patience and skill to connect with people who have diverse perspectives, but the benefits are worth the effort.

  > List and explain three communication strategies people might use to learn about different cultures and to share their own ideas with others.

  > Think of a time when you met someone who seemed very different from you at first but eventually became a close friend or colleague. Create a timeline that illustrates turning points in your relationship when you learned more about each other and developed rapport.

  > How do you rate yourself in terms of intercultural sensitivity and open-mindedness? What steps might you take to keep growing in this regard?

KEY TERMS

coculture p. 78

collectivistic cultures p. 81

culture p. 78

ethnicity p. 87

ethnocentrism p. 95

hegemony p. 95

high-context cultures p. 83

imaginary audience p. 91

in-groups p. 80

individualistic cultures p. 81

intergroup communication p. 79

intersectionality theory p. 87

low-context cultures p. 83

organizational culture p. 92

out-groups p. 80

personal fable p. 91

power distance p. 85

prejudice p. 95

race p. 87

salience p. 80

social media bots p. 90

social media trolls p. 90

uncertainty avoidance p. 84

ACTIVITY

Examine the comments posted in response to a blog or video online. Analyze how well the people involved have followed tips in the checklist "Discussing Politics Responsibly on Social Media" in this chapter. Rewrite several of the posts to be more respectful of diverse viewpoints.

Language

CHAPTER OUTLINE

5.1 The Nature of Language 102
  Language Is Symbolic
  Meanings Are in People, Not in Words
  Language Is Rule Governed

5.2 The Power of Language 107
  Language *Shapes* Values, Attitudes, and Beliefs
  Language *Reflects* Values, Attitudes, and Beliefs

5.3 Troublesome Language 114
  The Language of Misunderstandings

5.4 Disruptive Language 117
  Confusing Facts and Inferences
  Presenting Opinions as Facts
  Personal Attacks

5.5 Gender and Language 120

  MAKING THE GRADE 124
  KEY TERMS 125
  ACTIVITIES 125

LEARNING OBJECTIVES

5.1
Explain how symbols and linguistic rules allow people to achieve shared meaning.

5.2
Identify ways in which language shapes people's attitudes and reflects how they feel about themselves and others.

5.3
Recognize and remedy confusing and inflammatory language.

5.4
Describe the ways in which traditionally masculine and feminine speech patterns are alike and how they differ.

Consider how language is intertwined with your identity and the way you communicate with others.

What words do people use to describe you? Which words best reflect how you feel about yourself? Which are inconsistent with who you are on the inside?

Describe examples from your own experience in which language helped you connect with another person in a powerful way.

Looking around you and in the media, do you see evidence of people discussing sensitive issues in a civil and respectful way? Do you see evidence of hostile and judgmental language? How would you rate your own communication skills in this regard?

IT WAS LIKE "an explosion in my mind" says actor Asia Kate Dillon, describing how it felt to see the term *nonbinary* for the first time.[1] Dillon plays Taylor, the first openly nonbinary character in a mainstream television program. Taylor joined the cast of *Billions* in season 2 with the introductory statement, "Hello, I'm Taylor. My pronouns are 'they,' 'theirs' and 'them.'"[2]

Pronouns are one example of how powerful language can be. Nonbinary individuals do not identify exclusively with one gender or another, so words such as *he* and *she* can feel confining and inaccurate in describing who they really are. Dillon, who identifies as nonbinary in real life as well, cried upon encountering the term, saying that "it finally helped me put language to a feeling that I'd had my entire life."[3]

Language affects our lives in many ways. It helps us express how we feel (as in cautiously optimistic or filled with dread), define our relationships with others (perhaps best friend, acquaintance, or fiancé), and much more. All the while, language is alive and changing.

This chapter explores how language shapes and reflects the way we see the world, other people, and ourselves. It considers:

- the evolving nature of language,
- language's power to shape perceptions,
- how words can create mutual understanding or, conversely, can confuse or hurt others, and
- how people of different genders may use language differently.

By the end of the chapter, you should be better equipped to use the tool of language to improve your everyday interactions.

5.1 The Nature of Language

A British person who has a *chinwag* (chat) with someone from the United States or Canada might feel *knackered* (exhausted) by how hard it is to share meaning, even within the same language. Sometimes you probably feel the same way when speaking with people much closer to home.

To understand what's going on, let's first define some basic terms. A **language** is a collection of symbols governed by rules and used to convey messages between individuals. A **dialect** is a version of the same language that includes substantially different words and meanings.[4] English includes dozens of dialects, as do the 7,000 or so other languages of the world. A closer look at the nature of language reveals why people often hear something different from what the speaker meant to say. Language is powerful and indispensable, but it is also imprecise and constantly evolving.

Language Is Symbolic

There's nothing natural about calling your loyal, four-footed companion a *dog*. Words are arbitrary symbols that have no meaning in themselves. To a Spanish speaker, the symbol would be *perro*; to a computer, it would be *01100100 01101111 01100111*.

These terms, like virtually all language, are **symbols** in that their meaning relies on how members of a particular speech community understand and use them. Symbols are different from *signs*, which are inherently linked to what they represent. For example, smoke is a sign of fire.

Not all linguistic symbols are spoken or written words. Sign language, as "spoken" by most Deaf people, is symbolic and not the pantomime it might seem to nonsigners. There are hundreds of different sign languages spoken around the world, including American Sign Language, British Sign Language, French Sign Language, Danish Sign Language, Chinese Sign Language—even Australian Aboriginal and Mayan sign languages. Each has its own way of representing ideas.[5]

Symbols are more than just labels: They shape the way people experience the world. You can prove this by trying a simple experiment:[6] Spit into a glass and then see if you can drink your spit. Most people find this proposal mildly disgusting. But ask yourself why this is so. After all, we swallow our own saliva all the time. The answer arises out of the symbolic labels we use. After the saliva is in the glass, we call it *spit* and think of it in a different way. In other words, our reaction is to the name, not the thing.

The naming process operates in virtually every situation. How you react to a stranger will depend on the symbols you use to categorize them: gay (or straight), religious (or not), attractive (or unattractive), and so on.

Meanings Are in People, Not in Words

Ask a dozen people what the same symbol means, and you are likely to get 12 different answers. Does an American flag bring up associations of patriots giving their lives for their country? Fourth of July parades? Cultural imperialism? How about a cross: What does it represent? The message of Jesus Christ? Fire-lit gatherings of Ku Klux Klansmen? Your childhood Sunday school? The necklace your sister always wears?

As with physical symbols, the place to look for meaning in language isn't in the words themselves but in the way people make sense of them. Linguistic theorists C. K. Ogden and I. A. Richards illustrated the notion that meanings are social constructions in their well-known "triangle of meaning" (Figure 5.1).[7] This model shows that there's only an indirect relationship—indicated by a broken line—between a word and what it is used to represent. Some references are fairly clear, at least to members of the same speech community. In other cases, though, interpretations can be quite different. Consider abstract concepts such as *feminism*, *environmentalism*, and *conservatism*. Problems arise when we mistakenly assume that other people use these words (and others) in the same way we do.

Part of the person-centered nature of language involves the difference between denotative and connotative meanings. **Denotative meanings** are formally recognized definitions of a term, whereas **connotative meanings** involve thoughts and feelings associated with words. There is usually little confusion about the denotative meaning of a word such as *chair*. But consider terms such as *survivor* and *victim*. In reference to violent assaults, these terms have nearly synonymous denotative meanings: one who has been harmed by another person. But for many

> **symbol** A representation such as a word that stands for something else based on shared rules for interpreting it.
>
> **denotative meanings** Formally recognized definitions for words, as in those found in a dictionary.
>
> **connotative meanings** Informal, implied interpretations for words and phrases that reflect the people, culture, emotions, and situations involved.

Thought or Reference

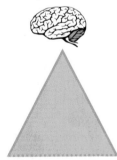

Symbol   Referent
(e.g., "dog," "perro")

FIGURE 5.1 Ogden and Richards's Triangle of Meaning

people, *survivor* connotes someone who manages to thrive despite adversity, and *victim* suggests a state of helplessness. The meanings people associate with words often have far more significance than do their dictionary definitions. In the end, words don't mean things; people do—and often in widely different ways.

Despite the potential for linguistic problems, the situation isn't hopeless. We do, after all, communicate with one another reasonably well most of the time. And with enough effort, we can clear up most of the misunderstandings that occur. The key to more accurate use of language is to avoid assuming that others interpret words the same way we do. In truth, successful communication occurs when we negotiate the meaning of a statement. As one French proverb puts it: *The spoken word belongs half to the one who speaks it and half to the one who hears.*

Language Is Rule Governed

Thirty years ago, the statement *"she texted a selfie"* would have been incomprehensible. The same goes for terms such as *blog* and *emoji.* Those words weren't in use back then. Keeping up to date with vocabulary and grammar is a lifelong process, as they are constantly evolving. Here are four types of rules that provide structure for languages as they evolve.

phonological rules Linguistic rules governing how sounds are combined to form words.

syntactic rules Rules that govern how symbols can be arranged (as opposed to the meanings of those symbols).

Phonological rules govern how words sound when pronounced. For instance, the words *champagne, double,* and *occasion* are spelled identically in French and English but are pronounced differently. Non-native speakers learning English are plagued by inconsistent phonological rules, as a few examples illustrate:

He could lead if he would get the lead out.

A farm can produce produce.

The dump was so full it had to refuse refuse.

The present is a good time to present the present.

I did not object to the object.

The bandage was wound around the wound.

I shed a tear when I saw the tear in my clothes.

In the film franchise starring the Minions, the little yellow creatures speak what may sound like gibberish or baby talk but is actually a complete language.

Can you name some syntactic rules that govern speech in your native language (for example, subject before verb)? What would it sound like to violate some of those rules?

Syntactic rules govern the structure of language—the way symbols can be arranged. Although most of us aren't able to describe the syntactic rules that govern our language, it's easy to recognize their existence by noting how odd a statement that violates these rules appears to be. For example, correct English syntax requires that every word contain at least one vowel, and it prohibits sentences such as *Have you the cookies brought?*, which is a perfectly acceptable word order in German. While syntactic rules differ among languages, linguist Noam Chomsky has suggested that all seem to be variations on a single model, which he calls Universal Grammar. Chomsky has argued that Universal Grammar is hardwired into all humans. This enables children to perform the astonishing feat of learning to speak competently at an early age.[8]

Technology has spawned versions of English with their own syntactic rules.[9] For example, people have devised a streamlined version of English for instant messages, texts, and tweets that speeds up typing in real-time communication (although it probably makes composition teachers grind their teeth in anguish):

A: Hey r u @ home?

B: ys

A: y

B: cuz i need to study for finals c u later tho bye

A: TTYL

Semantic rules are guidelines about the meaning of specific words. They make it possible for us to agree that *bikes* are for riding and *books* are for reading. It's possible, however, for the same word to represent different things. For example, in German *gift* means poison.[10] A popular brand of detergent in the Middle East is called *Barf*, which means *snow* in Hindi and Farsi.[11] *Sale* means *dirty* in French. Clearly, it's essential to know the semantic rules of the language or dialect you are using.

Even in the same language, semantic misunderstandings occur when words can be interpreted in more than one way, as the following humorous church notices demonstrate:

> The peacemaking meeting scheduled for today has been canceled due to a conflict.
>
> For those of you who have children and don't know it, we have a nursery downstairs.
>
> The ladies of the Church have cast off clothing of every kind. They may be seen in the basement on Friday afternoon.
>
> Sunday's sermon topic will be "What Is Hell?" Come early and listen to our choir practice.

Pragmatic rules govern how people use language in everyday interactions. Communication theorists use the term **speech act** to describe the purpose or intention of a communication episode.[12] For example, an utterance might be treated as a question, an apology, a complaint, or a joke. Pragmatic rules help people understand what sort of communication is occurring. This goes beyond simple grammar. If a friend says, "Don't you look great today?" they probably don't really mean it as a

> **semantic rules** Rules that govern the meaning of language as opposed to its structure.
>
> **pragmatic rules** Rules that govern how people use language in everyday interaction.
>
> **speech act** The purpose or intention of a communication episode (e.g., greeting, breaking up, kidding around).

Words that look the same but have different meanings depending on the language or coculture are sometimes called "false friends." This image might make English speakers chuckle, but in Sweden it's an everyday sign indicating an exit on the roadway.

Which words do you use whose meaning might be interpreted differently by different people?

question. It's more of a compliment. Like all rules governing language, there are personal and cultural variations of pragmatic rules. Someone who overhears a conversation between you and your best friend may think you're putting each other down, but based on shared understanding, the two of you may know you're just teasing.

You won't typically find pragmatic rules written anywhere, but people familiar with the language and culture rely on them to make sense of what is going on. Their usefulness is especially obvious when you find yourself wondering about them, as when a new acquaintance says "hope to see you soon" and you wonder if they really mean it or are just being polite. The challenge is that the people involved in a communication episode may not interpret it in the same way. Consider the example of a male boss saying "You look very pretty today" to a female employee. He may interpret the statement very differently than she does based on a number of factors.

EACH PERSON'S SELF-CONCEPT

Boss: Views himself as a nice guy.
Employee: Determined to succeed on merits, not appearance.

THE EPISODE IN WHICH THE COMMENT OCCURS

Boss: Casual remark at the start of the work day.
Employee: A possible come-on?

PERCEIVED RELATIONSHIP

Boss: Views employees like members of the family.
Employee: Depends on boss's goodwill for advancement.

CULTURAL BACKGROUND

Boss: Member of generation in which comments about appearance were common.
Employee: Member of generation sensitive to sexual harassment.

As this example shows, pragmatic rules don't involve semantic issues because the words themselves are clear. For example, consider the question "What do you do for a living?" The denotative meaning of this question is straightforward, but the pragmatic implications differ by culture. Americans typically ask and answer this question readily, but it might cause offense in France, where people may consider it an indirect way of asking, "How much money do you make?" Conversely, Americans typically consider it rude to ask adults how old they are, but it's more socially acceptable in Japan, where the oldest person in a conversation is typically given the most respect.

Pragmatic rules also govern language that some people find offensive but that others regard positively. Consider the word *queer*. In earlier generations, this term was a slur directed against homosexual men. More recently, however, some people have adopted it as a proud expression of their nontraditional gender identity. Researchers use the term **reappropriation** to describe how members of marginalized groups sometimes reframe the meaning of a term that has historically been used in a derogatory way.[13] Apparently unaware that "queer" has been reappropriated among some groups of people, Twitter gatekeepers launched an effort in 2018 to ban its use on the platform, leading one person to tweet: "Twitter thinks queer is a slur, which means basically everyone I know is going to be locked out."[14]

The use of profanity is another example of how pragmatic rules govern communication. These days, you're likely to encounter swear words regularly. American books published in the mid-2000s include about 28 times more curse words than those released in the 1950s.[15] Even book titles include expletives, with bestsellers such as *The Subtle Art of Not Giving a F—k* and *How to Get Sh-t Done*. Similar trends are visible in everyday communication. A study released in 2016 showed

reappropriation The process by which members of a marginalized group reframe the meaning of a term that has historically been used in a derogatory way.

that 1 in 4 Americans uses the f-word daily, a steep increase from 10 years earlier.[16] As with all forms of communication, the challenge is to gauge how likely the people around you are to consider an expletive appropriate or offensive. Among like-minded individuals, cursing can indicate a level of comfort and acceptance,[17] and mild swear words such as *damn* can sometimes make a public speaker more persuasive.[18] On the other hand, of course, people may be shocked or offended by profanity. Almost everyone agrees that it's rude to curse in the presence of children, and it may come off as disrespectful to swear in some situations—such as during a family dinner, a class, or a church service. Although swearing is acceptable in some workplaces, 8 of 10 employers in one survey said that they would question the professionalism of employees who use profanity on the job.[19] The best practice is to observe the language habits of people around you and to err on the side of caution when in doubt.

On the TV show *The Good Place*, Eleanor (Kristen Bell, shown here with costar Ted Danson) has joined the afterlife, where she is prohibited from cursing. To her frustration, her attempts at profanity come out as *motherforker, bullshirt, son of a bench*, and so on.

What pragmatic rules govern the use of profanity in your speech communities? What are the consequences for violating them?

5.2 The Power of Language

Imagine that you have just been accepted for a year-long study abroad program. You call your parents, who congratulate you and predict that it will be one of the best experiences of your life. Then you Facetime a friend, who says, "You're so brave! I'd be terrified to go someplace if I didn't speak the language." These encounters and your previous experiences are likely to affect how you feel about the opportunity ahead of you.

Coordinated management of meaning (CMM), a concept originally proposed by Barnett Pearce and Vernon Cronen, suggests that people co-create meaning in the process of communicating with each other.[20] Is your study abroad opportunity a cause for joy or fear? The sense you make of it is influenced by an array of factors, including your parents' excited reaction, your friend's statement of apprehensions, and exposure to messages throughout your life that have influenced your ideas about travel, independence, education, and more. From the perspective of CMM, you are influenced by others, but you also play an active role in the meaning-making process. For example, you might reply to your friend, "Not worried! I'll learn the language in no time." Your friend may accept your confident statement at face value or challenge it, perhaps by observing, "You can't fool me. You're excited, but also nervous. Am I right?"

Thinking about your communication exchanges in a single day will help you appreciate that language is a valuable tool in the ongoing process of negotiating meaning with others. In this section, we focus on how the words we use, and the way we say them, affect our lives.

coordinated management of meaning (CMM) The notion that people co-create meaning in the process of communicating with one another.

Language *Shapes* Values, Attitudes, and Beliefs

Let's return for a moment to the study abroad example. Perhaps you describe your initial reaction as "happy but scared." These are broad and rather vague terms for emotions. But pause to reflect. The ability to identify and describe nuances

Language and Worldview

Mudita: Taking delight in the happiness of others (Sanskrit).

Voorpret: The sense of delightful anticipation before a big event (Dutch).

Koi No Yokan: The sense upon first meeting a person that the two of you are going to fall in love (Japanese).

Cavoli Riscaldati: The result of attempting to revive an unworkable relationship; translates to "reheated cabbage" (Italian)[21]

Witzelsucht: The tendency to excessively make puns and tell inappropriate jokes and pointless stories (German).[22]

If you feel that life would be a little bit different if you had these phrases in your vocabulary, score one for linguistic relativism, the notion that words influence the way we experience the world.[23]

If not, score one for the idea that language *reflects* experiences more than it shapes them.

Whichever position you take, you're in good company. The best-known declaration of linguistic relativism is the *Sapir–Whorf hypothesis.*[24] Philosophers and scientists have been debating it and theories like it for about 150 years.

Words may be especially powerful in defining abstract concepts such as time. "For English speakers—people who read from left to right—time tends to flow from left to right," says cognitive scientist Lera Boroditsky.[25] But it's the opposite for people who speak languages such as Hebrew or Arabic that are read from right to left. Asked to put a series of images in chronological order, they are likely to put the first one on the right and the last one on the left.

A similar effect is noted for gender. Some languages refer to objects as either male or female. When researchers asked Germans to describe bridges, they tended to use words such as beautiful, fragile, pretty, and slender. As you might guess, their language defines bridges as feminine. By contrast, Spanish speakers, who refer to bridges as masculine, were more likely to describe them as sturdy, strong, and towering.[26,27]

Some evidence suggests that language may affect perceptions of the physical world. In one experiment, researchers asked English speakers and Himba speakers (from a region of southeast Africa) to distinguish between colors. The Himba speakers, whose language includes words for many shades of green, were able to distinguish between green hues that English speakers perceived as being all one color. Conversely, English speakers were able to distinguish between blue and green, whereas Himba speakers, who use the same word for both, often didn't see a difference.[28]

People who doubt linguistic relativity argue that, when we don't have a specific word for something, we use others in its place. They also point out that perception is not strictly in vocabulary's cage. That is, people often invent things for which they previously had no words. (Think Internet, microwave oven, and telephone.[29])

Although the debate about language relativity isn't likely to be resolved any time soon, it reveals a great deal about the power and limitations of words. Keep this in mind the next time words fail you and you think of a witty comeback only after the moment has passed. The Germans have a word for that: *treppenwitz.*

"You'll have to phrase it another way. They have no word for 'fetch.'"

Source: Drew Dernavich The New Yorker Collection/The Cartoon Bank

of how you feel is called *emotional granularity*. It requires self-awareness and a well-developed vocabulary. For example, you might realize that, rather than simply feeling "happy," you feel "ecstatic," "victorious," or "gratified." And perhaps "nervous excitement" or "terror" more accurately describes how you feel than the simple term "scared." Evidence suggests that using more specific language can help you understand your feelings better, express yourself more clearly, and even enjoy better health.[30] Indeed, the words at your disposal may have an impact on your perceptions in general. **Linguistic relativism** proposes that the language people speak shapes the way they view the world.[31] For an explanation of this phenomenon, see "Understanding Diversity" on page 108. Then read on to explore some of the ways that language shapes people's attitudes and worldview.

Names   "What's in a name?" Juliet asked rhetorically. A lot, it turns out. Research has demonstrated that names are more than just a simple means of identification: They shape the way others think of us, the way we view ourselves, and the way we act.

At the most fundamental level, the phonetic sound of a person's name may affect the way people regard them, especially when little other information is available. For example, voters tend to favor names that are simple, familiar, and easily pronounced. In one series of local elections, candidates Sanders, Reilly, Grady, and Combs attracted more votes than Pekelis, Dellwo, Schumacher, and Bernsdorf. Names don't guarantee victory, but in 78 elections, 48 outcomes supported the value of having a more common name.[32]

Names may also shape and reinforce a sense of personal identity. Naming a baby in reference to a family member (such as Junior or Trey) can create a connection between the child and the child's namesake. Names can also make a powerful statement about cultural identity. Some names may suggest a "Black" identity, whereas others sound more "White."[33] The same could be said for Latinx, feminine/masculine, Jewish, and other names.

Unfortunately, names can be used as the basis for discrimination. When researchers posted more than 6,000 AirBnB requests that were identical except for the users' names, they found that would-be guests with African American–sounding names were 15% more likely to be declined lodging than those whose names sounded White.[34] A similar pattern is evident in employment decisions. In the United States, job applicants with names such as Mohammed and Lakisha typically receive fewer calls from employers than equally qualified candidates whose names sound more European.[35,36] Because of this potential for discrimination, some people advocate for applications in which job candidates' names are masked during the review process.[37] (See the "@Work" box for more about names in the workplace.)

Accents   In the classic musical *My Fair Lady,* Professor Henry Higgins transforms Eliza Doolittle from a lowly flower girl into a high-society woman by helping her replace her cockney accent with an upper-crust speaking style. It's not a far-fetched idea. An **accent** involves pronunciation perceived as different from the local speech style.[38] When we say another person speaks with an accent, we usually mean they either have a non-native manner of speech (i.e., a foreign accent) or a so-called nonstandard speaking style. (In the United States, think classic Brooklynese, African American, or Cajun).[39] Taken this way, *everyone* has an accent. To your ear, most of the people you encounter every day might sound accent-free, but to an English-speaking person from, say, India, they would sound decidedly American.

An accent can either enhance or detract from speakers' social status. Among English speakers around the world, British accents are typically considered most pleasing to the ear, followed by American, Irish, and Australian accents.[40] But some non-native accents carry a stigma.[41] In the United States, employers are more likely to hire candidates who sound as if English is their first language than those who have Asian or Hispanic accents, even when the candidates speak English clearly and proficiently.[42]

linguistic relativism   The notion that language influences the way we experience the world.

accent   Pronunciation perceived as different from the locally accepted speech style.

ASK YOURSELF

Imagine for a moment that the words *love, disappointment,* and *hate* did not exist. Would you be aware of those feelings in the same way? Could you talk about them as effectively?

@WORK

What's in a Name?

When it comes to career success, names matter more than you might imagine.

Research suggests that many people pass judgment on job applicants based on their names. For example, prospective employers tend to rate applicants with common first names more highly than those with unique or unusual ones.[43] This bias presents challenges for people with unique names and for those from cultures with different naming practices.

People may also predict career success based on how closely a person's name matches the gender associated with their job. College students surveyed predicted that people with feminine names like Emma or Marta were more likely than others to be successful in traditionally female occupations such as nursing. By contrast, they estimated that people with masculine names like Hank or Bruno would do better in traditionally male jobs such as plumbing.[44]

Findings like these are worth noting if you hope to succeed in a field in which your identity doesn't match traditional expectations. For example, in the field of law, research suggests that your chances of success are greatest if you have a gender-neutral or traditionally male name. Researchers in one study found that a woman named Cameron is roughly three times more likely to become a judge than one named Sue. A female Bruce is five times more likely.[45]

Most people aren't willing to change their name to further their career goals, but it's possible to choose variants of a name that have a professional advantage. For example, Christina Jones might use the nickname Chris for gender neutrality on job applications. Someone with a hard-to-pronounce name might choose a nickname for work purposes. For example, it's customary in China for businesspeople and students of English to choose a Western name that sounds similar to theirs. For example, Junyuan may go by Joanna, at least at first. In a competitive job market, little differences can mean a great deal.

Based on the research described here, do you think your name may affect your career? Have others' names ever shaped your perceptions of them?

Listeners often assume that accents are linked with particular abilities and personality traits. When researchers asked women to listen to audiotaped voices and then say which speaker they would ask for help if their handbag were stolen, the women favored people with New York accents. By contrast, the women were more likely to ask people with Midwestern accents for directions. The researchers speculate that people with New York accents seemed tougher and more aggressive, and those with Midwestern accents seemed more approachable and friendlier.[46]

Gender References   By now it should be clear that the power of language to shape attitudes goes beyond individual cases and influences how we perceive entire groups of people. Children exposed to terms such as *fireman* and *businessman* are typically less likely than other children to think that women can pursue those occupations.[47,48] This assumption is far less prevalent among children exposed to gender-neutral terms such as *firefighter* and *businessperson*. Based on such evidence, many people argue that gender-neutral language is not merely "politically correct" but a powerful force in shaping opportunities and identities.

Using the right pronoun can be tricky.

When in doubt, how can you tell what pronoun others prefer?

ETHICAL CHALLENGE Freedom of Speech

One of the most treasured civil liberties is freedom of speech. At the same time, most people would agree that some forms of speech are hateful and demeaning.

- Do you think laws and policies can and should be made that limit certain types of racist and sexist communication?

- If not, how would you justify treating language that some people find personally degrading?

- If so, how should those limits be drafted to protect open debate?

Some languages emphasize gender less than others. In Finnish, for example, the pronoun *hän* refers to both males and females. Finnish speakers sometimes puzzle over whether to call individuals he or she in English because they aren't accustomed to categorizing people that way.[49] The implications of such language differences are notable. Gender equality is greater in countries such as Finland, where language is nongendered, than in regions where the predominant language (such as Spanish, German, or Russian) attributes a gender to nearly every noun.[50] In Spanish, a spoon (*la cuchara*) is feminine, a fork (*el tenedor*) is masculine, a napkin (*la servilleta*) is feminine, and so on. First names in those languages also tend to be distinctly masculine or feminine.

Although English is not entirely gender neutral, it's fairly easy to use nonsexist language. For example, the term *mankind* may be replaced by *humanity, human beings, human race,* or *people; manmade* may be replaced by *artificial, manufactured,* and *synthetic; manpower* by *human power, workers,* and *workforce;* and *manhood* by *adulthood.*

Language *Reflects* Values, Attitudes, and Beliefs

Besides shaping the way we view ourselves and others, language reflects our values, attitudes, and beliefs. Feelings of control, attraction, commitment, and responsibility—all these and more are reflected in the way we use language.

Power   Americans typically consider language powerful when it is clear, assertive, and direct. By contrast, language is often labeled powerless when it suggests that a speaker is uncertain, hesitant, intensely emotional, deferential, or nonassertive.[51]

"Powerful" speech can be an important tool. In employment interviews, for example, people who seem confident and assertive usually fare better than those who stammer and seem unsure of themselves.[52]

It doesn't pay to overdo it, however. Just as an extremely "powerless" approach can feel weak, an overly "powerful" statement can come off as presumptuous and bossy. A lot depends on the context. Consider the following statements a student might make to a professor:

"Excuse me, sir, I hate to say this, but I won't be able to turn in the assignment on time. I had a personal emergency and . . . well . . . it was just impossible to finish it by today. Would that be okay?"

"I wasn't able to finish the assignment that was due today. I'll have it in your mailbox on Monday."

If you were the professor in this situation, which approach would you prefer? The first sounds tentative and not as fluent, which, on the face of it, seems powerless. In some situations, however, less assertive speakers seem friendlier, more sincere, and less coercive than more assertive ones.[53] You might appreciate the second approach, which is more direct and "powerful," or you may find it presumptuous and disrespectful.

TABLE 5.1

"Powerless" Language

TYPE OF USAGE	EXAMPLE
Hedges suggest that a speaker is tentative or unsure.	"I'm *kinda* disappointed . . ." "I think *maybe* we should . . ." "*I guess* I'd like to . . ."
Hesitations prolong the length of an utterance.	"*Uh*, can I have a minute of your time?" "*Well, let's see*, we could try this idea . . ." "I wish you would—*er*—try to be on time."
Intensifiers indicate powerful emotion.	"I *really* loved the movie . . ." "You did a *fabulous* job!!!"
Polite forms display deference or consideration.	"Excuse me, sir . . ." "I hope I'm not bothering you . . ."
Tag questions transform statements into questions.	"It's about time we got started, *isn't it*?" "We should give it another try, *don't you think*?"
Disclaimers display that a statement is atypical for the speaker or is a preliminary conclusion open to revision.	"I don't usually say this, but . . ." "I'm not really sure, but . . ."

Your reaction to each approach is likely to reflect a number of factors: *Do you know the student well? Is it typical for the student to miss deadlines? Do you share the same cultural expectations?* People in some cultures admire self-confidence and direct speech. However, in many cultures, helping others save face is a higher priority, so communicators tend to speak in ambiguous terms and use hedge words (such as *maybe*) and disclaimers (as in *"I may be wrong about this. but . . ."*). This is true in many Japanese and Korean cultures. Similarly, in traditional Mexican culture, it's considered polite, rather than powerless, to add *"por favor?"* (*"if you please?"*) to the end of requests, such as when ordering food in a restaurant. By contrast, "powerful" declarative statements, such as "I'll have the fish," are likely to seem bossy, rude, and disrespectful.[54]

As you have probably gathered, the terms *powerful* and *powerless* can be misnomers. In the United States, women's traditional speech patterns have often been described as less powerful than men's. However, there is considerable diversity among people of every gender identity. And people who do not seem powerful in traditional or obvious ways may have a great deal of influence, such as through their relationships with others. Since we tend to trust and cooperate most with people who build supportive, friendly relationships with us, sharing power *with* others can be more effective than exercising power *over* them.[55] The best communicators, regardless of their gender, read situations well and combine elements of both "powerful" and "powerless" speech.

Table 5.1 provides examples of speech often categorized as powerless. As you review the list, keep two factors in mind: First, statements may serve several purposes. For example, a phrase may serve as both a hedge and a hesitation. Second, since meaning is personal and situational, these examples are effective in some situations but less so in others.

Since emerging in the 1970s, hip-hop has entered the mainstream, as evidenced by the blockbuster success of the musical *Hamilton*.

Do you ever adapt your linguistic style when moving between different groups?

Affiliation One means of building and demonstrating solidarity with others is through language. Close friends and romantic partners often use nicknames and personal references that signify the nature of their bond.[56] Using the same vocabulary sets these people apart from others, reminding themselves and the rest of the world of their relationship. The same process works among members of larger groups, ranging from street gangs to military personnel. Fans of the same sports team may share specialized cheers, greetings, and other linguistic rituals that make it clear they are on the same side. Communication researchers call this linguistic accommodation **convergence**.

Communicators can experience convergence online as well as in face-to-face interactions. Members of online communities often develop a shared language and conversational style, and their affiliation with one another can be seen in increased use of the pronoun *we*.[57] On a larger scale, instant message and email users create and use shortcuts that mark them as Internet-savvy. If you know what ILYSM, OOMF, and HMU mean, you're probably part of that group. (Those acronyms stand for "I like [or love] you so much," "one of my followers," and "hit me up" with a picture or message.)

The principle of speech accommodation works in reverse, too. Communicators who want to set themselves apart from others may adopt the strategy of **divergence**, speaking in a way that emphasizes their difference from others. For example, people may use dialect to show solidarity with one another and distinguish themselves from other groups. Philadelphia natives use the word *jawn* to mean almost anything from a place, to a thing, to a person. Use of the word marks them as insiders, while newcomers are almost certainly baffled by statements such as "Pass me the jawn" and "They're building a new jawn downtown."[58] (We'll talk more about slang terms such as this later in the chapter.) Divergence also operates in other settings. A physician or attorney, for example, who wants to establish credibility with a client might speak formally and use professional jargon to create a sense of distance. The implicit message here is "I'm different from (and more knowledgeable than) you."

Convergence and divergence aren't the only ways to express affiliation. **Linguistic intergroup bias** reflects whether or not you regard others as part of your in-group. A positive bias leads people to describe the personality traits of in-group members in favorable terms and those of out-group members negatively.[59] For example, if an in-group member gives money to someone in need, you are likely to describe that member as a generous person. If an out-group member (someone with whom you don't identify) gives to the same person in need, you are likely to describe the behavior as a one-time act. The same in-group preferences are revealed when people describe undesirable behaviors. If an in-group member behaves poorly, you are likely to describe the behavior using a concrete action verb, such as "John cheated in the game." In contrast, if the person you are describing is an out-group member, you are more likely to use general disposition adjectives such as "John is a cheater."

Responsibility   Language can also reveal a speaker's willingness to accept responsibility for a message, as the following examples illustrate.

- **"It" versus "I" statements.** "*It's* not finished" (less responsible) versus "*I* didn't finish it" (more responsible).

- **"You" versus "I" statements.** "Sometimes *you* make me angry" (less responsible) versus "Sometimes *I* get angry when you do that" (more responsible). "I" statements are more likely than accusatory ones to generate positive reactions from others.[60]

- **"But" statements.** "It's a good idea, *but* it won't work." "You're really terrific, *but* I think we ought to spend less time together." (*But* cancels everything before it.)

- **Questions versus statements.** "Do *you* think we ought to do that?" (less personal responsibility) versus "*I* don't think we ought to do that" (more personal responsibility).

---

convergence Accommodating one's speaking style to another person, usually a person who is desirable or has higher status.

divergence A linguistic strategy in which speakers emphasize differences between their communicative style and that of others to create distance.

linguistic intergroup bias The tendency to label people and behaviors in terms that reflect their in-group or out-group status.

ON YOUR FEET

Recall a time when you adapted your language to converge with the norms of a particular speech community. What changes did you make? How successful were you in converging?

🛈 5.3 Troublesome Language

Besides being a blessing that enables us to live together, language can be something of a curse. We all have known the frustration of being misunderstood, and most of us have been baffled by another person's overreaction to an innocent comment. In the following pages we look at several kinds of troublesome language, with the goal of helping you communicate in a way that makes matters better instead of worse.

The Language of Misunderstandings

Language problems are semantic. They result when we simply don't understand others completely or accurately. Most of these misunderstandings arise from some common problems that may be easily remedied—after you recognize them.

> **equivocal words** Words that have more than one dictionary definition.
>
> **equivocation** A deliberately vague statement that can be interpreted in more than one way.
>
> **relative words** Terms that gain their meaning by comparison.

Equivocal Language   Misunderstandings can occur when words are **equivocal**, meaning that they are open to more than one interpretation. For example, a nurse once told a patient that he "wouldn't be needing" the materials he requested from home. He interpreted the statement to mean he was near death, but the nurse actually meant he would be going home soon.

Some words are equivocal as a result of cultural or cocultural differences in language usage. While teaching in Ireland, an American friend of ours asked a male colleague if he would give her a ride to the pub. After a few chuckles, he said, "You mean a lift." Our friend was surprised to learn that the word *ride* has sexual connotations in Ireland.

Equivocal misunderstandings can have serious consequences. Equivocation at least partially explains why men may sometimes persist in attempts to become physically intimate when women have expressed unwillingness to do so.[61] Interviews and focus groups with college students revealed that rather than saying "no" outright to a man's sexual advances, women often use ambiguous phrases such as, "I'm confused about this," "I'm not sure that we're ready for this yet," and "Are you sure you want to do this?" Whereas women viewed indirect statements as meaning "no," men were more likely to interpret them as meaning "maybe." As the researchers put it, "male/female misunderstandings are not so much a matter of males hearing resistance messages as 'go,' but rather their not hearing them as 'stop.'" Under the law, "no" means precisely that, and anyone who argues otherwise can be in for serious legal problems.

In contrast to equivocal words, which may confuse people unintentionally, an **equivocation** is a *deliberately* vague statement that can be interpreted in more than one way. If your date asks how you like their new haircut, you might equivocate by saying, "Your hair looks so shiny," rather than admitting that you don't like the style. Equivocations can spare people the embarrassment that might come from a bluntly truthful answer. But they can also be underhanded, as when an employee calls in sick, saying, "I'm not feeling well" when the whole truth is that she's exhausted from partying all night.

No doubt this sign draws a laugh from many who see it.

What are some examples of words with multiple meanings?

Relative Words   Is the school you attend large or small? Compared to a campus such as Ohio State University, with an enrollment of more than 60,000 students, it probably seems small, but in reference to a smaller institution, it might seem quite large. **Relative words** gain their meaning by comparison.

In the same way, relative words such as *fast* and *slow, smart* and *stupid,* and *short* and *long* depend on comparison for their meaning.

Some relative words are so common that we mistakenly assume that they have a clear meaning. For example, how much is "a few," and how much is "a lot"? An inquisitive blogger who posted those questions received dozens of replies. Definitions of "a few" varied from one to a dozen. Most people said "a lot" is at least 20. One respondent suggested that people use "a horde" to describe items that are more numerous than "a lot" but less plentiful than a "swarm" and far less than "zounds."[62]

Using relative words without explaining them can lead to communication problems. Have you been disappointed to learn that classes you've heard were "easy" turned out to be hard, that trips you were told would be "short" were long, that "hilarious" movies were mediocre? The problem in each case came from failing to anchor the relative word to more precise measures for comparison.

Slang   Language used by a group of people whose members belong to a similar coculture or other group is called **slang**. For instance, cyclists who talk about "bonking" are referring to running out of energy. And social media enthusiasts probably recognize that *hundo p* is slang for 100%, as in: "i'm hundo p into this new app." Other slang consists of *regionalisms*—terms that only people from a relatively small geographic area use and understand. Residents of the largest U.S. state know that when a fellow Alaskan says, "I'm going outside," they are leaving the state. Slang defines insiders and outsiders, creating a sense of identity and solidarity among members, while outsiders are likely to be mystified or to misunderstand.[63] (This relates to the concept of communication divergence we discussed earlier in the chapter.)

Jargon   Almost everyone uses some sort of **jargon**—specialized vocabulary that functions as a linguistic shorthand for people with common backgrounds and experience. Whereas slang tends to be casual and changing, jargon is typically more technical and enduring. Some jargon consists of *acronyms*—initials used in place of the words they represent. In finance, P&L (pronounced P-N-L) translates as "profit and loss," and military people label failure to serve at one's post as being AWOL (absent without leave). The digital age has spawned its own vocabulary of jargon. For instance, UGC refers to user-generated content.

Jargon can be an efficient way to use language for people who understand it. The trauma team in a hospital emergency room can save time, and possibly lives, by speaking in shorthand, referring to "GSWs" (gunshot wounds), "chem 7" lab tests, and so on. But the same specialized vocabulary that works so well among insiders can bewilder and confuse a patient's family members who don't understand the jargon. The same sort of misunderstandings can arise in less critical settings, such as classrooms and business meetings, where people may be reluctant to admit that they don't understand the jargon being used.

Euphemisms   A **euphemism** is a mild or indirect term substituted for a more direct but potentially less pleasant one. We use euphemisms when we say *restroom* instead of *toilet* or *full-figured* instead of *overweight.* Airline pilots rely on euphemism when they tell passengers to expect *bumpy air* rather than *turbulence.*

Euphemisms often seem more polite and less anxiety-provoking than other words. However, they can be vague and misleading. Terms such as *domestic disturbance* and *battle fatigue* are easy to hear, but they downplay the harsh realities involved. In the same way, being *excessed, decruited,* or *graduated* doesn't make the reality of losing one's job any easier.[64]

slang Language used by a group of people whose members belong to a similar coculture or other group.

jargon Specialized vocabulary used as a kind of shorthand by people with common backgrounds and experience.

euphemism A mild or indirect term or expression used in place of a more direct but less pleasant one.

"Be honest with me Roger. By 'mid-course correction' you mean divorce, don't you."

Source: Leo Cullum The New Yorker Collection/The Cartoon Bank

ETHICAL CHALLENGE "Telling It Like It Is"

For most Americans, "telling it like it is" is generally a virtue, whereas "beating around the bush" is a minor sin. You can test the function of indirect speech by following these directions.

- Identify an example of someone using a euphemism and equivocation.

- Imagine how matters would have been different if the speaker or writer had used direct language instead.

- Based on your observations, discuss situations in which equivocation and euphemisms are helpful and those in which they make effective communication difficult.

abstract language Language that lacks specificity or does not refer to observable behavior or other sensory data.

abstraction ladder A range of more to less abstract terms describing a person, object, or event.

Overly Abstract Language **Abstract language**—speech that refers to events or objects only vaguely—serves two main functions. First, it's often faster, easier, and more useful to talk in terms of higher-level abstractions than to be highly specific. For example, it's less cumbersome to talk about *Europe* than to list all of the countries on that continent. In the same way, using relatively abstract terms such as *friendly* or *smart* can make it easier to describe people than listing their specific actions. A second, less obvious function is that abstract language allows people to avoid confrontations by avoiding specifics. Suppose, for example, your boss is enthusiastic about a new approach to attracting clients that you think is a terrible idea. Telling the truth might seem risky, but lying—as in saying, "I think it's a great idea"—wouldn't feel right either. Instead, you might comment on the general idea rather than the specific proposal, perhaps by saying "I agree that we need to brainstorm a new approach" and hoping that other ideas will emerge.

Most things can be described with varying degrees of specificity. An **abstraction ladder** depicts this by describing the same idea in graduated levels of specificity, with the most general and abstract description at the top and the least abstract and most specific description at the bottom.[65] For example, an abstraction ladder for the material you are reading might read something like this:

A book

A textbook

A communication textbook

Understanding Human Communication

Chapter 5 of *Understanding Human Communication*

Page 116 of Chapter 5 of *Understanding Human Communication*

Although vagueness has its uses, highly abstract language can cause several types of problems. The first is *stereotyping*. Consider claims such as "All whites are bigots," "Men don't care about relationships," "The police are a bunch of pigs," or "Professors around here care more about their research than they do about students." Each of these claims ignores the very important fact that abstract descriptions are almost always too general; they say more than we really mean.

Besides creating stereotypical attitudes, abstract language can lead to the problem of *confusing others*. Imagine the lack of understanding that results from imprecise language in situations like the following:

A: We never do anything that's fun anymore.
B: What do you mean?
A: We used to do lots of unusual things, but now it's the same old stuff, over and over.
B: But last week we went on that camping trip, and tomorrow we're going to that party where we'll meet all sorts of new people. Those are new things.

A: That's not what I mean. I'm talking about really unusual stuff.

B: *(becoming confused and a little impatient)* Like what? Taking hard drugs or going over Niagara Falls in a barrel?

A: Don't be stupid. All I'm saying is that we're in a rut. We should be living more exciting lives.

B: Well, I don't know what you want.

The best way to avoid this sort of overly abstract language is to use **behavioral descriptions** instead. (See Table 5.2.) Behavioral descriptions move down the abstraction ladder to identify the specific, observable phenomenon being discussed. A thorough description should answer three questions:

1. **Who is involved?** Are you speaking for just yourself or for others as well? Are you talking about a group of people ("the neighbors," "women") or specific individuals ("the people next door with the barking dog," "Lola and Lizzie")?

2. **In what circumstances does the behavior occur?** It probably doesn't occur all the time. In order to be understood, pin down what circumstances set this situation apart from other ones. For example, does it happen in specific places (at parties, at work, in public) or at particular times, such as when you're tired or stressed?

3. **What behaviors are involved?** Though terms such as *more cooperative* and *helpful* might sound like concrete descriptions of behavior, they are usually too vague to clearly explain what's on your mind. Behaviors must be observable, ideally both to you and to others. For instance, moving down the abstraction ladder from the relatively vague term *helpful,* you might come to behaviors such as "does the dishes every other day," "volunteers to help me with my studies," or "fixes dinner once or twice a week without being asked." It's easy to see that terms like these are easier for both you and others to understand than are more vague abstractions.

Behavioral descriptions can improve communication in a wide range of situations, as Table 5.2 illustrates. Research also supports the value of specific language. One study found that well-adjusted couples had just as many conflicts as poorly adjusted couples, but instead of blaming each other, well-adjusted couples expressed their complaints in behavioral terms.[66] For instance, rather than saying, "You're a slob," an enlightened partner might say, "I wish you wouldn't leave your dishes in the sink."

The checklist on this page can help you prevent misunderstandings by choosing your words carefully.

5.4 Disruptive Language

Misunderstandings are a fact of life. But irresponsible and hostile use of language is avoidable. Sadly, there seems to be an epidemic of incivility. More than 3 out of 4 American workers say they are treated rudely by others at least once a week.[67] And although nearly 100 percent of Americans say it's important for a politician to act in a civil manner, 4 out of 5 felt that the presidential election involving Hillary Clinton and Donald Trump was uncivil.[68]

CHECKLIST ✓

Avoiding Misunderstandings

Impetuous word choices can be confusing or hurtful. Here are some tips to help clarify your language and avoid mix-ups.

☐ **Use slang and jargon with caution.** For example, If you say "that person needs to get woke" (educated about social issues) or "the SCOTUS is completely out of touch" (referring to the Supreme Court of the United States), make sure others understand what you mean.

☐ **Explain your terms.** Abstract words such as *good, bad, helpful,* and *happy* mean different things to different people.

☐ **Clarify whom you represent.** If you say "We think . . ." be clear about who "we" is. Otherwise, use "I" statements.

☐ **Be careful with euphemisms and equivocations.** If you say, "He went to a better place," listeners may wonder if he died, got a better job, or found a higher-quality restaurant for dinner.

☐ **Focus on specific behaviors.** "It's important that you arrive by 9 o'clock every morning" is clearer than "Be punctual."

behavioral description An account that refers only to observable phenomena.

TABLE 5.2

Abstract Versus Behavioral Descriptions

		BEHAVIORAL DESCRIPTION			
	ABSTRACT DESCRIPTION	**WHO IS INVOLVED**	**IN WHAT CIRCUMSTANCES**	**SPECIFIC BEHAVIORS**	**THE DIFFERENCE**
PROBLEM	I talk too much ...	... around people I find intimidating	... when I want them to like me.	I talk (mostly about myself) instead of giving them a chance to speak or asking about their lives.	Behavioral description more clearly identifies behaviors to change.
GOAL	I want to be more constructive ...	... with my roommate	... when we talk about household duties	... instead of finding fault with her ideas, and suggesting alternatives that might work.	Behavioral description clearly outlines how to act; abstract description doesn't.
APPRECIATION	"You've really been helpful lately."	Coworkers	When I've had to take time off work for personal reasons	"You graciously took on my workload in my absence."	Behavioral descriptions show gratitude more specifically.
REQUEST	"Behave!"	Romantic partner	"When we're around my family ...	... please don't tell jokes that involve sex."	Behavioral description specifies desired behavioral change.

"Does being a jerk make you more authentic?" asks history teacher Scott Culpepper.[69] The answer is no. The data are clear: People who force their opinions on others and treat them disrespectfully are ultimately less successful than their more agreeable and respectful peers.[70]

Of course, not all disagreements can, or should, be avoided. Rather, the goal is to eliminate ineffective linguistic habits from your communication repertoire to minimize clashes that don't need to happen and to allow you to communicate about differences in a respectful way. Here are some communication pitfalls to avoid.

Confusing Facts and Inferences

To express yourself responsibly, it's important to consider the difference between **factual statements**, which can be objectively shown to be true, and **inferential statements**, which are conclusions based on an interpretation of evidence. Consider a few examples:

factual statement A statement that can be verified as being true or false.

inferential statement A conclusion arrived at from an interpretation of evidence.

FACT	INFERENCE
He hit a lamppost while driving down the street.	He was probably texting when he drove into the lamppost.
You interrupted me before I finished what I was saying.	You don't care about what I have to say.
You haven't paid your share of the rent on time for the past 3 months.	You're trying to weasel out of your responsibilities.
I haven't gotten a raise in almost a year.	The boss is exploiting me.
She believes that health care should be available to everyone.	She's a socialist.

There's nothing wrong with making inferences as long as you identify them as such: "She stomped out and slammed the door. It looked to me as if she were furious." The danger comes when you confuse inferences with facts and make them sound like the absolute truth.

One way to avoid fact–inference confusion is to use the perception-checking skill described in Chapter 3 to test the accuracy of your inferences. Recall that a perception check has three parts: a description of the behavior being discussed, your interpretation of that behavior, and a request for verification. For instance, instead of saying "Why are you laughing at me?" you could say "When you laugh like that *[description of behavior]*, I get the idea you think something I did was stupid *[interpretation]*. Are you making fun of me *[question]*?"

Presenting Opinions as Facts

Listen to the people around you and in the media. Consider how often they present opinions as if they are facts. **Opinion statements** are based on the speaker's beliefs.

Opinion statements cannot be proved or disproved, so they aren't facts. Consider a few examples of the difference between factual statements and opinion statements:

OPINION	FACT
The climate in Portland is better than in Seattle.	It rains more in Seattle than in Portland.
Kareem is the greatest basketball player in the history of the game.	Kareem Abdul-Jabbar is the all-time leading scorer in the National Basketball Association.
The United States is the best model of economic success in the world.	Per capita income in the United States is higher than in many other countries.

> **opinion statement** A statement based on the speaker's beliefs.
>
> **emotive language** Language that conveys an attitude rather than simply offering an objective description.

When factual statements and opinion statements are set side by side in this way, the difference between them becomes clear. In everyday conversation, however, we often present opinions as if they were facts, and in doing so we invite an unnecessary argument. For example:

"That was a dumb thing to say!"

"Spending that much on a car is a waste of money!"

"You can't get a fair shake in this country unless you're a white male."

Notice how much less antagonistic each statement would be if it were prefaced by a qualifier such as *"In my opinion . . ."* or *"It seems to me. . . ."*

Particularly troublesome are opinion statements meant to incite strong emotional reactions, sometimes called **emotive language**. For example, a statement such as *"Worthless bums are ruining our town"* is emotionally inflammatory. A more responsible statement might sound something like this: *"I worry that the rising number of homeless people will cause families to move away from this area."*

One problem with emotive language is that it tends to inspire reactions based more on emotional fervor than on rational thought. This may lead us to believe an emotionally charged speaker even if the person presents no solid evidence. Or it may cause us to strike out in anger against people whose arguments are different from our own. "Overly strong emotional language antagonizes

Donald Trump's unapologetic use of disruptive language upended notions of political civility. Supporters have called it candor, while detractors have accused him of being a hateful demagogue.

Do you ever use emotive language? If so, what are the consequences?

CHECKLIST ✓

Distinguishing Between Facts and Opinions

There is a place for personal opinions, but it's troublesome when personal opinions are misrepresented as facts. Use the following guidelines to determine whether information—in a social media post, news item, or personal conversation—is fact, opinion disguised as fact, or opinion.[78,79]

☐ **If statements meet the following criteria, they are probably facts.**

- The evidence presented can be objectively proven or verified.
- The information is current and relevant.
- Valid sources of information are provided.
- An effort is made to encourage additional and emerging information.

☐ **Beware of opinions masquerading as facts. Following are tell-tale signs.**

- Statements seem designed mostly to stir up people's emotions.
- Claims are not supported with objective information.
- The argument is based on an isolated or unusual case.
- Assertions are overgeneralized or out of date.

☐ **Give thoughtful consideration to responsible opinions, which meet the following criteria.**

- Statements are clearly acknowledged to be perceptions (*I feel that . . .* ").
- Assertions are supported with trustworthy information.
- Respect is shown for other opinions.

the receiver and wipes away impulses to listen, to stay friends, or even to talk together any further," reflects psychologist Susan Heitler.[71]

At a time when facts are sometimes written off as "fake" and inaccurate information masquerades as objective truth, the difference between fact and opinion can seem nebulous. The checklist on this page offers experts' tips for distinguishing between facts and opinions in the news and in everyday conversations.

Personal Attacks

The "sticks and stones" nursery rhyme got it wrong. Words *can* hurt, and it's unfair to engage in personal attacks and name calling. The **ad hominem fallacy** (translated as *to the person*) involves attacking a person's character rather than debating the issues at hand. For example, during his candidacy for president of the United States, Donald Trump referred to opponents by derogatory names such as "crooked" Hillary Clinton,[72] "whack job" Bernie Sanders,[73] "lyin'" Ted Cruz,[74] and "low energy" Ben Carson. Name calling has a long and ugly past. Throughout history, derogatory terms such as the *n-word*, *gooks*, and *fags* have been used to set some people apart and stigmatize them.

Social media can be conducive to personal attacks by creating what some theorists call "webs of hate."[75] Online, it's easy to network with people who have similar prejudices, bolstering the sense that those attitudes are more prevalent than they really are, and creating speech communities in which hateful speech is tolerated or encouraged.

In contentious political contexts, it can be especially important to avoid language that alienates or offends. Instead, use words that encourage respectful discourse about diverse perspectives. See the checklist on page 121 for ways to make sure that your use of language is clear, thoughtful, and constructive.

🔈 5.5 Gender and Language

"Why do women say they're fine when they're not?"

"Why do men want to talk about sex so much?"

"Why do women talk on the phone for hours?"

"Why won't he tell me how he feels?"

These are common questions on websites with titles such as "I Don't Understand Women"[76] and "8 Things We Don't Understand about Men."[77] Gendered perspectives on language fascinate everyday people as well as researchers. The tricky part can be differentiating between stereotypes and current realities. This section explores 10 true and false statements about how men and women use language.

1. **Metaphorically speaking, men are from Mars, and women are from Venus.** *False.* There's no denying that gender differences *can* be perplexing. But the sexes aren't actually "opposite" or nearly as different as the Mars–Venus metaphor suggests. As you read in Chapter 3, gender is a matrix that includes a nearly infinite array of identities. People are more alike than they are different, and there is immense diversity within every identity group. As you read the following points, keep in mind that generalities don't describe every person, and terms such as

men and *women*, though common in the literature, don't begin to describe the true diversity among people.

2. **Women talk more than men.** *False.* Men and women speak roughly the same number of words per day, but women tend to speak most freely when talking to other women, whereas men usually do most of the talking in professional settings.[80]

3. **Men and women talk about different things.** *True . . . sometimes.* This is most true when women talk to women and men talk to men. Among themselves, women tend to spend more time discussing relational issues such as family, friends, and emotions. Male friends, on the other hand, are more likely to discuss recreational topics such as sports, technology use, and nightlife.[81] That is not to say that people of different genders always talk about different things. Nearly everyone reports talking frequently about work, movies, and television.[82]

4. **Where romance is concerned, it's complicated.** *Undeniable.* Although social roles have changed dramatically over the decades, powerful vestiges of traditional gender roles persist when it comes to romance. For example, when researchers recently studied an online chat site for teens, they found that young men were more likely than young women to post flirtatious comments and bold sexual invitations, and females were more likely to post friendly comments and to ask about and share their feelings.[83]

   Research about first dates and speed dating shows that men are more inclined than women to bring up the topic of sex, and women are more likely than men to think that a date is successful if conversation flows smoothly and the tone is friendly.[84,85] Take heart: These differences between gender roles tend to moderate as dating partners get to know each other better.

5. **Men and women communicate for different reasons.** *Often true.* People of all genders share a desire to build and maintain social relationships through communication. Their ways of accomplishing this are often different, however. In general, men are more likely than women to emphasize making conversation fun. Their discussions involve a greater amount of joking and good-natured teasing. By contrast, women's conversations focus more frequently on feelings, relationships, and personal problems.[86] Consequently, women may wonder why men "aren't taking them seriously," and men may wonder why women are so "emotionally intense."

6. **Women are emotionally expressive.** *Often true.* In the United States, because women frequently use conversation to pursue social needs, they are often said to have an *affective* (emotionally based) style, meaning that their language focuses on emotions. Female speech typically contains statements showing support for the other person, demonstrations of equality, and efforts to keep the conversation going. Because of these goals, traditionally female speech often contains statements of sympathy and empathy: *"I've felt just like that myself," "The same thing happened to me!"* Instant messages written by women tend to be more expressive than those composed by men.[87] They are more likely to contain laughter ("hehe"), emoticons (smiley faces), emphasis (italics, boldface,

ON YOUR FEET

Identify an instance in the news or social media in which someone presents an opinion as if it's a fact or uses emotive language. Prepare a brief oral presentation in which you describe the example and suggest more responsible ways to convey the information.

CHECKLIST ✓

Choose Your Words Carefully

It may be tempting sometimes to blurt out statements without thinking, but impetuous word choices can be confusing, hurtful, and downright wrong. Here are some tips to help clarify your language and avoid mix-ups.

☐ **Don't jump to conclusions.** *"I was disappointed when you missed the deadline"* is better than *"You don't care about the job."*

☐ **Don't present opinions as facts.** For example, instead of *"Online classes are best"* say *"I prefer online classes."*

☐ **Avoid emotive language.** Statements such as *"She's a phony"* imply broad or hurtful value judgments. Instead, reference a specific behavior, as in, *"Her resume says she graduated, but she hasn't finished her senior year yet."*

☐ **Be polite.** Even if you disagree with someone, address them respectfully rather than resorting to insults or name calling.

repeated letters), and adjectives. Women are also inclined to ask lots of questions that invite the other person to share information: *"How did you feel about that?"* *"What did you do next?"* However, people of all genders are typically hesitant to share their feelings when they think others will consider them weak or moody.[88] This is particularly true in professional settings, where women often perceive that they will be judged more harshly than men if they express emotions. This is one rationale behind saying *"Nothing's wrong"* when something actually is.

7. **Men don't show their feelings.** *Partly true, partly false.* In contrast to women, men in some cultures have traditionally been socialized to use language *instrumentally*—that is, to accomplish tasks. This is one reason that, when someone shares a problem, instead of empathizing, men are prone to offer advice: "That's nothing to worry about . . ." or "Here's what you need to do. . . ." To compound the issue, men in many cultures are discouraged from crying or showing feelings of sadness. Consequently, they often cope with difficult situations by using humor or distractions to avoid showing grief or sorrow, especially in public.[89] That's not to say that men never express emotions. In some cultures, such as in traditional Arab communities, men are very emotionally expressive.[90] And whereas men in the United States have traditionally been discouraged from showing sadness or fear, they are encouraged to show emotions in some situations. Expressive behavior at sporting events is a case in point.[91] And in their private lives, men typically do express emotion, but they may do so through their *actions* (such as physical affection and favors) more than their words.[92]

8. **Women's speech is typically powerless, and men's is more powerful.** *Not necessarily.* As you read earlier in the chapter, while it is true that traditionally feminine speech includes more hedges and hesitations than traditionally male speech, those less assertive speech patterns can be a powerful means of building relationships and collaborating with others.

9. **Men and women are hardwired to communicate differently.** *Partly true, but less so than many people assume.* Research shows that men with high testosterone levels are more competitive than those with lower levels of the hormone, and they respond with more emotional language than other men when faced with setbacks.[93,94] Estrogen is associated with heightened emotional experiences and expression of emotion.[95] However, hormones do not correlate perfectly with biological sex. While men typically have more testosterone and women more estrogen, both hormones are present in everyone, and their presence varies by individual. Moreover, the influence of hormones is less intense than most people think. For example, only 3 to 8% of women experience hormonal mood swings beyond the range of everyday emotions.[96]

10. **Men and women are socialized to communicate differently.** *True.* One of the main reasons men and women communicate differently is that society expects them to. When researchers asked people to pretend they were either male or female while describing a painting, the differences were pronounced, reflecting the way people *think* men and women communicate, even more than the way they actually do.[97] Because of preconceived notions, men and women encounter different conversational climates. People in one study interrupted female speakers more than male speakers, even though all the speakers were trained to say much the same thing.[98] Similar effects are notable in the written word. Health news aimed at female audiences typically includes more hedges, hesitations, and tag questions than articles in men's magazines.[99]

By now it's probably clear that neither characteristically male nor female styles of speech meet all communication needs. You can improve your linguistic

ASK YOURSELF

Research shows that male and female speech is different in some respects and similar in others.

How would you characterize your use of language on a stereotypically male–female spectrum?

UNDERSTANDING YOUR COMMUNICATION

How Do You Use Language?

Answer the following questions to see what orientation is suggested by the way you use language.

1. **Your best friend is upset upon learning that he was not accepted into graduate school. What are you most likely to say?**
   a. "You seem discouraged. Tell me what's going through your head."
   b. "Grad school is overrated. Tons of successful people don't have master's degrees."
   c. "There are other great schools. How might I help you apply to them?"
   d. "You are a good student. Don't let this get you down. The school that accepts you will be very lucky."

2. **You are planning a sales pitch that could earn your company millions of dollars. What is your pitch most likely to include?**
   a. A focus on the client's most deeply held values
   b. A list of the reasons your company is better than the competition
   c. Specific features that make your product highly useful and effective
   d. Jargon and other language that shows you understand the client's business

3. **You hope to meet with a professor to learn more about a topic covered in class. How would you word a meeting request?**
   a. "I'm excited about the ideas you shared in class. Could I meet with you to learn more?"
   b. "This topic is critical to my long-term success. Can I meet with you to learn more about it?"
   c. "I'd like to hear what steps you think I should follow to be successful at this. Can we meet?"
   d. "I like what you said in class. I know you're busy, but would it be possible to meet and talk more about it?"

4. **Your family is planning a holiday celebration, but you'd like to go skiing with friends instead. How are you most likely to broach the topic with your family?**
   a. "You've always been so supportive of me. I think you'll understand. . . ."
   b. "Going skiing with my friends is a once-in-a-lifetime opportunity."
   c. "The ski trip is an opportunity to make new friends and maybe even some future business contacts."
   d. "I'd love to go skiing. But I won't go unless you're 100% okay with it."

INTERPRETING YOUR RESPONSES

Read the following explanations to learn more about your use of language. (More than one explanation may apply.)

Affective
If you answered "a" to two or more questions, your language tends to focus on emotions—yours and other people's. This affective approach (page 121) can make you a sensitive listener and a motivational speaker. Just be sure to balance this strength with awareness of practical concerns.

Opinionated
If you answered "b" to more than one question, you tend to voice strong viewpoints. Educated opinions can be useful, but review the discussion about inferences and opinions (pages 118–119) to make sure you present your ideas with respect for others. Presenting your views as facts can squelch open communication and blind you to alternative ways of understanding the world around you.

Instrumental
If you answered "c" to two or more questions, you are inclined to adopt an instrumental approach to language (page 122). Your focus on strategies and goals can be highly effective, but you may come off as headstrong in some situations. Make sure you don't lose sight of the emotional (affective) aspects of the issue at hand.

Affiliative
If you answered "d" to more than one question, you are disposed toward an affiliative language style. You tend to display convergence (alignment) with other people and avoid actions that might place you at odds. Your thoughtfulness is no doubt appreciated. At the same time, make an effort to take a stand when it's important to do so. The advice on assertive communication in Chapter 9 may be helpful.

competence by switching and combining styles. If you reflexively take an instrumental approach that focuses on the content of others' remarks, consider paying more attention to the unstated relational messages behind their words. If you generally focus on the feelings part of a message, consider being more task oriented. If your first instinct is to be supportive, consider the value of offering advice; and if advice is your reflexive way of responding, think about whether offering support and understanding might sometimes be more helpful. Research confirms what common sense suggests: Balancing a task-oriented approach with a relationship-oriented approach is usually the most effective strategy. Choosing the approach that is right for the other communicator and the situation can create satisfaction far greater than that which comes from using a single stereotypical style.

As we finish this chapter, take the "Understanding Your Communication" quiz to see how you typically use language.

MAKING THE GRADE

For more resources to help you understand and apply the information in this chapter, visit the *Understanding Human Communication* website at www.oup.com/he/adler-uhc14e.

OBJECTIVE 5.1   Explain how symbols and linguistic rules allow people to achieve shared meaning.

- A language is a collection of symbols governed by a variety of rules and used to convey messages between people.
- Because of its symbolic nature, language is not a precise tool. Meanings are in people, not in words themselves.
- For effective communication to occur, it is necessary to negotiate meanings for ambiguous statements.
  - > Test your understanding by explaining the symbolic, rule-governed nature of language to someone unfamiliar with the concepts in this chapter.
  - > How has the principle that meanings are in people, not words, created challenges in your life?
  - > How can you do a better job expressing yourself and understanding the other person in an important relationship?

OBJECTIVE 5.2   Identify ways in which language shapes people's attitudes and reflects how they feel about themselves and others.

- People collaborate with one another using communication to make sense of the world around them.

- Language features such as names, accents, and gendered language influence people's values, attitudes, and beliefs.
- The words people use reflect a sense of power, affiliation with others, and responsibility.
  - > How might the language people use affect and reflect their social status? Give examples of statements often categorized as "powerless" and statements considered to be "powerful."
  - > Record or transcribe from memory a recent conversation in which you were involved. Identify statements that illustrate concepts described on pages 107–113.
  - > How might you use language to impress a prospective employer during a job interview? To show support for a friend?

OBJECTIVE 5.3   Recognize and remedy confusing and inflammatory language.

- Vague and specialized language has the potential to create misunderstandings, either intentionally or not.
- Language can be disruptive when people assert their inferences and opinions as if they are facts or verbally attack others.
  - > Compare and contrast equivocal language and equivocations.
  - > When do you consider it acceptable to use euphemisms? When do you think such language is confusing or unfair?
  - > Identify examples of troublesome language in a movie or television show. How might you rewrite the script to reflect more effective ways of communicating?

OBJECTIVE 5.4 **Describe the ways in which tradition-ally masculine and feminine speech patterns are alike and how they differ.**

- Although there are differences in the ways men and women speak, people often assume the differences are greater than they are.

- In some situations, men and women tend to talk about different topics and to do either more of the talking or more of the listening.

- Despite social changes and evolving ideas about gender, traditional expectations are still detectable in many situations.

  > In your own words, describe the similarities and differences between characteristically male and female language as outlined on pages 120–124.

  > How do these similarities and differences compare to your use of language? To the linguistic style of people you know well?

  > If the language you use is closest to the "masculine" styles described, when and how might you use "feminine" strategies to expand your repertoire? If your style is more "feminine," when and how might you use "masculine" styles effectively?

KEY TERMS

abstraction ladder p. 116

abstract language p. 116

accent p. 109

ad hominem fallacy p. 120

behavioral description p. 117

connotative meanings p. 103

coordinated management of meaning (CMM) p. 107

convergence p. 113

denotative meanings p. 103

dialect p. 102

divergence p. 113

emotive language p. 119

equivocal words p. 114

equivocation p. 114

euphemism p. 115

factual statement p. 118

inferential statement p. 118

jargon p. 115

language p. 102

linguistic intergroup bias p. 113

linguistic relativism p. 109

opinion statement p. 119

phonological rules p. 104

pragmatic rules p. 105

reappropriation p. 106

relative words p. 114

semantic rules p. 105

slang p. 115

speech act p. 105

symbol p. 103

syntactic rules p. 104

ACTIVITIES

1. Increase your ability to achieve an optimal balance between powerful speech and polite speech by rehearsing one of the following scenarios:

    a. Describe your qualifications to a potential employer for a job that interests you

    b. Request an extension on a deadline from one of your professors

    c. Explain to a merchant why you want a cash refund on an unsatisfactory piece of merchandise when the store's policy is to issue credit vouchers

    d. Ask your boss for 3 days off so you can attend a friend's out-of-town wedding

    e. Approach your neighbors whose dog barks while they are away from home

2. Note differences in the language use of three men and three women you know. Your analysis will be most accurate if you record the speech of each person you analyze. Consider the following categories:

    > Conversational content

    > Conversational style

    > Reasons for communicating

    > Use of powerful/powerless speech

    Based on your observations, answer the following questions:

    > How much does gender influence speech in these episodes?

    > What role might other variables play? Consider occupational or social status, cultural background, social philosophy, competitive–cooperative orientation, and other factors in your analysis.

Listening

CHAPTER OUTLINE

🧍 6.1 The Value of Listening 128

🧍 6.2 Misconceptions About Listening 129
    Myth: Listening and Hearing Are the Same Thing
    Myth: Listening Is a Natural Process
    Myth: All Listeners Receive the Same Message

🧍 6.3 Overcoming Challenges to Listening 133
    Message Overload
    Noise
    Cultural Differences

🧍 6.4 Faulty Listening Habits 135

🧍 6.5 Listening to Connect and Support 137
    Be Sensitive to Personal and Situational Factors
    Allow Enough Time
    Ask Questions
    Listen for Unexpressed Thoughts and Feelings
    Encourage Further Comments
    Reflect Back the Speaker's Thoughts
    Consider the Other Person's Needs When Analyzing
    Reserve Judgment, Except in Rare Cases
    Think Twice Before Offering Advice
    Offer Comfort, If Appropriate

🧍 6.6 Gender and Supportive Listening 143

🧍 6.7 Listening to Accomplish, Analyze, or Critique 144
    Task-Oriented Listening
    Analytical Listening
    Critical Listening

    MAKING THE GRADE 151
    KEY TERMS 152
    ACTIVITIES 152

LEARNING OBJECTIVES

6.1

Summarize the benefits of being an effective listener.

6.2

Outline the most common misconceptions about listening, and assess how successfully you avoid them.

6.3

Recognize and develop strategies to overcome barriers to listening.

6.4

Identify and minimize faulty listening habits.

6.5

Describe and practice listening strategies for effectively connecting with and supporting others.

6.6

Describe how gender may influence listening behavior.

6.7

Know when and how to listen to accomplish a task, analyze a message, and critically evaluate another's remarks.

AFTER WAITING MONTHS to get a 20-minute meeting with a high-powered CEO, Mark Goulston was frustrated when the executive seemed preoccupied and wasn't listening. Rather than muddle through a disappointing meeting, Goulston acknowledged aloud, "There is something on your mind."[1] He suggested they reschedule, allowing the CEO time to focus on what was bothering him.

To Goulston's surprise, the CEO started to cry. He said his wife was undergoing a biopsy that day, and he feared a bad outcome. "I'm at work, but I'm not really here," the CEO said. "You've known me for 5 minutes and there are people that have known me for 20 years . . . and they don't know what you know."[2]

The CEO was so moved by Goulston's perceptiveness and compassion that he then began to listen. Goulston, the author of *Just Listen*,[3] is an expert on paying attention and reading between the lines, but anyone can become a better listener by building skills and avoiding bad habits.

Really listening involves a level of discipline and skill few people master. Yet the payoffs are enormous, as you will see. Masterful listening can help you make wise decisions, make a positive impression on others, and enrich your relationships. In this chapter, we consider both why and how to be a good listener. You will learn more about:

- listening misconceptions,
- factors that often make it difficult to listen well,
- bad listening habits to avoid,
- how and why to connect with and support others as a listener,
- common gender differences in listening, and
- types of listening by goal.

6.1 The Value of Listening

Imagine you are at a career fair where you have access to talent scouts and employers who can help you land the job of your dreams. You might approach these people with a mental list of everything you want to say—a description of your talents, experience, goals, and so on. Or you might take a different approach and listen.

If you chose the listen-more-than-you talk option, pat yourself on the back. And if you are a good listener in other situations as well, the benefits multiply. In his best-selling book *The 7 Habits of Highly Effective People*, Stephen Covey observes that most people only pretend to listen while they actually rehearse what they want to say themselves.[4] Rare (and highly effective) is the person who listens with the sincere desire to *understand,* observes Covey. An impressive body of evidence backs up this claim, as you will see in the following list of reasons to become a better listener:

- **People with good listening skills are more likely than others to be hired and promoted.**[5] "Listening is more important than speaking," advises a spokesperson for one of the largest career networks sites in the United States. In fact, she ranks the importance of listening among the "top 5 things recruiters wish you knew."[6] (The other four involve dressing appropriately, handling rejection well, being proactive, and being polite and considerate.) Listening skills are also important once you get the job. Because good listeners are typically judged to be appealing and trustworthy,[7] they are especially popular with employers and with customers and clients.[8,9]

- **Listening is a leadership skill.** Leaders who are good listeners typically have more influence and stronger relationships with team members than less attentive leaders do.[10] In fact, leaders' listening skills are even more influential than their talking skills.[11] As columnist Doug Larson puts it, "Wisdom is the reward you get for a lifetime of listening when you'd have preferred to talk."[12]

- **Good listeners are not easily fooled.** People who listen carefully and weigh the merits of what they hear are more likely than others to spot what some researchers call "pseudo-profound bullshit"—statements that sound smart but are actually misleading or nonsensical.[13] Mindful listening (a topic we'll discuss more in a moment) is your best defense.

- **Asking for and listening to advice makes you look good.** "Many people are reluctant to seek advice for fear of appearing incompetent," observes a research team who studied the issue.[14] What they found was the opposite— that people think more *highly* of people who ask them for guidance about challenging issues than those who struggle through a difficult challenge on their own. Of course, that's just the first step. Making the most of that advice requires good listening skills and follow-through.

- **Listening makes you a better friend and romantic partner.** Friends and partners who listen well are considered to be more supportive than those who don't.[15] That probably doesn't surprise you, but this may: Listening well on a date can significantly increase your attractiveness rating.[16,17] The caveat is that you can't *pretend* to listen. Effective listeners are sincerely interested and engaged.

Despite the importance of listening, experience shows that much of the listening we (and others) do is not very effective. We misunderstand others and are misunderstood in return. We become bored and feign attention while our minds wander. We engage in a battle of interruptions in which each person fights to speak without hearing the other's ideas. Some of this poor listening is inevitable, perhaps even justified. But in other cases, we can be better receivers by learning more about listening.

6.2 Misconceptions About Listening

Julio was nervous about meeting his wife's coworkers at a party. After fretting about what he might say to impress them, he decided to focus on what *other* people were saying. He paid close attention to people's words and body language and responded with statements such as "I understand what you're saying. You feel strongly that . . ." and "Let me see if I understand what you mean . . ."[18] Throughout the party, people

Source: © 2003 Zits Partnership Distributed by King Features Syndicate, Inc.

hearing The process wherein sound waves strike the eardrum and cause vibrations that are transmitted to the brain.

listening The process wherein the brain recognizes sounds and gives them meaning.

attending The process of focusing on certain stimuli in the environment.

understanding The act of interpreting a message by following syntactic, semantic, and pragmatic rules.

listening fidelity The degree of congruence between what a listener understands and what the message sender was attempting to communicate.

responding Providing observable feedback to another person's behavior or speech.

told Julio's wife how much they enjoyed talking to him, calling him articulate and charismatic. Perhaps without knowing it, Julio didn't succumb to the following myths about listening.

Myth: Listening and Hearing Are the Same Thing

In Chapter 1, we introduced the term *receiving* to describe the process by which a message is decoded. In fact, the process of receiving a message involves multiple stages. **Hearing** is the physiological ability to perceive the presence of sounds in the environment. If you have that ability, hearing occurs automatically when sound waves strike your eardrums and cause vibrations that are transmitted to your brain. People with physical hearing disorders lose some or all of the ability to detect sounds. Beyond hearing, **listening** occurs when the brain recognizes sounds and gives them meaning. Unlike hearing, listening requires conscious effort and skill. Even when people cannot hear, they can be attentive listeners in other ways. The phrase "I listen with my eyes" is common in the Deaf community.[19] It refers not only to sign language but to the ability to gain meaning by using all of the senses.

Although hearing is automatic, listening is another matter. It's common to hear others speak without listening. Sometimes we deliberately tune out speech—for example, a friend's boring remarks or a boss's unwanted criticism.

Successful listening consists of several stages. After hearing, the next stage is **attending**—the act of paying attention to a signal. Your needs, wants, desires, and interests influence what you attend to as a listener.

The next step in listening is **understanding**—the process of making sense of a message. Communication researchers use the term **listening fidelity** to describe the degree of congruence between what a listener understands and what the sender was attempting to communicate.[20] Chapter 5 discussed many of the ingredients that make it possible to understand language: syntax (how words are ordered), semantics (what words mean), and pragmatic rules about using and interpreting language. Taking all of this into account, it's clear that listening isn't the passive activity you might have imagined. It relies on a sophisticated combination of effort, knowledge, skills, and physical ability.[21] It's hard work.

The next listening stage involves **responding** to a message—offering observable feedback to the speaker. Feedback may include eye contact, facial expressions, questions and comments, posture, and more.

Feedback serves two important functions: It helps you clarify your understanding of a speaker's message, and it shows that you care about what that speaker is saying. Listeners don't always respond to a speaker in obvious ways—but research suggests that they should. When people are asked to evaluate the listening skills of people around them, the number-one trait they consider is whether the listener offers feedback.[22] Conversely, it's easy to see how discouraging it is when audience members yawn, slump, or display bored expressions. Adding responsiveness to the listening model demonstrates that communication is transactional (Chapter 1). As listeners, we are active participants in a communication transaction. While we receive messages, we also send them. For example, although some people insist that they are listening even when they seem distracted and unresponsive, their demeanor probably puts a damper on the conversation.

You probably think of listening as something you do only with your ears. But a similar version of the process also occurs when using social media. If you text a friend "We're on for tonight. Meet at home or restaurant?" and they text back "Sounds good!," you might think they didn't give much effort to understanding your message. Social media professionals also use the term "listening" when they monitor posts to see what people say about them. See *Understanding Communication Technology* for more about the practice of social media eavesdropping.

The final step in the listening process is **remembering**.[23] It has long fascinated scientists that people remember every detail of some messages but very little of

> **remembering** The act of recalling previously introduced information. The amount of recall drops off in two phases: short term and long term.

UNDERSTANDING COMMUNICATION TECHNOLOGY

Who Is Listening to You Online?

During a lecture, Australian university student Jonathan Pease tweeted the following message: "Sitting in the back row at syndey uni carving 'Rooney eats it' on the desk . . . feels good to be a rebel." He soon discovered that his online audience included Sydney University officials, who soon posted a tweet of their own: "Defacing our desks, Jonathan? Hope you enjoyed your course."[24] Luckily for Pease, the authorities had a sense of humor. Their response also reinforced an important point: When you go online, your audience may include more viewers than you imagine.

University officials aren't the only ones who track people online. The practice known as "social media listening" has become a sensation among marketing professionals who want to understand consumers' preferences better and develop relationships with them. Robert Caruso realized this when he saw a commercial for Total Bib on TV and tweeted to his friends that it reminded him of a *Saturday Night Live* sketch. "A few hours later, I received a reply tweet from Total Bib thanking me for the mention and engaging me in conversation," Caruso says. "I was pretty amazed since I was not following them previously, they were simply monitoring the stream." Not only that, but Total Bib representatives looked up Caruso online and learned that he was a single father with a toddler, so they sent him a complimentary bib. Caruso said he was touched by the personal attention.[25]

So-called listening software can monitor websites, Facebook, Google searches, blogs, Twitter, and more. That means that marketing professionals monitor people's online search terms and the websites they visit, in addition to the messages and photos they post.

At its best, monitoring technology allows marketers to listen avidly to consumers. At its worst, it can be "unethical and creepy," in the words of Tom Petrocelli, a technology blogger and marketing specialist.[26] Petrocelli is a fan of media listening when it's done well, but he counts the following actions in the creepy column: spying on employees, requiring them to share their "friends" list with the marketing team, and sending out mass emails or tweets uninvited.

One thing is for sure: Whether you consider online monitoring to be an invasion of privacy or marketing genius, the odds are that you are subject to this type of listening.

TING

EAR ... MIND

EYE

HEART

The Chinese word *ting* refers to deep, mindful listening. In its written form, the word combines the symbols for ears, eyes, heart, and mind.

Have you ever felt that someone listened to you with an open mind and heart, attentive to your words and your feelings? If so, how did you respond?

residual message The part of a message a receiver can recall after short- and long-term memory loss.

mindful listening Being fully present with people—paying close attention to their gestures, manner, and silences, as well as to what they say.

others. For example, you may remember many specifics about gossip you heard but forget what your roommate asked you to buy at the store. By some accounts, people tend to forget about half of what they hear *immediately after* hearing it, suggesting that they did not truly listen to and store the information.[27]

Given the amount of information we process every day—from instructors, friends, social media, TV, and other sources—it's no wonder the **residual message** (what we remember) is only a small fraction of what we hear. However, with effort, we can increase our ability to remember what is important to us. We'll explore ways of doing that later in the chapter.

Myth: Listening Is a Natural Process

Another common myth is that listening is like breathing—a natural activity that people usually do well. The truth is that listening is a skill much like speaking: Everybody does it, though few people do it well. In the workplace, good listeners are typically more influential than their peers because they are perceived to be more agreeable, open, and approachable than people who listen poorly.[28]

However, most of us are not the good listeners we think we are. In one survey, 96% of professionals rated themselves good listeners, but 80% of them admitted to multitasking while on the phone, not giving callers their full attention.[29]

Sometimes it's okay to be mindless about what you hear. Paying attention to every song on the radio or every commercial on TV would distract you from more important matters. But problems arise when people are lazy about listening to things that really matter. For example, a college student hurt by his girlfriend's poor listening skills wrote in an online forum, "I have opened up to her about really, really personal things and then two weeks later or within the week . . . she's like, 'oh, you never mentioned it to me.' I just find this really really rude and insulting."[30]

Mindful listening involves being fully present with others—paying close attention to their gestures, manner, and silences, as well as to what they say.[31] It requires a commitment to understanding the other's perspectives without being judgmental or defensive. Consistent with the idea of mindful listening, the Chinese concept of *ting* describes listening with open ears and eyes as well as an open mind and open heart.[32] This type of listening can be difficult, especially when we are busy or when we feel vulnerable ourselves, yet the investment is worthwhile. For more tips on mindful listening, see the checklist on page 133.

Myth: All Listeners Receive the Same Message

When two or more people listen to a speaker, we tend to assume they all hear and understand the same message, but it's not true. Recall the discussion of perception in Chapter 3, in which we pointed out the many factors that cause each of us to perceive an event differently. Perhaps you're hungry, thinking about something else, or just not interested. Your friend might find a joke funny, whereas you consider it silly or even offensive. For any number of reasons, you may attach different meaning or significance to the same words.

Misunderstandings are especially likely when remarks are interpreted out of context. When Fifth Harmony member Normani Kordei called one of her groupmates "very quirky" in an interview, she says she meant it in a good way, but some fans interpreted her remark as an insult and attacked Kordei on social media.[33]

6.3 Overcoming Challenges to Listening

Given the number of people and devices clamoring for your attention, listening can seem harder than ever before.[34] Consider these examples:

- Your phone vibrates while you're listening to a lecture or talking to a friend. You peek at the screen.

- You're binge-watching your favorite TV drama. A neighbor drops by to warn you about some car break-ins nearby. You know the issue may be important, but you are irritated by the interruption.

- Over coffee, a friend complains about having a bad day. You want to be supportive, but you're preoccupied with problems of your own, and you need to get back to work soon to meet a deadline.

- Your boss critiques your work. You think her comments are unfair, and you find yourself responding defensively.

- A family member tells the same story you've heard dozens of times before. You feel obliged to act interested, but your mind is far away.

As we will explore, a range of factors can affect your ability to listen in cases such as these.

Message Overload

It's impossible to listen carefully to all the information that bombards you daily, not only in face-to-face messages but in texts, calls, emails, and social media posts. Aside from personal messages, most people are awash in programming from mass media. This deluge of communication has made paying close attention more difficult than ever before.[35] Experts suggest turning off distracting communication devices while you work on complex tasks, sending only clear and brief emails with specific subject lines and thinking twice before sharing trivial information with everyone you know.[36] The @*Work* feature on page 134 considers what happens when we multitask.

Noise

Effective listening often requires overcoming various forms of noise.

Physical Noise   External distractions can make it hard to pay attention to others. The sounds of traffic, construction, alarms, music, or others' speech or media use can interfere with our ability to hear well, let alone listen. You can listen better by removing the sources of distracting noise whenever possible: turning off the television, closing the window, and so on. In some cases, you may need to find a more hospitable place to have a conversation.

Psychological Noise   People are often wrapped up in personal concerns that seem more important to them than the messages others are sending. It's hard to pay attention to someone else when you're anticipating an upcoming test or mentally replaying what you did last night. Good listeners develop the discipline to set aside personal concerns and give others' messages the attention they deserve.

CHECKLIST ✅

Tips for Mindful Listening

Mindful listening takes effort, but it pays off in terms of self-awareness and stronger connections with others. Here are some tips for becoming a more mindful listener.[37,38,39]

☐ **Commit to being fully present.**

☐ **Minimize distractions, including extraneous thoughts and worries.**

☐ **Listen for underlying messages as well as surface meanings.**

☐ **Pay attention to cues about how the speaker feels.**

☐ **Mentally acknowledge your own feelings.** ("I'm feeling defensive. I'll set that aside for now and try to understand more fully what this person is sharing with me.")

☐ **Acknowledge or ask about the other person's feelings.** ("Are you feeling discouraged?")

☐ **Ask questions and check your understanding.** ("I heard you say that you're confused. Is that a good reflection of how you feel? What are the pros and cons of the situation as you see them?")

☐ **Be patient. Don't interrupt.**

☐ **Become comfortable with silence.**

☐ **Don't rush the speaker.** The goal is to understand.

Multitasking Can Make You Less Productive

Multitasking may be a fact of life on the job, but research suggests that dividing your attention has its costs. In one widely reported study, volunteers tried to carry out various problem-solving tasks while being deluged with phone calls and emails.[40] Even though experimenters told the subjects to ignore these distractions, the average performance drop was equivalent to a 10-point decline in IQ.

You might expect that greater exposure to multiple messages would improve multitasking performance, but just the opposite seems to be the case. Heavy media multitaskers perform worse on task switching than light media multitaskers.[41] Although chronic multitaskers believe they are competent at processing information, in fact they're worse than those who focus more on a single medium.[42]

You may not be able to escape multiple demands at work, but don't hold any illusions about the cost of information overload. When the matter at hand is truly important, the most effective approach may be to turn off the phone, close down the email program or browser, and devote your attention to the single task before you.

Source: DILBERT © 2009 Scott Adams. Used by permission of UNIVERSAL UCLICK. All rights reserved.

ON YOUR FEET

In one of your most important relationships, which bad listening habits do you find most annoying? What would the other person say are your most annoying listening faults?

Physiological Noise Another listening distraction may arise from physiology, or the way our bodies function. You've probably noticed that it's harder to listen if you are tired, hungry, too hot, or too cold. If this psychological noise is too great, it may be better to have an important conversation at another time.

Cultural Differences

The behaviors that define a good listener may vary by culture. Many Americans are impressed by listeners who ask questions and make supportive statements.[43] By contrast, Iranians tend to judge people's listening skills based on more subtle indicators such as posture and eye contact. That's probably because the Iranian culture relies more on context.[44] (As you may remember from Chapter 4, members of high-context cultures are particularly attentive to nonverbal cues.) Meanwhile, Germans are most likely to think people are good listeners if they show continuous attention while others speak.[45] Expectations vary by generation as well. If you grew up texting and tweeting, you may think that glancing at your phone during a meal is fine. But someone from an older generation may perceive it as a lack of attention and respect. One lesson is that, whereas people in some cultures and age groups may overlook a quick glance at a cell phone or TV screen, others may interpret that behavior as rudely inattentive. See the checklist in the following section for ways to minimize the influence of social media on your listening habits.

❶ 6.4 Faulty Listening Habits

Shasta was at dinner with a group of friends when she proposed that they take turns sharing one thing about the friendship that they liked and one thing they would like to improve or expand. "Everyone shared really beautiful things," Shasta recalls. "It was super touching." Then, right before it was her turn, the conversation shifted to a different topic. No one seemed to realize that she never got a turn to weigh in. "I felt hurt," she says, looking back.[46] Here are seven bad habits to control if you want to be fully present with the people around you.

1. **Pretending to listen.** When people **pseudolisten,** they give the appearance of being attentive, but they aren't really. They may look people in the eye, nod and smile at the right times, and may even answer occasionally. That appearance of interest, however, masks thoughts that have nothing to do with what the speaker is saying.

2. **Tuning in and out.** When people respond only to the parts of a speaker's remarks that interest them, they are engaging in **selective listening.** All of us are selective listeners from time to time, but it's a habit that can lead to confusion, misunderstandings, and hurt feelings.

3. **Being defensive.** People who perceive that they are being attacked—even when they aren't—can become **defensive listeners,** more interested in justifying themselves than in understanding the other person's point of view. Guilt or insecurity is often at the root of defensive listening, with the effect that casual remarks may be taken as threats (*"How dare you ask me if I enjoyed lunch! It's my business and no one else's if I cheat on my diet!"*). A common defensive response is counterattacking the perceived critic, but that's not likely to make you a lot of friends. Instead of proving your point, you will probably cause the other person to feel defensive and angry. By contrast, non-defensive listeners are sincerely interested in understanding the other person's perspective, even when the topic makes them uncomfortable. See the checklist on page 136 for tips on listening nondefensively.

CHECKLIST ✓

Ways to Limit Social Media Distractions

By some estimates, most people swipe, stroke, tap, or click their smartphones hundreds or even thousands of times a day.[47] Psychologist and social media analyst Sherry Turkle offers the following tips for staying present:[48]

☐ **Don't reach for a device every time you get a free moment.** Instead, take stock of what you are seeing, feeling, smelling, and hearing. It will make you more attentive to the people and things around you.

☐ **Create "device-free zones."** By committing to setting devices aside at the dinner table, in the car, or in the living room, you can be alone with your thoughts or carry on a conversation without distractions.

☐ **Share your feelings in person every so often.** Rather than posting or tweeting your thoughts, consider sharing them face to face.

pseudolistening An imitation of true listening.

selective listening A listening style in which the receiver responds only to messages that interest them.

defensive listening A response style in which the receiver perceives a speaker's comments as an attack.

Calvin and Hobbes   by Bill Watterson

Source: CALVIN and HOBBES © 1995 Watterson. Reprinted by permission of UNIVERSAL UCLICK. All rights reserved.

Techniques for Listening Nondefensively

It's natural to feel uncomfortable when the boss wants to talk about the deadline you missed or when your roommate is upset because you left a mess. However, nondefensive listeners resist the impulse to fight back. Here are some tips for listening nondefensively, even when the heat is on.[55,56]

- [ ] **Take a deep breath and remind yourself that you are a likable person.**

- [ ] **Let go of the idea that you (or anyone) can be perfect.**

- [ ] **Accept that the other person's feelings are real, even if their interpretation is different from yours.**

- [ ] **Ask questions.** ("Did you feel that I was taking you for granted when you saw my dishes in the sink?")

- [ ] **Acknowledge the other person's feelings.** ("I can understand why you were disappointed.")

- [ ] **Avoid displaying nonverbal cues, such as eye rolls and heavy sighs, that seem defensive or dismissive.**

- [ ] **Use "I" language to express your feelings.** ("I wanted to edit the report one more time. I didn't realize the delay would be so costly.")

- [ ] **If warranted, apologize for your behavior.**

- [ ] **Learn from the encounter and move on.** And at the very least, congratulate yourself for handling a difficult situation with sincerity and openness.

4. **Avoiding the issue**. People who avoid difficult subjects are **insulated listeners**. It's understandable that someone might tune out when the topic is touchy. Perhaps you really don't want to hear your boss outline the reasons you didn't get a promotion. But even so there are advantages to listening well. You are likely to earn the boss's respect and learn some valuable tips if you are a receptive listener. When the tables are turned and you're the one who has to deliver touchy information, you can help listeners by being upfront about the sensitive nature of what you are about to disclose. One blog for young people offers this example: "Mom, I have something to tell you. I'm not proud of what I've done, and you might be mad. But I know I need to tell you. Can you hear me out?"[49]

5. **Missing the underlying point**. Rather than looking below the surface, **insensitive listeners** tend to take remarks at face value. An insensitive listener might miss the warble in a friend's voice that suggests they are more upset than their words let on. Or when a partner complains, "I always take out the trash," an insensitive listener might miss that what's wanted is a thank-you.

6. **Being self-centered.** The next time you're engaged in conversation, consider who has control. **Conversational narcissists** focus on themselves and their interests instead of listening to and encouraging others.[50] One type of conversational narcissist is the **stage hog**, who actively claims more than the fair share of the spotlight. Stage hogs tend to interrupt a lot and switch topics to suit their interests. They may assume that their ideas are better or more important than others' or that they can guess what other people are going to say.[51] Other conversational narcissists are more passive. They may not interrupt, but neither do they encourage others. A **passive narcissist** is unlikely to make supportive comments such as *"Uh-huh," "That's funny,"* and *"What happened next?"*[52] Whatever their approach, conversational narcissists tend to discourage the equal give-and-take that is the hallmark of mutually satisfying conversation.

7. **Assuming that talking is more impressive than listening**. As comedian Paula Poundstone jokes, "It's not that I'm not interested in what other people have to say, it's that I can't hear them over the sound of my own voice."[53] It's often tempting to talk more than we listen. However, there is no calculating the esteem and wisdom people earn by listening well rather than talking all the time. The playwright Wilson Mizner put it this way: "A good listener is not only popular everywhere, but after a while he [or she] gets to know something."[54]

You may recognize the egotism behind many of these bad habits but still feel tempted to engage in them sometimes. As we said, listening is hard. It's worth the effort, however. Next we take a closer look at some of the functions listening serves.

6.5 Listening to Connect and Support

When a pale and disheveled man approached Samarnh Pang at the university where he works, Pang didn't walk away. Instead, he listened as the man described his hardships and explained that many people shun him because his is old and unclean. "I sensed that this man just needed someone to listen to his stories," Pang says. He was right. After the older man had talked for a while, he smiled and said to Pang, "Son, you are the first person to listen to my stories with kind attention. Thank you for being kind hearted."[57]

There's no doubt that supportive listening can make a difference. Research shows that it can reduce loneliness and stress and build self-esteem.[58] Relational and supportive listeners are typically viewed by others to be friendly, honest, understanding, and encouraging.[59] And the benefits go both ways. People who provide social support often feel an enhanced sense of well-being themselves.[60]

In this section, we explore the role of **relational listening**, which involves emotionally connecting with others; and **supportive listening**, which goes a step further, with the goal of helping a speaker deal with personal dilemmas, whether they be minor stressors or life-changing situations.

Despite the benefits of relational and supportive listening, there can be drawbacks. It's easy to become overly involved with others' feelings. When that happens, listeners may feel overwhelmed and unable to offer an objective perspective.[61] And even when they are trying to be supportive, listeners can come off as opinionated and intrusive. Following are 10 strategies to avoid the pitfalls and be an effective listener when the goal is to connect with another person and/or help them through a difficult time.

Be Sensitive to Personal and Situational Factors

Before committing yourself to helping another person—even someone in obvious distress—make sure your support is welcome. Even then, there is no single best way to provide it. There is enormous variability in which style will work with a given person in a given situation. This explains why communicators who are able to use a wide variety of helping styles are usually more effective than those who rely on just one or two styles.[62]

You can boost the odds of choosing the best helping style in each situation by considering three factors.

- *Personal preference:* Some people prefer to handle difficult situations on their own. They may benefit from the opportunity simply to voice their thoughts. By contrast, other people welcome advice and analysis.[63]

- *The situation:* Timing is everything. Sometimes the most supportive thing you can do is listen quietly. At other times, listening well may help you realize that running an errand or offering to help with a task may be more helpful.[64]

- *Your own strengths and weaknesses:* You may be best at listening quietly, offering a prompt from time to time. Or perhaps you are especially insightful and can offer useful analysis of problems. These can be listening strengths. Just be careful that you don't use an approach that is comfortable for you even when a different one might be more effective. In some situations, speakers may assume that quiet listeners aren't paying attention or that "helpful" listeners are overly judgmental or eager to tell others what to do.

insulated listening A style in which the receiver ignores undesirable information.

insensitive listening The failure to recognize the thoughts or feelings that are not directly expressed by a speaker, and instead accepting the speaker's words at face value.

conversational narcissists People who focus on themselves and their interests instead of listening to and encouraging others.

stage hogs People who are overly invested in being the center of attention.

passive narcissists People who are so wrapped up in themselves that they fail to be supportive or encouraging of others.

relational listening A listening style that is driven primarily by the desire to build emotional closeness with the speaker.

supportive listening The reception approach to use when others seek help for personal dilemmas.

ON YOUR FEET

Think of a time when good listening helped you make a tough decision or navigate a stressful time in your life. Prepare a brief oral presentation in which you share the story with others. Be sure to describe the specific listening behaviors that made a difference for you.

In a world full of distractions, it can require effort to stop what you're doing and listen to others.

What steps might you take to be a more focused, patient listener?

In most cases, the best way to help is to use a combination of responses in a way that meets the needs of the other person and suits your personal communication style.[65]

Allow Enough Time

Connecting with and supporting others can take time. If you're in a hurry, it may be best to reschedule important conversations for a better time. You might say, *"I want to give you my undivided attention. Can we meet at 5 o'clock for coffee?"*

In other situations, brief interactions can be meaningful. To a harassed customer service representative you might say sympathetically, *"Busy day, huh?"* Or you might thank an especially patient salesperson by saying, *"I really appreciate you taking time to explain this so patiently."*

The gift of attention often speaks for itself, even when you don't know what to say. Medical studies show that, even when doctors cannot offer a cure, patients' coping skills are positively linked to the amount of time their doctors spend listening to them.[66]

Ask Questions

Asking questions can help a conversational partner define vague ideas more precisely. You might respond to a friend by asking, *"You said Greg has been acting 'differently' toward you lately. What has he been doing?"* or *"You told your roommates that you wanted them to be more helpful in keeping the place clean. What would you like them to do?"* Questions can also encourage people to keep talking, which is particularly helpful when you are dealing with someone who is quiet or fearful of being judged. Use caution, however. Questions may sometimes confuse or distract a speaker. See the checklist on the next page for experts' tips about effective questioning.

Listen for Unexpressed Thoughts and Feelings

People often don't say what's on their minds or in their hearts, perhaps because they're confused, fearful of being judged, or trying to be polite. However, these unstated messages can be as important as the spoken ones. It can be valuable to listen for unexpressed messages. Consider a few examples:

STATEMENT	POSSIBLE UNEXPRESSED MESSAGE
"Don't apologize. It's not a big deal."	"I'm angry (or hurt, disappointed) by what you did."
"You're going clubbing tonight? That sounds like fun!"	"I'd like to come along."
"Check out this news story. That's my little sister!"	"I'm proud of what she did."
"That was quite a party you [neighbors] had last night. You were going strong at 2 a.m."	"The noise bothered me."
"You like gaming? I do too!"	"Perhaps we can be friends."

There are several ways to explore unexpressed messages. You can *paraphrase* by restating the speaker's thoughts and feelings in different words, as in *"It sounds like that really surprised you"* or *"So you aren't sure what to do next, right?"* (We'll talk more about paraphrasing in a few pages.) You can *prompt* the speaker to volunteer more information with questions such as *"Really?"* and *"Is that right?"* Or you can *ask questions* such as *"What are the pros and cons, as you see them?"*

Encourage Further Comments

Sometimes you can strengthen relationships and support others simply by encouraging them to say more. Blogger Sarah Q remembers an unexpected source of support when she was being bullied in high school. Her English teacher noticed Sarah's distress and began spending lunch periods with her. "I told her everything," Sarah says. "She knew what was going on, and she helped me through it all. . . . She sat there and listened. And I needed that." Sarah has since transferred to a different school where she has good friends, but she says she will always be grateful to the teacher who truly listened.[67]

Whereas questioning requires a great deal of input from the respondent, another approach is more passive. **Prompting** involves using silences and brief statements of encouragement to draw others out and in so doing to help them solve their own problems. Consider this example:

> **Pablo:** Julie's dad is selling a complete computer system for only $1,200, but if I want it I have to buy it now. He's got another interested buyer. It's a great deal. But buying it would wipe out my savings. At the rate I spend money, it would take me a year to save up this much again.
>
> **Tim:** What a dilemma.
>
> **Pablo:** I wouldn't be able to take that ski trip over winter break . . . but I sure could save time with my schoolwork . . . and do a better job, too.
>
> **Tim:** That's for sure.
>
> **Pablo:** Do you think I should buy it?
>
> **Tim:** I don't know. What do you think?
>
> **Pablo:** I just can't decide.
>
> **Tim:** *(silence)*
>
> **Pablo:** I'm going to do it. I'll never get a deal like this again.

Prompting works best when it's done sincerely. Your nonverbal behaviors—eye contact, posture, facial expression, tone of voice—must show that you are concerned with the other person's problem but are not advocating for one outcome over another. Great teachers harness this power regularly. They know that students often learn more when they are asked questions and are encouraged to work through problems than when they are given the answers up front.[68]

When encouraging others to talk, be careful to not redirect the conversation back to yourself. Avoid responses like these:

> **Abel:** "I don't know whether to quit or stay in a job I hate."
>
> **Briana:** "You think your job is bad? Let me tell you about one I had last year . . . "
>
> **Carlo:** "My grandma is having health problems and I'd love go visit her, but midterms are coming up and I'd hate to miss them."
>
> **Danielle:** "Family always comes first. When my grandfather had an accident . . .

Temptations to Avoid When Asking Questions

Supportive listeners follow these principles:

☐ **Don't ask questions just to satisfy your curiosity.** You might be tempted to ask *"What did they say then?"* or *"What happened next?"* just because you'd like to know. In some cases, however, questions such as these might distract the speaker from the real issue. As a general rule, let the speaker decide what details are most important from their perspective.

☐ **Don't get specific before you know the whole story.** Asking *"When did the problem begin?"* or *"What are you going to do now?"* might speed up the speaker—but such questions may also lead them to think you are impatient and uninterested in hearing the whole story. It's important to be sure you're on the right track before asking about specifics.

☐ **Don't use questions to disguise suggestions or criticism.** We've all been questioned by parents, teachers, or others who seemed to be trying to trap or guide us through questions, as in *"Do you think it's wise to put off your homework until Sunday night?"* Used this way, questioning becomes a strategy that implies that the questioner already has an opinion about what should happen but isn't willing to say so outright.

☐ **Don't violate the speaker's privacy.** When you consider exploring unexpressed feelings and thoughts, be careful not to pry.

prompting Using silence and brief statements to encourage a speaker to continue talking.

You can show that you understand and help a speaker clarify their thoughts by listening carefully and reflecting back what a speaker seems to be saying.

When can you understand better by reflecting the speaker's thoughts?

reflecting Listening that helps the person speaking think about the words they have just spoken.

Reflect Back the Speaker's Thoughts

Supportive listeners often **reflect** aloud about the thoughts and feelings they have heard a speaker express. This is akin to paraphrasing, but the goal of reflecting isn't so much to clarify your understanding as to help the other person reflect on their thoughts. The following conversation between two friends shows how reflecting can offer support and help a person find the answer to their own problem:

Jill: I've had the strangest feeling about my boss lately.

Mark: What's that? *(A simple question invites Jill to go on.)*

Jill: I'm starting to think maybe he has this thing about women—or maybe it's just about me.

Mark: You mean he's coming on to you? *(Mark paraphrases what he thinks Jill has said.)*

Jill: Oh no, not at all! But it seems like he doesn't take women—or at least me—seriously. *(Jill corrects Mark's misunderstanding and explains herself.)*

Mark: What do you mean? *(Mark asks another simple question to get more information.)*

Jill: Well, whenever we're in a meeting or just talking around the office and he asks for ideas, he always seems to pick men. He gives orders to women—men, too—but he never asks the women to say what they think. But I know he counts on and acknowledges some women in the office.

Mark: Now you sound confused. *(Reflects her apparent feeling.)*

Jill: I am confused. I don't think it's just my imagination. I mean I'm a good producer, but he has never—not once—asked me for my ideas about how to improve sales or anything. And I can't remember a time when he's asked any other woman. But maybe I'm overreacting.

Mark: You're not positive whether you're right, but I can tell that this has you concerned. *(Mark paraphrases both Jill's central theme and her feeling.)*

Jill: Yes. But I don't know what to do about it.

Mark: Maybe you should . . . *(Starts to offer advice but catches himself and decides to ask a sincere question instead.)* So what are your choices?

Jill: Well, I could just ask him if he's aware that he never asks women's opinions. But that might sound too aggressive and angry.

Mark: And you're not angry? *(Tries to clarify how Jill is feeling.)*

Jill: Not really. I don't know whether I should be angry because he's not taking women's ideas seriously, or whether he just doesn't take my ideas seriously, or whether it's nothing at all.

Mark: So you're mostly confused. *(Reflects Jill's apparent feeling again.)*

Jill: Yes! I don't know where I stand with my boss, and not being sure is starting to get to me. I wish I knew what he thinks of me. Maybe I could just tell him I'm confused about what is going on here and ask him to clear it up. But what if it's nothing? Then I'll look insecure.

Mark: *(Mark thinks Jill should confront her boss, but he isn't positive that this is the best approach, so he paraphrases what Jill seems to be saying.)* And that would make you look bad.

Jill: I'm afraid maybe it would. I wonder if I could talk it over with any-body else in the office and get their ideas . . .

Mark: . . . see what they think . . .

Jill: Yeah. Maybe I could ask Brenda. She's easy to talk to, and I do respect her judgment. Maybe she could give me some ideas about how to handle this.

Mark: Sounds like you're comfortable with talking to Brenda first.

Jill: *(Warming to the idea.)* Yes! Then if it's nothing, I can calm down. But if I do need to talk to the boss, I'll know I'm doing the right thing.

Mark: Great. Let me know how it goes.

Reflecting a speaker's ideas and feelings can be surprisingly helpful. First, reflecting helps the other person sort out the problem. In the dialogue you just read, Mark's paraphrasing helped Jill consider carefully what bothered her about her boss's behavior. The clarity that comes from this sort of perspective can make it possible to find solutions that weren't apparent before. Reflecting also helps the person unload more of the concerns they have been carrying around, often leading to the relief that comes from catharsis. Finally, listeners who reflect the speaker's thoughts and feelings (instead of judging or analyzing, for example) show their involvement and concern.

Consider the Other Person's Needs When Analyzing

There are many reasons to analyze what you hear—to interpret it and consider it from multiple perspectives. In a few minutes, we'll consider the importance of analytical listening when assessing whether or not information is trustworthy. But when the goal is supportive listening, analysis serves another function—to offer an interpretation of a speaker's message that may help them achieve more clarity. Analysis statements meant to be supportive are probably familiar to you:

> *"I think what's really bothering you is . . ."*
> *"She's doing it because . . ."*
> *"I don't think you really meant that."*
> *"Maybe the problem started when she . . ."*

Interpretations are sometimes effective in helping people with problems consider alternatives meanings. Under the right circumstances, an outside perspective can make a confusing problem suddenly clear, either suggesting a solution or at least providing an understanding of what is occurring.

There can be two problems with analyzing, however. First, your interpretation may not be correct, in which case the speaker may become even more confused upon hearing it. Second, even if your interpretation is correct, saying it aloud might not be useful. There's a chance that it will arouse defensiveness in the speaker because analysis can imply superiority and judgment, and even if it doesn't, the person may not be willing or able to understand your view of the problem.

How can you know when it's helpful to offer an analysis? The checklist on this page suggests several guidelines to follow.

Reserve Judgment, Except in Rare Cases

Judgments can be helpful, but for the most part, they are a risky way to respond to someone in distress. A **judging response** evaluates the sender's thoughts or behaviors in some way. The evaluation may be favorable (*"That's a good idea"* or

ASK YOURSELF

If a friend is having trouble deciding between two job offers, how might you use supportive listening to help them work through the decision?

CHECKLIST ✓

What to Consider Before Offering a Judgment

Judgments have the best chance of being well received when the following conditions are present:

☐ **The person with the problem has requested an evaluation from you.** Occasionally, an unsolicited judgment may bring someone to their senses, but more often an uninvited evaluation will trigger a defensive response.

☐ **Your judgment is genuinely constructive and not designed as a putdown.** If you are tempted to use judgments as a weapon, don't fool yourself into thinking that you are being helpful. Often the statement *"I'm telling you this for your own good"* simply isn't true.

If you follow these two guidelines, your judgments will probably be less frequent and better received than if you share opinions without weighing the risks involved.

judging response Feedback that indicates a listener is evaluating the sender's thoughts or behaviors.

advising response Helping response in which the receiver offers suggestions about how the speaker should deal with a problem.

comforting A response style in which a listener reassures, supports, encourages, or distracts the person seeking help.

"You're on the right track now") or unfavorable (*"An attitude like that won't get you anywhere"*). But in either case, it implies that the listener is qualified to pass judgment on the speaker, which can cause hurt feelings.

Sometimes negative judgments are purely critical. How many times have you heard responses such as *"Well, you asked for it!"* or *"I told you so!"* or *"You're just feeling sorry for yourself"*? Statements like these can sometimes serve as a verbal wake-up call, but they often make matters worse. At other times, negative judgments involve constructive criticism, which is less critical and more intended to help a person improve in the future. Friends may offer constructive criticism about everything from the choice of clothing to jobs and to friends, and teachers may evaluate students' work to help them master concepts and skills. Whether or not it's justified, even constructive criticism may make the other person feel defensive. See the checklist on this page for factors to consider before offering a judgmental response.

Think Twice Before Offering Advice

When someone shares a concern with you, you might offer a solution, which scholars call an **advising response**. Although advice is sometimes valuable, often it isn't as helpful as you might think. For one thing, it can be hard to tell when someone actually wants your opinion. Statements such as *"What do you think of Jeff?,"* *"Would that be an example of sexual harassment?,"* and *"I'm really confused"* may be designed more to solicit information than to get advice. And even when someone asks for advice outright (as in *"What do you think I should do?"*), offering it may not be helpful for several reasons. For one, what's right for one person may not be right for another. If your suggestion is not the best course to follow, it can be harmful. Another downside is that advice can seem like a putdown in that it casts the advice giver as wiser or more experienced than the recipient. And finally, advice may discourage others from making their own decisions or feeling accountable for them. For the most part, advice is most welcome when it has been clearly requested and when the advisor respects the face needs of the recipient (Chapter 3).[69] Before offering advice, consider the checklist on page 143.

Offer Comfort, If Appropriate

Sometimes a listener's goal is to offer **comfort** by reassuring, supporting, encouraging, or distracting the person seeking help. Comforting responses can take several forms:

Agreement	"You're right—the landlord is being unfair."
	"Yeah, that class was tough for me, too."
Offer of help	"I'm here if you need me."
	"Let me try to explain it to him."
Praise	"I don't care what the boss said, I think you did a great job!"
	"You're a terrific person! If she doesn't recognize it, that's her problem."

Reassurance	"The worst part is over. It will probably get easier from here."
	"I know you'll do a great job."
Diversion	"Let's catch a movie and get your mind off this."
	"That reminds me of the time we . . ."
Acknowledgment	"I can see that really hurts."
	"I know how important that was to you."
	"It's no fun to feel unappreciated."

Even if you mean well, some responses may fail to help. Telling someone who is obviously upset that everything is all right, or joking about a serious matter, can leave the other person feeling worse, not better. They might see your comments as a putdown or an attempt to trivialize their feelings. It's also usually hurtful to be judgmental: *"There are people worse off than you are"* or *"No one ever said life was fair."*[70] And it may also be frustrating to hear *"I understand how you feel"* from someone who can't really know what they are going through.[71] An American Red Cross grief counselor explains that simply being present can be more helpful than anything else when people who are distressed or grief-stricken:

> Listen. Don't say anything. Saying *"it'll be okay,"* or *"I know how you feel"* can backfire. . . . Be there, be present, listen. The clergy refer to it as a "ministry of presence." You don't need to do anything, just be there or have them know you're available.[72]

See the checklist on page 144 for some factors to consider when offering comfort as a listener.

🎧 6.6 Gender and Supportive Listening

Men and women have traditionally defined supportive communication somewhat differently. Linguist Deborah Tannen, who is famous for her research on gender and communication, offers the example of telling one's troubles to another person.

When women share their troubles with other women, Tannen says, the response is often a matching disclosure. For example, a woman might say, *"I understand. My partner never remembers my birthday!"* Such a response is usually understood between women as a sign of their connectedness and solidarity. Indeed, women may even dig deep to find a matching experience or emotion to share,[73] which is one reason that happiness (as well as dissatisfaction) often feels contagious.

Men have traditionally been socialized to focus less on emotional connection and more on competition and emotional control. Consequently, if a woman responds to a man's troubles talk with a matching experience, it may feel to him like a one-up, as if she is implying *"your problems are not so remarkable"* or *"mine are even worse."* In short,

Factors to Consider Before Offering Advice

Before offering advice, make sure the following conditions are present:[74]

☐ **Be confident that the advice is correct.** Resist the temptation to act like an authority on matters you know little about. Furthermore, be aware that just because a course of action worked for you doesn't guarantee that it will work for everybody.

☐ **Ask yourself whether the person seeking your advice seems willing to accept it.** In this way you can avoid the frustration of making good suggestions, only to find that the person with the problem had another solution in mind all the time.

☐ **Be certain that the receiver won't blame you if the advice doesn't work out.** You may be offering the suggestions, but the choice and responsibility for accepting them are up to the recipient of your advice.

☐ **Deliver your advice supportively, in a face-saving manner.** Advice that is perceived as being offered constructively, in the context of a solid relationship, is much better than critical comments offered in a way that signals a lack of respect for the receiver.

CHECKLIST ✅

Ways to Offer Comfort

For occasions when comforting is an appropriate response, here are some helpful guidelines.

☐ **Make sure your comforting remarks are sincere.** Phony agreement or encouragement is probably worse than no support at all because it adds the insult of your dishonesty to the pain the other person is already feeling.

☐ **Be sure the other person can accept your support.** Sometimes we become so upset that we aren't ready or able to hear anything positive.

☐ **Avoid remarks that seem to make light of the person's feelings.** Saying *"There's nothing to worry about"* or *"You are overreacting"* is likely to make the person feel foolish and misunderstood rather than comforted.

All in all, the most comforting response is often a sincere commitment to stand by the person through hard times and to listen without judgment as they describe what they are feeling.[80]

task-oriented listening A response style in which the goal is to secure information necessary to get a job done.

what feels like empathy to her may feel like a putdown to him. And because men are often discouraged from expressing intense emotions, they may consider it supportive to offer a solution or a distraction such as *"Don't worry about it"* or *"Here's what you should do . . ."* As you might predict, women who are accustomed to a different style of social support may feel that people who respond this way are brushing off their concerns or belittling their problems.

The result of these different perspectives, Tannen observes, is often a mutual sense of frustration:

> She blames him for telling her what to do and failing to provide the expected comfort, whereas he thinks he did exactly what she requested and cannot fathom why she would keep talking about a problem if she does not want to do anything about it.[78]

Of course, it's important to avoid overgeneralizing. Gender roles continually evolve, and a number of factors interact with gender to shape how people provide social support—including cultural background, personal goals, expressive style, and cognitive complexity. All the same, understanding traditional patterns and social mores may help us avoid the assumption that our way is the only way or the right way to offer comfort.

6.7 Listening to Accomplish, Analyze, or Critique

"No matter how informed humans are, no matter how well educated, well traveled, or experienced, it is both a blessing and a curse that we are forever tethered to a single set of eyeballs through which to see the world," writes writer Francesca Moroney. The antidote to that limitation, she proposes, is to listen to one another.

Listening well offers numerous benefits, including the opportunity to learn something new, develop a more nuanced view of the world, and sometimes, to clarify your own beliefs in comparison to other people's.[79] In this section, we consider the value and techniques of task-oriented, analytical, and critical listening.

Task-Oriented Listening

The purpose of **task-oriented listening** is to secure information necessary to complete the job at hand. This might involve following your boss's instructions at work, following a friend's tips for mastering a new game or app, getting tips from a sports coach—the list goes on and on.

Task-oriented listeners are often concerned with efficiency. They may view time as a scarce and valuable commodity, and they may grow impatient when they think others are wasting it. A task orientation can be an asset when deadlines and other pressures demand fast action. Such listeners keep a focus on the job at hand and encourage others to be organized and concise.

Despite its advantages, a task orientation can put off others when it seems to disregard their feelings. Don't forget that emotional issues and concerns can be an important part of business and personal transactions. Also, an excessive focus on getting things done quickly can hamper the kind of thoughtful deliberation that some jobs require.

You can become more effective as a task-oriented listener by approaching others with a constructive attitude and by using some simple but effective skills. The following guidelines should help you be more effective.

Listen for Key Ideas It's easy to lose patience with long-winded speakers, but good task-oriented listeners stay tuned in. They are able to extract the main points, even from a complicated message. If you can't figure out what the speaker is driving at, you can always ask in a tactful way by using the skills of questioning and paraphrasing, which we will examine now.

Ask Questions Questioning involves asking for additional information to clarify your understanding of the sender's message. If you are heading to a friend's house for the first time, typical questions might be, *"How is the traffic between here and there?"* or *"Is there anything you'd like me to bring?"* One key element of these types of questions is that they ask the speaker to elaborate.

Not all questions are genuine requests for information. **Counterfeit questions** are disguised attempts to send a message. There are four main types of counterfeit questions.

- *Assertions disguised as questions:* When someone says, *"Are you going to stand up and give him what he deserves?"* they are clearly voicing an opinion rather than seeking a yes or no answer.

- *Lurking hidden agendas:* If someone asks *"Why don't we reorganize our work space?"* and what they really want is the corner office for themselves, the question is merely a vehicle for advancing a personal goal.

- *Leading questions:* Leading questions imply that there is a "correct" answer. For example, if a controlling leader asks *"Don't you think my idea is great?"* savvy listeners know that a brutally honest answer may not be well received.

- *Unchecked assumptions:* If someone asks *"Why aren't you listening to me?"* they are advancing the unchecked assumption that the other person isn't paying attention. Likewise, *"What's the matter?"* assumes that something is wrong. As Chapter 3 explains, perception checking is a much better strategy: *"When you kept looking out the window during our meeting, I thought you weren't listening to my idea, but maybe you were considering the implications of what I suggested. What was on your mind?"*

Paraphrase Another type of feedback can also help you confirm your understanding. **Paraphrasing** involves restating in your own words the message you think the speaker has just sent. For example:

(*To a direction giver*) "You're telling me to drive down to the traffic light by the high school and turn toward the mountains, is that it?"

> **counterfeit question** A question that is not truly a request for new information.
>
> **paraphrasing** Feedback in which the receiver rewords the speaker's thoughts and feelings.

COMMUNICATION TOOLS

Source: Courtesy of Ted Goff.

CHECKLIST

How and When to Paraphrase

If you decide that paraphrasing is a good option, you can make your response sound more natural by taking any of the following approaches, depending on the situation:

☐ **Change the speaker's wording.** *Speaker: Bilingual education is just another failed idea. Paraphrase: You're mad because you think bilingual ed sounds good, but it doesn't work? (Reflects both the speaker's feeling and the reason for it.)*

☐ **Offer an example of what you think the speaker is talking about.** When the speaker makes an abstract statement, you may suggest a specific example or two to see if your understanding is accurate. *Speaker: Lee is such a jerk. I can't believe the way he acted last night. Paraphrase: You think those jokes were pretty offensive, huh? (Reflects the listener's guess about the speaker's reason for objecting to the behavior.)*

☐ **Reflect on the underlying theme of the speaker's remarks.** When you want to summarize the theme that seems to have run through another person's conversation, a complete or partial perception check is appropriate. *Speaker: Remember to lock the door. Paraphrase: You keep reminding me to be careful. It sounds like you're worried that something bad might happen. Am I right? (Reflects the speaker's thoughts and feelings and explicitly seeks clarification.)*

(*To the boss*) "So you need me both this Saturday *and* next Saturday—right?"

(*To a professor*) "When you said, 'Don't worry about the low grade on the quiz,' did you mean it won't count against my grade?"

In other cases, a paraphrase will reflect your understanding of the speaker's feelings:

"[To a boss] You said not to worry about the customer's complaint, but I get the feeling it's a problem. Am I mistaken, or is this cause for concern?"

"You said you've got a minute to talk, but I'm not sure whether it's a good time for you."

"You said, 'Forget it,' but it sounds like you're mad. Are you?"

In each case, the key to success is to restate the other person's comments in your own words as a way of cross-checking the information. If you simply repeat the speaker's comments verbatim, you will sound foolish—and you still might misunderstand what has been said and why. Notice the difference between simply parroting (repeating without understanding) a statement and really paraphrasing:

Speaker:	I'd like to go to the retreat, but I can't afford it.
Parroting:	You'd like to go, but you can't afford it.
Paraphrasing:	So if we could find a way to pay for you, you'd be willing to come. Is that right?

Speaker:	What's the matter with you?
Parroting:	You'd like to know what's wrong with me?
Paraphrasing:	You think I'm mad at you?

As these examples suggest, effective paraphrasing is a skill that takes time to develop. It can be worth the effort, however, because it offers two very real advantages. First, it boosts the odds that you'll accurately and fully understand what others are saying. We've already seen that using one-way listening or even asking questions may lead you to think that you've understood a speaker when, in fact, you haven't. Second, paraphrasing guides you toward sincerely trying to understand another person instead of using faulty listening styles such as stage hogging, selective listening, and so on. Listeners who paraphrase to check their understanding are judged to be more socially adept than listeners who do not.[81] (See the checklist on this page for paraphrasing approaches.)

Take Notes   Understanding others is crucial, of course, but it doesn't guarantee that you will remember everything you need to know. As you read earlier in this chapter, listeners usually forget about half of what they hear immediately afterward.

Sometimes recall isn't especially important. You don't need to retain many details of the vacation adventures recounted by a neighbor or the childhood stories told by a relative. At other times, though, remembering a message—even minute details—is important. The lectures you hear in class are an obvious example. Likewise, it can be important to remember the details of plans that involve you: the time of a future appointment, advice for managing your finances, or the orders given by your boss at work.

At times like these, it's smart to take notes instead of relying on your memory. Sometimes these notes may be simple and brief: a name and phone number jotted on a scrap of paper, or a list of things to pick up at the market. In other cases—a lecture, for example—your notes should usually be much longer. See the checklist on this page for note-taking strategies when the details are essential.

Analytical Listening

The goal of **analytical listening** is to fully understand a message. Your own experience will prove that full understanding is rare. Just think about the times when others fail to understand *you*. You also know from experience how hard it can be to understand complicated ideas—in your studies, at work, and about the world at large.

Perhaps the biggest challenge in good listening is to avoid judging an idea before you fully grasp it. Pay attention to your own thoughts and you're likely to catch yourself wanting to rebut an idea before you've tried to understand.

Nobel Prize-winning physicist Richard Feynman knew a thing or two about grasping complicated information. The four-part process he shared with the world can be useful whether you are trying to understand a political candidate's position, a chemistry lecture, a new policy at work, or anything else that is both complicated and important.[82] To use the **Feynman Technique**, follow these four steps:

- Listen carefully to information about the new concept and then describe it as best you can on a sheet of paper. Feel free to use words, images, arrows, or anything else. (It's okay if your understanding is not perfect at this stage.)

- Explain the concept as if you were talking to a child or a new student, using words and graphics as if you are a teacher.

- Consider which aspects of the concept seem clear to you and which are still a little foggy.

- Review the original information (using your notes, follow-up questions, a recording, or printed material) to better understand details you haven't mastered yet.

Repeat the process, if necessary, until you can explain the concept in simple language. This gradual approach should help you analyze even highly complex information.

Sometimes the challenge isn't the complexity of a message, but the emotional reaction you feel as a listener. When you encounter messages that stir up strong feelings, here are some tips from the experts to avoid tuning out or jumping to conclusions.

Listen for Information Before Evaluating   Although it's tempting to avoid, unfriend, and unfollow people whose beliefs are different than your own, "it's worth listening to people you disagree with," urges Zachary R. Wood, who wrote the book *Uncensored* about difficult conversations surrounding race, free speech, and dissenting viewpoints.[83] Wood has made it a practice to bring people with diverse beliefs together simply to listen to one another.

The principle of listening to information with an open mind seems almost too obvious to mention, yet all of us are guilty of judging a speaker's ideas before we completely understand them. The tendency to make premature judgments is

analytical listening A response style in which the primary goal is to understand a message.

Feynman Technique A process proposed by physicist Richard Feynman that involves depicting a complex concept as best one can on paper, describing the concept as if teaching it to a child, considering what aspects of the idea are still unclear, and then reviewing information further to achieve even deeper understanding of it.

especially strong when the idea you are hearing conflicts with your own beliefs. As one writer put it:

> The right to speak is meaningless if no one will listen. . . . It is simply not enough that we reject censorship . . . we have an affirmative responsibility to hear the argument before we disagree with it.[84]

You can avoid the tendency to judge before understanding by following the simple rule of paraphrasing a speaker's ideas before responding to them. The effort required to translate the other person's ideas into your own words will keep you from arguing, and if your interpretation is mistaken, you'll know immediately.

Separate the Message from the Speaker The first recorded cases of blaming the messenger for an unpleasant message occurred in ancient Greece. When messengers reported losses in battles, their generals sometimes responded to the bad news by having the messengers put to death. This sort of irrational reaction is still common (though fortunately less violent) today. Consider a few situations in which there is a tendency to get angry with a communicator bearing unpleasant news: An instructor tries to explain why you did poorly on a major paper; a friend explains what you did to make a fool of yourself at the party last Saturday night; the boss points out how you could do your job better. At times like these, becoming irritated with the bearer of unpleasant information may not only cause you to miss important information but can also harm your relationship with the other person.

There's a second way that confusing the message and the messenger can prevent you from understanding important ideas. At times you may mistakenly discount the value of a message because of the person who is presenting it. Even the most boring instructors, the most idiotic relatives, and the most demanding bosses occasionally make good points. If you write off everything a person says before you consider it, you may be cheating yourself out of valuable information.

Search for Value Even if you listen with an open mind, sooner or later you will end up hearing information that is either so unimportant or so badly delivered that you're tempted to tune out. Although making a quick escape from such tedious situations is sometimes the best thing to do, there are times when you can profit from paying close attention to apparently worthless communication. This is especially true when you're trapped in a situation in which the only alternatives to attentiveness are pseudolistening or downright rudeness.

Once you try, you probably can find some value in even the worst situations. Consider how you might listen opportunistically when you find yourself locked in a boring conversation with someone whose ideas are worthless. Rather than torture yourself until escape is possible, you could keep yourself amused—and perhaps learn something useful—by listening carefully until you can answer the following (unspoken) questions:

> "Is there anything useful in what this person is saying?"
> "What led the speaker to come up with ideas like these?"
> "What lessons can I learn from this person that will keep me from sounding the same way in other situations?"

Listening with a constructive attitude is important, but not all information (or all speakers) are trustworthy. We turn next to a listening approach that involves evaluating the merit of what you hear.

Critical Listening

The goal of **critical listening** is to go beyond trying to understand the topic at hand and, instead, to assess its quality. At their best, critical listeners apply the

critical listening Listening in which the goal is to evaluate the quality or accuracy of the speaker's remarks.

techniques described here to consider whether an idea holds up under careful scrutiny.

Examine the Speaker's Evidence and Reasoning   Trustworthy speakers usually offer support to back up their statements. A car dealer who argues that domestic cars are just as reliable as imports might cite frequency-of-repair statistics from *Consumer Reports* or refer you to satisfied customers. A professor arguing that students are more community-oriented than they used to be might tell stories about then and now.

Chapter 12 describes several types of supporting material that can be used to prove a point: definitions, descriptions, analogies, statistics, and so on. Also see the checklist on this page for questions to consider when evaluating a speaker's message.

Evaluate the Speaker's Credibility   The acceptability of an idea often depends on its source. If your longtime family friend, a self-made millionaire, encouraged you to invest in jojoba fruit futures, you might welcome the tip. If your deadbeat brother-in-law made the same offer, you would probably laugh off the suggestion.

Chapter 14 discusses credibility in detail, but two questions provide a quick guideline for deciding whether to accept a speaker as an authority:

- **Is the speaker competent?** Does the speaker have the experience or the expertise to qualify as an authority on this subject? Note that someone who is knowledgeable in one area may not be well qualified to comment in another area. For instance, your friend who can answer any question about computer programming might be a terrible advisor when the subject turns to romance.

- **Is the speaker impartial?** Knowledge alone isn't enough to certify a speaker's ideas as acceptable. People who have a personal stake in the outcome of a topic are more likely than others to be biased. The unqualified praise a commission-earning salesperson gives a product may be more suspect than the mixed review you get from a user. This doesn't mean you should disregard all comments you hear from an involved party—only that you should consider the possibility of intentional or unintentional bias.

Examine Emotional Appeals   Sometimes emotion alone may be enough reason to persuade you. You might lend your friend $20 just for old times' sake even though you don't expect to see the money again soon. In other cases, it's a mistake to let yourself be swayed by emotion when the logic of a point isn't sound. The excitement or fun in an ad or the lure of low monthly payments probably isn't good enough reason to buy a product you can't afford. Again, the fallacies described in Chapter 14 will help you recognize flaws in emotional appeals.

As you read about various approaches to listening, you may have noted that you habitually use some more than others. Researchers are still trying to determine how much we rely on different approaches—and how much we should.[85] There's no question that you can control the way you listen and use approaches that best suit the situation at hand. Take the *Understanding Your Communication* quiz for insights about your listening skills.

CHECKLIST ✓

Evaluating a Speaker's Message

Whatever form of support a speaker uses, you can ask several questions to determine the quality of the evidence and reasoning:[86]

☐ **Is the evidence recent enough?** In many cases, old evidence is worthless. Before you accept even the most credible evidence, be sure it isn't obsolete.

☐ **Is enough evidence presented?** One or two pieces of support may be exceptions and not conclusive evidence. Be careful not to generalize from limited evidence.

☐ **Is the evidence from a reliable source?** Even a large amount of recent evidence may be untrustworthy if the source is weak.

☐ **Can the evidence be interpreted in more than one way?** Evidence that supports one claim might also support others. Alternative explanations don't necessarily mean that the one being argued is wrong, but they do raise questions that should be answered before you accept an argument.

ASK YOURSELF

Audiences may be enthralled by the rhetoric in political campaigns. But careful listening is essential to sorting out truths and evaluating candidates' logic. How can you apply critical thinking to political discourse?

What Are Your Listening Strengths?

Answer the following questions to gauge which listening approaches best describe you.

1. Which of the following best describes you?

   a. I'm a quick learner who can hear instructions and put them into action.

   b. I have an intuitive sense, not just of what people say, but how they are feeling.

   c. I'm a good judge of character. I can usually tell whether people are trustworthy or not.

   d. I'm a rapid thinker who is often able to jump in and finish people's sentences for them.

2. Imagine you are tutoring an elementary school student in math. What are you most likely to do?

   a. Focus on clearly articulating the steps involved in solving simple equations

   b. Begin each tutoring session by asking about the student's day

   c. Pay close attention to what the student says to see if they really understand

   d. Feel frustrated if it seems the student isn't listening or isn't motivated

3. A friend launches into a lengthy description of a problem with a coworker. What are you most likely to do?

   a. Offer some ideas for discussing the issue with the coworker

   b. Show that you are listening by maintaining eye contact, leaning forward, and asking questions

   c. Read between the lines to better understand what is contributing to the problem

   d. Pretend to listen but tune out after 5 minutes or so

4. If you had your way, which of the following rules would apply to team meetings?

   a. Chit-chat would be limited to 5 minutes, so we can get to the point at hand

   b. Everyone would get a turn to speak

   c. People would back up their opinions with clear data and examples

   d. There would be no meetings; they're usually a waste of time

INTERPRETING YOUR RESPONSES

Read the explanations below to see which listening approaches you frequently take. (More than one may apply.)

Task Oriented

If you answered "a" to more than one question, you tend to be an action-oriented listener. You value getting the job done and can become frustrated with inefficiency. Your task orientation (page 144) can help teams stay on track. Just be careful that you don't overlook the importance of building strong relationships, which are essential for getting the job done. Tips for group work in Chapter 12 may be especially interesting to you.

Relational/Supportive

If you answered "b" to more than one question, you tend to be a relational and/or supportive listener (pages 137–143). It's likely that people feel comfortable sharing their problems and secrets with you. Your strong listening skills make you a trusted friend and colleague. At work, however, this can make it difficult to get things done. Make an effort to set boundaries so people don't talk your ear off.

Analytical/Critical

If you answered "c" to more than one question, you often engage in analytical and/or critical listening (pages 147–149). You tend to be a skeptical listener who isn't easily taken in by phony people or unsubstantiated ideas. Your ability to synthesize information and judge its merits is a strength. At the same time, guard against the temptation to reach snap judgments. Take time to consider people and ideas thoughtfully before you write them off. The tips for mindful listening (pages 132–133) can help.

Impatient

If you answered "d" to more than one question, you have a tendency to be an impatient or distracted listener. Your frustration probably shows more than you think. Review the tips throughout this chapter for ways to become more focused and active in your listening approach.

MAKING THE GRADE

For more resources to help you understand and apply the information in this chapter, visit the *Understanding Human Communication* website at www.oup.com/he/adler-uhc14e.

OBJECTIVE 6.1 Summarize the benefits of being an effective listener.

- Listening—the process of giving meaning to an oral message—is a vitally important part of the communication process.

- There are many advantages to being a good listener. People with good listening skills are more likely than others to be hired, promoted, and respected as leaders. In addition, listening improves relationships.

- Identify people from your own experience whose listening ability illustrates the advantages outlined in this chapter.

  > Listen, *really* listen, to someone important in your life. Note how that person responds and how you feel about the experience.

  > How could better listening benefit your life?

OBJECTIVE 6.2 Outline the most common misconceptions about listening, and assess how successfully you avoid them.

- Listening and hearing are not the same thing. Hearing is only the first step in the process of listening. Beyond that, listening involves attending, understanding, responding, and remembering.

- Listening is not a natural process. It takes both time and effort. Mindful listening requires a commitment to understand others' perspectives without being judgmental or defensive.

- It's a mistake to assume that all receivers hear and understand messages identically. Recognizing the potential for multiple interpretations and misunderstanding can prevent problems.

  > Give examples of how common misconceptions about listening create problems.

  > How can you use the information in this section to improve your listening skills?

OBJECTIVE 6.3 Recognize and develop strategies to overcome barriers to listening.

- A variety of factors contribute to ineffective listening, including message overload, rapid thought, physical noise, psychological noise, physiological noise, and cultural differences.

  > Which factors described on pages 133–134 interfere with your ability to listen well?

  > Develop an action plan to minimize the impact of these listening challenges.

OBJECTIVE 6.4 Identify and minimize faulty listening habits.

- Faulty listening habits include pseudolistening, selective listening, defensiveness, avoiding difficult issues, insensitivity, conversational narcissism, and the assumption that talking is more impressive than listening.

  > From recent experience, identify examples of each type of ineffective listening described in this section.

  > Focus on the bad listening habit you are most guilty of and brainstorm ways to overcome it.

OBJECTIVE 6.5 Describe and practice listening strategies for effectively connecting with and supporting others.

- Relational and supportive listening involve the willingness to spend time with people to better understand their feelings and perspectives. The goal is to help the speaker, not the receiver.

- These approaches are most effective when the listener considers the people and situation involved, is patient, sensitive to underlying meanings, asks questions and encourages the speaker to continue, reflects back the speaker's thoughts, and is careful to offer analysis, judgment, advice, and comfort when they are likely to help the listener.

- Listeners can be most helpful when they use a variety of styles, focus on the emotional dimensions of a message, and avoid being judgmental.

  > What listening behaviors do you find most helpful when you want to connect with another person or share a problem with them? Which do you find least helpful?

  > Think of someone close to you. What listening behaviors do you think that person most appreciates? Which do they find least helpful? How might you increase your use of listening behaviors they find most helpful?

OBJECTIVE 6.6 Describe how gender may influence listening behavior.

- When women engage in supportive listening with each other, they often share experiences that are similar to those of the speakers. They may find this comforting, but men may interpret these responses as one-ups implying that the listener's problems are even worse than theirs.

- Men are likely to offer listening responses that seek to solve the speaker's problem or distract them from their worries. Men may appreciate this, but women may prefer that they express empathy instead.

  > Do your listening habits reflect traditional gender norms or not? Explain how.

  > Think of an example from your own experience or a movie or TV show in which a misunderstanding arose from a listener's attempt to offer someone support.

OBJECTIVE 6.7 Know when and how to listen to accomplish a task, analyze a message, and critically evaluate another's remarks.

- Task-oriented listening helps people accomplish mutual goals. It involves an active approach in which people often identify key ideas, ask questions, paraphrase, and take notes.

- Analytic listening involves the willingness to suspend judgment and consider a variety of perspectives to achieve a clear understanding. This type of listening requires people to distinguish between what is true and what is not.

- Critical listening is appropriate when the goal is to judge the quality of an idea. A critical analysis is most successful when the listener ensures correct understanding of a message before passing judgment, when the speaker's credibility is taken into account, when the quality of supporting evidence is examined, and when the logic of the speaker's arguments is examined carefully.

  > Identify situations in which each type of listening described in this section is most appropriate.

  > On your own or with feedback from others who know you well, assess your ability to listen for each of the following goals: to accomplish a task, analyze ideas, and critically evaluate messages. Which types of listening are your strongest and weakest?

KEY TERMS

advising response p. 142

analytical listening p. 147

attending p. 130

comforting p. 142

conversational narcissists p. 136

counterfeit question p. 145

critical listening p. 148

defensive listening p. 135

Feynman Technique p. 147

hearing p. 130

insensitive listening p. 136

insulated listening p. 136

judging response p. 142

listening p. 130

listening fidelity p. 130

mindful listening p. 132

paraphrasing p. 145

passive narcissist p. 136

prompting p. 139

pseudolistening p. 135

reflecting p. 140

relational listening p. 137

remembering p. 131

residual message p. 132

responding p. 130

selective listening p. 135

stage hogs p. 136

supportive listening p. 137

task-oriented listening p. 144

understanding p. 130

ACTIVITIES

1. To consider how misconceptions about listening affect your life, identify important situations in which you have made each of the following mistaken assumptions. In each case, describe the consequences:

   - Thinking that because you were hearing a message you were listening to it.
   - Believing that listening effectively is natural and effortless.
   - Assuming that other listeners were understanding a message in the same way as you.

2. Imagine that a friend says, "I'm so overwhelmed by work and school demands that I just don't know what to do."

   - Experiment with listening responses by writing a separate response for each of the following that illustrates what it might sound like.

     > Advising

     > Judging

     > Analyzing

     > Questioning

     > Comforting

     > Prompting

     > Reflecting

- Do the same for each of the following.

  > At a party, a guest you have just met says, "Everybody seems like they've been friends for years. I don't know anybody here. How about you?"

  > Your best friend has been quiet lately. When you ask if anything is wrong, they snap, "No!" in an irritated tone of voice.

  > A fellow worker says, "The boss keeps making sexual remarks to me. I think it's a come-on, and I don't know what to do."

  > It's registration time at college. One of your friends asks if you think they should enroll in the communication class you've taken.

  > Someone with whom you live remarks, "It seems like this place is always a mess. We get it cleaned up, and then an hour later it's trashed."

- Discuss the pros and cons of using each response style considering the situation and who is involved.

Nonverbal Communication

7

CHAPTER OUTLINE

7.1 Characteristics of Nonverbal Communication 157
Nonverbal Communication Is Unavoidable
Nonverbal Communication Is Ambiguous
Nonverbal Cues Convey Emotion
Nonverbal Cues Influence Identities and Relationships

7.2 Functions of Nonverbal Communication 161
Repeating
Substituting
Complementing
Accenting
Regulating
Contradicting
Deceiving

7.3 Types of Nonverbal Communication 166
Body Movements
Voice
Appearance
Touch
Space

7.4 Influences on Nonverbal Communication 174
Culture
Gender

MAKING THE GRADE **177**
KEY TERMS **178**
ACTIVITIES **178**

LEARNING OBJECTIVES

7.1
Explain the characteristics of nonverbal communication and the social goals it serves.

7.2
Describe key functions served by nonverbal communication.

7.3
List the types of nonverbal communication, and explain how each operates in everyday interaction.

7.4
Explain the ways in which nonverbal communication reflects culture and gender differences.

Amy Cuddy's

research and TED talk on body language highlight the power of nonverbal communication. As you read this chapter, consider these questions:

What do your nonverbal cues suggest to other people?

How might you alter your nonverbal behavior to feel and appear more confident and powerful?

What nonverbal cues displayed by others catch your attention most? What do you tend to overlook?

WONDER WOMAN AND SUPERMAN might be on to something. Their feet-apart, hands-on-hips stance is a powerful example of nonverbal communication. Social psychologist Amy Cuddy calls it a power pose. According to her research, it may change the way people think about you and may help improve your performance at school and work.[1]

Cuddy became interested in body language as a business professor at Harvard University. She noticed that some students engage in highly assertive nonverbal behaviors. "They get right into the middle of the room before class even starts, like they really want to occupy space. When they sit down, they're sort of spread out," she says.[2] Those are the students, she soon realized, who are most likely to get noticed. They raise their hands high and take part in discussions. By contrast, other students "are virtually collapsing as they come in. They sit in their chairs and they make themselves tiny, and they go like this," Cuddy says, raising her hand no higher than her head, with her arm close to her face.

Cuddy has also noticed that women in her classes are more likely than men to demonstrate low-power behaviors. Slightly built herself, she began to apply the idea to her own behavior and found that people took her more seriously when she demonstrated a confident presence.

Cuddy's research suggests that nonverbal communication can affect not just how others perceive you, but how you feel about *yourself*. Try this quick exercise: The next time you are entering a stressful situation, adopt a power pose in private for two minutes beforehand. It may make you feel more confident, Cuddy predicts. Of course, the effect is not guaranteed.[3] Adopting a powerful pose may be akin to a placebo: It works if you *think* it will.

Nonverbal communication is important in many ways. The ability to effectively encode and decode nonverbal messages is a strong predictor of well-being.[4] In general, people with strong nonverbal communication skills are more persuasive than those who are less skilled, and they have a greater chance of success in settings ranging from careers to poker to romance. Nonverbal sensitivity is a major part of emotional intelligence (Chapter 3), and researchers have come to recognize that it is impossible to study spoken language without paying attention to its nonverbal dimensions.[5]

Although people commonly assume that nonverbal communication is silent, that's not always the case. By definition, **nonverbal communication** is the process

nonverbal communication
Messages expressed without words, as through body movements, facial expressions, eye contact, tone of voice, and so on.

TABLE 7.1		
Types of Communication		
	VOCAL COMMUNICATION	**NONVOCAL COMMUNICATION**
VERBAL COMMUNICATION	Spoken words	Written words and gestures with specific verbal meanings
NONVERBAL COMMUNICATION	Tone of voice, sighs, screams, vocal qualities (loudness, pitch, and so on)	Gestures, movement, appearance, facial expression, and so on

Source: Adapted from Stewart, J., & D'Angelo, G. (1980). *Together: Communicating interpersonally* (2nd ed.). Reading, MA: Addison-Wesley, p. 22. Copyright © 1993 by McGraw-Hill. Reprinted/adapted by permission.

of conveying messages without using words. Nonverbal cues include facial expressions, touch, use of space, clothing, gestures (if they don't convey specific words), and other cues you can't hear. But nonverbal communication also includes audible cues that aren't linguistic—such as humming, sobs, how loudly or quickly someone speaks, and so on. To test your understanding of what qualifies, answer the following questions:

- Is American Sign Language mostly verbal or nonverbal?
- What about an email?
- How about laughter?

If you answered verbal, verbal, and nonverbal, respectively, you are correct. American Sign Language doesn't require sound, but it is word-based, thus verbal.[6] Emails are usually verbal for the same reason. (If you imagined an email filled with nothing but emojis, that would indeed be nonverbal.) Laughter involves vocal chords but doesn't rely on words, so it's nonverbal. These distinctions only begin to convey the richness of nonverbal messages. Table 7.1 illustrates these differences.

This chapter will help you become more mindful about the nonverbal cues you display to others and more sensitive to the messages others convey without words. We will explore:

- defining characteristics of nonverbal communication,
- the functions served by nonverbal cues,
- types of nonverbal communication, and
- how culture and social ideas about gender affect nonverbal communication.

Reading about these topics won't transform you into a mind reader, but it should make you a far more accurate observer of others—and yourself.

7.1 Characteristics of Nonverbal Communication

You have probably noticed that there is often a gap between what people say and how they actually feel. An acquaintance says, *"I'd like to get together again"* in a way that leaves you suspecting the opposite. A speaker tries to appear confident but acts in a way that almost screams out, *"I'm nervous!"* You ask a friend what's wrong, and the *"Nothing"* you get in response rings hollow. Then there are other times when a message comes through even though there are no words at all. A look of irritation, a smile, or a sigh can say it all. Situations like these have one element in common: Messages were sent nonverbally. Although you have certainly recognized nonverbal messages before, this chapter should introduce you to a richness of information you have never noticed as well as qualities that reveal the nature and richness of nonverbal communication.

What about languages that don't involve spoken words? For example, is American Sign Language considered verbal or nonverbal communication? Most scholars would say sign language is verbal because it largely uses gestures to express particular words.[7] This means that sign language and written words are verbal, but messages transmitted by vocal means that don't involve language—sighs, laughs, and other utterances—are nonverbal. We will discuss this in more detail later in the chapter.

Nonverbal Communication Is Unavoidable

Even if you try *not* to send nonverbal cues—perhaps by closing your eyes, leaving the room, or being silent—those behaviors transmit messages to others.[8] Think about the last time you blushed, stammered, or cried even when you didn't want to. Chances are that people around you attached meaning to your behavior. For example, a shaky voice in a job interview might communicate that you're nervous.

Why is sending nonverbal cues sometimes beyond our control? Scientists think it's because of the way our brains work. One part of the brain (the cerebrum) governs speech, but a different part (the limbic system) processes emotional reactions and awareness of many elements of the environment.[9] The limbic system allows you to respond to situations automatically. You jump when something pops out of the bushes, you get a burst of adrenaline in a stressful situation, and you cry or laugh when something moves you.

Of course, you can control some aspects of your nonverbal communication, such as eye contact and posture. But even when you send nonverbal cues deliberately, it's hard to be aware of everything you are doing. In one study, participants were asked to show nonverbally that they either liked or disliked a partner.[10] Immediately afterward, less than a quarter of the participants could describe all the nonverbal behaviors they used to send those messages. That's probably because we send messages in many ways simultaneously (e.g. through facial and bodily movements, use of space, and touch). Although 100% awareness isn't possible, the checklist in this section provides some strategies for communicating more mindfully.

Just because communicators are nonverbally expressive doesn't mean that others pick up on their cues. One study comparing the richness of email to in-person communication confirmed that far more nonverbal information is available in face-to-face conversations, but it also showed that some communicators (especially men in the study) failed to recognize many of the cues offered.[11]

The fact that you and everyone around you are constantly sending nonverbal clues is important because it means that you have a constant source of information available about yourself and others. If you tune in to these signals, you will be more aware of how those around you are feeling and thinking, and you will be better able to respond to them. As you will see next, however, it's a mistake to assume that nonverbal behaviors have clear or obvious meanings.

Nonverbal Communication Is Ambiguous

You may run across a book, podcast, or video that promises to reveal the hidden meaning of nonverbal behavior. Arms crossed? *She's angry.* Eyes looking up? *He's skeptical.* As you probably realize by now, simplistic interpretations such as these

Signing is a language because it uses gestures to express particular words. But, like proficient speakers, good signers use facial expressions to add depth to a message.

How expressive are you as a communicator? How do you think others would respond if you were more expressive? Less expressive?

are bogus. The reality is that nonverbal cues are difficult to interpret accurately because cues can have multiple meanings. Consider an example: You text someone you met recently, saying "Let's get together soon." The reply: A single Thinking Face emoji. What's that supposed to mean? Could it be *"Let me figure out when I'm free?" "Thinking about whether I want to see you?" "Do I know you?"* In situations like this, there's no way to be certain.[12] Your best guess about the meaning of this nonverbal message, like all others, will depend on several factors:

- The *context* in which the nonverbal behavior occurs. (Have you been in touch since the first meeting, or is this your first contact since then?)

- The *history and tone* of your relationship with the sender. (Was your first conversation meeting at church? In a dive bar? Did the other person express a desire to see you again then?)

- The *sender's mood* at the time. (Difficult to determine with no face-to-face information; easier if you are asking in person.)

- Your *own feelings*. (If you're feeling insecure, almost anything can seem like a threat.)

No matter the meaning you think you detect, you should consider nonverbal behaviors, not as facts, but as clues to be checked out.

Nonverbal communication can be ambiguous even in seemingly innocuous situations. Years ago, employees of the Safeway supermarket chain discovered this fact firsthand when they tried to follow the company's new "superior customer service" policy that required them to smile and make eye contact with customers. Twelve employees filed grievances over the policy, reporting that several customers had propositioned them, misinterpreting their actions as come-ons.[13]

Some people are more skillful than others at accurately sending and decoding nonverbal behavior.[14] The ability to interpret nonverbal cues tends to increase with age and training, although there are still differences in ability owing to personality and occupation. For the most part, extroverts are more accurate judges of nonverbal behavior than are introverts.[15] And, in general, women are better than men at decoding nonverbal messages.[16] (We'll talk about the reasons for that later in the chapter.) The challenge is especially difficult when neurology comes into play. People born with a syndrome called nonverbal learning disorder (NVLD) have trouble making sense of nonverbal cues because of a processing deficit in the right hemisphere of the brain.[17] They often take humorous or sarcastic messages literally because those cues are based heavily on nonverbal signals. People with NVLD also have trouble figuring out how to behave appropriately in new social situations, so they rely on rote formulas that often don't work. For example, a child who has learned the formal way of meeting an adult for the first time by shaking hands and saying *"Pleased to meet you"* might try this approach with a group of peers. The result, of course, is typically regarded as odd or nerdy. And their disability may lead them to miss nonverbal cues sent by the other children that this isn't the right approach.[18] It's not easy. Even for people who don't live with NVLD, the nuances of nonverbal behavior can be confusing.

CHECKLIST ✅

Three Ways to Convey Nonverbal Cues More Mindfully

Nonverbal communication often escapes conscious awareness. For example, you might come across as mad or tired without meaning to. Here are three ways to become more mindful of the nonverbal cues you send to others.

☐ **Aim for the right amount of eye contact.** Sometimes, locking eyes can come off as aggressive. On the other hand, lack of eye contact can make you seem unconfident or uninterested. In most situations, people find moderate eye contact to be pleasant and appealing (e.g., glancing away briefly every 7 to 10 seconds).

☐ **Monitor your tone of voice.** Even a simple remark such as, "That's not what I expected" can be interpreted in many ways based on the tone.[19] Thinking about potential interpretations can help you choose a tone that supports the message you hope to convey.

☐ **Observe yourself.** Ask a friend to record you giving a speech or interacting with others, and then study your nonverbal behavior in the video. Most of us have blind spots when it comes to our own communication.[20] What messages are suggested by your posture, gestures, and facial expressions?

expectancy violation theory The proposition that nonverbal cues cause physical and/or emotional arousal, especially if they deviate from what is considered normal. People who consider a particular nonverbal violation to be positive may accommodate (respond similarly), but they are likely to compensate (respond with opposite or distancing behavior) if they find it unpleasant.

affect displays Facial expressions, body movements, and vocal traits that reveal emotional states.

Expectancy violation theory helps explain how we manage ambiguity using nonverbal communication.[21,22] The theory is composed of three main propositions:

- People experience a degree of physical and/or psychological arousal when processing nonverbal cues, especially if those cues violate what they normally expect from others. For example, you might have a strong reaction if a stranger stares at you, touches you, or gets very close to you.

- A violation may be perceived either positively or negatively. If the person is attractive, their prolonged gaze may create positive feelings. But in other situations, you may feel afraid or uncomfortable if someone stares at you.

- People either *accommodate* or *compensate* based on how they feel about the perceived violation. If someone you like invades your personal space, you might not mind, and you might even lean closer toward them. That's accommodating. On the other hand, if the violation is unwelcome, you might compensate by moving away, avoiding eye contact, or ignoring the person.

Expectancy violation theory illustrates how nonverbal communication can create a sense of closeness or distance between people. It presents several implications. One is that nonverbal communication is highly potent. It can stimulate strong reactions in others, even when that isn't the sender's intention. Another implication is that a violation can be perceived in many ways. One person may consider a gentle touch on the arm to be sweet and thoughtful, whereas someone else may interpret the same gesture as aggressive or overly familiar. By the same token, a friend may not mind if you put your phone on the table during dinner (a nonverbal cue), but a date may consider it insulting, as if they aren't important enough to have your undivided attention.[23] The best advice is to think about the implications of your nonverbal cues, proceed slowly, and pay close attention to people's reactions.

Nonverbal Cues Convey Emotion

A friend who knows you well might recognize that you are shocked, happy, stressed, or sad, even when you're trying to hide those feelings or when you haven't fully acknowledged them within yourself.

One reason nonverbal cues are so powerful is that they convey some meanings better than words can. You can prove this for yourself by imagining how you could express each item on the following list nonverbally:

- You're bored.
- You are opposed to capital punishment.
- You are attracted to another person in the group.
- You want to know if you will be tested on this material.
- You are nervous about trying this experiment.

The first, third, and fifth items in this list all involve feelings. You could probably imagine how each could be expressed nonverbally through what social scientists call **affect displays**—facial expressions, body movements, and vocal traits that convey emotion. (*Affect* is another word for emotion.) By contrast, the second and fourth items involve ideas more than emotions, and they would be difficult to convey without using words. The same principle holds in everyday life: Nonverbal behavior offers many cues about the way people feel—often more than we get from their words alone.

Nonverbal Cues Influence Identities and Relationships

In Chapter 3 we explored the notion that people strive to create images of themselves as they want others to view them. A great deal of this occurs nonverbally. Consider what happens when you attend a party where you are likely to meet strangers you would like to get to know better. Instead of projecting your desired image verbally (*"Hi! I'm attractive, friendly, and easygoing"*), you behave in ways that support identity. You might smile a lot and adopt a relaxed posture. It's also likely that you will dress carefully—even if you are trying to create the illusion that you haven't given a lot of attention to your appearance.

Consider the wide range of ways you could behave when greeting another person. Depending on the nature of your relationship (or what you want it to be), you could wave, shake hands, nod, smile, pat the other person on the back, give a hug, or avoid all contact. Even trying to *not* communicate can send a message, as when you avoid talking to someone.

What emotions do you imagine this couple is feeling? Grief? Anguish? In fact, they just learned that they won $1 million in the New Jersey state lottery.

Have you ever made mistaken assumptions about others' nonverbal behaviors? Have others misinterpreted yours? When should you be more cautious about jumping to conclusions?

🧭 7.2 Functions of Nonverbal Communication

Although verbal and nonverbal messages differ in many ways, the two forms of communication operate together on most occasions. The following discussion explains the many functions of nonverbal communication and shows how nonverbal messages relate to verbal ones.

Repeating

If someone asks you for directions to the nearest drugstore, you could say *"north of here about two blocks"* and then point north with your index finger. This sort of repetition isn't just decorative. People remember comments accompanied by gestures more than those made with words alone.[24]

Substituting

When a friend asks you what's new, you might shrug your shoulders instead of answering in words. In this way, your nonverbal behavior is a substitute for a verbal response. In other situations, you might raise your eyebrows after a colleague makes an off-color comment or hug someone far longer than usual to show that you missed them. Sometimes a nonverbal cue says it all.

Some gestures substitute for specific words. **Emblems** are deliberate nonverbal behaviors that have precise meanings known to everyone within a cultural group. For example, most Americans consider that a head nod means *"yes,"* a head shake means *"no,"* a wave means *"hello"* or *"good-bye,"* and a hand to the ear means *"I can't hear you."* Keep in mind that the meaning of many emblems varies by culture. (See the *Understanding Your Communication* feature in this section.)

Complementing

Sometimes nonverbal behaviors reinforce the content of a verbal message. Consider, for example, a friend who apologizes for forgetting an appointment with you. You will be most likely to believe your friend if they use a sincere-sounding tone of voice and show an apologetic facial expression. We often recognize the significance of complementary nonverbal behavior when it is missing. If your friend's apology is delivered with a shrug, a smirk, and a light tone of voice, you will probably doubt its sincerity, no matter how profuse the verbal explanation is.

ASK YOURSELF ❓

What might other people assume about you based on your nonverbal cues right this moment? Would their assumptions be mostly correct or incorrect? In what way?

emblems Deliberate gestures with precise meanings, known to virtually all members of a cultural group.

UNDERSTANDING YOUR COMMUNICATION

How Worldly Are Your Nonverbal Communication Skills?

 Answer the following questions to test your knowledge about nonverbal communication in different cultures.

1. **In the United States, touching your index finger to your thumb while your other fingers point upward means "OK." But what does it mean elsewhere? (Two of the following are correct.)**

   a. It signifies money in Japan.

   b. People in Greece and Turkey interpret it to mean 30.

   c. In France, it means "you're worth zero."

   d. It's a compliment in Russia, implying that "you and I are close friends."

2. **People around the world recognize the sign for "peace" or "victory"—two fingers up, thumb holding down the other fingers, palm facing out. But in many places, the same gesture means something different if you show the back of your hand instead. What does that gesture mean to people in England, New Zealand, and Australia?**

   a. "May I have seconds?"

   b. "I'll be right back."

   c. "Goodbye."

   d. "Up yours!"

3. **In the United States, people convey "come closer" by alternately extending and curling their index finger in someone's direction. Where might the same gesture be considered a serious insult?**

   a. Egypt

   b. The Philippines

   c. Spain

   d. Saudi Arabia

4. **The "thumbs up" sign that means "yes" or "job well done" in the United States means something else in other cultures. Two of the following are true. Which ones?**

   a. It means "it's my turn to talk" in Tahiti and neighboring South Pacific islands.

   b. It means "up yours" in Australia, Greece, and the Middle East.

   c. It means the number 5 in Japan.

   d. In Myanmar, it's a symbol of mourning, meaning "someone has died."

INTERPRETING YOUR RESPONSES
Read the explanations below to see how many answers you got right.

Question 1
The gesture that means "OK" in the United States means "money" in Japan. But it has a darker meaning in France, where it conveys "you're worth zero." It's risky to use this gesture in others parts of the world as well. In Brazil, Germany, and Russia, it depicts a private bodily orifice; and in Turkey and Greece it's taken as a vulgar sexual invitation. The correct answers are "a" and "c."

Question 2
The palm-forward V sign is popular around the world, especially in Japan, where it's customary to flash a peace sign while being photographed.[25] But a slight variation makes a big difference. Winston Churchill occasionally shocked audiences during World War II by "flipping them off" (knuckles forward) when he really meant to flash a victory sign (palm forward).[26] Years later, U.S. president Richard Nixon made the same mistake in Australia, essentially conveying "f--- you" to an Australian crowd when he got the gesture wrong.[27] The correct answer is "d."

Question 3
The gesture Americans use to mean "come closer" is offensive in many places, including The Philippines, Slovakia, China, and Malaysia. People in those cultures summon a dog that way, so it's a putdown to use it with a person. Answer "b" is correct.

Question 4
"Thumbs up" has a positive connotation in the United States. Meanwhile, people in Germany and Hungary interpret the gesture to mean the number 1, and people in Japan use it for the number 5. However, the gesture is taken as an insult (akin to "Up yours!") in the Middle East. Both "b" and "c" are correct.

Much complementing behavior consists of **illustrators**—nonverbal behaviors that accompany and support spoken words. Scratching your head when searching for an idea and snapping your fingers when it occurs are examples of illustrators that complement verbal messages.

> **illustrators** Nonverbal behaviors that accompany and support verbal messages.

Accenting

Just as we use *italics* to emphasize an idea in print, we use nonverbal devices to emphasize oral messages. Pointing an accusing finger adds emphasis to criticism (and probably creates defensiveness in the receiver). Stressing certain words with the voice ("It was *your* idea!") is another way to add nonverbal accents. See *Understanding Communication Technology* in this section for more about accenting in online messages.

Regulating

Nonverbal behaviors can control the flow of verbal communication. For example, conversational partners send and receive turn-taking cues.[28] The speaker may use nonverbal fillers, such as *"um"* or an audible intake of breath, to signal that they would like to keep talking.[29] Or they may hold up a finger to suggest that the listener wait to speak. Conversely, long pauses are often taken as opportunities for others to speak. Nonverbal regulators also include signs that you would like to wrap up the conversation, such as an extended *okaaaay* or a glance at your phone.

When you are ready to yield the floor, the unstated rule is as follows: Create a rising vocal intonation pattern, then use a falling intonation pattern, or draw out the final syllable of the clause at the end of your statement. Finally, stop speaking. If you want to maintain your turn when another speaker seems ready to cut you off, you can suppress the attempt by taking an audible breath, using a sustained intonation pattern (because rising and falling patterns suggest the end of a statement), and avoiding any pauses in your speech. Other nonverbal cues exist for gaining the floor and for signaling that you do not want to speak.

Contradicting

It's not unusual for people to say one thing but display nonverbal cues that suggest the opposite. A classic example is when someone with a red face and bulging veins yells, *"I'm not angry!"* As you have no doubt experienced, when verbal and nonverbal messages are at odds, people are more likely to believe the nonverbal cues than the words.[30]

Many contradictions between verbal and nonverbal messages are unintentional. You may say you understand something, not realizing that your puzzled expression suggests otherwise. Not all mixed messages are unintentional. For example, if you become bored with a conversation while your companion keeps rambling on, you probably wouldn't bluntly say, *"I want you to stop talking."* Instead, you might nod politely and murmur *"uh-huh"* and *"no kidding?"* at the appropriate times but subtly signal your desire to leave by looking around the room, turning slightly away from the speaker, or picking up your bag or backpack. These cues may be enough to end the conversation without the awkwardness of expressing what you are feeling in words.

Deceiving

After reading this chapter, you might assume that you'll be able to recognize deception by focusing on nonverbal cues. But it's not that easy. Decades of research have revealed that there are no sure-fire nonverbal cues that indicate deception.[31]

UNDERSTANDING COMMUNICATION TECHNOLOGY

Nonverbal Expressiveness Online

Communication scholars characterize face-to-face interaction as rich in nonverbal cues that convey feelings and attitudes. Even telephone conversations carry a fair amount of emotional information via the speakers' vocal qualities. By comparison, most text-based communication online is relatively lean in relational information. With only words, subtlety can be lost. This explains why hints and jokes that might work well in person or on the phone often fail when communicated online.

Ever since the early days of email, online correspondents have devised emoticons, using keystrokes to create sad expressions such as :-(, surprised looks such as :-0, and more. Even now, when graphic emojis are readily available, it's difficult to keep up with society's craving for more nonverbal cues in cyberspace. Recognizing the need for more than a thumbs-up or thumbs-down image, Facebook added a series of new graphics to expand on the familiar but overused "Like" button.

Asterisks Enclosing a statement in asterisks can add emphasis to particular words. Notice how this nonverbal cue affects the tone of the following messages:

I really want to hear from you.

I *really* want to hear from you.

I really want to hear from *you*.

Capitalization Capitalizing a word or phrase can also emphasize a point:

I hate to be a pest, but I need the $20 you owe me TODAY.

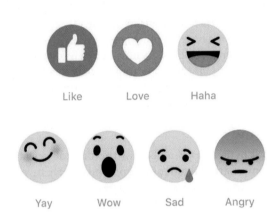

Like   Love   Haha

Yay   Wow   Sad   Angry

Overuse of capitals can be offensive, however. Be sure to avoid typing messages in all uppercase letters, which creates the impression of shouting:

HOW ARE YOU DOING? WE ARE HAVING A GREAT TIME HERE. BE SURE TO COME SEE US SOON.

Multiple Methods of Emphasis Just as in-person interactions include many forms of nonverbal expression, when you want to emphasize a point via technology you can use multiple methods:

I can't believe you told the boss that I sleep with a teddy bear! I wanted to *die* of embarrassment. Please don't *EVER* **EVER** do that kind of thing again.

We seem to be worse at catching deceivers when we participate actively in conversations than when we observe from the sidelines. This is probably because of the mental energy it takes to manage our own participation in the encounters.[32] In real life, most people have only a 50% chance of accurately identifying a liar.[33] As one writer put it, "There is no unique telltale signal for a fib. Pinocchio's nose just doesn't exist, and that makes liars difficult to spot."[34]

You might imagine that it's easier to tell when young children are lying. That might be true for preschoolers, but older children develop the ability to lie convincingly—just as they develop other communication skills. In a series of experiments, an adult told elementary-school-aged children (one at a time) that they could win a prize by guessing the numbers on cards lying face down on a table. The adult then left each child alone with the cards for a few minutes. As you might predict, some youngsters couldn't resist taking a peek. When the experimenter returned and asked the children if they had looked at the cards, some answered truthfully and others lied. When a group of randomly selected adults viewed the children's statements, they couldn't tell which of them were telling the truth. The participants' guesses were no better than chance. Even child care professionals and

Source: DILBERT © 2006 Scott Adams. Used by permission of UNIVERSAL UCLICK. All rights reserved.

parents who interact with children every day did no better at identifying deception than if they had simply flipped a coin.[35]

Even with people you know well, detecting honesty and deception isn't as accurate as you might imagine. For the most part, close friends and romantic partners are little better than strangers at detecting when their partners are lying.[36,37] This may be partly because we *want* to believe those we know well, and we are reluctant to believe others. If you have a **truth bias** regarding someone you know, you probably assume they are being honest unless you have a compelling reason to suspect otherwise.[38] Conversely, you may harbor a **deception bias**, assuming that people you don't favor are likely to be untruthful. A deception bias might arise from awareness of previous lies or bad behavior, feelings of insecurity, past experiences with other liars, or a history of telling lies yourself.[39]

By now, you probably realize that "common-sense" notions about lying are faulty. Liars might avoid eye contact, stutter, stammer, sweat, and so on. But some of them don't. And many people who display these nonverbal cues are telling the truth. With this awareness in mind, it *is* slightly easier to catch someone in a lie when the deceiver hasn't had a chance to rehearse, feels strongly about the information being hidden, and/or feels anxious or guilty about lying.[40] Table 7.2 lists situations in which deceptive messages are most noticeable.

ASK YOURSELF

How good are you at lying? Invite a friend to guess whether you are lying when you describe something you have (or pretend to have) in your bag or backpack. Was your friend's conclusion correct?

truth bias The tendency to assume that others are being honest.

deception bias The tendency to assume that others are lying.

TABLE 7.2

Nonverbal Clues to Deception

DECEPTION CLUES ARE MOST LIKELY WHEN THE DECEIVER...	DECEPTION CLUES ARE LEAST LIKELY WHEN THE DECEIVER...
Wants to hide emotions being experienced at the moment.	Wants to hide information unrelated to their emotions.
Feels strongly about the information being hidden.	Has no strong feelings about the information being hidden.
Feels apprehensive about the deception.	Feels confident about the deception.
Feels guilty about being deceptive.	Experiences little guilt about the deception.
Gets little enjoyment from being deceptive.	Enjoys the deception.
Has to construct the message in real time without mentally rehearsing it first.	Knows the deceptive message well and has thought it through in advance.

Source: Based on material from Ekman, P. (1981). Mistakes when deceiving. In T. A. Sebok & R. Rosenthal (Eds.), *The Clever Hans phenomenon: Communication with horses, whales, apes and people* (pp. 269–278). New York: New York Academy of Sciences. See also Samhita, L., & Gross, H. J. (2013). The "Clever Hans phenomenon" revisited. *Communicative and Integrative Biology*, 6(6), e27122. doi:10.4161/cib.27122

7.3 Types of Nonverbal Communication

Now that you understand how nonverbal messages operate as a form of communication, we can explore the various types of nonverbal behavior. The following pages explain how people send messages with their bodies, artifacts (such as clothing), environments, and the way they use time.

Body Movements

Stop reading for a moment and notice how you are sitting and what movements and facial expressions you have been displaying in the last few minutes. What do these say about how you feel? Are there other people near you now? What messages do you get from their posture and movements? Watch a randomly selected clip on YouTube without sound to see what messages are suggested by the movements of people on the screen. These simple experiments illustrate the communicative power of **kinesics**, the study of how people use their bodies and faces to communicate with others. In this section, we explore three types of kinetic nonverbal behaviors: posture, fidgeting, and facial expressions.

kinesics The study of body movement, facial expression, gesture, and posture.

manipulators Movements in which a person grooms, massages, rubs, holds, pinches, picks, or otherwise manipulates an object or body part.

Posture   There's a reason parents often tell their children to stand up straight: Good posture sends a message to others that you are confident and capable.[41] As you read in the opening to this chapter, standing tall may also make you *feel* more confident.[42]

Sometimes postural messages are obvious. If someone shuffles in with their head down or slumps over while sitting in a chair, it's apparent that something significant is going on. But most postural cues are subtler. For instance, mirroring the posture of another person can have positive consequences. One experiment showed that career counselors who used "posture echoes" while communicating with clients were considered more empathic than those who did not.[43] Researchers have also found that partners in romantic relationships tend to mirror each other's behaviors.[44]

Because posture sends messages about how alert and vulnerable someone is, criminals may use it to identify people they think will be easy targets. One study revealed that rapists sometimes use postural clues to select victims they believe will be easy to intimidate.[45] Walking slowly and tentatively, staring at the ground, and moving your arms and legs in short, jerky motions can suggest to others that you lack confidence and aren't paying attention to your surroundings.

Fidgeting   You have probably seen a public speaker before who looked visibly nervous. Perhaps they fidgeted with their clothing, clasped and unclasped their hands, or played with a pen or paperclip while they were speaking. Social scientists call these behaviors **manipulators**, in the sense that they involve ma-

nipulating or fiddling with things.[46] Research confirms what common sense suggests—that greater than normal use of manipulators is often a sign of discomfort. The good news is that you can appear more confident by avoiding the temptation to fidget. If you anticipate a stressful situation, follow these tips: Wear clothing that makes you feel confident and comfortable, don't hold anything in your hands, and try to relax your body and use gestures naturally.

Despite people's best efforts to conceal their emotions, their faces may reveal clues about how they feel.

What can you learn by paying more attention to facial expressions?

Eye Contact   How important is eye contact? A look at the cereal aisle in the grocery store will suggest an answer. Check out the old favorites—the Quaker Oats man, the Trix rabbit, the Sun-Maid girl, and Chef Boy-Ar-Dee. The odds are that all of them will be looking back at you. Researchers at Cornell University found that people were more likely to choose Trix over competing brands if the rabbit was looking at them rather than away.[47] "Making eye contact, even with a character on a cereal box, inspires powerful feelings of connection," said Brian Wansink, one of the study's authors.[48]

Our need for eye contact begins at birth. Newborns instinctively lock eyes with their caregivers.[49] But the meaning people give eye contact throughout their lives varies by culture. In Euro-American culture, meeting someone's glance with your eyes is usually a sign of involvement or interest, whereas looking away signals a desire to avoid contact. However, in some cultures—such as traditional Asian, Latin American, and Native American—it may be considered aggressive or disrespectful to make eye contact with a stranger or authority figure.[50] It's easy to imagine the misunderstandings that occur when one person's "friendly gaze" feels rude to another, and, conversely, how "politely" looking away can feel like a sign of indifference.

Facial Expressions   In the Pixar movie *Inside Out*, 11-year-old Riley tries to manage a range of emotions—fear, joy, sadness, disgust, and anger—after she moves to San Francisco with her family and must adapt to new situations. The movie creators clearly did their homework. Around the world, people are remarkably similar in how they express the five emotions that Riley experiences—as well as a sixth emotion, surprise. For example, exposed to something funny, people everywhere display an expression that others can usually identify as a smile, even if they don't speak the same language. The same is true for expressions of disgust, sadness, and so on.

> **affect blend**   The combination of two or more expressions, each showing a different emotion.

This doesn't mean that facial expressions are always easy to understand. For one thing, people can produce a large number of expressions and change them very quickly. A split-second frown can be replaced by a look of affection or surprise. For another, people can "say" one thing with part of their face and something different with another. A smile might be accompanied by lowered eyebrows that suggest disapproval along with mirth. **Affect blends** are simultaneous expressions that show two or more emotions, such as fearful surprise or angry disgust.

Despite the universality of some emotional expressions, culture is another factor that makes it complicated to interpret facial expressions (and other forms of nonverbal communication) with certainty. For example, in much of the United States and Europe, people who smile are regarded as friendlier, less aggressive, and more confident than those who don't.[51,52] But people in some cultures are put off by smiling, especially when whole-face grins are involved. In Russia, wearing a serious expression can be a status booster, since it suggests that one is serious, reliable, and powerful.[53] As with all nonverbal communication, it pays to be mindful of the situations, cultures, and relationships involved.

In the movie *Inside Out*, a young girl experiences a range of emotions as she adjusts to life in a new city. Her feelings are depicted as animated characters (from left to right) anger, fear, joy, disgust, and sadness. Nonverbal cues for these emotions are the same in virtually every culture.

Watch a video in a language you don't understand. What cues give you the impression of anger, disgust, joy, fear, and sadness?

Voice

paralanguage Nonlinguistic means of vocal expression: rate, pitch, tone, and so on.

disfluencies Vocal interruptions such as stammering and use of *"uh," "um,"* and *"er."*

"That kind of joke can get you in trouble," says the boss with a chuckle. The employee receiving the message may wonder if the boss is serious or only kidding. Social scientists use the term **paralanguage** to describe nonverbal cues that are audible. These include tone, speed, pitch, volume, and laughter, to name just a few. You might use your voice to emphasize particular words, or you might pepper your speech with **disfluencies** (such as stammering and use of "uh," "um," and "er").

The impact of paralinguistic cues is powerful. When asked to determine a speaker's attitude, listeners typically pay more attention to vocal nuances than to the speaker's words.[54] And when vocal cues contradict a verbal message, listeners usually trust the cues more than the words.[55] To try this idea out, say each of these statements as if you really mean them:

- Thanks for waking me up.
- I really had a wonderful time on my blind date.
- There's nothing I like better than waking up before sunrise.

Now say them again in a sarcastic way. You'll probably find that paralanguage can change the meaning of the statements, even suggesting that they mean the opposite of what they say on the surface. That makes it important to match a compliment, apology, command, or any other statement with vocal cues that support the message you mean to convey.

Some vocal factors influence the way a speaker is perceived by others. For example, communicators who speak loudly and without hesitations are viewed as more confident than those who pause and speak quietly.[56] People who speak somewhat slowly are often judged to have greater conversational control than fast talkers.[57] Along with vocal qualities, accents can shape perceptions. In one study, employers responded favorably to non-native English speakers with French accents but were put off by some other accents, such as Japanese, perhaps because the employers considered them more unfamiliar or difficult to understand.[58] See the *@work* feature for more about the role of vocal quality in creating a positive impression.

Appearance

How we appear can be just as revealing as how we sound and move. Here we explore the communicative power of physical attractiveness, clothing, and body art.

Physical Attractiveness   Some observers have coined the phrase "the Tinder trap" (after the popular dating app) to describe the tendency, especially online, to judge people primarily or even solely on how they look. As experience has no doubt taught you, looks are not the most important factor in judging relationship potential, but they contribute significantly to first impressions.

People who are considered physically appealing are more likely than others to be

"Wow . . . We could really fill this room with uncomfortable silences."

Source: Alex Gregory The New Yorker Collection/The Cartoon Bank

Vocal Cues and Career Success

Vocal cues "show that we are alive inside—thoughtful, active," says researcher Nicholas Epley."[59] Without vocal cues, Epley observes, it's hard to judge a speaker's personality or intent.

Epley was part of a research team that asked MBA candidates to prepare a two-minute pitch to a prospective employer.[60] The researchers recorded these pitches and asked 162 people to evaluate them. Some of the evaluators watched videotapes of the pitches, others listed them on audio, and a third group read the pitches in transcript form. Those who heard the pitches as audio or video recordings rated the candidates as more intelligent than those who only read the transcripts.

One implication is that, whenever possible, it's a good idea to meet with prospective employers in person. Résumés and cover letters, though indispensable, probably don't have the impact of a nonverbally rich interaction.

satisfied with life.[61] And no wonder. Attractive people are more likely than their peers to win elections,[62] be granted a not-guilty verdict in court,[63] influence their peers,[64] and succeed in business. More than 200 managers in one survey admitted that attractive people get preferential treatment both in hiring decisions and on the job.[65] One study even found that people with attractive avatars enjoy an advantage in online job interviews.[66]

What some researchers call the *beauty premium* involves a reciprocal cause-and-effect pattern. People tend to assume that attractive people are more confident and capable than others, so they treat them as if they are—which can lead attractive people to actually be more confident (though not necessarily more talented) than others.[67] And because people tend to smile at attractive people, they tend to smile back, which can boost their attractiveness rating even more.[68] One lesson is that presenting yourself in a confident and friendly way can improve others' opinion of your appearance, even if you don't consider yourself the best looking person in the room. See the *Ethical Challenge* in this section to explore the implications of judging people on their looks.

All this being said, if you aren't extraordinarily gorgeous or handsome by society's standards, don't despair. Evidence suggests that, as we get to know people and like them, we start to regard them as better looking, and the reverse is true for people with unappealing habits or personalities.[69] Moreover, we view others as beautiful or ugly not just on the basis of their "original equipment" but also on how they use that equipment. As we mentioned, posture, gestures, facial expressions, and other behaviors can increase the attractiveness of an otherwise unremarkable person. And occasionally, physical attractiveness has a downside. Employers sometimes turn down especially good-looking candidates because they perceive them to be threats.[70] All in all, while attractiveness generally gets rewarded, over-the-top good looks can be intimidating.[71]

Clothing   As a means of nonverbal communication, clothing can be used to convey economic status, educational level, athletic ability, interests, and more.

Most college-age men say they consider whether various clothing choices would cast them as members of a particular group such as goth, gay, alternative, or gangster.[72] And even when they give the impression that they haven't spent a lot of time choosing what they wear, 9 in 10 surveyed said they deliberately select clothing that reflects their interests and hobbies, such as t-shirts with sports emblems or bands, or clothing that makes them seem destined for professional success.[73]

Women often face several challenges in clothing selection: For one, the latest fashions might not look good on them or be designed to fit them.[75] For another, they receive mixed messages about how they should appear. Media images tend to depict women in skimpy clothing, but dressing that way can lead others to take them less seriously. In one study, university students judged women in photos to be more intelligent, powerful, organized, professional, and efficient when they were dressed in business clothes rather than in form-fitting or revealing clothing.[76] Third, clothing and shoes that society expects women to wear can be uncomfortable and impractical.[77] Clothing options may be especially problematic for women who identify as queer.[78] They often say that few role models or clothing styles reflect their identity. As a result, they may feel alienated and ignored whether they shop in the "women's" or the "men's" section. The majority of queer women surveyed said they would appreciate shopping in clothing sections that are not designated for one gender or another.

Even the colors people wear may influence how others perceive them. In one study, experimenters asked participants to rate the attractiveness of individuals shown in photos. The photos were identical except that the experimenters had digitally changed the color of the models' shirts. Participants consistently rated both male and female models to be more attractive when they were shown wearing red or black shirts compared to yellow, blue, or green.[79] Before you load your wardrobe with red and black, however, consider that perceptions vary by context. For example, patients are significantly more willing to share their social, sexual, and psychological problems with doctors wearing white coats or surgical scrubs than those wearing business dress or casual attire.[80]

Clothing is especially important in the early stages of a relationship, when making a positive first impression encourages others to get to know you better. This is equally important in personal situations and in employment interviews. In both cases, your style of dress (and personal grooming) can make the difference between the chance to progress further and outright rejection.

Body Art   The percentage of Americans with tattoos has increased from 3% to 37% in the last 30 years, making indelible body art a relatively common means of self-expression.[81] The most popular reasons to get a tattoo are to commemorate an important relationship, life event, or philosophy.[82] When it comes to ink, beauty is in the eye of the beholder *and* the identity of the design wearer. College students tend to consider tattoos more attractive on younger women than on older ones.[83] When researchers showed people images of the same men with or without tattoos, the respondents tended to rate the tattooed versions as more masculine and aggressive.[84] That can be both good and bad. Women in the study found the tattooed men appealing but not necessarily good husband or father material. On the other hand, men did not consider the tattooed males to be any less appealing as long-term partners. But beware: So far, tattoos are still taboo in many workplaces. Overall, people with visible tattoos have more difficulty getting hired than others.[85]

Touch

A supportive pat on the back, a high five, or even an inappropriate graze can be more powerful than words. **Haptics**, the study of touch, has revealed that physical contact is even more powerful than you might think.

Experts propose that one reason actions speak louder than words is that touch is the first language we learn as infants.[86] In many ways, touch is essential to our lives. Mortality rates in orphanages decreased two-thirds in the early 20th century when people recognized that babies need to be touched and carried in order to thrive.[87,88] As children grow, touch seems to increase their mental functioning as well as their physical health. Children given plenty of physical stimulation tend to develop significantly higher IQs and less communication apprehension than those who have less contact.[89]

Touch can also be persuasive. When people were asked to sign a petition, they were more likely to cooperate when the person asking touched them lightly on the arm. In one variation of the study, 70% of those who were touched complied, whereas only 40% of the untouched people agreed to sign.[90]

Touch can be welcome in other scenarios as well. Fleeting touches on the hand and shoulder often result in larger tips for restaurant servers.[91] And romantic partners who frequently touch each other are typically more satisfied with their relationships than other couples are.[92] Even athletes reflect the power of touch. One study of the National Basketball Association revealed that the touchiest teams had the most successful records, while the lowest scoring teams touched each other the least.[93]

Although touch may be appreciated in some situations, it's downright annoying or frightening in others. In one study, shoppers touched by other shoppers (particularly males) bought less and left the store more quickly than shoppers who were not touched, perhaps because, in that situation, being touched had the unpleasant connotation of being jostled or crowded.[94] See @*work* for more about the benefits and potential drawbacks of using touch as a means of expression.

Space

The use of space can create nonverbal messages in two ways: the distance we put between ourselves and the territory we consider our own. We'll now look at each of these dimensions.

Distance   The study of the way people use space is called **proxemics**. Preferred spaces are largely a matter of cultural norms. For example, people living in hyperdense Hong Kong manage to live in crowded residential quarters that most North Americans would find intolerable. Anthropologist Edward T. Hall defined four

Most clothing and body art are conscious choices, crafted to support a desired identity.
What does your physical appearance say about you? How might you change the way others perceive you by modifying your appearance?

haptics The study of touch.

proxemics The study of how people and animals use space.

Touch and Career Success

The old phrase "keeping in touch" takes on new meaning once you understand the relationship between haptics and career effectiveness.

Some of the most pronounced benefits of touching occur in the health and helping professions. For example, patients are more likely to take their medicine as prescribed when physicians give a slight touch while prescribing it.[95] In counseling, touch often increases self-disclosure and verbalization of psychiatric patients.[96]

Touch can also enhance success in sales and marketing. When retail sales personnel touch customers without making them feel rushed or crowded, the customers tend to increase their shopping time, evaluate the store more highly, and buy more.[97] When an offer to sample a product is accompanied by a touch, customers are more likely to try the sample and buy the product.[98]

Of course, touch should be culturally appropriate. Whereas appropriate contact can enhance your success, it isn't welcome in all cultures or in all situations. Too much contact can be bothersome, annoying, or even creepy.

---

intimate distance A distance between people that ranges from touching to about 18 inches; usually used by those who are emotionally close.

personal distance A distance between people that ranges from 18 inches to 4 feet; common among most relational partners.

social distance A distance between people that ranges from 4 to 12 feet; commonly used by people who work together or interact during sales transactions.

public distance A distance between people that exceeds 12 feet; common in public speaking.

---

distances used in mainstream North American culture.[99] He says that we choose a particular distance depending on how we feel toward the other person at a given time, the context of the conversation, and our personal goals.

Intimate distance begins with skin contact and ranges to about 18 inches. The most obvious context for intimate distance involves interaction with people to whom we're emotionally close—and then mostly in private situations. Intimate distance between individuals also occurs in less intimate circumstances such as visiting the doctor or getting your hair cut. Allowing someone to move into the intimate zone usually is a sign of trust.

Personal distance ranges from 18 inches at its closest point to 4 feet at its farthest. The closer range is the distance at which most relational partners stand in public. We are often uncomfortable if someone else "moves in" to this area without invitation. The far range of personal distance runs from about 2.5 to 4 feet. This is the zone just beyond the other person's reach—the distance at which we can keep someone "at arm's length." This term suggests the type of communication that goes on at this range: Interaction is still reasonably personal, but less so than communication that occurs a foot or so closer.

Social distance ranges from 4 to about 12 feet. Within it are the kinds of communication that usually occur in business situations. Its closer range, from 4 to 7 feet, is the distance at which conversations usually occur between salespeople and customers and between people who work together. We use the far range of social distance—7 to 12 feet—for formal and impersonal situations. This is the distance apart we generally sit from the boss.

Public distance is Hall's term for the farthest zone, running outward from 12 feet. The closer range of public distance is the one most teachers use in the classroom. In the farther range of public space—25 feet and beyond—two-way communication becomes difficult. In some cases, it's necessary for speakers to use public distance owing to the size of their audience, but we can assume that anyone who voluntarily chooses to use it when they could be closer is not interested in having a dialogue.

Choosing the optimal distance can have a powerful effect on how others respond to us. For example, students are more satisfied with teachers who reduce the distance between themselves and the class. They also are more satisfied with the course itself, and they are more likely to follow the teacher's instructions.[100] Likewise, medical patients are often more satisfied with physicians who don't "keep their distance."[101]

Allowing the right amount of personal space is so important that scientists who create interactive robots take proxemics into account. They have found that people tend to be creeped out by robots who get too close, especially if the robot makes consistent "eye contact" with them. However, people don't mind as much if robots whose nonverbal behaviors seem friendly and nonthreatening move into their personal space.[102]

Territoriality   **Territoriality** involves the tendency to claim places and spaces you consider to be more or less your own. These might include your bedroom, house, or the chair you usually occupy in class or at your favorite coffee hangout. People tend to use nonverbal markers to declare which territory is theirs. You might erect a fence around your yard or spread a blanket on the beach to mark the area as yours, at least temporarily. You may feel annoyed or disrespected if someone encroaches on your territory without permission. Indeed, honoring boundaries is one way of showing respect. People with higher status are typically granted more personal territory and privacy. You probably knock before entering the boss's office if the door is closed, whereas a boss can usually walk into your work area without hesitating. In the military, greater space and privacy usually come with rank: Privates typically sleep 40 to a barracks, sergeants have their own private rooms, and high-ranking officers have government-provided houses.

The rules of personal space are revealed most powerfully when they're broken.

Have you ever invaded someone's personal space? Have you ever been unexpectedly standoffish? What were the consequences?

territoriality The tendency to claim spaces or things as one's own, at least temporarily.

chronemics The study of how humans use and structure time.

Environment   Google, Microsoft, and some other employers have begun to design communication-friendly work environments in which the furniture can be moved at will and there is plenty of space for conversation and group interaction. They know what social scientists have long found to be true: People's physical environments shape the communication that occurs within them. Natural light, flexible seating configurations, and views of nature have been shown to enhance learning in the classroom[103] and productivity at work.[104]

Especially appealing are work spaces that feature both areas for team collaboration and private spaces in which people can work without interruptions.[105] Perhaps mindful of this, when members of the Barbarian Group marketing firm redesigned their New York City headquarters, they had architects create a 1,000-foot-long high-tech "superdesk." The desk looks a bit like a curving white ribbon that weaves through headquarters, offering spaces where people can easily collaborate with one another as well as sheltered "grottoes" where they can work or converse quietly.[106]

Time   **Chronemics** is the study of how people use and structure time.[107] The way we handle time can express both intentional and unintentional messages.[108] Social psychologist Robert Levine describes several ways that time can communicate.[109] For instance, in a culture that values time highly, like the United States, waiting can be an indicator of status. "Important" people (whose time is

Airbnb office space offers areas for collaboration, relaxation, and individual work.

If you could design the environment for your dream job, what would it look like? How would it shape communication at work?

monochronic The use of time that emphasizes punctuality, schedules, and completing one task at a time.

polychronic The use of time that emphasizes flexible schedules in which multiple tasks are pursued at the same time.

supposedly more valuable than that of others) may be seen by appointment only, whereas it is acceptable to intrude without notice on "less important" people. To see how this rule operates, consider how natural it is for a boss to drop in to a subordinate's office unannounced, whereas some employees would never intrude into the boss's office without an appointment. A related rule is that low-status people must not make more important people wait. It would be a serious mistake to show up late for a job interview, although the interviewer might keep you cooling your heels in the lobby. Important people are often whisked to the head of a restaurant or airport line, whereas the presumably less exalted are forced to wait their turn.

Similar principles apply beyond work. For example, when singer Justin Bieber took the stage almost two hours late at a sold-out concert in London, his fans were outraged and took to Twitter and other social media to complain. Bieber's management of time sent an unspoken message to many fans: "My life is more important than yours."

The use of time depends greatly on culture.[110] Some cultures (e.g., North American, German, and Swiss) tend to be **monochronic**, emphasizing punctuality, schedules, and completing one task at a time. Other cultures (e.g., South American, Mediterranean, and Arab) are more **polychronic**, with flexible schedules in which people pursue multiple tasks at the same time. One psychologist discovered the difference between North and South American attitudes when teaching at a university in Brazil.[111] He found that some students arrived halfway through a two-hour class and most of them stayed put and kept asking questions when the class was scheduled to end. A half hour after the official end of the class, the professor finally closed off discussion because there was no indication that the students intended to leave. This flexibility of time is quite different from what is common in most North American colleges.

Even within a culture, rules of time vary. Sometimes the differences are geographic. In New York City, the party invitation may say "9 P.M.," but nobody would think of showing up before 9:30. In Salt Lake City, guests may be expected to show up on time, or perhaps even a bit early.[112] Even within the same geographic area, different groups establish their own rules about the use of time. Consider your own experience. In school, some instructors begin and end class punctually, whereas others are more casual. With some people you feel comfortable talking for hours in person or on the phone, whereas with others time seems to be precious and not meant to be "wasted."

7.4 Influences on Nonverbal Communication

Sometimes nonverbal cues are easy to interpret no matter who is involved. As we mentioned before, facial expressions that reflect emotions such as happiness and sadness are similar around the world.[113] Other nonverbal cues can be culture-specific or influenced by gender identity.

Cultures have different nonverbal languages as well as verbal ones. Fiorello La Guardia, the legendary mayor of New York from 1934 to 1945, was fluent in

English, Italian, and Yiddish. Researchers who watched films of his campaign speeches with the sound turned off found that they could tell which language he was speaking by the changes in his nonverbal behavior.[114] Here are some of the ways that culture influences nonverbal communication.

Culture

Use of Space   Edward Hall points out that, whereas Americans are comfortable conducting business at a distance of roughly 4 feet, people from the Middle East stand much closer.[115] It's easy to visualize the awkward advance-and-retreat pattern that might occur when two diplomats or businesspeople from these cultures meet. The Middle Easterner would probably keep moving forward to close a gap that feels wide to them, whereas the American would probably continually back away. Both would feel uncomfortable, although they might not know why.

Eye Contact   Like distance, patterns of eye contact vary around the world.[116] A direct gaze is considered appropriate for speakers in Latin America, the Arab world, and southern Europe. On the other hand, Asians, Indians, Pakistanis, and northern Europeans typically gaze at a listener peripherally or not at all. In either case, deviations from the norm are likely to make listeners who are unfamiliar with the culture uncomfortable.

Nonverbal Focus   Culture also affects how nonverbal cues are monitored. In Japan, for instance, people tend to look to the eyes for emotional cues, whereas Americans and Europeans focus more on the mouth.[117] These differences can often be seen in the text-based emoticons commonly used in these cultures. (Search for "Western and Eastern emoticons" in your browser for examples.)

Paralanguage   Even within a culture, various groups can have different nonverbal rules. For instance, younger Americans often use "uptalk" (statements ending with a rise in pitch) and "vocal fry" (words ending with a low guttural rumble). Celebrities such as Kim Kardashian and Zooey Deschanel popularized these vocalic styles. It's therefore not surprising that females use them more than males do,[118] although age of the speaker (i.e., Millennial or younger) has more of an impact than gender. There is some debate about whether vocal fry diminishes one's credibility[119] or enhances it.[120] Either way, vocal mannerisms are one way that speakers affiliate with their communities.

Affect Displays   Although some nonverbal expressions are more or less universal, the way they are used varies widely around the world. In some cultures, display rules discourage the overt demonstration of feelings such as happiness or anger. In other cultures, displaying the same feelings is perfectly appropriate. Thus, a person from Japan may appear to be much more controlled and placid than an Arab, when in fact their feelings are nearly identical.

It has become fashionable in some circles to imitate the throaty rumble known as vocal fry, made popular by Katy Perry and other celebrities.

Do you ever change the sound of your voice to make a particular impression on people? If so, how?

The same principle operates closer to home among co-cultures. For example, observational studies have shown that, in general, black women in all-black groups are more nonverbally expressive and interrupt one another more than do white women in all-white groups.[121] This doesn't mean that black women always feel more intensely than their white counterparts. A more likely explanation is that the two groups tend to follow different cultural rules. The researchers found that, in racially mixed groups, both black and white women moved closer to the others' style. This nonverbal convergence shows that skilled communicators can adapt their behavior when interacting with members of other cultures or cocultures to make the exchange smoother and more effective.

All in all, communicators are likely to be more tolerant of others after they understand that nonverbal behaviors they consider unusual may be the result of cultural differences. In one study, American adults were presented with videotaped scenes of speakers from the United States, France, and Germany.[122] When the sound was cut off, viewers judged foreigners more negatively than their fellow citizens. But when the speakers' voices were added (allowing viewers to recognize that they were from a different country), participants were less critical of the foreign speakers.

Gender

Although differences between the sexes are often smaller than people think, women in general are more nonverbally expressive than men, and women are typically better at recognizing others' nonverbal behavior.[123] More specifically, research shows that, compared with men, women tend to:

- Smile more
- Use more facial expressions
- Use more head, hand, and arm gestures (but fewer expansive gestures)
- Touch others more
- Stand closer to others
- Be more vocally expressive
- Make more eye contact

Most communication scholars agree that these differences are influenced more by social conditions than by biological differences. Of particular influence are media depictions and social power structures, which we discuss next.

Media Influences   Media portrayals tend to reinforce stereotypes about gender and nonverbal communication. Television programs for children and teenagers commonly define girls in terms of their appearance and their ability to attract the attention of boys, and boys in terms of rugged independence and a preoccupation with pursuing girls.[124] The depictions are overwhelmingly heterosexual and consistent with the stereotype that one should be either a "hot girl" or a "cool dude," as one research team put it.[125] Television commercials and video games for all ages feature similar, even more extreme, versions of these themes.[126,127]

LGBTQ individuals are given disproportionally little screen time, especially in the movies, and gay characters are often treated as comical figures,[128] as when LeFou in Disney's *Beauty and Beast* clumsily flirts with the hypermasculine Gaston.[129]

Exposure to media messages is associated with stereotypical beliefs. For example, men who frequently play video games in which women are portrayed as "damsels in distress" are more likely than other men to think women are weak and helpless.[130] Stereotypes affect people's self-image as well. Selfies (photos of oneself) on social media tend to correspond closely with images in the media—females suggestively dressed and smiling and men emphasizing their physiques and emotional control.[131]

Social Structure   Women may demonstrate greater sensitivity than men to nonverbal cues because of their social status. Because women have historically had less power, they have had greater incentive to read men's nonverbal cues than the other way around.[132] You have probably noticed a similar dynamic in work settings, where people are usually more tuned into the boss's moods than the boss is to theirs. It may also be that, because women have traditionally been responsible for child care, they learn to display and decipher nonverbal cues so that they can better communicate with youngsters who are not yet proficient using language.

Despite these differences, men's and women's nonverbal communication patterns have a good deal in common.[133] You can prove this by imagining what it would be like to use radically different nonverbal rules. Standing only an inch away from others, sniffing strangers, or tapping people's foreheads to get their attention would mark you as bizarre no matter your gender.

Moreover, according to a recent study, nonverbal differences are less pronounced in conversations involving gay and lesbian individuals than in those involving heterosexuals, presumably because heterosexuals feel less constrained by gender-related stereotypes.[134] All in all, gender and culture have an influence on nonverbal style, but the differences are often a matter of degree and cultural influence.

As noted in the opening feature about Amy Cuddy's work, people of different genders often present themselves in more or less powerful ways, but a great deal of that is under their control. Likewise, as you become more aware of the signals you (and others) send nonverbally, your communication ability may improve beyond words.

MAKING THE GRADE

At www.oup.com/he/adler-uhc14e, you will find a variety of resources to enhance your understanding, including video clips, animations, self-quizzes, additional activities, audio and video summaries, interactive self-assessments, and more.

OBJECTIVE 7.1   Explain the characteristics of nonverbal communication and the social goals it serves.

- Nonverbal communication helps people manage their identities, define their relationships, and convey emotions.
- It's impossible to avoid communicating nonverbally. Humans constantly send messages about themselves that are available for others to receive.
- Nonverbal communication is ambiguous. There are many possible interpretations for any behavior. This ambiguity makes it important for the receiver to verify any interpretation before jumping to conclusions.

- Nonverbal communication is different from verbal communication in complexity, flow, clarity, impact, and intentionality.

   > Describe three messages that qualify as nonverbal and one message that is verbal. Explain the difference between these two types of communication.

   > Considering that people cannot *not* communicate, what messages do you think you send nonverbally to strangers who observe you in public?

   > If a friend were preparing for a job interview and asked your advice about appearing and feeling confident, what advice would you give about managing their nonverbal communication?

OBJECTIVE 7.2 **Describe key functions served by nonverbal communication.**

- Nonverbal communication serves many functions: repeating, substituting, complementing, accenting, regulating, and contradicting verbal behavior, as well as deceiving.

- Detecting deception based on nonverbal cues is much harder than most people think because people tend to have truth or deception biases and because nonverbal behavior is highly ambiguous.

  > Give an example of each of the following nonverbal behaviors: repeating, substituting for, complementing, accenting, regulating, and contradicting verbal messages.

  > Try interacting with people for an hour without using words, then reflect on the experience. What were you able to convey easily through nonverbal means? What was most difficult?

OBJECTIVE 7.3 **List the types of nonverbal communication, and explain how each operates in everyday interaction.**

- We communicate nonverbally in many ways: through posture, fidgeting, eye contact, facial expressions, voice, physical attractiveness, clothing, touch, distance and territoriality, environment, and time.

- Members of some cultures tend to be monochronic, whereas others are more polychronic.

  > Describe the difference between a monochronic time orientation and a polychronic orientation.

  > Pause to look at your surroundings. How conducive are they to a positive state of mind? To social interaction? To contemplation? How does your environment influence the way you feel right now?

  > Keep a tally of how many disfluencies you utter in one day. Consider whether you are happy with the results. If not, what would you like to change?

  > Notice the nonverbal cues of someone around you right now. Write down two or three interpretations of how that person is feeling. Ask if any of them are accurate.

OBJECTIVE 7.4 **Explain the ways in which nonverbal communication reflects culture and gender differences.**

- Based on their culture, people may focus on different nonverbal cues and have different expectations about how to use space, eye contact, paralanguage, and affect displays.

- In some cultures, eye contact and physical closeness are interpreted as signs of attentiveness, in others as challenges or indications of disrespect.

- In general, women are socialized to be more attentive to nonverbal cues than men are.

- Media depictions and social roles mostly presume that everyone is heterosexual and that women should be beautiful and display helplessness and men should present themselves as rugged, strong, and independent.

  > Name three rules of appropriateness for making or avoiding eye contact that are familiar to you but may be unfamiliar to someone from a different culture.

  > What advice would you offer someone who is packing clothing for a trip during which they will encounter people from many different cultures?

  > How would you explain the reasons that men and women may use and interpret nonverbal cues differently?

KEY TERMS

affect blend p. 167
affect displays p. 160
chronemics p. 173
deception bias p. 165
disfluencies p. 168
emblems p. 161
expectancy violation theory p. 160
haptics p. 171
illustrators p. 163
intimate distance p. 172
kinesics p. 166
manipulators p. 166
monochronic p. 174
nonverbal communication p. 156
paralanguage p. 168
personal distance p. 172
polychronic p. 174
proxemics p. 171
public distance p. 172
social distance p. 172
territoriality p. 173
truth bias p. 165

ACTIVITIES

1. You can practice conveying and interpreting vocal messages by following these directions.

   a. Join with a partner and designate one person A and the other B.

   b. Partner A should choose a passage of 25 to 50 words from a newspaper or magazine, using his or her voice to convey one of the following attitudes:

   > Egotism

   > Friendliness

> Insecurity

> Irritation

> Confidence

c. Partner B should try to detect the emotion being conveyed.

d. Switch roles and repeat the process. Continue alternating roles until each of you has both conveyed and tried to interpret at least four emotions.

e. After completing the preceding steps, discuss the following questions:

> What vocal cues did you use to make your guesses?

> Were some emotions easier to guess than others?

> Given the accuracy of your guesses, how would you assess your ability to interpret vocal cues?

> How can you use your increased sensitivity to vocal cues to improve your everyday communication competence?

2. To consider the influence of culture on nonverbal communication, do the following:

a. Identify at least three significant differences between nonverbal practices in two cultures or cocultures (e.g., ethnic, age, or socioeconomic groups) within your own society.

b. Describe the potential difficulties that could arise out of the differing nonverbal practices when members from the cultural groups interact. Are there any ways of avoiding these difficulties?

c. Now describe the advantages that might come from differing cultural nonverbal practices. How might people from diverse backgrounds profit by encountering one another's customs and norms?

Understanding Interpersonal Communication

8

CHAPTER OUTLINE

8.1 Characteristics of Interpersonal Communication 182

What Makes Communication Interpersonal?
How People Choose Relational Partners
Content and Relational Messages
Metacommunication
Self-Disclosure
Online Interpersonal Communication

8.2 Communicating with Friends and Family 191

Unique Qualities of Friendship
Friendship Development
Gender and Friendship
Family Relationships

8.3 Communicating with Romantic Partners 197

Stages of Romantic Relationships
Love Languages
Male and Female Intimacy Styles

8.4 Relational Dialectics 203

Connection Versus Autonomy
Openness Versus Privacy
Predictability Versus Novelty

8.5 Lies and Evasions 205

Altruistic Lies
Evasions
Self-Serving Lies

MAKING THE GRADE 207

KEY TERMS 208

ACTIVITIES 208

LEARNING OBJECTIVES

8.1

Explain how content and relational meaning, metacommunication, self-disclosure, and online communication influence interpersonal relationships.

8.2

Identify common communication patterns in friendships, parent–child dynamics, and sibling relationships.

8.3

Describe stages of romantic relationships and options for conveying intimate messages.

8.4

Compare dialectical continua and strategies for managing them.

8.5

Analyze the functions served by altruistic lies, evasions, and self-serving lies.

Reality television may not be especially realistic, but it's a reminder that interpersonal relationships aren't always easy. As you read this chapter, consider the following questions in light of your own relationships:

When has communication felt most personal, and most impersonal to you? What made the difference?

When is the last time you told someone a deeply personal secret about yourself? How was it received?

Have you ever grappled with wanting to be with someone but also craving time for yourself? If so, how did you handle those competing needs?

"WILL YOU ACCEPT THIS ROSE?" This line from *The Bachelor* and *The Bachelorette* invites a contestant to be part of another round in the reality television game of love. To date, viewers have watched 29 couples say "I will" in marriage proposals at the shows' conclusions. However, only four of those couples have actually said "I do."[1] What happens after the show to turn "I love you forever" into "Maybe not"?

This chapter explores the role of communication in important relationships. As you have no doubt experienced, close relationships—whether with romantic partners, friends, or family members—sometimes seem effortless and rewarding. At other times, they are challenging and downright confusing.

Reality shows barely scratch the surface, of course. Sean Lowe, one of the few who found lasting love on *The Bachelor*, describes his wake-up moment: "You leave the show, then you get into the real world and find out like, 'Oh crap! Being in a relationship isn't always easy and it actually takes work.'"[2]

Communicating effectively when powerful emotions are involved requires know-how and awareness. In this chapter, we will explore:

- communication patterns that help people develop close relationships,
- the different ways we communicate with friends, family members, and romantic partners,
- how couples manage the give-and-take of competing relational and personal needs, and
- the role of deception, whether well meaning or self-serving.

The section on lies, one of the most sensitive aspects of interpersonal communication, leads into Chapter 9's coverage of effective conflict management.

8.1 Characteristics of Interpersonal Communication

Pause for a moment to consider all the people you encounter in a typical day. Some may be close friends and loved ones; others may be acquaintances; and still others are probably strangers. Now, consider how you communicate with these different people. Are there things you share about yourself with some but not with others? Which of these people make you smile or, conversely, frustrate you on a regular basis? In this unit, we define interpersonal communication, consider how we choose relational partners, and explore some of the key dimensions of communicating in close relationships.

What Makes Communication Interpersonal?

To clarify what communication is and isn't interpersonal, consider a few of the people you're likely to encounter on a typical day. Which of the following encounters do you think qualify as interpersonal communication?

- In your morning class, you strike up a conversation with a classmate you've just met.

- Later, you take part in a group project meeting in which everyone contributes.

- At midday, you pick up lunch and chat briefly about the weather with the cashier.

- After work, you enjoy a conversation with your housemate, who you know will understand what a crazy day you've had and will share some personal experiences and feelings in return.

- Before bed, you get better acquainted by texting with someone you recently met online.

In a truly interpersonal relationship, partners consider one another irreplaceable and treat one another as unique individuals.

Think about your truly interpersonal relationships. What makes each person irreplaceable and unique? How does communication in your interpersonal relationships differ from other, more superficial ones?

If you picked the scenario with your roommate, you're right. If you also selected the online relationship, score that one as a maybe for now. The other encounters are important in their own ways, but they probably don't qualify as interpersonal.

To understand why, think of the opposite of interpersonal communication as *impersonal.* Chatting with strangers and interacting in a meeting are not very personal. As we'll use the term, **interpersonal communication** involves interaction between people who are part of a *close* and *irreplaceable relationship* in which they *treat each other as unique individuals.* Let's consider the implications of that definition by returning to our examples.

Truly interpersonal relationships aren't always happy, but there will never be other ones exactly like them. The conversation with your housemate is interpersonal because you're invested in listening to and sharing personal information with each other as unique individuals. The relationship with your online friend may or may not be interpersonal. It depends—you guessed it—on how much personal information you each share and whether you treat one another as unique individuals.

Of course, it isn't always so easy to categorize relationships. On a continuum between impersonal and interpersonal, some relationships fall in the middle. Our main point here is that interpersonal relationships are marked by emotional closeness and awareness of each partner's unique qualities.

interpersonal communication Two-way interactions between people who are part of a close and irreplaceable relationship in which they treat each other as unique individuals.

How People Choose Relational Partners

Considering the number of people with whom we communicate every day, truly interpersonal interaction is rather scarce. That isn't necessarily unfortunate. Most of us don't have the time or energy to create personal relationships with everyone we encounter—or even to act in a personal way all the time with the people we know and love best. In fact, the scarcity of qualitatively interpersonal communication contributes to its value. Like precious jewels and one-of-a-kind artwork, interpersonal relationships are special because of their scarcity.

Sometimes we don't have a choice about our relationships. Children can't select their parents, and most workers aren't able to choose their bosses or colleagues. In many other cases, though, we seek out some people and actively avoid others. Following are seven common reasons why people establish close relationships with particular people.

They Have a Lot in Common   In most cases, people gravitate to those whose temperament, values, and life goals are similar to their own. For example, the more alike a married couple's personalities are, the more likely they are to report being happy and satisfied in their marriage.[3] Similarity is also a factor in the early stages of friendship. When strangers are given a choice where to sit, they tend to settle near people whose most obvious features are similar to theirs, for example, those who also wear glasses or have the same color hair.[4] Researchers suggest that similarities are comfortable and may reduce people's fear of rejection. As you will see, however, superficial characteristics such as appearance aren't a good indicator of long-term compatibility.

They Balance Each Other Out   The folk wisdom that "opposites attract" seems to contradict the similarity principle just described. In truth, both are valid. Differences strengthen a relationship when they are *complementary*—when each partner's characteristics satisfy the other's needs. For example, when introverts and extroverts pair up as friends, they typically report that the quieter person serves as a steady anchor for the friendship, and the more gregarious partner propels the other to take part in activities they might otherwise avoid.[5]

They Like and Appreciate Each Other   Of course, we aren't drawn toward everyone who seems to like us, but to a great extent, we like people who like us and shy away from those who dislike us or seem indifferent.

They Admire Each Other   Forming relationships with talented and accomplished people can inspire you and offer mutual validation.[6]

They Open Up to Each Other   People who reveal important information about themselves often seem more likable, provided that what they share is appropriate to the situation and the stage of the relationship.[7] Self-disclosure is appealing when it reveals similarities in experiences (*"I broke off an engagement, myself"*) or attitudes (*"I feel nervous with strangers, too"*). And when people share private information, it suggests that they respect and trust each other. (We'll talk more about self-disclosure later in the chapter.)

They Interact Frequently   In many cases, proximity leads to liking.[8] People are more likely to develop friendships with close neighbors than with distant ones. Proximity allows them to learn more about each other and to engage in relationship-building behaviors. Plus, people in close proximity may be more similar to one another than those who live, work, and play in different places. A sense of community can be achieved online as well. The Internet provides a means of creating closeness through "virtual proximity."[9]

The Relationship Is Rewarding   Relationships reflect a semi-economic model called **social exchange theory**, which suggests that people stay in relationships if the rewards are greater than or equal to the costs.[10] The rewards of being in a relationship may be tangible (a nice place to live, a high-paying job) or intangible (prestige, emotional support, companionship). By contrast, costs are undesirable outcomes, such as a sense of obligation, putting up with annoying habits, emotional pain, and so on. According to social exchange theory, people use this formula (usually

ASK YOURSELF (?)

Have you ever been attracted to someone based on physical similarities and then found out you were incompatible in other ways?

social exchange theory The idea that relationships seem worthwhile if the rewards are greater than or equal to the costs involved.

unconsciously) to decide whether dealing with another person is a "good deal" or "not worth the effort."

There Is a Good Balance   Having just pointed out common considerations when choosing relational partners, a few caveats are in order. Read the following before you fall into the trap of thinking you must be supermodel stunning, Mensa smart, or Olympic-level talented for people to find you appealing.

- *First impressions can mislead.* Evidence shows that we befriend people whose interests and attitudes *seem* similar to our own. This is partly an illusion, however. We tend to overestimate how similar we are to our friends and underestimate how similar we are to people we don't know well.[11] In addition, there is strong evidence that superficial similarities such as appearance do not predict long-term happiness with a relationship,[12] and when we are willing to communicate with a range of people, our differences are not usually as great as we thought.[13]

- *Our priorities change.* For example, physical factors that catch our eye at first glance aren't necessarily what we want in the long run.[14] As one social scientist put it, "attractive features may open doors, but apparently, it takes more than physical beauty to keep them open."[15]

- *Perfection can be a turn-off.* We like people who are attractive and talented, but we are uncomfortable around those who are *too* perfect. Let's face it: No one wants to look bad by comparison. And it's more important to be friendly than to be flawless. In one study, researchers concluded that, if people had to choose, most would rather spend time with a "lovable fool" than a "competent jerk."[16]

- *It's not all about communication, but it's a lot about communication.* The online dating service eHarmony matches couples based on "29 dimensions of compatibility," and other online dating sites make similar promises. We can imagine mathematical formulas for finding friends as well. However, the long-term success of people matched by computer algorithms is no greater than that of people who meet on their own.[17] That's because long-term compatibility relies less on superficial similarities and more on how people interact with each other once they start a relationship and encounter stressful issues.

A few lessons emerge from these observations. One is to break free of your comfort zone and give new people a chance. Another is that, when you are the person who seems different from others, you can help reduce the stranger barrier by being friendly and approachable and letting people get to know the real you. Finally, don't be discouraged if you don't seem perfect by society's standards. Being perfect is overrated. Being nice matters more.

Content and Relational Messages

Lucien texted "Free for lunch today?" to a good friend, Haris, and received the reply, "Sorry, can't. Next time?" Assuming that Haris was busy at work or home studying, Lucien went to a restaurant alone and saw Haris there having lunch with a mutual friend of theirs.

Communicating in close relationships is anything but simple. Will Lucien be glad to see Haris and their mutual friend? Will he feel disrespected that Haris didn't invite him? Will Haris feel guilty? Glad? Indifferent? We don't know. In the context of a close relationship, a gesture, text, statement, or silence can mean many different things. And because our emotions are involved, our reactions may surprise even us. One thing we do know: Virtually every verbal statement contains both a

ON YOUR FEET

Describe the qualities that first attracted you to a special person in your life. Did the importance you placed on those characteristics change as the relationship developed? If so, how? Prepare to share your experiences in a brief oral presentation.

"She's texting me, but I think she's also subtexting me."

Source: Leo Cullum The New Yorker Collection/The Cartoon Bank.

content message A message that communicates information about the subject being discussed.

relational message A message that expresses the social relationship between two or more people.

affinity The degree to which people like or appreciate one another, whether or not they display that outwardly.

immediacy Expression of interest and attraction communicated verbally and/ or nonverbally.

respect The degree to which a person holds another in esteem, whether or not they like them.

control The amount of influence one has over others.

content message, which focuses explicitly on the subject being discussed, and a **relational message**, which makes a statement (often implied) about how the parties feel toward one another. As the cartoon in this section shows, even text messages can contain subtexts that imply how the message should be interpreted. Following are some of the dimensions communicated on a relational level.

Affinity The degree to which people like or appreciate others is called **affinity**. Nonverbal cues often reveal more about affinity than words do. Think of the different tones of voice you could use to say "Thanks a lot." Imagine how this simple message could convey gratitude and appreciation. Now visualize how the same words could express irritation and dislike.

Immediacy Whereas affinity involves attraction, **immediacy** reflects the level of engagement. Is the communicator interested and involved with you, or detached and distant? Affinity and intimacy can interact in several ways. For example, a communicator might feel favorably toward others (high affinity) and be fully engaged with them (high immediacy) or rarely engaged with them (low immediacy).

Respect The degree to which we admire others and hold them in esteem is known as **respect**. While respect and affinity might seem similar, they are actually different dimensions of a relationship.[18] You might not like your boss or teacher but you still might respect their talents. Respect is an important but often overlooked ingredient in satisfying relationships. It is a better predictor of relational satisfaction than liking, or even loving.[19] Lucien may have been disappointed that Haris had other plans but respected him as a busy adult with the right to spend time as he chooses.

Control In every conversation and every relationship, there is some distribution of **control**, that is, the amount of influence exercised by the people involved. Control can be distributed evenly among relational partners, or one person can have more or less than the other. An uneven distribution of control in one way isn't necessarily problematic if it balances out in other ways. Perhaps you do most of the cooking and your roommate does most of the cleaning. But relationships suffer if control is lopsided, as when one partner orders the other around or always gets their way.[20] In our example, perhaps Lucien invites Haris to lunch nearly every day and Haris sidestepped the invitation today to have more control over his own schedule.

We can speculate on Lucien and Haris's feelings, but we are only guessing. Keep this in mind when you evaluate the relational messages of people around you. The impatient tone of voice you take as a sign of anger might be due to fatigue, and the interruption you consider belittling might arise from enthusiasm about your idea. Ultimately, relational meanings are ambiguous and situational. It's a good idea to check your understanding before making assumptions about them.

Metacommunication

Most of the time, relational messages aren't put into words. You can get an idea about whether someone likes or respects you, and how much control they seek, by the way they act rather than what they say. But there *are* ways to bring relational dynamics into the open.

Social scientists use the term **metacommunication** to describe messages that refer to other messages.[21] In other words, metacommunication is communication about communication. Whenever you discuss a relationship with others, you're metacommunicating: *"It sounds like you're angry at me." "I appreciate how honest you've been." "You keep insisting, even after I've said 'no.'"* Metacommunication occurs online, too, with initials jk (*just kidding*) or emojis such as 😠 and 😣 . Following are some key points to know about metacommunication.

Metacommunication Provides a Look Below the Surface   Consider friends arguing because one wants to watch TV while the other wants to talk. Imagine how much better the chances of a positive outcome would be if they used metacommunication: *"Look, it's not the TV watching itself that bothers me. It's that I think you're watching because you're mad at me. Am I right?"* Talking about communication is a way to address issues that might otherwise fester or cause misunderstandings. One friend might say to another, *"You looked surprised when you saw me at the restaurant. Were your feelings hurt?"*

Metacommunication Isn't Just for Handling Problems   It's also a way to reinforce the good aspects of a relationship, as in *"Thank you for praising my work in front of the boss."* Comments such as this let others know that you value their behavior. They also boost the odds that the other person will continue that behavior in the future.

Metacommunication Can Be Risky   Calling attention to relational dynamics, especially if they are problematic, might expose vulnerabilities. Some people assume that "our relationship isn't working if we have to keep talking about it." Furthermore, metacommunication involves a certain degree of analysis (*"It seems like you're angry at me."*), which can lead to resentment (*"Don't presume to know how I feel."*). This doesn't mean metacommunication is a bad idea, just that it's a tool you should use carefully. One strategy for cautiously and respectfully metacommunicating is perception checking, introduced in Chapter 3.

Self-Disclosure

"We don't have any secrets," some people claim. **Self-disclosure** is the process of deliberately revealing significant information about oneself that would not normally be known by others. Under the right conditions, self-disclosure is rewarding.[22] Talking about your feelings and experiences can yield a greater sense of clarity and emotional closeness. It can be validating to know that others know and like the real you and are willing to reveal something about themselves in return. At the same time self-disclosure, like metacommunication, involves an element of vulnerability. Disclosing too soon or too much can be used against you and make others uncomfortable.

Here we consider two models—the social penetration model and the Johari Window—that offer insights into the types and reasons for self-disclosure.

Social Penetration Model   Social psychologists Irwin Altman and Dalmas Taylor describe two ways in which communication can be more or

metacommunication Messages that refer to other messages; communication about communication.

self-disclosure The process of deliberately revealing information about oneself that is significant and that would not normally be known by others.

ON YOUR FEET

Think of a recent conversation with someone who is important to you. What about the conversation felt satisfying (or not)? If you were a relationship coach using this conversation as a teaching tool, what lessons might you share in terms of what worked well and what could have been more effective?

"There's something you need to know about me, Donna. I don't like people knowing things about me."

Source: Leo Cullum The New Yorker Collection/The Cartoon Bank

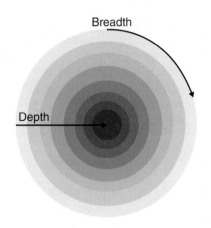

FIGURE 8.1 Social Penetration Model

social penetration model A theory that describes how intimacy can be achieved via the breadth and depth of self-disclosure.

Johari Window A model that describes the relationship between self-disclosure and self-awareness.

less disclosive.[23] Their **social penetration model** (Figure 8.1) proposes that communication occurs within two dimensions: (1) *breadth*, which represents the range of subjects being discussed and (2) *depth*, how in-depth and personal the information is.

For example, as you start to reveal information about your personal life to coworkers—perhaps what you did over the weekend or stories about your family— the breadth of disclosure in your relationship expands. The depth may also expand if you shift from relatively nonrevealing messages (*"I went out with friends."*) to more personal ones (*"I went on this awful date set up by my mom's friend. . . ."*).

What makes the disclosure in some messages deeper than in others? Some revelations are more *significant*. Consider the difference between saying, *"I love my family"* and *"I love you."* Other statements qualify as deep disclosure because they are *private.* Sharing a secret you've told to only a few close friends is certainly an act of self-disclosure, but it's even more revealing to divulge information that you've never told anyone.

Each of your personal relationships has a different combination of breadth and depth. As relationships become more intimate, disclosure increases (usually gradually) in terms of both factors. However, it can be difficult to know how much to share how soon since every relationship is different. As a rule of thumb, theorists recommend the "Goldilocks principle": Judge by the other person's reaction if you are offering "too much" or "too little" and aim instead for "just right."[24]

The Johari Window   The **Johari Window** is another model that describes the value of self-disclosure.[25] Imagine a frame that contains everything there is to know about you: your likes and dislikes, your goals, your secrets, your needs—everything.

Of course, you aren't aware of everything about yourself. Like most people, you're probably discovering new things about yourself all the time. To represent this, we can divide the frame containing everything about you into two parts: the part you know about (on the left in Figure 8.2) and the part you don't know about (on the right).

We can also divide this frame in another way: The top row contains the things about you that others know, and the bottom row things about you that you keep to yourself. Altogether, the Johari Window presents *everything about you* divided into four parts. It's worth considering each block in the model separately.

- Quadrant 1 represents your *open area*, information about you that both you and your relational partner are aware of. For example, you may both know that you aspire to be a CEO one today.

- Quadrant 2 represents the *blind area*—information you are unaware of but others know. Do you get snarky after a couple of beers? Do you talk about yourself too much? Do your jokes hurt others' feelings?

- Quadrant 3 represents your *hidden area*—information that you know but aren't willing to reveal to others. Do you have romantic feelings for a good friend? Are you shyer than you let on? Items in this hidden area may eventually become known to others if you self-disclose them.

- Quadrant 4 represents information that is *unknown* to both you and others. For example, it's not unusual to discover that you have an unrecognized talent, strength, or weakness. Recognizing that you still have plenty to learn about yourself and others can foster an attitude of humility that is likely to serve you well.

Interpersonal relationships of any depth are virtually impossible if the individuals involved have little open area. You have probably found yourself in situations in which you felt frustrated trying to get

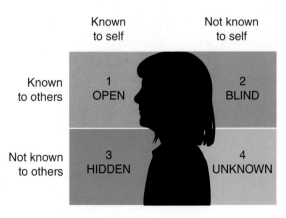

FIGURE 8.2 The Johari Window

to know someone who was too reserved. On the other hand, if you are a *Bachelor* or *Bachelorette* fan, you know the cringeworthy sensation of self-disclosure pushed to its limits. Reality TV manufactures such moments by pressuring strangers to develop intimate relationships very quickly with almost no privacy. But you needn't fall prey to the same traps in real life. As you can imagine, sharing every detail of your personal life with people you barely know isn't usually effective. The checklist in this section presents eight questions you can ask yourself to determine when and how self-disclosing may be beneficial to you and others.

Online Interpersonal Communication

Maya and Jad live in different countries and know each other only through online communication. Maya acknowledges that many people might be skeptical about the quality of their friendship, but she says, Jad is "one of my closest friends, as odd as that may seem." The two have shared personal experiences, including some they haven't shared with their so-called RL (real life) friends. "I trust him enough to ask things I can't ask face-to-face," Maya says, adding, "I like to think I have found a safe little corner where I can talk and joke with someone who I would not have met otherwise."[26]

Is the communication between Maya and Jad interpersonal? Early definitions specified that interpersonal communication had to take place in person.[27] That was presumably because theorists considered face-to-face interactions to have a richness lacking in other channels. Today, few scholars dispute the idea that interpersonal communication can occur via texts, emails, video chats, and other technical channels.[28] Based on Maya's description, her online relationship with Jad does seem to be interpersonal. In this section, we consider five of the most rewarding aspects of online interpersonal communication and two of its less appealing qualities.

Online Communication Helps People Stay Connected   Even when relational partners' routines don't mesh, technology can help them stay in touch. Perhaps for this reason, adolescents who use online communication (in moderation) typically have more cohesive friendships than other teens,[29] and couples who talk on the phone frequently often feel more loving, committed, and confident about their relationships than couples who don't.[30]

There's More Diversity Online   Most face-to-face communication networks are limited to people in a relatively small geographic region. But the number of people you can befriend online is virtually endless. "I grew up in a fairly small town, so being a sci-fi and comics nerd who loved makeup, '80s and '90s pop music, fancy cake, and sushi pretty much made me a peer group of one," reflects Rachel. Online, however, she has access to "an entire world's worth of people," including many who share her interests and passions.[31]

Online Communication Can Feel Nonthreatening   When a researcher in Turkey interviewed college students from four different parts of the world, they said they would like to have more international friends, but they felt anxious about approaching people from different cultures in person.[32] They were afraid they would say the wrong thing or not know

---

CHECKLIST ✓

Questions to Ask Yourself Before Self-Disclosing

No single style of self-disclosure is appropriate for every situation. However, there are some questions you can ask to determine when and how self-disclosure may be beneficial for you and others.[33]

☐ **Is the other person important to you?**

☐ **Is the disclosure appropriate?**

☐ **Is the risk of disclosing reasonable?**

☐ **Are the amount and type of disclosure appropriate?**

☐ **Is the disclosure relevant to the situation at hand?**

☐ **Is the disclosure reciprocated?**

☐ **Will the effect be constructive?**

☐ **Is the self-disclosure clear and fair?**

what to say. The students were less anxious about communicating *online* with people from different cultures. After the chance to strike up cross-cultural friendships in cyberspace, most students in the study said they would like to meet their new international friends in person, evidence that mediated relationships can be a gateway to face-to-face interaction.

Online Communication Can Be Validating   One appealing quality of online communication is its potential to convey social support. Posting news of the A+ you earned in English is likely to be rewarded almost instantly with "likes" and congratulations. University students who use social networking sites typically experience less stress than their peers, especially when they consider their online friends to be supportive, interpersonally attractive, and trustworthy.[34]

Online Communication Has a Pause Option . . . Sometimes   Many forms of online communication are asynchronous. They allow you to think about messages and then reply when you are ready (see Chapter 2 for more on asynchronicity). This is an advantage because you can catch mistakes or avoid blurting out something you might regret later.[35]

Online Communication Can Be Distracting   On the downside, social media use can interfere with relationships in person. Episodes in which people are inattentive to those around them because they are paying attention to their devices instead are known as **phubbing** (a combination of phoning and snubbing).[36] Researchers in one study found that the mere presence of mobile devices can have a negative effect on closeness, connection, and conversation quality during face-to-face discussions of personal topics.[37] Cyber relationships can threaten in-person ones in other ways as well. See the *Ethical Challenge* feature to consider what online activities you consider to be "cheating" on a romantic partner.

Too Much Online Communication Can Be Problematic   In moderation, social media can boost feelings of connection and identity. However, it's possible to have thousands of superficial online "friends" but few you can count on during hard times. As a consequence, communicators who spend excessive time online tend to be lonelier than their peers.[38] Considering the pros and cons of online communication, most experts agree that moderation is key and the best relationships often include both.[39]

phubbing A mixture of the words *phoning* and *snubbing*, used to describe episodes in which people pay attention to their devices rather than to the people around them.

ASK YOURSELF

Do you feel that social media use ever interferes with your ability to get things done and to be fully present with the people around you? If so, what might you do to cut down a bit?

ETHICAL CHALLENGE Is It Cheating?

Tara was furious when she found out that Michael had been looking up past girlfriends online. She saw it as cheating on their relationship. Michael said he was just curious and had no intention of connecting with any of his old flames.[40]

As with many ethical considerations, there's no hard and fast rule about what qualifies as "digital infidelity." The best bet is for romantic partners to discuss their expectations. Here are some questions to consider together:

- Does looking up previous romantic partners online constitute cheating to you? Why or why not?

- If someone in a committed relationship engages in romantic talk online with someone they will never meet in person, do you think that is cheating? Why or why not?

- How about posting sexually provocative comments or photos to no one in particular?

🔵 8.2 Communicating with Friends and Family

"You are more than just my best friend—you are the sister I never knew I needed—you are family." With these words, poet Marisa Donnelly acknowledges the powerful influence of both friends and family.[41] In this unit, we explore communication in friendship and family relationships, which are some of the longest and most influential of our lives.

From the outset, it's important to note that some people are both friend and family. Sometimes, a parent, uncle, or sibling is also a friend. At other times, as poet Marisa Donnelly reflects, "people with whom we don't share DNA or a roof over our heads" become so close that they are family.[42]

Unique Qualities of Friendship

Friendships can take many forms. Consider Jacob, who happens to be straight, and Anthony, who happens to be gay. Aware that his best friend was discouraged about not having a date to their high school prom, Jacob held up a huge homemade banner in the school hallway that read, "you're hella gay, I'm hella str8, but you're like my brother, so be my d8?" Martinez called it "the sweetest, coolest thing that has ever happened."[43] Their remarkable friendship became the toast of social media, and the friends appeared on the *Ellen DeGeneres* show to share their story.[44]

Everyone has a friend who has said or done something that made a powerful difference to them. Friendships are unique for a number of reasons:

- Friends typically treat each other as equals, unlike parent–child, teacher–student, or doctor–patient relationships, in which one partner has more authority or higher status than the other.[45]

- While you can only have a limited number of family and romantic relationships, you can have a much larger circle of friends. For example, you may have different circles of work friends, traveling friends, online friends, and so on.

Good friends help people stay healthy, boost their self-esteem, and make them feel loved and supported.[46] Friends can also help each other adjust to new challenges and uncertainty.[47] It's not surprising, then, that people with strong and lasting friendships are happier than those without them.[48]

Friendship Development

What's a recipe for a good friendship? The simple answer is time, talk, and shared activities. In one study, a researcher tracked the development of friendships in both people who had relocated to a new city and first-year college students.[49] It took an average of 50 hours of time with someone to create a casual friendship, 90 hours before they became real friends, and about 200 hours to become close friends.

How and where you spend time is another element of developing friendships. You might spend hours each day with acquaintances at work or school, but you wouldn't consider them friends. One key to developing a friendship is time spent together in different settings—like going out for coffee, seeing a concert together, or visiting each other's homes. Shared leisure activities are often a sign that a friendship is deepening.

Communication also plays an important role. Close friendships are characterized by self-disclosure and discussing personal issues. But everyday talk is also

ASK YOURSELF (?)

Think of a close friend you have known a long time. Do you communicate differently with them than with friends you have made recently? If so, how?

important: Catching up, checking in, and joking around are important ways to keep a friendship strong.

A quick survey of your social network will confirm that friendships come in many forms. See how yours compare on the dimensions described here.

Short-Term Versus Long-Term   Short-term friends tend to change as our lives do. We say goodbye because we move, graduate, or switch jobs. Or perhaps we spend less time at parties or the ball field than we used to. It's natural for our social networks to change as a result. However, long-term friends are with us even when they aren't.[50] Particularly today, with so many different ways to stay in touch, people report that—as long as trust and a sense of connection are there—they feel as close to some of their long-term friends who live far away as to those who are nearby.[51]

Low Disclosure Versus High Disclosure   Some of your friends know more about you than others. Self-disclosure is associated with greater levels of intimacy such that only a few confidants are likely to know your deepest secrets. One interesting exception occurs among people who are highly self-disclosive online. They might announce personal news to hundreds of friends and acquaintances with a single post or tweet. This isn't necessarily bad.[52] However, it's easy to cross the line and go public with information you might later wish you had kept private. As one blogger points out, you may have several hundred online "friends," but not all of them need or want to hear that you were cheated on last night.[53]

Doing-Oriented Versus Being-Oriented   Some friends experience closeness "in the doing." That is, they enjoy performing tasks or attending events together, and they feel closer because of those shared experiences.[54] In these cases, different friends are likely to be tied to particular interests—a golfing buddy or shopping partner, for example. Other friendships are "being-oriented." For these friends, the main focus is on being together, and they might get together just to talk or hang out.[55] Even long-distance friends may send texts or photos to "be" together when they are miles apart.

Low Obligation Versus High Obligation   There are probably some friends for whom you would do just about anything—no request is too big. For others, you may feel a lower sense of obligation. Cultural elements may play a role. Friends raised in low-context cultures such as in the United States are more likely than those raised in high-context cultures such as China's to express their appreciation for a friend out loud (see Chapter 4). Chinese friends are more likely to express themselves indirectly—most often by doing favors for friends and by showing gratitude and reciprocity when friends do favors for them.[56] It's easy to imagine the misunderstandings that might occur when one friend puts a high value on words and the other on actions.

Frequent Contact Versus Occasional Contact   You probably keep in close touch with some friends. Perhaps you work out, travel, socialize, or Skype daily with them. But you might connect with other friends only at reunions or via occasional phone calls or text messages. Of course, infrequent contact doesn't always correlate with levels of disclosure or emotional closeness. Some friends go years without seeing each other and then reconnect as if they were never apart.

In-Person Versus Mediated   The average person has many more online friends than physical ones—double the amount, according to one report.[57] Quantity isn't the only difference between mediated and offline friendships, however. Here are some other differences:[58,59,60]

- Online-only friends are more likely than in-person friends to lie or express hostility.

- Face-to-face friends typically depend on each other more than online friends do, especially during the early stages of their relationships.

- In-person friends are more likely to talk about topics in depth, and they typically share a deeper level of mutual understanding and commitment than online friends do.

- Not surprisingly, in-person friends tend to have more similar social networks.

These differences may become less evident as friendships progress, however. As time goes on, some online relationships become even more personal than those in person.[61]

To enhance your communication skills, no matter what type of friends you have, see the checklist on this page about being a good friend, and then read on to learn more about gender differences and friendship.

Gender and Friendship

"Thank you for adventures—late-night doughnut trips, walks around the park, food and ice cream runs, and trips to the beach. I can always count on you to be up for anything, and I know it'll always be a blast."[62] In this open letter to her best friend, Samantha Smith sings the praises of male friends. Here we explore answers to some common questions about friendship and gender.

Do men and women do friendship differently? Although there are many similarities, there are some common differences too, especially in same-sex friendships. Male–male friendships typically involve good-natured competition and a focus on tasks and events, whereas female friends tend to treat each other more as equals and to engage in emotional support and self-disclosure.[63]

Can heterosexual men and women be just friends? Women typically say yes, but men often give a decidedly iffy answer. In a study of 88 college-age male–female friendship partners, most of the women said the friendship was purely platonic, with no romantic interest on either side.[64] However, heterosexual men in the study were more likely to say that they secretly harbored romantic fantasies about their female friends, and they suspected (often wrongly, it seems) that the feeling was mutual.

Researchers speculate that men and women get their wires crossed in part because they tend to communicate differently. As we mentioned earlier, women usually expect friends to be emotionally supportive and understanding.[65] From the male perspective, this may feel more like romance than friendship. By contrast, men tend to emphasize independence and friendly competition.[66] Those behaviors may not strike women as particularly romantic.

Are there advantages to other-sex friendships? Decidedly so. For example, men often say that they find it validating when female friends encourage them to be more emotionally expressive than usual, and women say they appreciate the opportunity to speak assertively with their guy friends.[67]

There also seems to be some truth to the idea that straight women and gay men make great friends.[68] This may be because gay men and straight women tend to trust each other's advice about love and romance. Both sides typically say they enjoy getting a different perspective without the complications of a hidden sexual agenda.[69]

CHECKLIST ✔

How to Be a Good Friend

Experts suggest the following communication strategies to keep your friendships strong.

☐ **Be a good listener.** Put aside distractions and pay close attention to your friend's words and nonverbal cues.

☐ **Give advice sparingly.** Despite your good intentions, offering advice, especially when it's not requested, can come off as insensitive and condescending.

☐ **Share your feelings appropriately.** Whether you feel upset or good about your friendship, say so.[70]

☐ **Be validating and appreciative.** Find ways to let your friends know they matter to you.[71] Depending on the relationship, this might involve a thoughtful phone call, spending time together, or remembering a birthday.

☐ **Apologize and forgive.** If you slip up— perhaps by forgetting an important date or saying something that embarrasses your friend—admit the mistake, apologize sincerely, and promise to do better in the future.[72, 73]

☐ **Keep confidences.** Two of the most dreaded violations of trust are revealing private information to others and saying unkind things about friends behind their back.[74]

☐ **Give and take equally.** The best friendships are characterized by equal give and take. One payoff of being a giver is the sense that you make a difference in someone's life.[75]

☐ **Stay true through hard times.** People who believe their friends will be there for them typically experience less everyday stress and more physical and emotional resilience than other people.[76]

What Kind of Friendship Do You Have?

 Think of a particular friend and select the answers below that most accurately describe your relationship.

1. **Which of the following best describes the time you spend with this friend?**

   a. We see each other a lot. I'd really miss our time together if something prevented that.

   b. Sometimes we spend time together and sometimes not. It's not a big deal either way.

   c. We don't see each other very often, but when we do, we're as in sync as if no time has passed.

   d. We haven't spent much time together yet, but I hope we will.

2. **If you were on a long car ride together, what would you most likely talk about?**

   a. Whatever is on my mind. I can tell this friend anything.

   b. Current events or what we've been up to at school or work.

   c. Funny memories. We've had many adventures together through the years.

   d. Where we raised, what we're studying in school, and other topics to get to know each other better.

3. **If your friend were in bed with the flu for several days, what would you most likely do?**

   a. Stop by to cheer them up and help out.

   b. Send a "get well soon" text.

   c. Call to say I wish I could be there in person.

   d. I probably wouldn't know about it until later.

4. **If this friend said something that hurt your feelings, what would you probably do?**

   a. Talk about it together and repair the rift.

   b. Avoid them for a while.

   c. Let it go. It's nothing compared to all we've been through together.

   d. Rethink my desire to be friends.

INTERPRETING YOUR RESPONSES

Read the following explanations to reflect on the qualities of your friendship. More than one may apply.

Loyal

If you answered "a" more than once, this is a close a friend you can count on. You are likely to disclose a great deal to each other (page 192) and to back each other up, even when things are difficult. This is likely to be a long-term relationship (page 192), with the rewards that come from knowing that someone knows you well and is there for you no matter what.

Independent

If you answered "b" two or more times, this is a friendship that doesn't require a great deal of commitment. You are able to get together when you like without feeling a strong sense of obligation to do so (page 192). Although not as close as some friendships, this one may be valuable, particularly if other obligations claim a lot of your time right now. Not everyone has to be a best friend. Just be careful not to let your desire for independence keep you from forming strong bonds with one or two friends you can always count on.

Far Yet Close

If you answered "c" more than once, this seems to be an enduring, long-term friendship (page 192) that remains strong even though you're not able to be together in person as much as you would like. You are likely to enjoy the benefits of being emotionally connected without much obligation to do things together or for each other (page 192). It can be a great feeling to know that distance and time cannot dim the memories you have shared together. At the same time, be sure not to take this friendship for granted. A thoughtful text or call may help you feel close even when you're not together physically.

Evolving

If you answered "d" two or more times, it's likely that your friendship is still developing. Your sense of obligation (page 192) is likely to be low at this stage, as you venture to disclose more about yourselves to each other (page 192). It remains to be seen if this will be a short- or long-term relationship (page 192), but the benefits of having strong friendships suggest that it may be worth the effort to find out.

How does gender diversity figure into friendship? Perhaps the greatest byproduct of friendship is a sense of mutual acceptance and respect. The **contact hypothesis** proposes that prejudice tends to diminish when we have personal contact with people we might otherwise stereotype.[77] For example, gay men who have close friends who are straight are less likely than their peers to perceive that society judges them harshly for being gay.[78]

Likewise, college students with at least one friend who is transgender are far less likely than their peers to harbor negative attitudes about transgender people.[79] For those who shy away from transgender or gender-transitioning individuals for fear of causing offense, the advice is clear: "Treat them like you usually do; they're still a person. Try to use their preferred pronouns and treat them like they're your friend—because they still are!"[80]

Family Relationships

In today's world, defining what makes a family isn't a simple matter. Theorist Martha Minnow proposes a solution: She suggests that people who share affection and resources as a family and who think of themselves and present themselves as a family *are* a **family**.[81] Your own experiences probably tell you that this concept of a family might encompass (or exclude) bloodline relatives, adopted family members, stepparents, honorary aunts and uncles, and blended families in which the siblings were born to different parents. This makes it easy to understand why people can be hurt by questions such as *"Is he your natural son?"* and *"Is she your real mother?"* Calling some family members "natural" or "real" implies that others are fake or that they don't belong.[82]

Parenting Relationships   Power and influence play a role in any relationship but especially in the communication dynamic between parents and children. Imagine establishing the curfew for teenage members of a family as we focus on three parenting styles.[83]

- **Authoritarian** parents are strict and demanding, and they expect unquestioning obedience. You might characterize this as a "do it because I said so" style consistent with a *conformity* approach. In our curfew example, teens would be expected to follow their parents' rules, beliefs, and values without challenging them.[84]

- **Authoritative** parents are also firm, clear, and strict, but they encourage children to communicate openly with them. These parents have high expectations, but they are willing to discuss them and to listen to children's input and even negotiate the rules. This family communication pattern emphasizes *conversation*. Teens and their parents would probably negotiate the curfew by talking openly about it and listening to each other.

- **Permissive** parents do not require children to follow many rules. Based on this approach, parents and children may communicate about other topics, but they probably don't spend a lot of time setting firm guidelines such as curfews.

contact hypothesis A proposition based on evidence that prejudice tends to diminish when people have personal contact with those they might otherwise stereotype.

family People who share affection and resources as a family and who think of themselves and present themselves as a family, regardless of genetics.

ASK YOURSELF

While you were growing up, were decisions such as teen curfews decided mostly through conversation or through conformity with rules set by your parents or guardians? Or were there few rules? How do you think the communication patterns you experienced as a child affect the way you communicate now?

The kind of communication a child learns growing up can set a pattern that lasts a lifetime. *How would you describe the parenting relationship in the household where you grew up? Is this a pattern you hope to continue into the next generation? Why or why not?*

Strengthening Family Ties

Communicating with family members can be a joy and a challenge. Following are some strategies for successful communication based on experts' advice.

☐ **Share family stories.** Family stories contribute to a shared sense of identity. They also convey the idea that adversity is an inevitable part of life, and they can suggest strategies for overcoming it.[96]

☐ **Listen to each other.** People who are involved in reflection and conversation learn how to manage and express their feelings better than people who don't. They tend to have better relationships as a result.[97,98]

☐ **Negotiate privacy rules.** Privacy violations among family members can have serious consequences.[99] At the same time, too much privacy can lead family members to overlook dangerous behavior and avoid distressing but important topics. Experts suggest that families talk about and agree on privacy expectations and rules.

☐ **Coach conflict management.** Effective conflict management doesn't just happen spontaneously.[100] Families should create safe environments for discussing issues and striving for mutually agreeable solutions.

☐ **Go heavy on confirming messages.** Supportive messages from family members can give people the confidence to believe in themselves.[101]

☐ **Have fun.** Happy families make it a point to minimize distractions and spend time together on a regular basis. They establish togetherness rituals that suit their busy lives,[102] and they share in adventures, both large and small.

Most evidence suggests that children who grow up with an authoritarian/conformity pattern do not get much experience sharing their emotions and negotiating with authority figures, so it may take work to develop these skills as adults.[85]

Children who grow up with authoritative parents are typically more adept at expressing their emotions confidently and effectively as they grow older.[86,87] Authoritative parents provide the dual benefits of structure and compassion. In the words of one researcher, they are "warm, responsive, assertive without being overly intrusive or restrictive."[88]

As you might imagine, children who don't engage in much give-and-take communication about the rules as they grow up are usually less comfortable negotiating expectations in other relationships.

Siblings   In the midst of what one theorist calls the "playing and arguing, joking and bickering, caring and fighting"[89] of sibling life, children learn a great deal about themselves and how to relate to others.

Just as people who think of themselves as family *are* family, a sibling relationship isn't limited to people who share biological parents. It's more about shared life experiences. As one person who grew up in a blended family puts it, "You feel beyond annoyed when explaining your family structure and someone says, 'Oh, so you're only half sisters.'" It's tempting, she says, to reply, "Only? ONLY? Well, you're my half-friend now."[90]

Whatever the origin, sibling relationships involve an interwoven, and often paradoxical, collection of emotions. Children are likely to feel both intense loyalty and fierce competition with their brothers and sisters and to be both loving and antagonistic toward them. Here are five types of sibling relationships people might settle into as they become adults.[91]

- **Supportive siblings** talk regularly and consider themselves to be accessible and emotionally close to one other. Supportive relationships are most common among siblings who are of similar ages, particularly if they come from large families.[92]

- **Longing siblings** typically admire and respect one other. However, they interact less frequently and with less depth than they would like. This can be especially difficult for younger siblings who watch older ones move out. One teen lamented when his brother left home: "[I] look at his empty desk, the table where we would sit and talk, and [I] start bawling. . . I know I'll see him again, but nothing will be the same."[93]

- **Competitive siblings** behave as rivals, vying for scarce resources such as their parents' time and respect.[94] It's not uncommon for siblings to feel competitive as they grow up. That feeling sometimes extends into adulthood, especially if they perceive that their parents continue to play favorites.[95]

- **Apathetic siblings** are relatively indifferent toward one another. They communicate on special occasions, such as holidays or weddings, or rarely at all.

- **Hostile siblings** harbor animosity toward each other and often stop communicating.[103] Unlike apathetic siblings, who may drift apart without hard feelings, hostile siblings usually feel a sense of jealousy, resentment, and anger toward one another.

See the checklist in this section for tips on communicating well with the people you call family.

8.3 Communicating with Romantic Partners

Romantic love is the stuff of songs, fairytales, and happy endings. So it might surprise you that the butterflies-in-your-belly sense of romantic bliss isn't a great predictor of happiness. A much better indicator is the effort that couples put into their communication. Factors such as trust, agreeableness, and emotional expressiveness are primarily responsible for long-term relationship success.[104,105] In this section, we explore the role of communication in forming and sustaining romantic relationships.

Stages of Romantic Relationships

Some romances ignite quickly, whereas others grow gradually. Either way, couples are likely to progress through a series of stages as they define what they mean to each other and what they expect in terms of shared activities, exclusivity, commitment, and their public identity.

One of the best-known explanations of how communication operates in different phases of a relationship was developed by communication scholar Mark Knapp. His **developmental model** depicts five stages of intimacy development (coming together) and five stages in which people distance themselves from each other (coming apart).[106] Other researchers have suggested that the middle phases of the model can also be understood in terms of keeping stable relationships operating smoothly and satisfactorily (relational maintenance).[107] Figure 8.3 shows how Knapp's 10 stages fit into this three-part view of communication in relationships. As you read on, consider how well these stages reflect communication in the close relationships you have experienced.

developmental model (of relational maintenance) Theoretical framework based on the idea that communication patterns are different in various stages of interpersonal relationships.

Initiating   The initiating stage occurs when people first encounter each other. Knapp restricts this stage to conversation openers, such as, *"It's nice to meet you"* and *"How's it going?"* In this stage, people form first impressions and have the opportunity to present themselves in an appealing manner.

Experimenting   People enter the experimental stage when they begin to get acquainted, with questions such as *"Where are you from?"* and *"What do you do?"* Though small talk might seem meaningless, it allows you to interact with a wide range of people to determine who is worth getting to know better. For couples who meet online, experimentation may involve meeting in person for the first time. See the checklist in this section for experts' tips on making that transition.

Communication in the initiating and experimenting stage doesn't qualify as interpersonal since the participants have not yet formed a close relationship, but it may lay the foundation for a closer relationship in the future.

Intensifying   In the intensifying stage, truly interpersonal relationships develop as people begin to express how they feel about each other. It's often a time of strong emotions and optimism that may lead either to a higher level of intimacy or to the

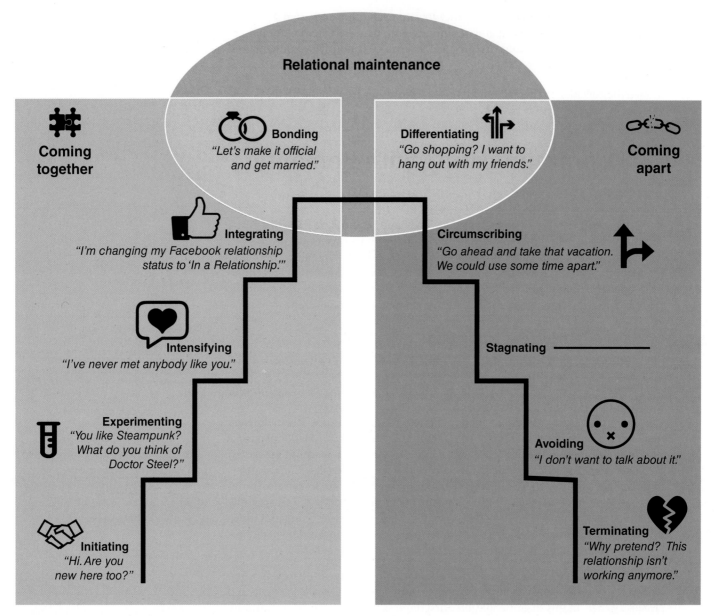

FIGURE 8.3 Knapp's Stages of Relational Development

end of the relationship if partners are not in sync. Couples often navigate this uncertainty by flirting, hinting around, asking hypothetical questions about their future, and being more affectionate than before. They might become bolder and more direct only if their partners seem receptive to these gestures.[108]

Integrating In the integration stage, couples begin to take on an identity as a social unit. Invitations may come addressed to the couple. And as it becomes a given that they will share resources and help each other, partners become comfortable making straightforward requests of each other.

Bonding The bonding stage is likely to involve a wedding, a commitment ceremony, or some other public means of communicating to the world that this is a relationship meant to last. Bonding generates social support for the relationship and demonstrates a strong sense of commitment and exclusivity.

Not all relationships last forever, however. And even when the bonds between partners are strong and enduring, it's sometimes desirable to create some distance. The following stages accomplish that.

Differentiating   In the differentiating stage, the emphasis shifts from "how we are alike" to "how we are different." For example, a couple who moves in together may find that they have different expectations about doing chores, sleeping late, what to watch on TV, and so on. This doesn't necessarily mean the relationship is doomed. Differences remind partners that they are distinct individuals. To maintain this balance, partners may claim different areas of the home for their private use and reduce their use of nicknames, gestures, and words that distinguish the relationship as intimate and unique.[109]

Circumscribing   In this stage, communication decreases significantly in quantity and quality. Rather than discuss a disagreement, partners may withdraw mentally by using silence, daydreaming, or fantasizing. They may also withdraw physically by spending less time together. Circumscribing entails a shrinking of interest and commitment.

Stagnating   If circumscribing continues, the relationship may begin to stagnate. Partners behave toward each other in old, familiar ways without much feeling. Like workers who have lost interest in their jobs yet continue to go through the motions, sadly, some couples unenthusiastically repeat the same conversations, see the same people, and follow the same routines without any sense of joy or novelty.

Avoiding   When stagnation becomes too unpleasant, partners distance themselves in more overt ways. They might use excuses, such as *"I've been busy lately,"* or direct requests, such as *"Please don't call. I don't want to talk to you."* In either case, the writing about the relationship's future is clearly on the wall.

Terminating   Characteristics of this final stage include conversations about where the relationship has gone wrong and the desire to break up. The relationship may end with a cordial dinner, a note left on the kitchen table, a phone call, or a legal document stating the dissolution. Depending on each person's feelings, this stage can be quite short, or it may be drawn out over time. See the *Understanding Communication Technology* box on page 200 for a discussion of breaking up online.

One key difference between couples who get together again after a breakup and those who go their separate ways is how well they communicate about their dissatisfaction and negotiate for a mutually appealing fresh start. Unsuccessful couples deal with their problems by avoidance, indirectness, and reduced involvement with each other. By contrast, couples who repair their relationships more often air their concerns and spend time and effort negotiating solutions to their problems.

You may assume that you can predict what will happen next if your relationship shows signs of coming together or coming apart. But couples differ in terms of how quickly they move through the stages, and just because you've drifted apart lately, the relationship isn't

CHECKLIST

Meeting an Online Date for the First Time

Following are some tips from communication researchers and dating experts to dial down the awkward and help put you and the other person at ease.

☐ **Be genuine from the beginning.** The computer-enhanced selfie may cause an online admirer to be disappointed in person. Ultimately, relationships fare best when your mediated self is a close reflection of your in-person self.[110]

☐ **Talk on the phone first.** If your online connection shows promise, see how a phone call goes before meeting in person.

☐ **Be safe.** Arrange to meet in a public setting rather than a private home, and provide your own transportation. (Avoid your favorite hangouts just in case you'd rather not run into each other again in the future.)[111]

☐ **Put romantic thoughts aside for now.** It may sound counterintuitive, but as dating advice columnist Jonathan Aslay observes, "most successful long-term relationships are built on a solid friendship."[112] Approaching the encounter as a new friendship can be less anxiety provoking and more realistic than expecting fireworks with someone you barely know.

☐ **Begin with a quick and easy meetup.** Rather than meeting for a meal on a first date, try meeting for coffee or a drink instead.[113]

☐ **Keep the tone light.** This isn't the time to make an over-the-top clothing selection, share your deepest secrets, or interrogate your date with highly personal questions. Instead, encourage conversation with casual questions, and share something (but not everything) about yourself.

☐ **Be playful.** Pay your date a sincere compliment, pose fun questions, share a light-hearted story about yourself, and don't forget to smile and laugh.[114]

UNDERSTANDING COMMUNICATION TECHNOLOGY

To End This Romance, Just Press "Send"

It was the middle of a workday two weeks ago, and Larry was in a meeting when a text message appeared on his phone screen. He glanced at it and thought: "You can't be serious."

It was no joke. His girlfriend was breaking up with him . . . again. And she was doing it by email . . . again. For the sixth time in eight months, she had ended their relationship electronically rather than face to face. He had sensed trouble—he had been opening his email with trepidation for weeks—so the previous day he had suggested that they meet in person to talk things over. But she nixed that, instead choosing to send the latest in what Larry had begun to consider part of a virtual genre: "the goodbye email."

Understandably, he'd like to say his own goodbye to that genre. "Email is horrible," says Larry, 36, a U.S. Air Force sergeant from New Hampshire. "You can't have dialogue. You don't have that person in front of you. You just have that black-and-white text. It's a very cold way of communicating."

Cold, maybe. Popular, though. The use of email and instant messaging to end intimate relationships is popular because technology makes it easy—some say too easy—to call the whole thing off. Want to avoid one of those squirmy,

awkward breakup scenes? Want to control the dialogue while removing facial expressions, vocal inflections, and body language from the equation? A solution is as near as your keyboard or cell phone.

Sometimes there is a legitimate reason for avoiding personal contact. Tara, a 32-year-old woman who lives near Boston, says her ex-husband was intimidating and emotionally abusive during their marriage. So, when she wanted to end the marriage several years ago, she felt more comfortable doing so via a text message.

Tara says that since then she has ended several other relationships by email. "I'm a softie, and I hate hurting people's feelings," she says. Recently, she laid the groundwork for breaking her engagement with a series of emails to her fiancé. After ending the engagement last week, she reached a moment of truth, she says, and has decided that from now on, if she wants to call it quits, "the email option is out."

Do you ever use online communication to say things you would be uncomfortable saying in person? When do you think this strategy is effective? When might it be unfair to the relationship or the other person?

ASK YOURSELF

The intensifying stage can be both exciting and unsettling. Is there a time when you felt vulnerable and wondered if you were setting yourself up to get hurt? Conversely, have you ever been so cautious that you lost the chance to get to know someone better? What did you learn from these experiences?

necessarily hopeless. Communication plays a significant role in the process. A number of practical lessons emerge from the developmental perspective:

- *Each stage requires different types of communication.* If you don't find yourself sharing secrets on your first date, don't worry. Partners may find that talking about highly personal issues deepens their bond in the intensifying stage but can be overwhelming sooner than that. Likewise, the polite behavior of the first two stages may seem cool and distant as intimacy increases. Every relationship, and every stage of involvement, has its own pace and rhythm.

- *Partners can change the direction a relationship is headed.* They may recognize the early signs of coming apart in time to reverse the trend. For example, if they realize they are differentiating or stagnating, they might refresh their relationship by doing more of the things they did while "coming together," such as going on dates, sharing feelings, and pursuing new experiences together.[115]

- *Relational development involves risk and vulnerability.* At any stage—even those associated with coming together—the relationship may falter. Intimacy only evolves if people are willing to take a chance of becoming gradually more self-disclosive.[116] Your knowledge of relational stages can help you understand whether the relationship is trending more toward the positive or the negative.

Love Languages

Relationship counselor Gary Chapman observes that people typically orient differently to five common love languages.[117]

Affirming Words   This language includes compliments, thanks, and statements that express love and commitment. Even when you know someone loves and values you, it's nice to hear it in words. The happiest couples continue to flirt with each other, even after they have been together for many years.[118]

Quality Time   Some people show love by completing tasks together, talking, or engaging in some other mutually enjoyable activity. The good news is that, even when people can't be together physically, talking about quality time can be an important means of expressing love. For example, partners separated by military deployments often say they feel closer to each other just talking about everyday activities and future plans.[119]

Acts of Service   People may show love by performing favors such as caring for each other when they are sick, doing the dishes, or making meals. Committed couples report that sharing daily tasks is the most frequent way they show their love and commitment.[120] Although each person need not contribute in exactly the same way, an overall sense that they are putting forth equal effort is essential to long-term happiness.[121]

Gifts   It's no coincidence that we buy gifts for loved ones on Valentine's Day and other occasions such as birthdays and anniversaries. For some people, receiving a gift—even an inexpensive or free one such as a flower from the garden or a handmade card—adds to their sense of being loved and valued.[122]

Physical Touch   Loving touch may involve a hug, a kiss, a pat on the back, or having sex. Researchers in one study asked couples to increase the number of times they kissed each other. Six weeks later, the couples' stress levels and relational satisfaction, and even their cholesterol levels, had significantly improved.[123]

A good deal of research supports the potency of love languages in promoting harmony.[124,125] Most people value all the languages to some degree, but they give some greater weight than others. Keep in mind that people differ in terms of what love languages they prefer. Good intentions may lead you astray if you assume that your partner feels the same way you do. The golden rule—do unto others as you would have them do unto you—isn't very useful when your partner's love language preferences differ from yours.[126]

You can learn more about love languages in your life by completing the *Understanding Your Communication* quiz on the next page.

Male and Female Intimacy Styles

The Internet is loaded with advice on understanding your romantic partner better. It ranges from "just because she says things are fine, don't assume she means it"[127] to "we [men] crave hugs and hand-holding too. And no, it doesn't always have to lead to sex."[128] Here's what research shows about the connection between gender and communication in romantic relationships.

Are women better than men at the lovey-dovey stuff? Until recently, most social scientists believed that women were better at developing and maintaining intimate relationships than men. This belief grew from the assumption that the most important ingredients of intimacy are sharing personal information and showing emotions. Most research *does* show that women (taken as a group, of course) *are* more willing than men to share their thoughts and feelings.[129] However, male–female differences aren't as great as they seem,[130] and emotional expression isn't the *only* way to develop close relationships. (Keep reading.)

ASK YOURSELF

Have you ever experienced a sense of drifting apart in a relationship? If so, what communication strategies did you use to either increase or decrease the emotional distance? Will you do anything differently if you find yourself in a similar situation again?

Sometimes the love language that resonates with one partner isn't meaningful to the other.

If you're in a romantic relationship, what's your love language? How well does it match your partner's? How can you communicate in ways that speak to your partner's love language?"

UNDERSTANDING YOUR COMMUNICATION

What Is Your Love Language?

 Answer the following questions to learn more about the love languages you prefer. If you're in a romantic relationship, consider inviting your partner to answer the same questions and then compare your responses. You may be surprised to find they aren't identical.

1. You have had a stressful time working on a team project. The best thing your romantic partner can do for you is:

a. Set aside distractions to spend some time with you

b. Do your chores so you can relax

c. Give you a big hug

d. Pamper you with a dessert you love

e. Tell you the team is lucky to have someone as talented as you

2. What is your favorite way to show that you care?

a. Go somewhere special together

b. Do a favor without being asked

c. Hold hands and sit close together

d. Surprise your romantic partner with a little treat

e. Tell your loved one how you feel in writing

3. With which of the following do you most agree?

a. The most lovable thing someone can do is give you their undivided attention.

b. Actions speak louder than words.

c. A loving touch says more than words can express.

d. Your dearest possessions are things loved ones have given you.

e. People don't say "I love you" nearly enough.

4. Your anniversary is coming up. Which of the following appeals to you most?

a. An afternoon together, just the two of you

b. A romantic, home-cooked dinner (you don't have to lift a finger)

c. A relaxing massage by candlelight

d. A photo album of good times you have shared

e. A homemade card that lists the qualities your romantic partner loves about you

INTERPRETING YOUR RESPONSES

For insight about your primary love languages, see which of the following best describes your answers.

Quality Time

If you answered "a" to one or more questions, you feel loved when people set aside life's distractions to spend time with you. Keep in mind that everyone defines quality time a bit differently. It may mean a thoughtful phone call during a busy day, a picnic in the park, or a few minutes every evening to share news about the day.

Acts of Service

Answering "b" means you feel loved when people do thoughtful things for you, such as washing your car, helping you with a repair job, bringing you breakfast in bed, or bathing the children so you can put your feet up. Even small gestures say *"I love you"* to people whose love language involves acts of service.

Physical Touch

Options labeled "c" are associated with the comfort and pleasure you get from physical affection. If your partner texts to say, "Wish we were snuggled up together!" they are speaking the language of touch.

Gifts

If you chose "d," you treasure thoughtful gifts from loved ones. Your prized possessions are likely to include items that look inconsequential to others but have sentimental value to you because of who gave them to you.

Words of Affirmation

Options labeled "e" refer to words that make us feel loved and valued, perhaps in a card, a song, or a text. To people who speak this love language, hearing that they are loved (and why) is the sweetest message imaginable.

My boyfriend considers it quality time when we do yard work or go fishing together, says one woman. Am I missing something? Whereas women typically value personal talk, men often demonstrate caring by doing things for their partners and spending time with them. It's easy to imagine the misunderstandings that result from different expectations. Indeed, women's most frequent complaint is that men don't stop to focus on "the relationship" enough.[131] Men, however, are more likely to complain about what women do or don't do in a behavioral sense. For example, they may consider it highly significant if a woman doesn't call when she says she will.

What about sex? Are men and women on the same page? Some are. But whereas many women think of sex as a way to express intimacy that has already developed, men are more likely to see it as a way to *create* that intimacy.[132] In this sense, the man who encourages sex early in a relationship or after a fight may view it as a way to build closeness. By contrast, the woman who views personal talk as the pathway to intimacy may resist the idea of physical closeness before the emotional side of the relationship has been discussed.

Much of the research talks about male–female couples. What happens when partners are of the same sex? Research is limited so far, but much of it suggests that, on average, long-term same-sex partners typically match up well in terms of supportive communication,[133] emotional closeness,[134] and the effort each partner puts into maintaining the relationship. Researchers speculate that, while same-sex couples face mostly the same challenges as anyone else, they have probably been socialized to communicate in similar ways and to have similar expectations.

Having considered the differences between friendship, family relationships, and romance, we conclude the chapter with an exploration of two dynamics that affect all relationships to one degree or another—dialectical tensions and deception.

8.4 Relational Dialectics

Relationships can feel like a balancing act. You want connection but also independence. You want to share your thoughts and feelings but also to have some privacy. You want the relationship to be fresh but still have predictability you can count on.

The model of **relational dialectics** suggests that partners in every close relationship constantly seek a balance between opposing forces such as togetherness versus independence, sharing versus privacy, and comfortable routines versus new adventures.[135] As you read about each set of opposing needs, consider how they operate in your life.

> **relational dialectics** The perspective that partners in interpersonal relationships must deal with simultaneous and opposing forces of connection versus autonomy, predictability versus novelty, and openness versus privacy.

Connection Versus Autonomy

The conflicting desires for togetherness and independence are embodied in the *connection–autonomy dialectic*. One of the most common reasons for breaking up is that one partner doesn't satisfy the other's need for connection:[136]

"We barely spent any time together."
"(S)he wasn't committed to the friendship."
"We had different needs."

But relationships can also split up for the opposite reason. One partner, or perhaps both, may feel stifled by what seem like excessive demands for staying connected.[137] In this case, complaints sound like this:

"She won't give me any space."
"He wants to be together all the time."
"(S)he is too needy."

Source: ©2006 Zits Partnership Distributed by King Features Syndicate Inc.

At different stages, the desire for connection or autonomy can change. Author Desmond Morris suggests that each of us repeatedly goes through three stages: *"Hold me tight," "Put me down,"* and *"Leave me alone."*[138] In marriages and other committed relationships, for example, the *"Hold me tight"* bonds of the first year are often followed by a desire for independence. This need for autonomy can manifest in many ways, such as making friends or engaging in activities that don't include one's partner or making a career move that might disrupt the relationship. Movement toward autonomy may lead to a breakup, but it can also be part of a cycle that redefines the relationship and allows partners to recapture or even surpass the closeness they had previously. For example, you might find that spending some time apart makes you miss and appreciate your partner more than ever.

Openness Versus Privacy

As you read earlier in the chapter, self-disclosure is one characteristic of interpersonal relationships. Yet, along with the need for intimacy, experience has probably shown you that you have an equally important need to keep some thoughts and experiences to yourself, even in the strongest of relationships. These sometimes-conflicting drives create the *openness–privacy dialectic*. When the drive for openness is strong, you might find yourself making demands for sharing personal information:

"What's on your mind?"
"How are you really feeling?"

But when you're on the receiving end of such demands, you might think or say things like the following:

"Stop pushing so hard!"
"Don't try to read my mind!"

Predictability Versus Novelty

Some relationships stay fresh through new experiences. At the same time, shared routines can create a sense of security. These opposing needs represent different ends of the *predictability–novelty dialectic*.

As the cartoon on this page shows, too much predictability can suck the excitement out of a relationship. But unpredictability can lead to shock and doubt about the relationship. *"I don't know who you are anymore,"* you might say or hear.

Dialectical tensions are a fact of life in close relationships, and there are a number of strategies people can use to manage them.[139]

- *Relationships involve continual change and negotiation.* Relational partners who understand dialectical tensions can give up the unrealistic notion that they will always be in sync or that negotiating relationship options should be effortless.

- *Partners can be in sync in some ways but not in others.* Recognizing different dialectical tensions may help you identify what's going on when you feel the tension of opposing drives.

- *It may be tempting to deny opposing tensions or over-correct.* For example, a romantic partner feeling that things have become stale might avoid talking about the issue, leave the relationship, or have an affair when it may have been healthier to pursue new hobbies together or plan an exciting vacation.

8.5 Lies and Evasions

People lie more than they probably realize—on average, once or twice per day[140] and even more when they meet someone new. Upon first meeting, the average is about three lies in the first 10 minutes, especially when romantic attraction is a factor.[141]

Not all lies are self-serving. At least some of the lies people tell are intended to be kind or polite. Lies can do the greatest damage when the relationship is intense, the importance of the subject is high, and there have been previous doubts about the deceiver's honesty. Of these three factors, the one most likely to cause a relational crisis is the sense that one's partner lied about something important.[142]

Experts suggest that, if you're considering deception, imagine how others would respond if they knew about it.[143] Would they accept your reasons for being untruthful, or would they be hurt by them? In light of that, we explore three types of lies here: altruistic lies, evasions, and self-serving lies.

Altruistic Lies

Some lies aim to protect other's feelings (see Table 8.1). For example, you might tell the host of a dinner party that the food was delicious even if it wasn't. Or you might compliment your significant other's new haircut or tattoo to avoid hurt feelings. **Altruistic lies** are defined—at least by the people who tell them—as

> **altruistic lies** Deception intended to be unmalicious, or even helpful, to the person to whom it is told.

TABLE 8.1

Types of Altruistic Lies

REASON	EXAMPLE
Acquire resources	"Oh, please let me add this class. If I don't get in, I'll never graduate on time!"
Protect resources	"I'd like to lend you the money, but I'm short myself."
Initiate and continue interaction	"Excuse me, I'm lost. Do you live around here?"
Avoid conflict	"It's not a big deal. We can do it your way. Really."
Avoid interaction or take leave	"That sounds like fun, but I'm busy Saturday night." "Oh, look what time it is! I've got to run!"
Present a competent image	"Sure, I understand. No problem."
Increase social desirability	"Yeah, I've done a fair amount of skiing."

Source: Adapted from categories originally presented in Camden, C., Motley, M. T., & Wilson, A. (1984, Fall). White lies in interpersonal communication: A taxonomy and preliminary investigation of social motivations. *Western Journal of Speech Communication, 48,* 315.

Altruistic lies are meant to spare people's feelings, but self-serving lies are often hurtful and manipulative.

Have you ever been caught telling a self-serving lie to a loved one or been hurt by a lie told by someone else? If so, how did it affect your relationship?

being harmless, or even helpful, to the person to whom they are told.[144] For the most part, "white lies" such as these fall in the category of being polite, and effective communicators know how and when to use them without causing offense.

Evasions

evasion The act of making a deliberately vague statement.

Unlike outright lies, **evasions** are vague statements that help speakers avoid telling the entire truth. Often motivated by good intentions, evasions are based on the belief that less clarity can be beneficial for the sender, the receiver, or sometimes both.[145] For instance, when your partner asks what you think of an awful outfit, you could *equivocate* by saying something truthful but vague, as in *"It's really unusual— one of a kind!"* Or you might *hint* when trying to escape a party by saying to the host, *"It's getting late,"* rather than, *"I'm bored and want to leave now."*

Self-Serving Lies

Self-serving lies are attempts to manipulate the listener into believing something that is untrue—not primarily to protect the listener but to advance the deceiver's agenda. For example, people might lie on their income tax returns or deny that they're under the influence if a cop pulls them over.

Self-serving lies may involve an *omission*—withholding information that another person deserves to know, or a *fabrication*—deliberately misleading another person for one's own benefit. For example, a romantic partner may keep a love affair secret or claim to be somewhere they weren't.

It's no surprise that self-serving lies can destroy trust and lead the deceived person to wonder what else their partner might be lying about. However, some couples rebound from serious deceptions, particularly if the lie involves an isolated incident and the wrongdoer's apology seems sincere.[146]

Although few of us will end up on a reality TV show, perhaps we can learn a few lessons from them about relationships: (1) Very often, it's not what you say but how you say it; (2) sharing either too much too soon or nothing about yourself can derail a relationship; and (3) the lies you tell today may be publicly revealed tomorrow. Former *Bachelorette* star Jillian Harris adds one more to the list: "It sounds cliché, but be yourself."[147]

MAKING THE GRADE

At www.oup.com/he/adler-uhc14e, you will find a variety of resources to enhance your understanding, including video clips, animations, self-quizzes, additional activities, audio and video summaries, interactive self-assessments, and more.

OBJECTIVE 8.1 Explain how content and relational meaning, metacommunication, self-disclosure, and online communication influence interpersonal relationships.

- Interpersonal communication involves two-way communication between people who are part of a close and irreplaceable relationship in which they treat each other as unique individuals.

- We typically gravitate to people who have a good deal in common with us, have characteristics that complement our own, like us back, open up to us, and offer rewards that are worth the costs required to maintain the relationship.

- Interpersonal communication consists of both content (literal) messages and relational (usually implied) messages that suggest how we feel about the other person in terms of affinity, respect, and control.

- Metacommunication involves interpersonal exchanges in which the parties talk about the nature of their interaction. It is communication about communication, as in *"Are you being serious?"*

- The social penetration model describes how intimacy can be achieved via the breadth and depth of self-disclosure.

- The Johari Window describes the relationship between self-disclosure and self-awareness in terms of what you know (and don't know) about yourself and what others know (and don't know) about you.

- Online communication can facilitate connections and social support that might otherwise be difficult or intimidating. However, people who overuse technology may find that it detracts from their in-person relationships and can lead them to feel lonely and isolated.

> List all the people you communicate with in one day. How many of these encounters are relatively impersonal? How many are interpersonal? What functions do both types of relationships serve in your life?

> Transcribe a recent conversation, including as much detail as you can about what people said and how they said it. See if you can identify examples of content and relational meaning, metacommunication, and self-disclosure.

> How might you maximize the benefits of online communication in relationships that are important to you? How might you minimize the potential drawbacks?

OBJECTIVE 8.2 Identify common communication patterns in friendships, parent–child dynamics, and sibling relationships.

- Friendships vary in terms of how long they last, how much the friends share with each other, what they do together, how obligated they feel toward one another, and how they communicate.

- Parents have an influence on whether children grow up to value *conversation* or *conformity* as a means of solving problems.

- Sibling relationships often involve a complex mixture of camaraderie and competition.

> Think of your closest friend. Are you mostly similar to each other, or are your characteristics complementary? What is most rewarding about the friendship? How do your differences influence the way you communicate together?

> Does an authoritarian, authoritative, or permissive parenting style appeal to you most? Why?

> If there is anyone in your life you consider to be a sibling, which of the styles described on page 196 best describes your relationship? How does that style influence the way you communicate with each other?

OBJECTIVE 8.3 **Describe stages of romantic relationships and options for conveying intimate messages.**

- Romantic relationships typically pass through stages of coming together (initiating, experimenting, intensifying), sustaining the relationship (integrating, bonding, differentiating, and circumscribing), and sometimes, of coming apart (stagnating, avoiding, and terminating).

- People typically value some love languages more than others. While all are important, it's a mistake to assume that everyone's preferences are the same as yours.

- Social ideas about gender and romance may influence how people express their emotions and how well their expectations match up with their partners'.

  > Create a timeline of a relationship (a friendship or romance), including key turning points. What did you say and do to get better acquainted? How did the way you communicated early on affect what happened between you? Does the relationship reflect any of the stages in the developmental model presented on pages 197–199?

  > Which of the love languages (affirming words, quality time, acts of service, gifts, or physical touch) are most meaningful to you? Which are most meaningful to the significant people in your life?

  > Do you identify more with using words or actions to convey how much you care?

OBJECTIVE 8.4 **Compare dialectical continua and strategies for managing them.**

- The relational dialectic perspective calls attention to the way relational partners negotiate a balance between opposing desires.

  > Draw three lines on a sheet of paper. On one line, write "autonomy" at one end and "connection" at the other end. On another line, write "openness" at "privacy" at opposite ends. On the third line, label one end "predictability" and the other end "novelty." Think of a close relationship, and draw a star on each continuum representing *your* personal preference (e.g., mostly prefer autonomy, closer to the middle, or mostly prefer connection). Then draw a circle on each continuum where you think your relational *partner* usually falls. In what ways are you similar? How are you different?

OBJECTIVE 8.5 **Analyze the functions served by altruistic lies, evasions, and self-serving lies.**

- People are likely to experience deceit even in their closest and most intimate relationships.

- Altruistic lies fall in the category of being polite, and effective communicators know how to use them without causing offense.

- Evasions are deliberately vague and include equivocation and hinting. They are generally meant to avoid hurting people's feelings.

- Self-serving lies are attempts to manipulate the listener into believing something that is untrue. They involve omissions or fabrications.

  > Keep a tally of how many white lies and evasive comments you make in a single day. What were your reasons for making each of them?

  > Consider a self-serving lie you have communicated in a close relationship. Looking back would you do anything differently? Why or why not?

KEY TERMS

affinity p. 186

altruistic lies p. 205

contact hypothesis p. 195

content message p. 186

control p. 186

**developmental models
(of relational maintenance)** p. 197

evasion p. 206

family p. 195

immediacy p. 186

interpersonal communication p. 183

Johari Window p. 188

metacommunication p. 187

phubbing p. 190

relational dialectics p. 203

relational message p. 186

respect p. 186

self-disclosure p. 187

social exchange theory p. 184

social penetration model p. 188

ACTIVITIES

1. Draw two columns on a sheet of paper.

   - In one column, list the people who are closest to you.

   - In the other column, list people you are acquainted with but whom you don't know well.

   - List several things you might say to people in one column that you probably wouldn't say to people in the other.

2. Answer the following questions as you think about a relationship with a person who is very important to you.

- Is the relationship more in a getting-to-know-you stage or more in an intensifying or sustaining stage? How does the stage you are in influence your communication with each other?

- What love languages do you use most? Are you satisfied with these, or could you do a better job showing how much you care?

Managing Conflict

9

CHAPTER OUTLINE

9.1 Understanding Interpersonal Conflict 213

Expressed Struggle
Interdependence
Perceived Incompatible Goals
Perceived Scarce Resources

9.2 Communication Climates 214

Confirming and Disconfirming Messages
How Communication Climates Develop

9.3 Conflict Communication Styles 220

Nonassertiveness
Indirect Communication
Passive Aggression
Direct Aggression
Assertiveness

9.4 Negotiation Strategies 225

Win–Lose
Lose–Lose
Compromise
Win–Win

9.5 Social Influences on Conflict Communication 232

Gender and Conflict Style
Cultural Approaches to Conflict

MAKING THE GRADE **236**

KEY TERMS **237**

ACTIVITIES **237**

LEARNING OBJECTIVES

9.1

Explain the key facets of interpersonal conflict, including expressed struggle, interdependence, and the perception of incompatible goals and scarce resources.

9.2

Describe the role of communication climate and relational spirals, and practice communication strategies for keeping relationships healthy.

9.3

Identify characteristics of nonassertive, indirect, passive aggressive, directly aggressive, and assertive communication, and explain how these conflict approaches vary.

9.4

Explain the differences between win–lose, lose–lose, compromising, and win–win negotiation strategies, and apply the steps involved in achieving win–win solutions.

9.5

Compare and contrast conflict management approaches that differ by gender and culture.

Consider the relevance
of relational climate and
conflict to your own life.

Is the emotional tone of your
most important relationships
warm and welcoming, stag-
nant, or chilly and unsatisfying? How so?

Recall a recent verbal or nonverbal message that made
you feel good about yourself.
Now think of one that made
you feel frustrated or unappreciated. What was different
about these episodes?

What happened the last time
you openly disagreed with
someone? Was your relationship with that person better
or worse afterward?

IT WAS A MAGIC MOMENT—a couple who met as lifeguards 25 years before, now happily married with children, sharing a nostalgic swim in a beautiful lake. As the couple paused to tread water, "our eyes met," remembers the wife. "I let my sentiments roam freely, tenderly telling Steve, 'I'm so glad we decided to do this together.'" She luxuriated in the moment, expecting "an equally gushing response." Instead, Steve said, "Yeah. Water's good," and starting paddling again.[1]

As quickly and unexpectedly as that, the seeds of conflict can emerge. It's no one's fault, necessarily. Goals and expectations differ. When they do, hurt feelings and frustration can quickly escalate into resentment or arguments.

The woman sharing a nostalgic swim with her husband was Brené Brown, a social work scholar and author of numerous books about embracing one's imperfections and daring to be vulnerable. That doesn't make her impervious to hurt feelings, of course. *Didn't he hear me?* she remembers thinking, as her husband swam away. She reflects, "My emotional reaction was embarrassment, with shame rising."[2]

Hurt but not defeated, Brené decided to try again when she and her husband reached the opposite shore of the lake. "I flashed a smile in hopes of softening him up and doubled down on my bid for connection," she recalls. She again looked Steve in the eyes, and this time said, "This is so great. I love that we're doing this. I feel so close to you." He replied, "Yep. Good swim," and swam away toward the original shore. Brené was left feeling not only disappointed but indignant.[3]

You've probably found yourself at odds with someone who is important to you. Conflict management is one of the biggest challenges we face—whether with romantic partners, friends, coworkers, or family members.

Although you may wish every conflict you experience could be resolved or disappear, that doesn't always happen. Sometimes, that's a good thing. Communicating about a conflict may stimulate creative thinking and deeper understanding. At the very least, it gives you and your relational partner the chance to put your thoughts into words. As you will see here, those goals can be even more valuable than finding a neat and tidy solution.

We don't control other people's behavior, so we can't guarantee particular outcomes. But we can approach conflict in a constructive and collaborative way. The main point of this chapter is that managing conflict skillfully can often lead to healthier, stronger, and more satisfying relationships. We will explore:

- factors that define and shape conflict communication,
- the importance of communication climates relevant to conflict management,
- conflict management styles,
- negotiation strategies that influence who wins and who loses, and
- conflict management patterns associated with gender and culture.

As famed problem-solver Bernard Meltzer once said, "If you have learned how to disagree without being disagreeable, then you have discovered the secret of getting along—whether it be business, family relations, or life itself."[4]

9.1 Understanding Interpersonal Conflict

Regardless of what we may wish for or dream about, a conflict-free world just doesn't exist. Even the best communicators, the luckiest people, are bound to find themselves in situations in which their needs don't match the needs of others. Money, time, power, sex, humor, and aesthetic taste, as well as a thousand other issues, arise and keep us from living in a state of perpetual agreement.

Whatever form it may take, every interpersonal **conflict** involves an expressed struggle between at least two interdependent parties who perceive incompatible goals, scarce resources, and/or interference from one another in achieving their goals.[5] A closer look at four parts of this definition helps illustrate the conditions that give rise to interpersonal conflict.

Expressed Struggle

There are times when we fume to ourselves rather than expressing our frustration. You may be upset for months because a neighbor's loud music keeps you from sleeping. That's most accurately described as internal conflict. Actual interpersonal conflict requires that both parties know a disagreement exists, such as when you let the neighbor know that you don't appreciate the decibel level. You might say this in words. Or you might use nonverbal cues, as in giving the neighbor a mean look, avoiding them, or slamming your windows shut. One way or another, once both parties know that a problem exists, it's an interpersonal conflict. In Brené Brown's swimming story, the conflict has yet to be expressed, but it will be.

Interdependence

However antagonistic they might feel toward each other, the parties in a conflict are usually dependent on each other. The welfare and satisfaction of one depend on the actions of another. After all, if they didn't need each other to solve the problem, they could solve it themselves or go their separate ways. Although this seems obvious from a distance, many people don't realize it in the midst of a disagreement. One of the first steps toward resolving a conflict is to take the attitude that "we're in this together."

Perceived Incompatible Goals

Conflicts often look as if one party's gain will be another's loss. If your neighbor turns down their loud music, they lose the enjoyment of hearing it the way they want, but if they keep the volume up, then you're awake and unhappy. It helps to

> **conflict** An expressed struggle between at least two interdependent parties who perceive incompatible goals, scarce rewards, and/or interference from the other party in achieving their goals.

ASK YOURSELF (?)

Think of a time when you and a relational partner experienced conflict. What goals and resources were involved? Were you able to express your feelings to each other and reach a mutually satisfying conclusion? Why or why not?

No matter how satisfying your relationships, some degree of conflict is inevitable.

When do you find yourself most at odds with the people who matter most? How do you handle conflicts when they arise?

realize that goals often are not as oppositional as they seem. Solutions may exist that allow both parties to get what they want. For instance, you might achieve peace and quiet by closing your windows and getting the neighbor to do the same. You might use earplugs. Or perhaps the neighbor could get a set of headphones and listen to the music at full volume without bothering anyone. If any of these solutions proves workable, then the conflict disappears.

Unfortunately, people often fail to see mutually satisfying answers to their problems. And as long as they *perceive* their goals to be mutually exclusive, they may create a self-fulfilling prophecy in which the conflict is very real.

Perceived Scarce Resources

In a conflict, people often believe that there isn't enough of the desired resource to go around. That's one reason conflict so often involves money. If a person asks for a pay raise and the boss would rather keep the money or use it to expand the business, then the two parties are in conflict.

Time is another scarce commodity. As authors, we constantly struggle with how to use the limited time we have to spend. Should we work on this book? Visit with our partners? Spend time with our kids? Enjoy the luxury of being alone? With only 24 hours in a day, we're bound to end up in conflicts with our families, editors, students, and friends—all of whom want more of our time than we have available to give. You probably know the feeling well.

Having laid out the ingredients for conflict and acknowledged that it's a fact of life, let's turn our attention to ways that people can manage conflict effectively and even use it to strength their relationships. Creating a healthy relational climate is a good first step.

🔵 9.2 Communication Climates

As Brené and Steve swam back across the lake, she envisioned the day unfolding in a pattern they had enacted many times before when they were frustrated with each other. She predicted that Steve would say, "What's for breakfast, babe?" and she would roll her eyes and say: "Gee, Steve. I forgot how vacation works. I forgot that I'm in charge of breakfast. And lunch. And dinner. And laundry. And packing and goggles. And . . ."[6]

communication climate The emotional tone of a relationship as it is expressed in the messages that the partners send and receive.

You get the point. Every relationship has a **communication climate**—an emotional tone. It's a lot like the weather. Some communication climates are fair and warm, whereas others are stormy and cold. Some are polluted and others healthy. Some relationships have stable climates, whereas others change dramatically—calm one moment and turbulent the next. Although the sun was shining, Brené predicted that a metaphorical dark cloud was brewing for her and her husband.

A communication climate doesn't involve specific activities as much as the way people feel about one another as they carry out those activities. Consider two communication classes, for example. Both meet for the same length of time and follow the same syllabus. It's easy to imagine how one of these classes might be a friendly, comfortable place to learn, whereas the other might be cold and tense—even hostile. The same principle holds for families, coworkers, and other relationships. Communication climates are a function more of the way people feel about one another than of the tasks they perform.

Communication climate influences how people respond when conflict emerges in a relationship. As you will see in the following section, some relationships involve trust and respect, whereas others are steeped in criticism and defensiveness.

Source: Ted Goff, North America Syndicate, 1994.

Confirming and Disconfirming Messages

What makes some relationship climates positive and others negative? A short but accurate answer is that the *communication climate is determined by the degree to which people see themselves as valued.* When you believe that the other person views you as important, you are likely to feel good about the relationship. By contrast, the relational climate suffers when you think others don't appreciate or care about you. Here we consider two types of messages that shape relational climates.

Disconfirming Messages   A message is considered **disconfirming** if it denies the value of another person.[7] Disagreeing with someone can be disconfirming, and it can be hurtful to point out something that bothers you about another person. That's not to say you will (or should) always agree with other people or find their behavior 100% acceptable. The point is more to handle those inevitable conflicts fairly and respectfully. The checklist on page 216 provides rules for fighting fair that should help in this regard.

Unfortunately, people sometimes handle conflict in ways that erode their relationships. John Gottman, who has spent more than four decades studying how people communicate, can predict with a rate of accuracy approaching 90% whether or not a married couple is headed toward divorce, mostly on the basis of their disconfirming behaviors.[8] Gottman calls the most hurtful of these the "Four Horsemen of the Apocalypse" because, when they are present on a regular basis, a relationship is usually is serious trouble and unlikely to survive.[9] Although Gottman studied married couples, the same types of messages can damage all types of relationships. As you read about the Four Horsemen, consider if you are ever guilty of any of them with partners, friends, roommates, family members, or anyone else you know.

- *Partners criticize each other.* Whereas it can be healthy for relational partners tactfully to point out specific behaviors that cause problems (*"I wish you would let me know when you're running late"*), **criticism** goes beyond that to deliver a personal, all-encompassing accusation such as *"You're lazy"* or *"The only person you think about is yourself."*

- *Partners show contempt.* **Contempt** takes criticism to an even more hurtful level by mocking, belittling, or ridiculing the other person (*"People laugh at you behind your back"* or *"You disgust me"*). Whereas criticism implies *"You are flawed,"* contempt implies *"I hate you."*[10] Expressions of contempt may be explicit, but they are more often expressed nonverbally—by sneering, eye rolling, and a condescending tone of voice. Gottman flatly states that the single best single predictor of divorce is contempt.[11] Experience probably shows you that it often spells doom in other relationships as well.

disconfirming messages Actions and words that imply a lack of agreement or respect for another person.

criticism A message that is personal, all-encompassing, and accusatory.

contempt Verbal and nonverbal messages that ridicule or belittle the other person.

Rules for Fighting Fair

Here are eight ways to engage in relationship-friendly conflict management, based on the work of Jack Gibb.[16]

☐ **Avoid judgment statements.** Don't make "you" statements, such as *"You don't know what you're talking about"* or *"You smoke too much,"* which are likely to cause defensiveness and escalate conflict.

☐ **Use "I" language.** Use statements such as *"I get frustrated when you interrupt me"* that focus on a specific behavior and the speaker's thoughts and feelings about it.

☐ **Be honest.** Think about what you want to say, and plan the wording of your message carefully so that you can express yourself clearly.

☐ **Show empathy.** Empathic messages show that you accept another person's feelings and can put yourself in their place. You might say, *"I can understand why you thought I was ignoring you at the party."*

☐ **Treat others as your equal.** Demonstrate that you are willing to listen to others and consider their needs and goals, not just your own.

☐ **Avoid attempts to control or manipulate others.** Be careful not to impose your preferences without regard for other people's needs or interests. For example, avoid guilt-provoking proclamations such as *"If you cared about me, you would…"* Instead, share your feelings and invite the other person to do the same.

☐ **Don't be a know-it-all.** Acknowledge that you don't have a lock on the truth. You might say, *"My impression is that the candidate has very little experience. What do you know about him?"*

☐ **Focus on mutually beneficial problem solving.** You can help build a healthy relational climate by being respectful, being a good listener, and seeking solutions that satisfy both your needs and the other person's.

- *Partners are defensive.* When faced with criticism and contempt, it's not surprising that partners react with **defensiveness**—protecting their self-worth by counterattacking (*"You're calling me a careless driver? You're the one who got a speeding ticket last month"*). Once an attack-and-defend pattern develops, conflict often escalates or partners start to avoid each other.

- *One or both partners engage in stonewalling.* **Stonewalling** is a form of avoidance in which one person refuses to engage with the other. Walking away or giving one's partner the silent treatment sends the message *"You aren't even worth my attention."* Disengagement may seem like a better option than arguing, but it robs partners of the chance to understand each other better.

These are the big offenders on Gottman's list, but people may engage in a number of other disconfirming messages as well. Table 9.1 lists a variety of tactics people use to create distance between themselves and others. It's easy to see how each of them is inherently disconfirming.

It's important to note that disconfirming messages, like virtually every other kind of communication, are a matter of perception. That's why it can be a good idea to engage in perception checking before jumping to conclusions: *"Were you laughing at my joke because you think I look stupid, or was it something else?"* You might find that a message you thought was disconfirming was actually delivered with good intentions.

Confirming Messages   Consider times when someone made you feel good with a compliment, a smile, or encouraging words. **Confirming messages** convey that you are valued by implying that "you exist," "you matter," "you are important."[12] Brené was trying to engage Steve in a confirming exchange when she told him she was glad to be there with him. She remembers how she felt when she didn't receive the validation she had expected in return: *"I thought What's going on? I don't know if I'm supposed to feel humiliated or hostile. I wanted to cry and I wanted to scream."*[13]

On the other side, put yourself in Steve's shoes. He might have been at a loss about how to respond. We can learn some valuable tools from scholars who have identified three main categories of confirming communication.[14] Here are those categories, in order from the most basic to the most powerful.

- *Show recognition.* Recognition seems easy and obvious, and yet there are many times when we don't respond to others on this basic level. Brené remembers that when Steve tossed off his "Yep. Good swim" response, "he seemed to be looking through me rather than at me."[15] Your friends may feel a similar sense of being invisible or ignored if you don't return phone messages or if you fail to say hi when you encounter each other at a party or on the street. Of course, this lack of recognition may simply be an oversight. You might not notice your friend, or the pressures of work and school might prevent you from staying in touch. Nonetheless, if the other person *perceives* you as avoiding contact, the message has the effect of being disconfirming. The *Understanding Communication Technology* feature on page 218 provides tips for

TABLE 9.1

Distancing Tactics

TACTIC	DESCRIPTION
Avoidance	Evading the other person
Deception	Lying to or misleading the other person
Disrespect	Treating the other person in a degrading way
Detachment	Acting emotionally uninterested in the other person
Discounting	Disregarding or minimizing the importance of what the other person says
Humoring	Not taking the other person seriously
Impersonal demeanor	Treating the other person like a stranger; interacting with that person as a role rather than a unique individual
Inattention	Not paying attention to the other person
Nonimmediacy	Displaying verbal or nonverbal clues that minimize interest, closeness, or availability
Reserve	Being unusually quiet and uncommunicative
Restraint	Curtailing normal social behaviors
Restriction of topics	Limiting conversation to less personal topics
Shortening of interaction	Ending conversations as quickly as possible

Source: Adapted from Hess, J. A. (2002). Distance regulation in personal relationships: The development of a conceptual model and a test of representational validity. *Journal of Social and Personal Relationships, 19*, 663–683.

making sure you aren't so preoccupied with high-tech gadgets that you neglect people around you.

- *Acknowledge the other person's thoughts and feelings.* Acknowledging the ideas and emotions of others is an even stronger form of confirmation than simply recognizing them. Attentive listening is probably the most common form of acknowledgment. Not surprisingly, leaders who are supportive of others and their ideas are more successful than those who are more concerned with promoting their own image and ideas.[17]

- *Show that you agree.* Whereas acknowledgment means you are interested in other people's ideas, endorsement means that you agree with them. Not surprisingly, we tend to be attracted to people who agree with us.[18] The message is: *We have a lot in common and are in sync.* You can probably find something in a message to endorse even if you don't agree with it entirely. You might say, *"I can see why you were so angry,"* to a friend, even if you don't approve of their outburst. Of course, outright praise is a strong form of endorsement and one you can use surprisingly often if you look for opportunities to compliment others.

It's hard to overstate the importance of confirming messages. People who offer confirmation generously are considered to be more appealing candidates for marriage than their less appreciative peers.[19] Confirming messages are just as important in other relationships. Family members are most satisfied when they regularly encourage each other, joke around, and share news about their day.[20] And in the classroom, motivation and learning increase when teachers demonstrate a genuine interest in and concern for students.[21]

How Communication Climates Develop

When we left Brené and Steve on their swim across the lake, she was feeling hurt and was already imagining the bickering that might lay ahead for them. One

ASK YOURSELF

How might you react if your partner is upset with you and you don't think you have done anything wrong? Do your answers suggest that you are ever guilty of criticism, contempt, defensiveness, or stonewalling? If so, how might you behave differently to avoid damaging your relationship?

defensiveness Protecting oneself by counterattacking the other person.

stonewalling Refusing to engage with the other person.

confirming messages Actions and words that express respect for another person.

UNDERSTANDING COMMUNICATION TECHNOLOGY

Can You Hear Me Now?

Thanks to technology, people have never been more connected—or more alienated.

I have traveled 36 hours to a conference on robotic technology in central Japan. The grand ballroom is Wi-Fi enabled, and the speaker is using the web for his presentation. Laptops are open, fingers are flying. But the audience is not listening. Most seem to be doing their email, downloading files, surfing the web, or looking for a cartoon to illustrate an upcoming presentation. Every once in a while audience members give the speaker some attention, lowering their laptop screens in a kind of digital curtsy.

In the hallway outside the plenary session, attendees are on their phones or using laptops and PDAs to check their email. Clusters of people chat with one another, making dinner plans, "networking" in that old sense of the term—the sense that

implies sharing a meal. But at this conference it is clear that what people mostly want from public space is to be alone with their personal networks. It is good to come together physically, but it is more important to stay tethered to the people who define one's virtual identity, the identity that counts.

We live in techno-enthusiastic times, and we are most likely to celebrate our gadgets. Certainly the advertising that sells us our devices has us working from beautiful, remote locations that signal our status. We are connected, tethered, so important that our physical presence is no longer required. There is much talk of new efficiencies; we can work from anywhere and all the time. But tethered life is complex; it is helpful to measure our thrilling new networks against what they may be doing to us as people.[22]

— *Sherry Turkle*

---

relational spiral A reciprocal communication pattern in which each person's message reinforces the other's.

escalatory conflict spiral A reciprocal pattern of communication in which messages, either confirming or disconfirming, between two or more communicators reinforce one another.

avoidance spiral Occurs when relational partners reduce their dependence on one another, withdraw, and become less invested in the relationship.

positive spiral Occurs when one person's confirming message leads to a similar or even more confirming response from the other person.

challenge of conflict management is that we tend to feel defensive and angry when our expectations are thwarted or we don't agree with our relational partners. One comment can escalate into hours of snide comments or tense silence.

A **relational spiral** is a communication pattern in which one person's behavior is followed by a similar or even more intense response by another person, which tends to inspire an even greater reaction in the first person, and so on.[23] There is a natural tendency to strike back when one's feelings are hurt, as captured in the old saying "what goes around comes around." But acting defensively can make a difficult situation even worse. The good news is that relational spirals aren't always negative.

Escalatory Conflict Spirals   In an **escalatory conflict spiral**, one perceived slight leads to another until the communication escalates into a full-fledged dispute.[24] Perhaps you feel that your friend said something unkind about you, so you say insulting things about your friend, and the cycle continues until you are both even more hurt and furious.

Avoidance Spirals   Not communicating can also be destructive. In **avoidance** spirals, rather than fighting, individuals slowly lessen their dependence on one another, withdraw, and become less invested in the relationship.[25] If you have ever said, *"I'm not calling them. If they want to talk, they can call me,"* you've been part of an avoidance spiral. Even the best relationships can go through periods of conflict and withdrawal. However, too much negativity may lead to a "point of no return" from which the relationship cannot be saved.

Positive Spirals   Fortunately, spirals can escalate in beneficial ways as well. In **positive spirals**, one person's confirming message leads to a similar response from the other person, and so on.[26] Offering a sincere compliment, apology, invitation, or simply one's undivided attention can inspire more of the same from your relational partner.

UNDERSTANDING YOUR COMMUNICATION

What's the Forecast for Your Communication Climate?

Think of an important person in your life—perhaps a friend, a roommate, a family member, or a romantic partner. Choose the option in each group below that best describes how you communicate with each other.

1. **When I am upset about something, my relational partner is most likely to:**

   a. Listen to me and provide emotional support

   b. Say I should have tried harder to fix or avoid the problem

   c. Ignore how I feel

2. **When we are planning a weekend activity and I want to do something my partner doesn't want to do, I tend to:**

   a. Suggest another option we will both enjoy

   b. Beg until I get my way

   c. Cancel our plans and engage in the activity with someone else

3. **When my partner and I disagree about a controversial subject, we usually:**

   a. Ask questions and listen to the other person's viewpoint

b. Accuse the other person of using poor judgment or ignoring the facts

   c. Avoid the subject

4. **If I didn't hear from my partner for a while, I would probably:**

   a. Call or text to make sure everything is okay

   b. Feel angry about being ignored

   c. Not notice

5. **The statement we are most likely to make during a typical conversation sounds something like this:**

   a. "I appreciate the way you . . ."

   b. "You always forget to . . ."

   c. "Were you saying something?"

INTERPRETING YOUR RESPONSES
Read the following explanations for a climate report about your relationship.

Warm and Sunny
If the majority of your answers are "a," your relational climate is warm and sunny, with a high probability of confirming messages (pages 216–217). You seem to be experiencing a positive spiral (page 218). Use suggestions throughout this chapter to strengthen and nurture your relationship even more.

Stormy
If you answered mostly "b," your relationship tends to be turbulent, with outbreaks of controlling or defensive behavior (page 216). Storm warning: You seem to be in a downward escalatory conflict spiral (page 216) that can damage your relationship. That's not to say it's hopeless, but you may want to consider underlying feelings—yours and the other person's. Guidance on self-disclosure (pages 187–189) may be helpful.

Chill in the Air
If most of your answers are "c," beware of falling temperatures. It's natural for people to drift apart sometimes, but your relationship shows signs of chilly indifference and avoidance spiraling (page 218). Consider whether you are guilty of the damaging patterns described on pages 215–216. You may be able to change the weather by engaging in more supportive communication (pages 216–217).

It often feels that relational spirals have a life of their own. People may be inclined, even without thinking about it, to mirror and escalate their partners' behaviors, even if they are harmful to the relationship. The best communicators recognize this tendency and make mindful choices instead. They may switch from negative to positive messages without discussing the matter. Or they may engage in metacommunication (Chapter 8). *"Hold on,"* one might say, *"This is getting us nowhere."*

Take the "What's the Forecast for Your Relational Climate?" quiz on this page and then follow along as we take a closer look at useful approaches and techniques for managing conflict.

9.3 Conflict Communication Styles

When Steve and Brené reached the dock where their swim had started, she decided to talk about her problem. Rather than blaming Steve for her hurt feelings, which was likely to escalate the conflict, she said to him instead, "I've been trying to connect with you and you keep blowing me off. I don't get it."[27]

Consider what you might have done in a similar situation. Are you prone to avoiding sensitive issues? Do you tend to hint around when something upsets you, or are you likely to say outright how you feel?

In this section, we consider five conflict communication styles, which are common patterns of behavior. As you will see, each style varies on two dimensions: concern for self and concern for others.[28] As you read, ask yourself which styles you use most often and how these styles affect the quality of your close relationships.

Nonassertiveness

nonassertion The inability or unwillingness to express one's thoughts or feelings.

The inability or unwillingness to express one's thoughts or feelings in a conflict is known as **nonassertion**. A nonassertive person may insist that "nothing is wrong" even when it is. Sometimes nonassertion comes from a lack of confidence. At other times, people lack the awareness or skills to use a more direct means of expression.

Nonassertion can take a variety of forms. One is *avoidance*—either steering clear of the other person or avoiding the topic, perhaps by talking about something else, joking, or denying that a problem exists. People who avoid conflicts usually believe it's easier to put up with the status quo than to face the problem head-on and try to solve it. *Accommodation* is another type of nonassertive response. Accommodators deal with conflict by giving in, thus putting others' needs ahead of their own.

COMMUNICATION TECHNOLOGY

You Can't Take It Back

Patricia is so furious with members of her project group that she sends out an angry group text calling them "lazy and irresponsible." Later, she wishes she had handled the situation differently.

Disagreements handled via texting, chatting, email, and blogging can unfold differently from those that play out in person. Some of the characteristics of mediated communication described in Chapter 2 are especially important during conflicts. Following are a few of the pros and cons of using online communication to address a conflict. Review these and then consider what advice you would offer Patricia and her team.

- For better or worse, it's often tempting to say things online that we might not say to people face to face.

- Mediated communication offers a chance to cool down and think carefully before posting a message or replying

to one. On the down side, people can easily ignore online posts and fail to respond to emails, texts, and IMs.

- Because emails and text messages come in written form, there's a permanent "transcript" that doesn't exist when communicators deal with conflict face to face. This can help clarify misperceptions and faulty memories. On the other hand, we may wish we could take back some messages that are now part of the record forever.

What advice do you have for Patricia now that she has already sent out the angry group text? What advice would you offer her teammates who receive it? What communication strategies might you suggest so the team can be deal with this conflict and others in the future?

While nonassertion won't solve a difficult or long-term problem, here are a few situations in which accommodating or avoiding is a sensible approach:

- *The conflict is minor or short-lived.* Avoidance may be the best choice if the matter isn't serious. You might let a colleague's occasional grumpiness pass without saying anything, knowing they are under a lot of stress.

- *The relationship is new or sensitive.* You might reasonably choose to avoid conflict with some people in your life. For example, you might not object if a new friend or your grandfather eats a snack you had put aside for yourself.

- *The risks are great.* You might choose to keep quiet if speaking up would put you at risk, as when it might get you fired from a job you can't afford to lose or provoke someone who might do you physical harm.

Nonassertion displays low concern for self and high concern for others. That can be a virtue. But in some cases, being "selfless" can damage your relationships. For one, you might begin to resent that people don't "listen to you" or honor your preferences, even though you may not be voicing your thoughts in a way others can understand. For another, relational partners may feel that they don't know the real you, and they may be frustrated that you don't give them honest feedback about what you like and don't like. As one analyst puts it, "trying to make everyone happy can make you [and others] miserable."[29] The *Ethical Challenge* feature explores the dilemma of downplaying one's preferences.

Indirect Communication

Whereas a nonassertive person resists dealing with conflict at all, someone using an indirect style addresses the conflict but in subtle ways. **Indirect communication** conveys a message in a roundabout manner. The goal is to get what you want without causing hard feelings. Consider the case of the neighbor's loud music. One indirect approach would be to strike up a friendly conversation with the neighbor and ask if anything you are doing is too noisy, hoping they get the hint.

Because indirect communication saves face for the other party, it is often kinder than blunt honesty. If your guests are staying too long, it's probably more polite to yawn and hint about your big day tomorrow than to bluntly ask them to leave. Likewise, if you're not interested in going out with someone who has asked you for a date, it may be more compassionate to claim that you're busy than to say, *"I'm not interested in seeing you."*

At other times, we communicate indirectly to protect ourselves. You might, for example, joke around about being underpaid rather than directly asking the boss

> **indirect communication** Hinting at a message instead of expressing thoughts and feelings directly.

ETHICAL CHALLENGE "It's Nothing!"

Anticipating their 10-year wedding anniversary, Alex asked his wife Danielle what she would like. She said, "Just surprise me!" So Alex planned a surprise get-away at a bed and breakfast Danielle likes. She didn't seem thrilled to be there, however. "I asked her what was wrong, and she kept saying, 'Nothing,'" Alex remembers. "Finally, the last day there, she told me she really had her heart set on a new wedding ring. How was I supposed to know that?"[30]

We might imagine that Danielle felt it would be unseemly to ask outright for a new ring. She might have dropped subtle hints but hoped that Alex would surprise her with a ring because he wanted to give her one, not because she asked him to.

If you have been on either side of this classic dilemma you know how hard it can be to navigate. Consider your views on the following:

- Is it ever justifiable to keep your desires to yourself, hoping that other people will pick up on them? If not, why? If so, when?

- Is it ever okay to behave as if you are angry with someone but refuse to share your feelings with that person? If so, when? How would you describe that style of conflict communication?

for a raise. At times like these, a subtle approach may get the message across while softening the blow of a negative response. The risk, of course, is that the other party will misunderstand you or will fail to get the message at all. There are times when an idea is so important that hinting lacks the necessary punch.

Indirect communication involves a moderate concern for self and others. Next we discuss two reactions to conflict that demonstrate low concern for others, and one approach that balances concern for self with concern for others.

Passive Aggression

passive aggression An indirect expression of aggression, delivered in a way that allows the sender to maintain a facade of innocence.

directly aggressive message Attacks the position and perhaps the dignity of the receiver.

assertive communication A style of communicating that directly expresses the sender's needs, thoughts, or feelings, delivered in a way that does not attack the receiver.

If you know someone who responds to conflict with unkind humor or snide comments and then acts like they didn't intend to hurt your feelings, you have experienced **passive aggression**, which occurs when a communicator expresses hostility in an ambiguous way. The hurtful but superficially "deniable" nature of passive aggression can be "crazymaking," in the words of scholar George Bach.[31] Bach describes five types of passive aggressive people who engage in "crazymaking" behavior:

- *Pseudoaccommodators* only pretend to agree with you. A passively aggressive person might commit to something (*"I'll be on time from now on"*) but not actually do it.

- *Guiltmakers* try to make you feel bad. A guiltmaker will agree to something and then make you feel responsible for the hardship it causes them (*"I really should be studying, but I'll give you a ride"*).

- *Jokers* use humor as a weapon. They might say unkind things and then insist they were "just kidding," insinuating that you are being too sensitive (*"Where's your sense of humor?"*).

- *Trivial tyrannizers* do small things to drive you crazy. Rather than express their feelings outright, they might "forget" to clean the kitchen or to put gas in the car just to annoy you.

- *Withholders* keep back something valuable. A withholder punishes others by refusing to provide thoughtful gestures such as courtesy, affection, or humor.

When passive aggression is present in the workplace, it can be frightening or demoralizing, even if it seems subtle on the surface. See the *@work* feature on page 223 for some response strategies if you ever feel harassed.

Direct Aggression

Directly aggressive people show little or no concern for others. A **directly aggressive message** confronts others in a way that attacks their position, or even their dignity. Aggressive people often use intimidation and insults to get their way. Many directly aggressive responses are easy to spot:

"You don't know what you're talking about."

"That was a stupid thing to do."

"What's the matter with you?"

Other forms of direct aggression come more from nonverbal messages than from words. It's easy to imagine a hostile way of expressing statements such as:

"What is it now?"

"I need some peace and quiet."

You may get what you want in the short run using verbal aggressiveness: Yelling *"Shut up"* might stop the other person from talking, and saying *"Get it yourself"*

Dealing with Sexual Harassment

Sexual harassment takes many forms. It can be a blatant sexual overture or a verbal or nonverbal behavior that creates a hostile work environment. Here are several options to consider if you or someone you care about experiences harassment:[32]

- **Tell the harasser to stop.** Assertively tell the harasser that the behavior is unwelcome, and insist that it stop immediately. Your statement should be firm, but it doesn't have to be angry.

- **Write a personal letter to the harasser.** A written statement may help the harasser understand what behavior you find offensive. Detail specifics about what happened, what behavior you want stopped, and how you felt. You might include a copy of your organization's sexual harassment policy. Keep a record of when you delivered your message.

- **Use company channels.** Report the situation to your supervisor, personnel office, or a committee that has been set up to consider harassment complaints.

- **File a legal complaint.** If all else fails or the incident is egregious, you may file a complaint with the federal Equal Employment Opportunity Commission or with your state agency. You have the right to obtain the services of an attorney regarding your legal options.[33]

---

may save you from some exertion. But the relational damage of this approach probably isn't worth the cost. Direct aggression can be hurtful, and the consequences for the relationship can be long lasting.[34]

In some cases, aggressive behavior is downright abusive. See the checklist in this section for experts' advice on how to stay safe.

Assertiveness

Winston Churchill is said to have proclaimed, "Courage is what it takes to stand up and speak. Courage is also what it takes to sit down and listen."[35] Assertiveness, which represents a balance between high self-interest and high concern for others, involves a good deal of both. **Assertive** people handle conflicts by expressing their needs, thoughts, and feelings clearly and inviting others to do the same. They communicate directly but without judging others or dictating to them. Assertive people have the attitude that most of the time it is possible to resolve problems to everyone's satisfaction. For example, a partner who has noticed that arguments erupt early in the week might approach the issue assertively by saying:

> I've noticed that we're often impatient with each other on Monday mornings. I think I'm especially tense because I dread the weekly staff meeting. I'd like to spend some time Sunday preparing for that meeting. I think that will make me less stressed, and maybe that will help us start the week together on a positive note. Is there something we can to do make Monday mornings less stressful for you?

As this scenario suggests, being assertive usually means talking about an issue when you have a cool head rather than in the heat of the moment. Assertive individuals avoid accusations and assumptions. Their motto might be: *We are good people with good intentions who can work this out together.* Being assertive requires self-awareness, patience, and good listening skills. It's not always easy. People who manage conflicts assertively may experience feelings of discomfort while they are

Protecting Yourself from an Abusive Partner

There are no magic communication formulas to prevent or stop the behavior of an abusive person, but there are steps you can take to help protect yourself.

☐ **Don't keep abuse a secret.** Abusers often isolate their partners from friends and loved ones because it's easier to control them if they don't have a strong network.[36] Avoid this trap by keeping close contact and open communication with people you trust. At the very least, tell someone what's happening and ask that person to assist you in getting help.

☐ **Have a plan for defense.** Program emergency numbers into your phone, and keep it handy. Agree on code words you can mention to trusted people. Avoid sharing passwords with the abuser.

☐ **Don't blame yourself.** Abused people often believe they are at fault in some way. Remember—*no one deserves abuse.* Abusive people make the choice to be abusive.[37] One source of information and assistance is www.healthyplace.com/abuse.

working through problems. However, they usually feel better about themselves and one another afterward. For example, romantic partners who approach conflict in a patient and caring way often feel closer to each other as a result.[38]

Here are five steps to follow when you want to approach a conflict assertively:

Describe the Behavior in Question   An assertive description is specific without being evaluative or judgmental.

> Behavioral description: *"You asked me to tell you what I thought of your new car, and when I told you, you said I was too critical."*

> Evaluative judgment: *"Don't be so touchy! It's hypocritical to ask for my opinion and then get mad when I give it to you."*

Judgmental words such as *touchy* and *hypocritical* invite a defensive reaction. The target of your accusation can reply, *"I'm not touchy or hypocritical!"* It's harder to argue with the facts stated in an objective, behavioral description. Furthermore, neutral language reduces the chances of a defensive reaction.

Share Your Interpretation of the Other Person's Behavior   This is where you can use the perception-checking process outlined in Chapter 3. Remember that, after referencing a specific behavior, a complete perception check includes two possible interpretations of the behavior and an invitation for the other person to respond:

> *"When it took me two days to call you back* [behavior], *perhaps you thought I didn't care* [one interpretation]. *Or you might have assumed our plans were off* [another interpretation]. *Did I hurt your feelings* [invitation to respond]?"*

The key is to label your hunches as such instead of suggesting that you are positive about what the other person's behavior means.

Describe Your Feelings   Expressing your feelings adds a new dimension to a message. For example, consider the difference between these two responses:

> *"When you call me in the middle of the day* [behavior], *I think you miss me and care about me* [interpretation], *and I feel special* [feeling]."*

> *"When you call me in the middle of the day* [behavior], *I think something must be wrong* [interpretation], *and I feel stressed* [feeling]."*

Applied to our previous example, an assertive message that conveys feeling might sound something like this:

> *"When you said I was too critical after you asked my opinion of your car* [behavior], *it seemed to me that you were disappointed* [interpretation], *and I felt bad for being so blunt* [feeling]."*

Describe the Consequences   A consequence statement explains what happens (or might happen) as a result of the behavior you have described. There are three kinds of consequences:

- What happens to you, the speaker: *"When you tease me, I'm tempted to avoid you."*
- What happens to the target of the message: *"When you drink too much, you start to drive dangerously."*
- What happens to others: *"When you play the radio so loud, it wakes up the baby."*

It's important that a consequence statement not sound like a threat or an ultimatum. It isn't meant to manipulate the other person. Instead, the goal is to explain what impact someone's behavior has, at least from your perspective.

State Your Intentions   Intention statements are the final element in the assertive format. They can communicate three kinds of messages:

- Where you stand on an issue: *"I wanted you to know how hurt I felt"* or *"I wanted you to know how much I appreciate your support."*

- Requests of others: *"I'd like to know whether you are angry"* or *"I hope you'll come again."*

- Descriptions of how you plan to act in the future: *"I've decided to stop lending you money."*

In our ongoing example, adding an intention statement would complete the assertive message:

> *"When you said I was too critical after you asked my opinion of your car* [behavior]*, it seemed to me that you were disappointed* [interpretation]*. That made me feel bad for being so blunt* [feeling]*. Now I realize that it hurt your feelings when I called your new car a gas guzzler* [consequence]*. I'm going to be more supportive and less critical in the future* [intention]*."*

It's good to know these steps in being assertive, but keep in mind that they are only a general guide. Depending on the situation, you may use a different order, combine steps, or return to some steps to make sure you both understand each other. In communication, as in many other activities, patience and persistence are essential. Take the *Understanding Your Communication* quiz to consider how assertive you usually are.

🔘 9.4 Negotiation Strategies

You may think of negotiating as a formal process—something people only do when buying a car or establishing the salary for a new job. But when you consider **negotiation** as "an interactive communication process" designed to help people reach agreement when one person wants something from another,[39] you probably realize that you negotiate more than you thought. You may negotiate to determine who will do specific household chores, what days you have off at work, where you go on a family vacation, what each group member will contribute to a shared project, and much more. In this unit, we consider four negotiation strategies. As you read about them, consider which ones you use now, and whether others might serve you better.

Win–Lose

Win–lose problem solving occurs when one party achieves their goal at the expense of someone else. People resort to this method of resolving disputes when they perceive a situation as being an "either–or" one: *Either I get what I want, or you get your way.* The most clear-cut examples of win–lose situations are games such as baseball or poker, in which the rules require a winner and a loser. Some interpersonal issues seem to fit into this win–lose framework: two coworkers seeking a promotion to the same job, or a couple who disagrees on how to spend their limited money.

Power is a distinguishing characteristic in win–lose problem solving, because it's necessary to defeat an opponent to get what you want. The most obvious kind of power is physical. Some parents threaten their children with warnings such as *"Stop misbehaving, or*

negotiation An interactive process meant to help people reach agreement when one person wants something from another.

win–lose problem solving An approach to conflict resolution in which one party reaches their goal at the expense of the other.

"It's not enough that we succeed. Cats must also fail."

Source: Leo Cullum The New Yorker Collection/The Cartoon Bank

UNDERSTANDING YOUR COMMUNICATION

How Assertive Are You?

    Choose the answer to each question that best describes you.

1. **You feel you deserve the corner office that has just become available. What would you do?**

    a. Stay quiet and hope the boss realizes that you deserve the office.

    b. Hint around that you have outgrown your cubicle.

    c. Tell your coworkers, "You deserve it more than I do," but secretly ask the boss if you can have it.

    d. Meet with your supervisor and lay out the reasons you think you deserve the office.

    e. Threaten to quit if you aren't assigned to the office.

2. **Your best friend just called to cancel your weekend trip together at the last minute. This isn't the first time your friend has done this, and you are very disappointed. What do you do?**

    a. Reassure your friend that it's okay and there are no hard feelings.

    b. Declare, "But I've already packed," hoping your friend will take the hint and decide to go after all.

    c. Resolve to cancel the next trip yourself to teach your friend a lesson.

    d. Say, "I feel disappointed, because I enjoy my time with you and because we have made nonrefundable deposits. Can we work this out?"

    e. Announce that the friendship is over. That's no way to treat someone you care about.

3. **During a classroom discussion, a fellow student makes a comment that you find offensive. What do you do?**

    a. Ignore it.

    b. Tell the instructor after class that the comment made you uncomfortable.

    c. Say nothing, but tell other people how much you dislike that person.

    d. Join the discussion, say that you see the issue differently, and invite your classmate to explain why they feel as they do.

    e. Announce that the statement is the most stupid thing you have ever heard.

4. **You are on a first date when the other person suggests seeing a movie you are sure you will hate. What do you do?**

    a. Say, "Sure!" How bad can it be?

    b. Lie and say you've already seen it.

    c. Say, "Okaaay," and raise your eyebrows in a way that suggests your date must be either stupid or kidding.

    d. Suggest that you engage in another activity instead.

    e. Proclaim that you'd rather stay home and watch old reruns than see that movie.

I'll send you to your room." Power can also involve rewards or punishments. In most jobs, supervisors have the power to decide who does what, when they will work, who is promoted, and even who is fired.

Even the usually admired democratic principle of majority rule is a win–lose method of resolving conflicts. However fair it may be, it results in one group getting its way and another group being unsatisfied.

There are some circumstances when win–lose problem solving may be necessary. For instance, if two people want to marry the same person, they can't both succeed at the same time. And it's often true that only one applicant can be hired for a job. Another circumstance in which you might engage in win–lose tactics is when someone is trying to manipulate you. See the *Ethical Challenge* box on page 227 to consider whether there are circumstances in which it is okay to fight dirty.

All the same, don't be quick to assume that your conflicts are necessarily win–lose. As you will soon read, many situations that seem to require a loser can actually be resolved to everyone's satisfaction.

EVALUATING YOUR RESPONSES

Based on your answers, see which of the following describes your usual conflict management style. (More than one may apply.)

Nonassertive

If you chose "a" multiple times, you tend to rank low on the assertiveness scale. The people around you may be unable to guess when you have a preference or hurt feelings. It may seem that "going with the flow" is the way to go, but research suggests that relationships flounder when people don't share their likes and dislikes with one another. Try voicing your feelings more clearly. People may like you more for it.

Indirect

If more than one of your answers is "b," you often know what you want, but you rely on subtlety to convey your preferences. This can be a strength because you aren't likely to offend people. However, don't be surprised if people sometimes fail to notice when you are upset. Research suggests that indirect communication works well for small concerns, but not for big ones. When the issue is important to you, step up to say so.

Passively Aggressive

Answering "c" multiple times is an indication that you are sometimes passively aggressive. Rather than taking the bull by the horns, you are more likely to seek revenge, complain to people around you, or use snide humor to make your point. These techniques can make the people around you feel belittled and frustrated. Plus, you are more likely to alienate people than to get your way in the long run. Try to break this habit by saying what you feel in a clear, calm way.

Assertive

If you chose "d" more than once, you tend to hit the bulls-eye in terms of healthy assertiveness. You say what you feel without infringing on other people's right to do the same. Your combination of respectfulness and self-confidence is likely to serve you well in relationships.

Aggressive

If "e" answers best describe your approach, you tend to overshoot assertive and land in the aggressive category instead. Although you may mean well, your comments are likely to offend and intimidate others. Try toning it down by stating your opinions (gently) and encouraging others to do the same. If you refrain from name calling and accusations, people are likely to take what you say more seriously.

ETHICAL CHALLENGE   Negotiating with a Bully

"Bullying doesn't end when you grow out of your playground days," observes negotiation specialist Alexandra Dickinson."[40] As an adult, how might you respond if someone with whom you are trying to negotiate raises their voice, makes inflammatory or insulting statements, threatens you, or refuses to listen to your perspective?

Experts suggest the following strategies:

- Remain calm. Don't try to "outbully the bully" by raising your own voice or making threats in return.[41]

- Take a break to let tempers cool.

- Try to identify the person's interests even if they aren't expressing them well. "I make a point of listening and taking notes . . . to reinforce that I'm paying close attention," says one professional.[42]

- Show sympathy and respect. Sometimes bullies just want to be heard and treated as if they matter. Meeting those needs might reduce their inclination to rant and rave.

- If nothing else works, consider leaving the room or discontinuing the conversation.

In your opinion, Is it ever justified to raise one's voice in a negotiation or argument? If not, why? If so, when?

The last time you lost your temper, what factors led you to feel that way? What might have made you feel better?

lose–lose problem solving An approach to conflict resolution in which neither party achieves its goals.

compromise An agreement that gives both parties at least some of what they wanted, although both sacrifice part of their goals.

win–win problem solving An approach to conflict resolution in which the parties work together to satisfy all their goals.

Lose–Lose

In **lose–lose problem solving**, neither side is satisfied with the outcome. Although it's hard to imagine that anyone would willingly use this approach, in truth, lose–lose is a fairly common way to handle conflicts. In many instances both parties strive to be winners, but as a result of the struggle, both end up losers. On the international scene, many wars illustrate this sad point. A nation that gains military victory at the cost of thousands of lives, large amounts of resources, and a damaged national consciousness hasn't truly won much. On a personal level the same principle holds true. Most of us have seen battles of pride in which both parties strike out and both suffer. Perhaps you are angry because your sister hasn't called in a while, so you resolve not to call her. The distance between you grows and you both lose.

Compromise

Unlike lose–lose outcomes, a **compromise** gives both parties at least some of what they wanted, although both sacrifice part of their goals. For example, imagine that one partner says, "You're either at work or school every night Monday through Thursday. I hardly ever get to see you." After talking about it, the couple might compromise by agreeing to devote two nights a week to work and two nights a week to each other.

Although a compromise may be better than nothing, it's often not the cure-all that people make it out to be. Conventional wisdom, such as "The key to a good marriage is compromise," overlooks the reality that better solutions may be available. In our example, the two-nights-a-week compromise isn't likely to be successful if one partner resents having to delay graduation by taking fewer night classes. It might be more gratifying to find a solution that allows the partners to meet their obligations *and* spend quality time together. If that sounds difficult to accomplish, you may be right. Negotiation isn't always easy. But win–win problem solving is often possible once people let go of the notion that problem solving always means that at least one person must lose or make concessions.

Win–Win

In **win–win problem solving**, the goal is to find a solution that satisfies both people's needs. Neither tries to win at the other's expense. Instead, both parties believe that, by working together, it's possible to find a solution that reaches all their goals. This is typically the most satisfying and relationship-friendly means of negotiating.

Roger Fisher and William Ury of the Harvard Negotiation Project are well-known proponents of win–win problem solving. They recommend that participants begin by focusing on their *interests* (*why* they want something) rather than their *positions* (what *solution* they think is best).[43] For example, suppose you'd like a quiet evening at home but your partner wants to go to a party. Those are your positions. Based on them, clearly one of you will win and the other will lose. However, after listening to each other and sharing your interests, you may realize that you can both get your way. You have an interest in spending a quiet evening rather than getting dressed up and talking to a room full of people. Your partner isn't crazy about going out tonight either but has an interest in connecting with two old friends who are going to be at the party. Once you understand both parties' underlying goals, a solution presents itself: Invite those two friends over for a casual dinner at your place before they head off to the party. This way, neither you nor your partner compromises on what you want to achieve. Indeed, the evening may be more enjoyable than either of you expected.

Following are the steps involved in win–win problem solving, which is consistent with being assertive. Here too, you'll probably find that, although it's good

ASK YOURSELF

Which of the conflict management styles we discussed previously (non-assertion, indirect communication, passive aggression, aggression, assertiveness) is most conducive to win–lose problem solving? Lose–lose? Compromise? Win–win?

to know all the recommended steps, you may choose to focus on some more than others, depending on the situation.

Identify Your Problem and Unmet Needs   Before you speak up, it's important to realize that if something bothers you, the problem is *yours*. Perhaps you're frustrated by your partner's tendency to yell at other drivers. Because *you* are the person who is dissatisfied, the problem is yours. Realizing this will make a big difference when the time comes to approach your partner. Instead of feeling and acting in an evaluative way, you'll be more likely to share your problem in a descriptive way, which will not only be more accurate but also will reduce the chance of a defensive reaction.

Pause to think about why your partner's behavior bothers you. Perhaps you're afraid they will offend someone you know. Maybe you're worried that their anger will lead to unsafe driving. Or maybe you would prefer to use the car ride as an opportunity for conversation.

If you feel vulnerable making your needs known, you aren't alone. Brené Brown points out that conflict can stir up deep-seated fears. "We don't know what to do with the discomfort and vulnerability," Brown says, adding that "emotion can feel terrible, even physically overwhelming. We can feel exposed, at risk, and uncertain."[44] Considering this, it's no wonder that many people avoid conflict, accommodate other's wishes, or disguise their vulnerability with aggression.

The irony, points out Brown, is that avoiding conflict and handling it badly usually make us feel worse and more disconnected from people, when what we usually want is to be understood and accepted. The good news, she says, is that we don't have to be experts at understanding emotions—ours or other people's. We need only to be curious about them in an open and nonjudgmental way. This might involve saying, *"I'm having an emotional reaction to what's happened and I want to understand."*[45]

Make a Date   Unconstructive fights often start because the initiator confronts someone who isn't ready. There are many times when a person isn't in the right frame of mind to face a conflict—perhaps owing to fatigue, being busy with something else, or not feeling well. At times like these, it's unfair to insist on having a difficult discussion without notice and expect to get the other person's full attention.

You might say, "Something's been bothering me. Can we talk about it?" If the answer is "yes," then you're ready to go further. If it isn't the right time for a serious discussion, find a time that's agreeable to both of you.

Describe Your Problem and Needs   Other people can't meet your needs without knowing why you're upset and what you want. It's up to you to describe your problem as specifically as possible. When you do, use terms that aren't overly vague or abstract, and express yourself in ways that don't cause the other person to feel judged and defensive. You might say:

> I look forward to riding home from work together because I like the chance to hear about your day and make plans for later [need/desire]. I know you get frustrated with city traffic [empathy], but I feel disappointed when you yell at other drivers instead of talking with me [problem].

Brené Brown told her husband that his brief responses had hurt her feelings, and she explained why this way:

> I feel like you're blowing me off, and the story I'm making up is either that you looked over at me while I was swimming and thought, *Man, she's getting old. She can't even swim freestyle anymore.* Or you saw me and thought *She sure as hell doesn't rock a Speedo like she did twenty-five years ago.*[46]

With this statement, Brené showed the self-awareness and courage to say outright why Steve's half-hearted responses were so painful to her. She recommends the phrase "the story I'm making up" as a way to express oneself without blaming the other person.

Check the Other Person's Understanding   After you have shared your problem and described what you need, it's important to make sure the other person has understood what you've said. As you may remember from the discussion of listening in Chapter 6, there's a good chance—especially in a stressful conflict situation—of your words being misinterpreted. If your partner says, *"You're telling me I'm a bad driver,"* you can take the opportunity to say something like, *"I'm not judging your driving. I know it's stressful. I'm just saying I'd love to have some quality time with you on the ride home."*

Ask About the Other Person's Needs   After you've made your position clear, it's time to find out what the other person needs in order to feel satisfied about the issue. There are two reasons why it's important to discover their needs. First, it's fair. After all, they have as much right as you do to feel satisfied, and if you expect help in meeting your needs, then it's reasonable that you behave in the same way toward them. Second, just as an unhappy relational partner will make it hard for you to become satisfied, a happy one will be more likely to cooperate in helping you reach your goals. Thus, it's in your own self-interest to discover and meet the other person's needs.

You might learn about the other person's needs simply by asking about them: *"Now that you know what I want and why, tell me what you need from me."* After they begin to talk, your job is to use the listening skills discussed in Chapter 6 to make sure you understand.

Back at the lake, Brené was surprised to learn that her husband was dealing with his own fears. He said he had suffered a vivid nightmare the previous night in which he had tried desperately to save all five of their children when a boat had come at them suddenly in the water. He told her:

> I don't know what you were saying to me today. I have no idea. I was fighting off a total panic attack during that entire swim. I was just trying to stay focused by counting my strokes.[47]

Brené understood. As a lifelong swimmer, she and Steve were aware of the dangers posed by sharing the waterway with motorboats. Such a nightmare would have unnerved her, too.

Check Your Understanding of the Other Person's Needs   Paraphrase or ask questions about your partner's needs until you're certain you understand them. The surest way to accomplish this is to use the paraphrasing skills you learned in Chapter 6. Perhaps the conversation reveals that your partner is frustrated because of work, or because the partner does all the driving or is hungry during the drive home.

Brainstorm Solutions and Choose One Together   Now that you understand each other's needs, the goal becomes finding a way to meet them. If your interest is quality time on the way home and your partner's interest is to avoid a stressful driving situation, think creatively about options that will allow you to meet both of those needs.

Brainstorm as many potential solutions as possible. The key word here is *quantity*. Write down every thought that comes up, no matter how unworkable it seems at first. Next, evaluate the solutions. This is the time to talk about which

ON YOUR FEET

Think of a conflict in your life. What were your needs and fears regarding the conflict? What were the other person's? Were you able to meet both of your needs? If so, how? If not, what might you have done differently? Create a brief oral presentation in which you share the lessons you learned about conflict management from this experience.

solutions are likely to meet both of your interests (win–win) and which probably wouldn't.

After evaluating the options, pick the one that looks best to both of you. Your decision doesn't have to be final, but it should seem potentially successful. It's important to be sure that you both understand and support the solution.

To go back to the driving example, perhaps you decide to meet for dinner after work and then drive home once the rush-hour traffic has subsided. Or perhaps you can take the train together and avoid rush-hour traffic, or you can alternate who drives each day. These solutions might satisfy your interest in spending time together and your partner's interest in avoiding a stressful commute. The beauty of win–win problem solving is that you will probably *both* benefit from quality time and less stress, which underscores the value of effective conflict management.

Follow Up on the Solution   You can't be sure the solution will work until you try it out. After you've tested it for a while, it's a good idea to set aside some time to talk over how things are going. You may find that you need to make some changes or even choose a different option. The idea is to keep on top of the problem and keep using creativity to solve it.

All of this being said, win–win solutions aren't always possible. There will be times when even the best-intentioned people simply won't be able to find a way of meeting all their needs. When that happens, compromising may be the most sensible approach. You will even encounter instances when pushing for your own solution is reasonable, and times when it makes sense to willingly accept the loser's role. Table 9.2 describes some factors to consider. But even when win–win problem solving isn't a perfect success, the steps we've discussed haven't been wasted. A genuine desire to learn what the other person wants and to try to satisfy those desires will build a climate of goodwill that can help you improve your relationship.

TABLE 9.2

Choosing the Most Appropriate Method of Conflict Resolution

1. Consider deferring to the other person:
   - When you discover you are wrong
   - When the issue is more important to the other person than it is to you
   - To let others learn by making their own mistakes
   - When the long-term cost of winning may not be worth the short-term gains

2. Consider compromising when:
   - There is not enough time to seek a win–win outcome
   - The issue is not important enough to negotiate at length
   - The other person is not willing to seek a win–win outcome

3. Consider competing when:
   - The issue is important and the other person will take advantage of your noncompetitive approach

4. Consider striving for a win–win solution when:
   - The issue is too important for a compromise
   - A long-term relationship between you and the other person is important
   - The other person is willing to cooperate

🌐 9.5 Social Influences on Conflict Communication

So far we have been talking about the dual goals of conflict management (concern for self and concern for others) as if they are of equal merit. But as you will see in this section, gender and cultural expectations sometimes privilege one of these interests over the other.

Gender and Conflict Style

Caroline recently experienced a difference of opinion with two different people. One, whom she calls Fran, stewed about the conflict for a week before emailing Caroline to say she was upset and wanted to resign from the project group in which the conflict took place. The other conflict was with Max, who immediately told Caroline when he was annoyed with her. Caroline and Max talked about the issue briefly and moved on. Caroline and Fran eventually made peace with one another, but only after a cooling down period and talking it through later.[48]

This example doesn't apply to all men and women, of course. You may relate more to Fran or more to Max regardless of your gender. However, it does illustrate some traditional differences in the way men and women respond to conflict. Let's explore the roots of those differences and how they may influence the way we communicate.

Physiological Differences   Biology explains some of the difference between the way males and females deal with conflict. During disagreements, men tend to experience greater physiological arousal than women, which comes in the form of increased heart rate and blood pressure. This may be why boys exhibit more aggressive behaviors than girls do, even when they are very young. About 1 in 20 male toddlers is frequently aggressive, compared with 1 in 100 female toddlers.[49]

Evolution may play a role as well. Because women are able to bear only a limited number of children, procreation has favored men who can successfully compete for their attention and demonstrate their superiority to other males. Moreover, in their traditional role as hunters and providers, men were challenged to be bold, physical risk takers.[50] It may be that men have evolved to be more physical and competitive than women because, at least in years gone by, that was an advantage. By contrast, women have traditionally nurtured children. In that role, there is an advantage to creating safe environments, working cooperatively with others, and understanding the nuances of nonverbal communication.[51] This may explain why women are often more sensitive to subtle cues and are more aligned to harmony and cooperation than to competition.

Of course, there are individual differences in physiology even within the same sex. And although biology and evolution have some influence, as we grow up, we learn to handle our emotions and mimic our role models. This means that culture plays an important role as well.[52]

As the Game of Thrones plot unfolded, Lady Sansa Stark (Sophie Turner) grew into a strong and canny noblewoman who held her own.

What social forces make it challenging for women to manage conflict on an equal footing with men?

Early Socialization   To understand common differences between the way men and women handle conflict, it helps to remember the playground. By the time boys are 6 years old, they tend to gravitate toward large groups in which there is a clear understanding of who outranks whom.[53] They have traditionally played games (think baseball, football, cops and robbers) in which competition is considered a way to earn respect and status. It's also common for boys to engage in physical tussles with each other, both as a form of play and as a means of settling disputes. Perhaps because males are taught to treat conflict and competition as natural, they are typically more likely than females to feel friendly toward their opponents after engaging them in competition.[54]

Young girls, on the other hand, tend to gravitate to interpersonal relationships. The emphasis is less on who outranks whom and more on who is closest to whom. Even in a group, girls typically know who has best-friend status. As a result, girls tend to engage in more prosocial behaviors (offering compliments, showing empathy, providing emotional support) than boys do, and girls more often shy away from direct confrontations.

Especially because children tend to have mostly same-sex friends, it's easy to imagine the misunderstandings that occur when they form other-sex relationships in their teen years and beyond. Girls may feel that boys are insensitive, boisterous, and emotionally distant, whereas boys may feel that girls are quick to get their feelings hurt but are reluctant to say outright what is bothering them. On the bright side, gender differences can be refreshing. Males typically say that they appreciate the emotional support of their female friends. And women say they enjoy the freedom to be more frank and assertive with their male friends than they usually are with their female friends.[55]

In Western cultures, females are typically expected to be accommodating and males to be competitive. But these expectations can lead to frustrating double binds, as we will discuss next.

Conflict Dilemmas   Women face a double standard: They may be judged more harshly than men if they are assertive, but they may be overlooked if they aren't. Typically, women are more likely than men to use indirect strategies instead of confronting conflicts head-on. They are also more likely to compromise and to give in to maintain relational harmony. This style works well in some situations, but not in others. In one study, men and women scored equally well on a set of mathematical challenges, but the men were twice as likely as the women to enter a tournament in which they could compete for cash prizes or raises based on their performance.[56] If they are reluctant to compete, women may be overlooked for raises, promotions, and other forms of recognition, even though they are highly qualified.

Cultural norms present a dilemma for men as well. They are typically rewarded for being competitive and assertive, but those behaviors can seem overly aggressive in close relationships. For example, when conflict arises with coworkers, friends, and loved ones, a competitive stance can make the situation worse.[57] That, coupled with men's high level of physical arousal in conflict conditions, can make interpersonal conflict particularly frustrating for them. They are more likely than women to withdraw if they become uncomfortable or fail to get their way.[58] Women may interpret this as indifference, but men often say they detach to avoid overreacting, physically and verbally, in the heat of the moment.

Gender differences that appear in face-to-face communication also persist online. When researchers compared messages posted by male and female teenagers, they found that the boys typically used assertive language, such as boasts and sexual invitations, whereas the girls used mostly cooperative language, such as compliments and questions.[59] The teens' adherence to traditional gender roles was also evident nonverbally. The girls tended to post photos of themselves in seductive, receptive poses, and the boys in more rugged, dominant poses. These personae are likely to influence how they behave when conflict arises.

Commonalities   General differences aside, it bears emphasizing that social expectations change over time and stereotypes do not always apply. The qualities men and women have in common far outnumber their differences. Put another way, although men and women differ on *average*, most of us live somewhere in the middle, where masculine and feminine styles overlap.[60] For example, men and women are roughly the same in terms of how much closeness they desire in relationships and the value they place on sharing ideas and feelings.[61]

One danger is that we may stereotype others and even ourselves. People who assume that men are aggressive and women are accommodating may notice behavior that fits these stereotypes (*"See how much he bosses her around? A typical man!"*). On the other hand, behavior that doesn't fit these preconceived ideas (accommodating men, pushy women) goes unnoticed or is criticized as "unmanly" or "unladylike."

What can we conclude about the influence of gender on conflict? Research has demonstrated that there are, indeed, some measurable differences in general. But although men and women may have characteristically different conflict styles, the individual style of each communicator is ultimately more important than a person's sex in shaping how they handle conflict.

Cultural Approaches to Conflict

How people communicate during conflicts varies widely from one culture to another. Here we consider the impact of two cultural variables we discussed in Chapter 4—individualism and collectivism, and high and low context—and a third variable, emotional expressiveness.

Individualism and Collectivism   In individualistic cultures, like many of those in the United States, the goals, rights, and needs of each person are considered important, and most people would agree that it is an individuals' right to stand up for themselves. People in such cultures typically value direct communication in which people say outright if something is bothering them.[62]

By contrast, people in collectivist cultures (more commonly in Latin America and Asia) usually consider the concerns of the group to be more important than those of any individual. Preserving and honoring the face of the other person are prime goals, and communicators go to great lengths to avoid communication that might embarrass a conversational partner. The kind of assertive behavior that might seem perfectly appropriate to an American or Canadian would seem rude and insensitive in these cultures.

High and Low Context   As you might imagine, people in low-context cultures, like that of the United States, often place a premium on being direct and literal. By contrast, people in high-context cultures, like that of Japan, more often value self-restraint and avoid confrontation. Communicators in high-context cultures derive meaning from a variety of unspoken cues, such as the situation, social conventions, and hints. For this reason, what seems like "beating around the bush" to an American might seem polite to an Asian. In Japan, for example, even a simple request like *"close the door"* may seem too straightforward. A more indirect statement such as *"it is somewhat cold today"* would be more appropriate. Or a Japanese person may glance at the door or tell a story about someone who got sick in a drafty room.[63] They may also be reluctant to simply say *"no"* to a request. A more likely answer would be, *"Let me think about it for a while,"* which anyone familiar with Japanese culture would

UNDERSTANDING DIVERSITY

They Seem to Be Arguing

"The Italian language brings out the passion in me," declares Ewa Niemiec, who is fluent in Italian, English, and Polish.[64] Like many people who are multilingual, she feels that each language evokes a different feeling. Speaking English makes her feel polite, but Italian makes her "mouthy and loud." That's not necessarily bad, she says, unless people mistake the passion for conflict or aggression.

Once, when Niemiec interviewed for a job in Italian, she found herself "bickering" with her would-be boss. She's not unusual in finding that experience enjoyable. Members of many cultures consider verbal disputes to be a form of intimacy and even a game.

Misunderstandings may arise, however, when people from emotionally reserved cultures observe or interact with people from emotionally expressive ones. Americans visiting Greece, for example, often think they are witnessing an argument when they are overhearing a friendly conversation.[65] A comparative study of American and Italian nursery school children showed that one of the Italian children's favorite pastimes was a kind of heated debating that Italians call *discussione*, which Americans would regard as arguing.

Niemiec encourages people to overcome the idea that emotional language is confrontational. "If you're planning to travel or live in Italy," she says, "be prepared for a land of strong feelings, loud voices, and even bigger hand gestures."

recognize as a refusal, but others may take to mean *"It's possible. I'll let you know my decision soon."*

Emotional Expressiveness   From the examples so far, you might expect the United States to top the charts in terms of directness when it comes to conflict management. However, a mediating factor is at play—emotional expressiveness. In this regard, the United States has a great deal in common with Asian cultures, namely, a preference for calm communication rather than heated displays of emotion.[66] From this perspective, it may seem rude, frightening, or incompetent to show intense emotion during conflict. Indeed, people who become passionate are warned about the danger of saying things they don't mean.

By contrast, in cultures that value emotional expressiveness, people who do *not* show passion are regarded as hiding their true feelings. African Americans, Arabs, Greeks, Italians, Cubans, and Russians are typically considered highly expressive.[67] To them, behaving calmly in a conflict episode may be a sign that a person is unconcerned, insincere, or untrustworthy. The *Understanding Diversity* feature in this section describes what happens when members of one culture enjoy heated, emotional discussions and members of another shy away from them.

With differences like these, it's easy to imagine how people from different cultural backgrounds might have trouble finding a conflict style that is comfortable for them both. Sometimes we don't even understand our own reactions to conflict. Many Americans find that, although they usually consider themselves to be direct and individualistic, they are fairly accommodating when it comes to conflict management.[68] This may surprise them as much as it surprises other people.

As for Brené Brown, she reflects that the conversation in the lake, which might have ended in bickering and withheld affection, instead resulted in a renewed sense of love and commitment. Expressing their fears and listening to each other brought her and Steve closer. Afterward, as the couple walked back up to the lake house, he popped her playfully with his wet towel and said, "Just so you know: You still rock a Speedo."[69]

MAKING THE GRADE

At www.oup.com/he/adler-uhc14e, you will find a variety of resources to enhance your understanding, including video clips, animations, self-quizzes, additional activities, audio and video summaries, interactive self-assessments, and more.

OBJECTIVE 9.1   Explain the key facets of interpersonal conflict, including expressed struggle, interdependence, and the perception of incompatible goals and scarce resources.

- Interpersonal conflict is an acknowledged struggle between at least two interdependent people who perceive that they have incompatible goals, scarce resources, and interference from one another in achieving their goals.

- Many people think that the existence of conflict means that there's little chance for happy relationships with others. Effective communicators know differently, however. They realize that although it's impossible to *eliminate* conflict, there are ways to *manage* it effectively.

- Managing conflict skillfully can lead to healthier, stronger, and more satisfying relationships.

  > What distinguishes interpersonal conflict from the frustration you may feel with the behavior of a stranger you will never see again?

  > Think of a significant interpersonal conflict in your own life. What goals and resources were involved?

  > Imagine that someone you care about has begun to say, *"You're right. I'm wrong"* any time conflict between the two of you emerges. Do you think this is a good or a bad sign for your relationship? Why?

OBJECTIVE 9.2   Describe the role of communication climate and relational spirals, and practice communication strategies for keeping relationships healthy.

- Communication climate refers to the emotional tone of a relationship.

- Confirming communication occurs on three increasingly positive levels: recognition, acknowledgment, and endorsement.

- Disconfirming messages deny the value of others and show a lack of respect. Four particularly damaging forms of disconfirming messages are criticism, contempt, defensiveness, and stonewalling.

- Relational spirals are reciprocal communication patterns that escalate in positive or negative ways.

- Communication strategies that enhance relational climates include using "I" language, striving for mutually satisfying options, being honest, showing empathy, and respecting other people's viewpoints.

  > Describe a confirming message you have sent to someone else. Then describe a disconfirming message you have sent, even if you didn't mean to. How did the other person react in each episode?

  > Describe a time when you and a relational partner were involved in an increasingly negative spiral. What did you do (or what might you have done) to help stop the downward spiral?

  > Identify several disconfirming messages from your own experience, and rewrite them as confirming ones using the tips for creating positive communication climates in this chapter (pages 216–217).

OBJECTIVE 9.3   Identify characteristics of nonassertive, indirect, passive aggressive, directly aggressive, and assertive communication, and explain how conflict approaches vary.

- Nonassertive behavior reflects a person's inability or unwillingness to express thoughts or feelings in a conflict, as when one engages in avoidance or accommodation.

- Indirect communication involves hinting about a conflict rather than discussing it directly.

- Passive aggressive behavior is somewhat indirect, but it has a hostile tone meant to make the recipient feel bad.

- Directly aggressive communicators seek to intimidate or belittle people rather than striving for productive solutions collaboratively.

- Assertive communicators share their feelings and goals and encourage others to do the same.

- Online communication offers people an opportunity to consider carefully before responding, but the relatively anonymous and permanent nature of online messages may escalate a conflict's intensity.

  > Describe the pros and cons of each of the following conflict styles: nonassertive, indirect, passive aggressive, directly aggressive, and assertive.

  > Think of a behavior that bothers you. Write out what you might say to the person involved, including the components of assertive messages described on pages 223–225.

  > Scan several Twitter feeds or the comments sections on a blog or YouTube video. Identify examples in which people said things they might not have in person.

OBJECTIVE 9.4 **Explain the differences among win–lose, lose–lose, compromising, and win–win negotiation strategies, and apply the steps involved in achieving win–win solutions.**

- Although win–lose and lose–lose conflicts do not sound appealing, in rare occasions they are the best option.

- Compromise is often heralded as effective conflict management, but it may not always be the best option, considering that it involves less-than-optimal results for everyone involved.

- Win–win outcomes involve goal fulfillment for everyone. This is often possible if the parties involved have the proper attitudes and skills.

  > Think of a time when you compromised to resolve a conflict. Were you satisfied with the result? Can you think of a way in which both of you might have met your goals completely?

  > List and describe the eight steps for win–win negotiating described in this chapter.

  > Imagine that you and a friend have just signed the lease on an apartment with one regular bedroom and one master suite. Using the principles of win–win negotiating, how might you work together to decide who gets which bedroom?

OBJECTIVE 9.5 **Compare and contrast conflict management approaches that differ by gender and culture.**

- Although individuals differ greatly, in broad terms, men are typically socialized to approach conflict in a competitive way and women in a cooperative way.

- In different cultures, it may be expected that people will approach conflict indirectly or that they will be expressive and direct about it.

  > Describe conflict management approaches that feel most comfortable to you. What has influenced you to respond in those ways?

  > Describe how you might adapt your conflict management approach to suit different situations, such as with your best friend, at work, with older adults in your family, and with your boss or professor.

KEY TERMS

assertive communication p. 223
avoidance spiral p. 218
communication climate p. 214
compromise p. 228
confirming messages p. 216
conflict p. 213
contempt p. 215
criticism p. 215
defensiveness p. 216
directly aggressive message p. 222
disconfirming messages p. 215
escalatory conflict spiral p. 218
indirect communication p. 221
lose–lose problem solving p. 228
negotiation p. 225
nonassertion p. 220
passive aggression p. 222
positive spiral p. 218
relational spiral p. 218
stonewalling p. 216
win–lose problem solving p. 225
win–win problem solving p. 228

ACTIVITIES

1. Deepen your understanding of how confirming and disconfirming messages create communication spirals by doing the following:

   a. Think of an interpersonal relationship. Describe several confirming or disconfirming messages that have helped create and maintain the relational climate. Be sure to identify both verbal and nonverbal messages.

   b. Show how the messages you have identified have created either negative or positive relational spirals.

2. Composing assertive responses for each of the following situations:

   a. A neighbor's barking dog is keeping you awake at night.

   b. A friend hasn't repaid the $20 she borrowed two weeks ago.

   c. Your boss made what sounded like a sarcastic remark about the way you put school before work.

   d. Now develop two assertive messages you could send to a real person in your life. Discuss how you could express these messages in a way that is appropriate for the situation and that fits your personal style.

Communicating for Career Success

CHAPTER OUTLINE

10.1 Setting the Stage for Career Success 241
Developing a Good Reputation
Managing Your Online Identity
Cultivating a Professional Network

10.2 Pursuing the Job You Want 244
Preparing Application Materials
Planning for a Job Interview
Participating in a Job Interview

10.3 Organizational Communication Factors 252
Culture in the Workplace
Patterns of Interaction
Power in the Workplace

10.4 Communicating in a Professional Environment 256
Avoiding Common Communication Mistakes
Communicating Well as a Follower
Communicating in a Professional Manner Online

MAKING THE GRADE **260**

KEY TERMS **261**

ACTIVITIES **261**

LEARNING OBJECTIVES

10.1

Describe communication behaviors of successful job candidates.

10.2

Practice communication strategies to make a good impression during employment interviews.

10.3

Explain how organizational communication patterns affect a work environment.

10.4

Describe ways to communicate effectively as a new employee in person and online.

Courtney Baxter

found a job that matches her qualifications by reaching out to someone she admired in the field.

What qualities describe your dream job?

Who might you contact to find out more about your chosen field?

How could you reach out to people who could help you advance in your career?

COURTNEY BAXTER WAS in a state of what she calls "postgraduation anxiety." She was eager to land a job relevant to her degree in gender and international studies but unsure where to begin.

On a whim, she emailed Courtney E. Martin, an author whose work about women's issues she had long admired. To her surprise, Martin wrote back with an invitation to meet in person. Baxter arrived with questions about the nature of the field and how she might break in.

Martin was so impressed with the recent grad's motivation and research that she arranged for Baxter to meet a colleague at the OpEd Project. That conversation led to a job offer. It's an entry-level position, but Baxter is thrilled to be have landed it. "I feel lucky that, at 23, I look forward to work every day," Baxter says of her big break.[1]

Employers across the board agree that communication is essential to getting a good job and excelling in the workplace. Communication skills consistently comprise 7 of the 10 qualities bosses value most highly. Communication ability even outranks technical knowledge on employers' wish lists.[2] That's no wonder: There are many reasons why good communicators flourish in the professional world:

- *Good communicators work well in teams.* "Life is a team sport," observes human resource specialist Robert Half.[3] This is especially true in the workplace, where effective teamwork is linked to successful outcomes and high morale, efficiency, problem solving, satisfaction, and loyalty to the organization.[4] Employers rank the ability to work well on teams second only to outstanding verbal communication skills.[5]

- *Good communicators enhance client satisfaction.* Pleasing communication is the number-one factor in consumer satisfaction.[6] Of particular importance are employees' listening skills, empathy, and cultural sensitivity.[7,8]

- *Good communicators build public awareness.* Skilled communicators not only promote the organization during one-on-one interactions; they can serve as brand ambassadors in front of audiences large and small. Team members might make sales pitches, inform audiences about the value of a product or service, or advocate for change or public policies. Analyzing audience needs and crafting ethical and effective messages are crucial at every level.[9] Chapters 12 through 15 will help you refine these skills.

- *Good communicators make good leaders.* Leaders spend most of their time communicating—mostly about organizational problems that boil down to poor communication, observes business analyst Matt Myatt.[10] The upshot is that communication is central to what leaders do and what they care about. Research shows that the best leaders are attentive listeners who focus on both tasks and relationships.[11]

- *Good communicators inspire others.* You have probably been inspired by a boss or colleague you would like to emulate. Phil Dourado remembers a boss who offered to resign rather than lay off employees. When the board of directors refused her resignation, his boss went to the team and said, "They've given me two months to come up with something else. So, what shall we do?" The staff worked together to cut costs by 20% and eliminated the need for layoffs. "We were all fiercely loyal to that boss for ever after," Dourado remembers.[12]

This chapter provides tips for communicating effectively in the career world. We will explore how to:

- monitor your online identity and network effectively,
- present yourself as an appealing candidate for a job of your choice,
- understand and adapt to a new work environment once you are hired, and
- communicate effectively on the job.

As you read experts' tips for impressing employers and coworkers, consider how you might use this knowledge to reach your career goals.

10.1 Setting the Stage for Career Success

Launching a successful career begins long before you apply for a job. "Ideally, college students should take steps to lay the foundation for an effective job search as early as the second semester of their freshman year," advises Mike Profita, a career development specialist with more than 20 years' experience.[13] Forward-thinking students realize that college is an audition for the work they will do later. They also know it's never too soon to invest in career-related networking. Here are some communication strategies you can use now to enhance your future success.

Developing a Good Reputation

Put yourself in the shoes of a professor, classmate, or current coworker. If you frequently show up late, they can only assume you will do the same in the future. If your work or appearance is sloppy, they probably can't imagine anything different from you. The opposite is also true. People who go above and beyond now will probably be first in line for glowing recommendations and opportunities later.

Jim Kellam, a biology professor at Saint Vincent College, devoted personal time over winter break to write an enthusiastic letter of recommendation for a student who had excelled in his classes. Would he do the same for a poor or mediocre student? No. There's too much at risk, Kellam says. For one thing, insincere praise won't help a graduate succeed, even if they get the job. For another, praising an underperformer may lead employers to believe that the program's "excellent" students aren't actually very good. Finally, Kellam says, "my reputation will be harmed because I lied or exaggerated how good the student would be for the job."[14]

The next time you do more than your share on a team project, manage a conflict skillfully, or show a great attitude even in a stressful situation, pat yourself on the back for building a good reputation.

Managing Your Online Identity

"Like it or not, your social media accounts are your brand," says one savvy media user.[15] Just as companies have brands that tell the public what makes them unique, your online identity has the power to shape how others see you. A digital presence

ASK YOURSELF

What would potential employers learn about you if they looked you up online? What might worry them? What might impress them?

The greater your online presence, the more vigilant you need to be to ensure potential employers won't find fault.

What would a potential boss think of your online identity?

networking The strategic process of meeting people and maintaining contacts to gain information and advice.

is critical in today's job market, but it's important to get it right. According to the *New York Times*, 70% of U.S. recruiters have rejected job candidates because of personal information online.[16] Here are experts' tips for creating an online presence that works in your favor.

Showcase Your Strengths and Goals   Creating an online identity isn't about fooling anyone—it's about portraying yourself in an authentic and favorable way. Take stock of your interests, talents, and goals. Make sure that people who encounter you online have a clear sense of them.[17]

Build a Professional Identity   Hoping to land a job in Washington, DC, Joseph Cadman posted a LinkedIn profile of himself in a suit and tie in front of the Capitol.[18] Make sure your online photos and information create a sense that you are ready for the career of your choice.

Avoid Embarrassing Posts   When employers review candidates' online presence (as most of them do), the most common deal breakers are evidence of drinking or partying, posts that are disrespectful toward others, and critical comments about a current or former employer.[19] Even if an off-color meme seems like a harmless joke, do you want it to represent you? Think carefully about everything you post.

Monitor Your Online Presence   Even information you think is private may be accessible online. Google yourself and see what comes up. Then expand your search to include other search engines because no individual search index will find everything on the Internet. Double-check the privacy settings on your social media accounts (although you shouldn't consider these settings foolproof), and sign up for Google Alerts to receive a notification when your name pops up online. (You might also create Google Alerts for potential employers to help stay current about them.)

Do Damage Control   Remove any incorrect, unfair, or potentially damaging information about yourself online, if you can. If you can't remove it, consider a service such as reputationdefender.com that will monitor your online identity, and ask the managers of offending websites to remove unflattering information. Of course, a far less expensive and burdensome approach is to minimize the chances of reputation damage by being on your best public behavior.

Beware of Mistaken Identities   You might find that unfavorable information pops up about someone with the same name as you. One job seeker Googled herself out of curiosity, only to find the first hit was the Facebook page of a person with the same name. Unfortunately, the profile was full of immature comments. To minimize the chance of being mistaken for someone else, you might distinguish yourself by including your middle name or middle initial on your résumé and online.

Don't Be Scared Off   With all these warnings, you may be tempted to avoid creating a digital footprint at all. That's probably not a good idea. About 7 in 10 employers check out candidates online, and nearly 6 in 10 are reluctant to interview those who have no online presence.[20] The odds are that cultivating an impressive and honest online identity will work in your favor.

Don't Stop When You Get Hired   Your social media conduct remains important once you land a job. Don't post information that disparages your employer or clients, reveals confidential information, or makes you (and hence your employer) look bad.[21] "People are looking," cautions a hiring manager.[22]

Cultivating a Professional Network

Courtney Baxter's story earlier in this chapter shows that good jobs go to proactive communicators who locate opportunities and work hard to get them. Finding and landing a job you love starts with **networking**—the process of meeting people and maintaining relationships that result in information and advice that enhance one's career. A whopping three out of four people in the workplace today obtained their jobs with the help of personal networking.[23]

A little reflection reveals why networking is such a valuable approach. If you're looking for a job, personal contacts can tell you about positions that may not even be public yet. And after you have identified a position you want, people you know can serve as references for you and give you tips on how to pursue the position most effectively.

Networking will only work, of course, if you are the kind of person others recognize as being worth endorsing. If you are willing to work hard and you have the necessary skills to do a job (or are willing to learn those skills), there are several steps you can take to create and benefit from a personal network.

Look for Networking Prospects   Besides the people in your everyday networks, you have access to a wealth of other contacts. These include former coworkers, fellow students (current or past), alumni from your school, teammates, people you've met at social and community events, professional people whose services you have used . . . the list can be long and diverse. The checklist in this section provides some tips for being in the right place to meet people who may be able to help with your career.

Engage in Online Networking   Numerous websites offer professional networking opportunities. The most popular is LinkedIn, which offers a basic membership for free. It's not too early to join such a site and set up a personal profile. It might include career-relevant projects you have been part of, volunteer work, awards, accomplishments, and interests. Above all, consider how your information will look to new contacts you want to impress. You may be proud of your membership in the National Rifle Association, Planned Parenthood, or a religious or political group, but a prospective employer might not find your affiliations so attractive.

Seek Referrals   Each contact in your immediate network has connections to other people who might be able to help you. Social scientists have verified the "six degrees of separation" hypothesis: the average number of links separating any two people in the world is indeed only half a dozen.[24] You can take advantage of this by seeking out people who are removed from your personal network by just one degree: If you ask 10 people for referrals, and each of them knows 5 others who might be able to help, you have the potential of support from 60 people.

Conduct Informational Interviews   Courtney Baxter, whose story begins this chapter, had the courage to reach out and ask for an

CHECKLIST ✓

Strategies to Meet Networking Prospects

☐ **Take part in volunteerism and service learning.** In addition to gaining experience you may use in a career, working in the community is likely to bring you in contact with civic and business leaders.

☐ **Attend lectures, forums, and networking events.** Make it your goal to speak personally with several people whose interests are similar to yours. Although it's natural to feel a little nervous and out of place at first, networking coach Darrah Brustein says, "Be yourself . . . . The people you connect with when you are authentic are the ones you'll want to stay in touch with."[25] After the event, send a personal note thanking the presenters and others you met there.

☐ **Keep up with local news.** There's no better way to know who is involved and what the latest issues are. This can help you identify potential mentors and speak knowledgeably about current issues when you meet them.

☐ **Join career-related organizations.** Many professional groups offer discounted membership fees for students. Take advantage of the opportunity to meet people in the career field you hope to join. And don't just sit in the back of the room. Volunteer to serve on committees or hold an office. Your hard work won't go unnoticed.

☐ **Use online resources.** In addition to cultivating a professional presence of your own, use online resources to identify people who can guide you. Search LinkedIn and similar sites for people who have something in common with you—perhaps you attended the same school or majored in the same discipline. Common interests can be great conversation starters.

informational interview A structured meeting in which a person seeks answers from someone whose knowledge can help them succeed.

informational interview, a structured meeting with someone whose knowledge might help you succeed. Informational interviews are usually conducted in person, but if necessary, you might also use the phone or an online meeting format. A good informational interview can help you achieve the following goals:

- learning more about a job, organization, or field
- making a positive impression on the person you are interviewing
- gaining referrals to other people who might be able to help you

Unless you know a prospective interviewer well, it's usually best to send your request for a meeting in a letter or email. Introduce yourself, explain your reason for wanting the interview, and emphasize that you're seeking information, not asking for a job. Request a specific, short amount of time for a meeting, such as 15 minutes.

In advance of an informational interview, prepare questions that focus on career-related information. For example, you might ask, "What are the three fastest-growing companies in this field?" or "What do you think about the risks of working for a startup company?" Most of the time, the best way to get information is to ask straightforward questions. But there are times when it's more gracious to be indirect. For instance, instead of asking "What's your salary?" you might ask, "What kind of salary might I expect if I ever held a position along this path?" Listen closely and ask follow-up questions. If the primary question is "Who are the best people to ask about careers in the financial planning field?," a secondary question might be, "What do you think is the best way for me to go about meeting them?"

Show Appreciation   Don't forget to thank the people in your network. Beyond a sincere thank you, take the time to maintain relationships and let your contacts know when their help has made a difference in your career advancement. Your thoughtfulness will distinguish you as the kind of person worth hiring or helping again in the future.

As the saying goes, "It isn't [just] what you know, it's who you know." Informational interviews can be a gateway to learning about your chosen career and expanding your personal network.

Who could help you gain valuable career information and contacts, and how might you approach them?

🔘 10.2 Pursuing the Job You Want

Once you've identified an interesting position, consider whether you are right for the job and vice versa. If it doesn't seem like the relationship will be mutually beneficial, keep looking. But if it seems like a good match, establish yourself as an interested and qualified candidate. Here are some strategies that should serve you well.

Preparing Application Materials

"First impressions count," says recruiter Susan Kihn, pointing out that potential employers often form their first impressions of you while skimming your résumé and cover letter. "The average employer only spends a few seconds looking at a résumé," Kihn says, so be sure that your most impressive and relevant qualifications are noticeable at a glance.[26] Here are several strategies for creating a great impression with your application materials.

Create a High-Quality Résumé   A good résumé provides a snapshot of your professional strengths and achievements and how to contact you. Since some employers keep applications on file for the future, be sure to include a permanent email address. You might use the sample in Figure 10.1 as a content guide, but keep in mind that there are different formats for different purposes. Type "create résumé" into your favorite search engine to see various options and tips for creating them.

Write a Confidence-Inspiring Cover Letter   As one expert puts it, a cover letter is "an introduction, a sales pitch, and a proposal for further action all in one."[27] Write a letter that provides a brief summary of your interests and experience. The checklist in this section and the sample cover letter in Figure 10.2 may be useful. Edit your

Rose Magnon

[mailing address]
[permanent email address]
[permanent phone number]
[URLs for online portfolio, LinkedIn page, or other
online presence]

PROFILE AND GOALS

Experienced in using social media to promote mutually beneficial collaborations in the community and with members of other cultures. Dedicated to a career in the nonprofit sector coordinating humanitarian efforts and securing funding for international partnerships.

EDUCATION

University of East Florida, Oceanview, FL
B.A. in Communication/Public Relations, 3.86 GPA

EMPLOYMENT EXPERIENCE

Communication Coordinator, University College of Arts, Social Sciences, and Humanities
2018 to present

- Responsible for internal and external college and departmental communication, including production of semiannual dean's report, social media management, creation and distribution of monthly newsletter, and university/community lecture series.

Director of Student Recruitment (temporary contract) January to August 2017

- Led nationwide marketing effort to recruit university accounting majors to take part in a new online employment platform. My efforts helped to attract 50 new students in 8 months.

ACTIVITIES & VOLUNTEER EXPERIENCE

Co-Chair of Cultural Team for United Way Global Resident Fellowship Program October to December 2018

- Developed partnerships with United Way organizations in Western Australia, France, and South Africa. Wrote influential whitepaper detailing lessons learned and recommendations for future programs.

Workplace Campaign Manager August 2009 to March 2013

- Managed $830,000 workplace campaign portfolio representing 120+ accounts with 130 volunteers. Secured new sponsorships and developed sponsorship campaign materials.

ASSOCIATION MEMBERSHIPS

- Oceanview Young Professionals, Government Affairs Council, 2017 to 2018
- Florida Public Relations Society of America, 2016 to present
- University of East Florida Forensics and Debate Team, 2017 to present

HONORS & AWARDS

- Top 3 finalist in Florida's State Collegiate Debate and Forensic Competition, 2017 and 2018
- Toastmasters Regional Impromptu Speaking Contest, 1st place 2018
- Dean's and President's List every semester since entering college in 2017

FIGURE 10.1 Sample Résumé

Rose Magnon
[mailing address]
[permanent phone number]
[permanent email address]

January 28, 2020

Renée Robinson, Executive Director
International Society for the Advancement of Children
2525 West 37th Avenue
Landersville, MD 55555

Dear Ms. Robinson:

I am interested in the public relations assistant position at the International Society for the Advancement of Children, as advertised in *The Philanthropy Newsletter*. I currently serve as a Communication Coordinator at the University of East Florida, in which position I have gained experience that would allow me to make significant contributions to your organization and its mission.

I believe I would be effective in the public relations assistant position at ISAC for three main reasons.

- I am a skilled communicator who crafts messages carefully and adapts well to different audiences and formats. I currently author three newsletters, each designed to reach a specific stakeholder group. I also curate content on numerous social media platforms including Facebook, Instagram, and Twitter. I believe my success in these endeavors would allow me to recruit sponsors for ISAC, coordinate and publicize ISAC events such as your annual Children's Festival, and share your success stories.

- I have demonstrated success in fundraising. As a volunteer workplace campaign manager, I coordinated fundraising efforts involving 100 volunteers. The result was a total contribution of more than $800,000. I would enjoy the chance to help coordinate and promote your semiannual Education for All event and other fundraisers.

- I am an accomplished speaker who enjoys interacting with audiences of all sizes. In the last few years, I have won a regional Toastmasters Competition and ranked in the Top 3 at state collegiate debate and forensics tournaments. I would like to use this skill at conferences and civic events to share stories about the good work that ISAC does.

I became interested in the International Society for the Advancement of Children while working on an international project with United Way. I particularly admire your efforts to provide education to children in impoverished areas of the world. My career goal is to coordinate humanitarian efforts, and I would be honored by the chance to help with the good work that you do.

I have attached my résumé, which includes links to my online portfolio and my LinkedIn profile, where you can see samples of my work. I hope you will consider me for the assistant public relations position. I am available for an interview at your convenience. Thanks very much for your consideration.

Sincerely,

Rose Magnon

FIGURE 10.2 Sample Cover Letter

materials thoroughly. Many employers immediately put cover letters and résumés that include typos or grammatical errors in the "no" stack. If possible, have a staff member at your school's career center critique your materials. The final documents should be clear, honest, succinct, and free of errors.

Follow Application Instructions While there may be stories of an unusual move that caught a hiring manager's attention, failing to follow instructions will most likely lead to rejection. If the posting asks you to submit your materials in PDF format, don't send them in Word. If it says "no phone calls," then don't call. As one recruiter puts it: "If you can't follow clear, simple directions, how can I trust that you will be able to give great attention to the details of your job?"[28]

Keep Organized Records Maintain a log of everyone you have communicated with (along with their contact information), when the message was sent or received, and what the exchange was about.

Planning for a Job Interview

A successful job interview can be a life-changing experience. It's worth preparing for it carefully. In this section, we consider experts' suggestions for engaging in a **selection interview**—a discussion in which you are considered for an opportunity. Employment interviews are the most common type, but the same principles apply if you are being considered for a promotion or reassignment, an award, a scholarship, or admission to a graduate program.

Do Your Research Displaying your knowledge of an organization in an interview is a terrific way to show potential employers that you're a motivated and savvy person. In some organizations, failure to demonstrate familiarity with the organization or job is an automatic disqualifier. Along with what you've learned from informational interviews, online research can reveal information about a prospective employer and the field in which you want to work. In your search engine, type the name of the organization and/or key people who work there. Use your research to prepare questions that demonstrate your familiarity with the organization and the field:

> *Do you think 5g will be replaced anytime soon by a faster data transfer protocol?*
>
> *Will the new employment regulations affect staffing?*

Prepare for Likely and Nontraditional Questions Regardless of the organization and job, most interviewers have similar concerns, which they explore with similar questions. Some also ask nontraditional questions to see how well candidates think on their feet, how they handle problems, and how creative they are.[29] Some examples include:

> *If you could have any superpower, what would it be?*
>
> *Name five uses for a stapler with no staples in it.*
>
> *If you were a sweater, what kind would you be?*

If asked one of these questions, it's important to maintain your composure and use your answers to demonstrate qualifications you would like to showcase. For example, one response to the sweater question might be, "I would be stylish and classy, but not flashy, so I would

---

CHECKLIST ✔

What to Include in a Cover Letter

☐ When possible, direct your letter to a specific person. (Be certain to get the spelling and title correct.)

☐ Indicate the position you are applying for, and introduce yourself.

☐ Briefly describe your accomplishments (be specific) as they apply to qualities and duties listed in the job posting.

☐ Demonstrate your knowledge of the company to show your initiative and interest in the job.

☐ State the next step you hope to take—usually a request for an interview.

☐ Conclude by expressing appreciation to the reader for considering you.

selection interview A formal meeting (in person or via communication technology) to evaluate a candidate for a job or other opportunity.

 ASK YOURSELF

Find a posting for a job you would like to have one day. What talents and experiences make you a good candidate for the position?

"What Is Your Greatest Weakness?"

It's a common question in job interviews and one of the toughest to answer. Whatever answer you give, try to show how awareness of your flaws makes you a desirable person to hire. Here are four ways you might respond, but there are endless possibilities. Ensure that your answers reflect your own experiences.

☐ *Discuss a weakness that can also be viewed as a strength.* "When I'm involved in a big project I tend to work too hard, and I can wear myself out."

☐ *Discuss a weakness that is unrelated to the job, and then relate it to the job.* "I'm not very interested in accounting. I'd much rather work with people selling a product I believe in."

☐ *Discuss a weakness the interviewer already knows about from your résumé, application, or the interview.* "I don't have a lot of experience in multimedia design at this early stage of my career. But, based on my experience in computer programming and my internship in graphic arts, I know that I can learn quickly."

☐ *Discuss a weakness you have been working to remedy.* "I know that being bilingual is important for this job. That's why I have enrolled in a Spanish course."

have enduring value as trends change. I would give great value for the price . . . not too cheap, but not too expensive. I would be flexible, so I could be used in many different situations." Notice that the attributes of the sweater would make the candidate a good person to hire.

The checklist in this section suggests strategies for responding to the common and challenging interview question, "What is your greatest weakness?"

Know the Law An interviewer who casually asks, "Are you married?" may be trying to make small talk, unaware that it's illegal to ask job candidates questions such as that. How do you respond if you fear your answer would jeopardize your chances of getting the job?

First, it's important to understand the law and your options as an applicant. It's illegal to ask about a person's race, color, religion, gender, sexual orientation, disabilities, national origin, marital status, or age. Questions must be related to what the U.S. government's Equal Employment Opportunity Commission (EEOC) calls *bona fide occupational qualifications.* In other words, prospective employers may only ask about topics that are related to the job at hand. For example, it's legal and useful for an interviewer to ask what languages you speak if you'll be dealing with non-native English speakers. The same question would probably be illegal if it had no relevance.

Despite the law, there's a good chance that interviewers will ask illegal questions. They are probably uninformed rather than malicious. Still, it's a good idea to prepare in advance for how you might respond. Here are several options:

- *Respond without objecting.* Answer the question, even though you know it's probably unlawful. Recognize, though, that this could open the door for other illegal questions—and perhaps even discrimination in hiring decisions.

- *Seek an explanation.* Ask the interviewer firmly and respectfully to explain why this question is related to the job: "I'm not sure how my marital status relates to my ability to do this job. Can you explain the connection?"

- *Redirect.* Shift the focus of the interview away from a question that isn't job related and toward the requirements of the position itself: "What you've said so far suggests that age is not as important for this position as knowledge of accounting. Can you tell me more about the kinds of accounting that are part of this job?"

- *Refuse.* Explain politely but firmly that you will not provide the information requested: "I'd rather not talk about my religion. That's a very private and personal matter for me."

- *Withdraw.* End the interview immediately and leave, stating your reasons firmly but professionally: "I'm uncomfortable with these questions about my personal life, and I don't see a good fit between me and this organization. Thank you for your time."

There's no single correct way to handle illegal questions. The option you choose will depend on several factors, including the apparent intent of the interviewer, the nature of the questions, your desire for the job—and finally, your level of comfort with the whole situation. Knowing your options going in may help you make an effective split-second decision if the need arises.

ETHICAL CHALLENGE Responding to Illegal Interview Questions

"You're sitting in the interview for your dream job, and it's going great," says human resources specialist Angela Smith. "You and the interviewer are really hitting it off. Then, out of the blue, she asks, 'Are you planning on having kids?'"[30] The question is illegal. What should you do?

The best bet is often a tactful dodge, Smith says. For example, you might say, "You know, I'm not quite there yet. But I am very interested in career paths at your company. Can you tell me more about that?"[31]

An interviewer may legally ask "Are you authorized to work in the U.S.?" and "Do you have a college degree?," but they are not allowed to ask where you are from or when you graduated. The first might hint at race or ethnicity, and the latter at your age.

Brainstorm how you might respond to the following questions in a way that redirects the interview to more productive (and legal) topics:

"That's a beautiful name. Where are you from?"

"I went to State College, too. When were you a student there?"

"You look familiar. Do you go to Olive Baptist Church?"

"How would you feel about working in an all-male (or all-female) work environment?"

Bring Copies of Your Résumé and Portfolio   Arrive at the interview with materials that will help the interviewer learn more about why you are ready, willing, and able to do the job. Bring extra copies of your résumé. If appropriate, also bring copies of your past work, such as reports you have helped prepare, performance reviews by former employers, drawings or designs you have created for work or school, and letters of recommendation. Besides showcasing your qualifications, items such as these demonstrate that you know how to sell yourself. Also bring the names, addresses, and phone numbers of any references you haven't listed in your résumé.

Know When and Where   Don't risk sabotaging the interview before it begins by showing up late. Be sure you are clear about the time and location of the meeting. Research parking or public transportation to be sure you aren't held up by delays. There's virtually no good excuse for showing up late. Even if the interviewer is forgiving, a bad start is likely to shake your confidence and impair your performance.

Reframe Anxiety as Enthusiasm   Feeling anxious about an employment interview is natural. Managing your feelings during an interview calls for many of the same strategies as managing your apprehension while giving a speech (Chapter 12). Realize that a certain amount of anxiety is understandable. If you can reframe those feelings as *excitement* about the prospect of landing a great job, the energy can work to your advantage.

Participating in a Job Interview

Once the day comes for your interview, it's your chance to shine. You are well prepared, and now it's time to present yourself in a positive and confident light. Here are some suggestions for doing your best.

Dress for Success   Most interviewers form opinions about applicants within the first four minutes of conversation,[32] so it makes sense to look your best. No matter what the job or company, be well groomed and neatly dressed, and

Being lost is no excuse for showing up late to an employment interview.

How can you be sure you'll arrive on time for important engagements?

Your choice of attire for an employment interview will depend on the field and organizational culture. If in doubt, the safe bet is to favor formality.

How would you dress when interviewing for your dream job?

ASK YOURSELF

Ask a friend to pose as a job interviewer. Practice introducing yourself to them while shaking hands and making eye contact. Are you happy with how clear and confident you seem? In what ways would you like to improve?

don't overdo it with makeup or accessories. The proper style of clothing can vary from one type of job or organization to another, so do some research to find out what the standards are for appropriate attire. When in doubt, particularly when applying for an office job, it's best to dress in formal and conservative business attire. It's unlikely that an employer will think less of you for being overdressed, but looking too casual can be taken as a sign that you don't take the job or the interview seriously.

Mind Your Manners   It's essential to demonstrate proper business etiquette from the moment you arrive for an interview. "You may be riding on the elevator with the head of your interview team," advises one business etiquette expert.[33] Turn off your phone before you enter the building, smile at people, put your shoulders back and head up, and don't fiddle with your clothing, hair, or belongings. In short, behave at all times as the sort of engaged, professional, and attentive coworker everyone wants on their team.

When you meet people, look them in the eye, shake hands firmly, and demonstrate an attentive listening posture—shoulders parallel to the speaker's, eyes focused on the speaker, and facial expressions that show you are paying attention. If multiple people are present, be sure to shake hands with all of them and include them in your comments and eye contact throughout the interview.

Follow the Interviewer's Lead   Let the interviewer set the tone of the session. Along with topics and verbal style, pay attention to nonverbal cues described in Chapter 7: the interviewer's posture, gestures, vocal qualities, and so on. If they are informal, you can loosen up a bit too, but if they are formal and proper, you should act the same way.

Keep Your Answers Succinct and Specific   It's easy to ramble in an interview, either out of enthusiasm, a desire to show off your knowledge, or nervousness. But in most cases, long answers are not a good idea. Generally, it's better to keep your responses concise, but provide specific examples to support your statements.

Describe Relevant Challenges, Actions, and Results   Most sophisticated employers realize that past performance can be the best predictor of future behavior. For that reason, you should be ready to answer questions like these:

Describe an experience in which you needed to work as part of a team.

Tell us about a time you had to think on your feet to handle a challenging situation.

Describe a situation in which you were faced with an ethical dilemma, and discuss how you handled it.

Think About the Question Behind the Question   An employment interview may seem like it's all about you, and in a way it is. But whatever form questions take, they all are exploring the employer's fundamental question: "What can you do for this organization?" For example, questions about your job history are meant to explore how well you will perform in the job for which you're interviewing. If the interviewer asks, "Were you happy in your previous jobs," the real concern is "How happy would you be working for us?" Table 10.1 offers strategies for constructing employer-related answers to other common questions.

TABLE 10.1

Responding to Common Interview Questions

QUESTION	TIPS
Tell me something about yourself.	Keep your answer focused on the job for which you're applying. This isn't the time to talk about irrelevant hobbies, your family, or pet peeves.
What makes you qualified to work for this company?	Prepare in advance by making a table with three columns: in one, list your main qualifications; in the next, list specific examples of each qualification; and in the last, explain how these qualifications would benefit the employer.
What accomplishments have given you the most satisfaction?	The accomplishments you describe should demonstrate qualities that would help you succeed in the job.
Why do you want to work for us?	Employers are impressed by candidates who have done their homework about the organization.
Where do you see yourself in five years?	Answer in a way that shows you understand the industry and the company. Share your ambitions, but also make it clear that you're willing to work hard to achieve them.
What major challenges have you faced, and how have you dealt with them?	The specific problems aren't as important as the way you responded to them. You may even choose to describe a problem you didn't handle well to show what you learned from the experience that can help you in the future.
What are your greatest strengths?	Link what you say to the job. "I'm a pretty good athlete" wouldn't be helpful for most jobs. Instead talk about being a team player, having competitive drive, or having the ability to persevere in the face of adversity.
What are your salary requirements?	Do advance research to determine the prevailing compensation rates in the industry and in your region. Shooting too high can knock you out of consideration, whereas shooting too low can cost you dearly.

Ask Good Questions of Your Own   Near the end of the interview, you'll probably be asked if you have any questions. You might feel as if you already know the important facts about the job, but asking questions based on your knowledge of the industry, the company, and the position can produce some information and show the interviewer that you have done your research and are realistically assessing the fit between yourself and the organization. In addition to generating questions specific to the company or organization, here are some that often work well:

What are the primary results you would like to see me produce?

What is the biggest challenge or opportunity facing the team now?

How would you describe the management style I could expect from supervisors?

Most experts say it's bad form to ask about salary or benefits during a selection interview, as this may seem presumptuous or arrogant. You'll have a chance to negotiate if an offer is made.

Follow Up After the Interview   Send a prompt, sincere, and personalized note of thanks. A thoughtful and well-written thank-you note can set you apart from other candidates, whereas failing to send a thank-you note within a day of your interview can eliminate you from the running.

- Express your appreciation for the chance to become acquainted with the interviewer(s) and the organization.
- Explain why you see a good fit between you and the job, highlighting your demonstrated skills.
- Finally, let the interviewer(s) know that the conversation left you excited about the chance of being associated with the organization.

ASK YOURSELF

Brainstorm a list of questions you might ask a prospective employer to show that you understand the industry and organization and are a good candidate for your dream job.

UNDERSTANDING COMMUNICATION TECHNOLOGY

Interviewing by Phone or Video

In an age when budgets are tight, communication technology is pervasive, and work teams are geographically distributed, it's no surprise that a growing number of interviews are conducted via phone and video conference.

Present a Professional Identity   Your screen name, if you are using one, should be professional and appropriate, not flirty or edgy. Likewise, pay attention to what you wear if it's a video conference. Even if you're at home, dress professionally for the interview. In addition, think about the background of the room that interviewers will see. A neutral backdrop without distractions is ideal. Be sure to minimize or eliminate background noise.

Practice with Technology in Advance

- Make sure you have the right software and are comfortable using it.
- Confirm that you have a solid Internet connection with enough speed to handle the conversation.
- Double-check your camera, microphone, and speakers to confirm that they function properly.
- Make sure lighting is sufficient to allow the interviewer to see you clearly.
- If you're using a phone or tablet, set it up on a tripod or other stable device to avoid distracting jiggles.
- If you're using a laptop, make sure the camera captures you head-on rather than at an unflattering angle.

Ensure You Have the Right Time for the Interview
Confirm the time in advance, especially when different time zones are involved. (If you're not sure, search the web for "world clock.") You might send a message such as, "I am looking forward to speaking with you Tuesday, March 12, at 8 a.m. Pacific/11 a.m. Eastern."

Ask in Advance How Long the Interview Will Last
"Long-distance interviews are sometimes meant to be a brief candidate introduction, not a thorough vetting session," says one job search coach. She advises, "If this is the case, be prepared to make the most of this brief first impression!"[34]

Look at the Camera, not at the Screen   Looking at your monitor may feel natural. However, it creates the impression you're not making eye contact with the interviewer. Instead, look directly at the camera on the device you're using for the interview, and remember to smile.

Conduct a Dress Rehearsal   Practicing is the best way to ensure you are prepared. Recruit a friend (or, even better, someone at your school's career center) to play the role of interviewer. Be sure to practice under the actual circumstances of the interview—remotely and with the same equipment and services you'll use for the real thing. Besides ironing out potential glitches, rehearsals should leave you feeling more confident.

Since employment decisions may be made quickly, send a gracious thank-you email the same day as your interview. One recruiter suggests sending the email after 5 p.m., because "by sending the note after working hours you are intimating that you go the extra mile no matter what."[35] This is one circumstance in which you might also send a handwritten message of thanks. Reread everything you write carefully several times before sending it, and if possible, have a skilled proofreader review it as well. One job seeker ruined her chances of employment by mentioning the "report" (instead of "rapport") that she felt with the interviewer.

If you are interviewing for a job via interactive technology, see the *Understanding Communication Technology* feature in this section.

10.3 Organizational Communication Factors

Understanding communication factors in your work environment is critical to success. In this section, we consider the importance of organizational culture, communication patterns, and power dynamics.

Culture in the Workplace

Chapter 4 explained that organizations have cultures of their own, made up of shared beliefs and patterns of behavior. Cultural expectations suggest what is frowned upon in the organization, and conversely, what is likely to garner approval. For example, colleagues in some organizations are rewarded for competing with one another, and personal conversations are largely considered a waste of time. In other organizations, people may consider friendly interactions to be productive and enjoyable. At the retail giant Zappos, one of the core values is to "create fun and a little weirdness."[36] Team members are encouraged to interact with one another, dress as they like, bring pets to work, take part in company talent shows, and play games as they strive for the mutual goal of "delivering happiness" to customers.[37,38]

Socializing (or not) with colleagues is just one dimension of organizational culture.

What qualities are important in your preferred organizational culture? How can you find out whether prospective organizations fit this profile?

Even when you have researched an organization in advance, expect a period of adjustment once you become part of it. **Acculturation** (also called socialization) is the process of adapting to a culture.[39] If you have ever adjusted to a new job, school, or city, you know that acculturation isn't a simple process.

Most experts agree that there is value in having consistent values and practices. A strong culture can unite organizational members in upholding key values and pursuing shared goals. Strong values can also be good for business. About two-thirds of consumers say they are willing to pay more for products and services offered by companies that are socially and environmentally conscious.[40] However, the danger in a strong culture is that people who challenge the status quo may be marginalized for not "fitting in," even if they have good ideas to share. Organizations in which people are only rewarded if they think and act alike tend to lack innovation and lose step with changes in the environment.[41] The best bet is a balance between core values and openness to new ideas.

In reality, acculturation is a two-way street. New members bring with them novel ways of thinking and behaving, which affect the existing organizational culture to some degree. For example, as members of the Millennial generation have embarked on careers, they have tended to favor employers who offer flexible working conditions and who are civically engaged and socially responsible.[42] Business articles abound that offer advice on attracting and keeping Millennial job candidates, mostly by being less rigid, less hierarchical, and more inclusive.[43,44]

Patterns of Interaction

One aspect of organizational culture involves who communicates with whom and how. Figure 10.3, which features a diagram usually called a **sociogram**, suggests the number and complexity of interactions that can occur. Arrows connecting members indicate remarks shared between them. Two-headed arrows represent two-way conversations, whereas one-headed arrows represent remarks that did not elicit a response. Arrows directed to the center of the circle indicate remarks made to members as a whole. A network analysis of this sort can reveal both the amount of participation by each member and the recipients of every member's remarks. For example:

- Person E appears to be connected to the group only through a relationship with person A.

acculturation The process of adapting to a culture.

sociogram A graphic representation of the interaction patterns in a group.

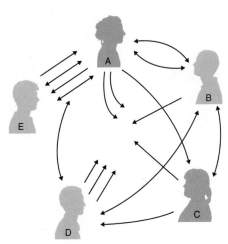

FIGURE 10.3 Patterns of Interaction in a Five-Person Group

all-channel network A communication network pattern in which group members are frequently together and share a great deal of information with one another.

chain network A communication network in which information passes sequentially from one member to another.

wheel network A communication network in which a gatekeeper regulates the flow of information from all other members.

gatekeeper Person in a small group through whom communication among other members flows.

power The ability to influence others' thoughts and/or actions.

- Person E never addressed any other members, nor did they address E.
- Person A is the most active and best connected member. They addressed remarks to the group as a whole and to every other member and were the object of remarks from three individuals as well.

Sociograms don't tell the whole story because they do not indicate the quality of the messages being exchanged. Nonetheless, they can be a useful tool for diagnosing group communication.

Figure 10.4 shows an **all-channel network** in which group members share the same information with everyone on the team. Emails are a handy way to accomplish this. As you probably know from experience, it's nice to be in the loop, but too much information can be overwhelming.

Another option is a **chain network** (also in Figure 10.4), in which information moves sequentially from one member to another. Chains are an efficient way to deliver simple verbal messages or to circulate written information when members can't attend a meeting at one time. However, chains aren't reliable for lengthy or complex verbal messages because the content often changes as it passes from one person to another. A simple statement such as "Cristiana is leaving Friday" might morph into "Cristiana has been fired," when the speaker really meant that she's going to a conference at the end of the week. The danger of message distortion along a chain means that you should be skeptical about rumors unless you can confirm the information.

Another communication pattern is the **wheel network** (Figure 10.4), in which one person acts as a clearinghouse, receiving and relaying messages to other members. Like chains, wheel networks are sometimes a practical choice, especially if the group is small and one member is available to communicate with others all or most of the time. The central person, called the **gatekeeper**, can become the informational hub who keeps track of messages and people. If the gatekeeper is a skilled communicator, these mediated messages may help the group function effectively. But if they consciously or unconsciously distort messages to suit personal goals or play members against one another, the group is likely to suffer.

Power in the Workplace

Power, the ability to influence others, comes in many forms. When Omar joined the staff, his great attitude and talent influenced everyone around him. His presence was a good reminder that power is not vested solely in leaders. Six types of power are common in the workplace.[45,46]

All-Channel Network

Wheel Network

Chain Network

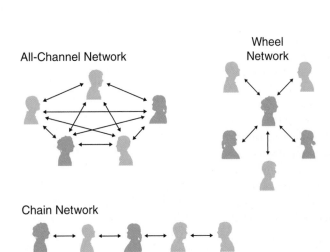

FIGURE 10.4 Organizational Communication Networks

Legitimate Power Sometimes called position power, **legitimate power** arises from the title one holds, such as supervisor, professor, or coach. People with legitimate power are said to be *nominal leaders*. *Nominal* comes from the Latin word for *name*, meaning that these leaders have been officially named to leadership positions. You can increase your legitimate power by becoming an authority figure, speaking up without dominating others, demonstrating competence, following group norms, and gaining the visible support of influential members.

Expert Power People have **expert power** when others perceive that they have valuable talents or knowledge. If you're lost in the woods, it makes sense to follow the advice of a group member who has wilderness experience. If your computer crashes at a critical time, you might turn to the team member with IT expertise. To gain expert power, make sure members are aware of your qualifications, be certain that what you convey is accurate, and don't act as if you are superior to others.

Connection Power   As its name implies, **connection power** comes from a member's ability to develop relationships that help the group reach its goals. A team seeking guest speakers for a seminar might rely on a well-connected member to line up candidates. To gain connection power, seek out opportunities to meet new people, nurture relationships through open and regular communication, and don't allow petty grievances to destroy valued relationships.

Reward Power   A person with the ability to grant or promise desirable consequences has **reward power**. Rewards come in a variety of forms, such as raises, promotions, and recognition. Even if you don't have a high-power job title, you can have reward power. For example, you might offer sincere, positive feedback to a classmate about a presentation they made in class. Your thoughtful words may ultimately be more treasured and memorable than the grade they receive from the instructor.

It's essential to respect the expertise of qualified team members, regardless of where they fit on an organizational chart.

What expertise does each member bring to the team to which you belong? What would be the consequences of failing to follow their advice?

Coercive Power   The threat or imposition of unpleasant consequences gives rise to **coercive power**. Bosses can coerce members via the threat of a demotion, an undesirable task, or even loss of a job. But peers also have coercive power. Working with an unhappy, unmotivated teammate can be punishing. For this reason, it's important to recruit well, build strong relationships, and insist on collegial behavior.

Referent Power   The basis of **referent power** is the respect, liking, and trust others have for someone. To gain referent power, listen to others' ideas and honor their contributions, do what you can to be liked and respected without compromising your principles, and present your ideas clearly and effectively.

After our look at various ways members influence one another, three important characteristics of power in groups become clearer.[47]

- *Power is conferred.* Power isn't something an individual possesses. Instead, it is conferred by the group. You may be an expert on the subject being considered, but if the other members don't think you are knowledgeable, you won't have expert power. By the same token, you might try to reward other people by praising their contributions, but if they don't value your compliments all the praise in the world won't influence them.

- *Power is distributed among group members.* Power rarely belongs to just one person. Even when a group has an official leader, other members usually have some power to affect what happens. This influence can be positive, as when it arises from information, expertise, or social reinforcement. It can also be negative, as when a member criticizes others or withholds contributions the group needs to succeed. You can appreciate how power is distributed among members by considering the effect just one member can have by not showing up for meetings or failing to carry out their part of the job.

- *Power isn't an either–or concept.* It's incorrect to assume that power is something that members either possess or lack. Rather, it is a matter of degree. Instead of talking about someone as "powerful" or "powerless," it's more accurate to talk about how much influence they have in various ways.

Table 10.2 summarizes types of power embodied in the workplace and how you might cultivate influence to help a group succeed.

ASK YOURSELF

What types of power do you have? How might you use that power responsibly to boost your chance of career success?

legitimate power The ability to influence group members based on one's official position.

expert power The ability to influence others by virtue of one's perceived expertise on the subject in question.

connection power The ability to influence others based on having relationships that might help the group reach its goal.

reward power The ability to influence others by granting or promising desirable consequences.

coercive power The ability to influence others by threatening or imposing unpleasant consequences.

referent power The ability to influence others by virtue of being liked or respected.

TABLE 10.2

Methods for Acquiring Power in Small Groups

LEGITIMATE AUTHORITY

1. Become an authority figure through election or appointment.

2. Speak up without dominating or antagonizing others.

3. Demonstrate competence.

4. Follow group norms.

5. Gain the support of other members.

INFORMATION POWER

1. Provide useful but scarce or restricted information.

2. Be certain the information you provide is accurate.

EXPERT POWER

1. Make sure members are aware of your qualifications.

2. Don't act superior.

REWARD AND COERCIVE POWER

1. Try to use rewards as a first resort and punishment as a last resort.

2. Make rewards and punishments clear in advance.

3. Be generous with praise.

REFERENT POWER

1. Gain the liking and respect of other members without compromising your principles.

2. Learn effective presentation skills. Present your ideas clearly to boost your credibility.

Source: Adapted from Rothwell, J. D. (1998). *In mixed company: Small group communication* (3rd ed.). Fort Worth, TX: Harcourt Brace, pp. 252–272. Reprinted with permission of Wadsworth, an imprint of the Wadsworth Group, a division of Thomson Learning. Fax 800-730-2215.

🛈 10.4 Communicating in a Professional Environment

The communication skills and strategies covered throughout this book will serve you well on the job and in life. The tips below are especially useful to keep in mind when starting a new position, and they will help you over the course of your career.

Avoiding Common Communication Mistakes

What you *should* do on the job is only part of the story. It's also important to consider what *not* to do. Here are some communication blunders to avoid at work.

Never Make Fun of Others   Some people learn this lesson the hard way. For example, when two women walked into the diner where he worked, Rik wisecracked to a coworker: "Oh look! Fat old ladies in baseball caps!"[48] Then he realized that they were his coworker's mother and aunt. Rik's colleague forgave him, but he learned a valuable lesson and swore off insensitive humor forever. Wisecracks at someone else's expense can be hurtful in any situation. At work, they can be cause for a reprimand, dismissal, or even a lawsuit.

Don't Overshare   It may seem important to "be yourself," but there are times when disclosing information about your personal life can damage your chances for professional success—or just flat out annoy people.[49] The "don't overshare" rule applies to online communication, too. Photos of your wild vacation aren't likely to impress the boss or potential employers. The best rule is to disclose cautiously, especially if the topic is a sensitive one such as religion, sexual relationships, or politics.[50] A trusted colleague may be able to offer advice about how much to share.

Respect Cultural Differences   Many Americans display what researchers call "instant intimacy."[51] They often address even new acquaintances, elders, and authority figures by first name. They also engage in a great deal of eye contact, touch their conversational partners, and ask personal questions. To people from different backgrounds, these behaviors may seem disrespectful. An Australian exchange student in the United States reflected on her experience: "There seemed to be a disproportionate amount of really probing conversations. Things I normally wouldn't chat about on a first conversation."[52] Review Chapter 4 for more guidance on being culturally sensitive.

When it comes to gossip, the golden rule applies: Don't say anything behind a colleague's back that you wouldn't want said about you.

How would you feel if you knew colleagues were talking behind your back?

Don't Gossip   Communicating with integrity isn't always easy. The culture in some organizations favors gossiping, bad-mouthing, and even lying about others. Nevertheless, principled communication means following your own set of ethics rather than relying on the approval of others. It may be helpful to know if someone was promoted, reprimanded, or fired, and why—but malicious gossip can mark you as untrustworthy and can damage team spirit.[53] Also be careful not to vent about work-related matters on social media. One executive proposed this test: Before you start talking, stop and ask yourself, "Is it kind?"[54]

Always Do Your Best   You may have heard the phrase *"don't sweat the small stuff."* In fact, making a good impression requires paying attention to every detail. Show up for work well groomed and professionally dressed. Another method for standing out is to do more than what is required.[55] You might finish a job sooner than anticipated or volunteer to work on a weekend or after hours (if it's allowed), offer to deliver a presentation, or tackle a project that keeps getting delayed. The time that jobs like this take may be well worth the good reputation they earn you.

Remain Calm   Losing control under pressure can jeopardize your career. Consider the extreme case of a JetBlue flight attendant who was fired after he became frustrated with a passenger, grabbed two beers, and jumped off the (parked) plane via an evacuation chute.[56] Even less dramatic freak-outs—raising your voice or dashing off an angry email or text message—can damage your career or land you among the unemployed. To stay collected when you feel yourself getting agitated, take a few deep breaths or a break, stop to listen and ask questions before responding, and vent your emotions to trusted friends while you are off the clock.[57]

Don't Fixate on Mistakes   What if you accidentally say "I love you" while ending a call with your boss? Or your eyes fill with tears during a stressful business meeting? Minor lapses in professionalism are bound to occur, even among people who have been in the workplace for many years. You can usually recover your dignity and your reputation by following these four steps: don't panic, acknowledge the gaffe, apologize, and return to life as usual.[58] For example, you might say, "Sorry about that! I'm in the habit of saying 'I love you' when I talk to my family. I'm sorry for the slip-up! I'll be more careful in the future." It's okay to laugh if the other person does, but don't dwell on the mistake. You want other people's opinion of you to focus on your impressive performance instead.

Communicating Well as a Follower

"What are you, a leader or a follower?" you're asked, and you know which position is generally considered the better one. But followers are far more important than you might have imagined. Consider the following examples.

ASK YOURSELF

What are some ways you might graciously exit a conversation involving gossip about a coworker?

UNDERSTANDING YOUR COMMUNICATION

How Good a Follower Are You?

Check all of the following that apply to you in your role as a follower.

- [ ] I think for myself.
- [ ] I go above and beyond job requirements.
- [ ] I am supportive of others.
- [ ] I am goal oriented.
- [ ] I focus on the end goal and help others stay focused as well.
- [ ] I take the initiative to make improvements.
- [ ] I realize that my ideas and experiences are essential to the success of the group.
- [ ] I take the initiative to manage my time.
- [ ] I frequently reflect on the job I am doing and how I can improve.
- [ ] I keep learning.
- [ ] I am a champion for new ideas.

If the majority of these statements describe you, pat yourself on the back. These are the qualities of an outstanding follower, according to Robert Kelley, author of *The Power of Followership*.[59]

- In a tense meeting, apartment dwellers are arguing about overcrowded parking and late-night noise. One tenant cracks a joke and lightens up the tense atmosphere.

- A project team is trying to come up with a new way to attract customers. The youngest member, fresh from a college advertising class, suggests a winning idea.

- Workers are upset after the boss passes over a popular colleague and hires a newcomer for a management position. Despite their anger, they accept the decision after the colleague persuades them that she is not interested in a career move anyway.

These examples show that influence isn't just the domain of leaders. Despite the common belief that leaders are the most important group members, good followers are also indispensable. Completing the *Understanding Your Communication* quiz in this section will help you appreciate the role you play as a follower.

The self-assessment makes it clear that good followers aren't sheep who blindly follow the herd. According to management consultant Robert Kelley, effective followers "think for themselves, are very active, and have very positive energy."[60] He points out that many leaders have a special term for followers such as these. They call them "my right-hand person" or my "go-to person."

Successful executives agree. In a study of more than 300 senior-level leaders, 94% said that followers help shape leaders, not just the other way around.[61] In their view, effective followers and leaders share many of the same qualities, including honesty, competence, intelligence, and character. The executives also said they appreciate followers who are loyal, dependable, and cooperative. But they didn't define those qualities in terms of blind obedience. Indeed, almost all of the executives disagreed with the statement that good followers "simply do what they are told." Overall, the lesson is that followership involves a sophisticated array of skills, a good measure of self-confidence, and a strong commitment to teamwork.

Not all followers communicate or contribute equally. Theorist Barbara Kellerman proposes that followers fall into five categories.[62] Which best describes you?

isolates Followers who are indifferent to the overall goals of the organization and communicate very little with people outside their immediate environment.

- **Isolates** are indifferent to the overall goals of the organization and communicate very little with people outside their immediate environment.

- **Bystanders** are aware of what's going on around them, but they tend to hang back and watch rather than play an active role. You may find yourself in a bystander role occasionally, especially when you are in a new situation. Because bystanders are usually not as emotionally involved as others, they can sometimes provide an objective, fresh perspective if they are encouraged to share their thoughts.

- **Participants** are moderately involved and attempt to have an impact. Some participants support leaders' efforts, whereas others work in opposition. (Opposition isn't necessarily a bad quality in followers. Good followers *should* object when leaders are unethical or ineffective.)

- **Activists** are more energetically and passionately engaged than participants. They, too, may act either in accordance with, or in opposition to, leaders' efforts. Their commitment is a plus in many ways. At the same time, activists sometimes have difficulty compromising and getting along with others.

- **Diehards** will, sometimes literally, sacrifice themselves for the cause. "Being a Diehard is all consuming. It is who you are. It determines what you do," Kellerman says.[63] Soldiers are a classic example, as are people who protest against oppressive rulers or fight for civil rights. Diehards may also work tirelessly in nonprofits or other organizations if they believe the services they provide are essential. Their commitment is unrivaled, but sometimes it's difficult to contain their enthusiasm, even when it runs counter to other peoples' goals.

bystanders Followers who are aware of what's going on around them but tend to hang back and watch rather than play an active role.

participants Followers who are moderately involved and attempt to have an impact either by supporting leaders' efforts or working in opposition to them.

activists Followers who are energetically and passionately engaged.

diehards Followers who will, sometimes literally, sacrifice themselves for a cause they believe in.

Communicating in a Professional Manner Online

Computer-mediated communication is indispensable in the workplace. Business professionals send or receive an average of 122 emails per day,[64] and videoconferencing is considered a vital means of communication in 86% of North American companies.[65] Experts offer the following tips for maintaining professionalism in an online environment.

ASK YOURSELF

Have you ever had an online moment you'd like to take back? If so, what advice would you give others?

Take Part in Training   Even if you're familiar with many types of communication technology, keep in mind that software and practices differ. A good strategy when you get a new job is to ask what communication platforms you will use so you can brush up on them in advance. If possible, also sign up for technology training once you are on the job.

Develop Camaraderie   Communication technology can make it easy to exchange information but harder to make a real connection. That's important to consider because people often feel less committed and less accountable if they don't know their teammates well.[66] Take time to get acquainted with online team members and, if possible, meet in person from time to time.

Use Correct Grammar, Spelling, and Punctuation   Messages such as "gotta go now" or "thanx tho" may be fine among friends, but they can cause clients, coworkers, and managers to doubt your professionalism and your literacy skills. Before hitting send, consider how you would feel if you received a message from your boss or a job applicant riddled with grammar and spelling errors.[67]

It's usually impossible to retrieve a message once it's entered cyberspace.

Have you ever sent a message you wish you had reviewed more carefully? If so, what were the consequences?

Don't Be Too Brief   You may mean a brief reply such as "OK" to sound upbeat, but with no other cues to go by, recipients may assume you are frustrated or impatient. It's all right to get to the point online, but do so in a courteous manner.

Think Before Hitting "Forward" or "Reply to All" Imagine being the boss who sent an email asking everyone to welcome a new team member—but accidentally included past correspondence that disclosed the new person's sign-on salary, which was higher than that of other, more experienced employees.[68] Take time to be sure that what you send is meant for all recipients.

Don't Convey Sensitive Information Bad news, criticism, and private information are best delivered in person—or if that's not possible, by phone. And never forget that information sent electronically may end up in the wrong inbox. If you wouldn't be comfortable having your words become public knowledge, don't send them electronically.[69]

Dress for the Camera Don't be the telecommuter who logged onto a videoconference wearing a business jacket and pajama pants and then realized once the cameras were on that she had left important papers on the other side of the room. Dress for video conferences as you would for in-person meetings.

Pay Attention Although it may be tempting to try to get other work done during a long-distance meeting, doing so may signal that you are a poor listener. In addition, avoid typing during a videoconference. "It's probably the biggest faux pas," according to Angie Hill, a Skype general manager.[70]

No matter what career you pursue, communication is indispensable for success. According to career coach Andrea Kay, "Communication is critical in any field, because every job entails working with other people: bosses, customers, clients, co-workers, vendors, you name it."[71]

Chapter 11 continues the theme of workplace communication by further exploring what it takes to succeed as a leader and team member.

MAKING THE GRADE

At www.oup.com/he/adler-uhc14e, you will find a variety of resources to enhance your understanding, including video clips, animations, self-quizzes, additional activities, audio and video summaries, interactive self-assessments, and more.

OBJECTIVE 10.1 Describe communication behaviors of successful job candidates.

- Keep in mind that life is an audition. The work habits you exhibit now will influence whether people let you know about career opportunities and recommend you for them.

- It's important to build an online identity even before you enter the job market. Present yourself in a favorable and honest light by showcasing your strengths and goals and avoiding behavior that might lead to embarrassing posts.

- To network effectively, view everyone as a networking prospect, engage in online networking, seek referrals, conduct informational interviews, and show appreciation to the people who help you.

  > Make a list of several people you might interview to learn more about your dream job and brainstorm a list of questions you would ask them.

> What might people conclude about you based on your work habits as a student? Are you prompt or tardy? Well prepared or not? Well groomed or sloppy? Committed to doing excellent work or just enough to get by? How well do your answers reflect what employers are looking for in a job candidate?

OBJECTIVE 10.2 Practice communication strategies to make a good impression during employment interviews.

- Create application materials that make it easy for prospective employers to see how your skills and aspirations match their needs.

- Succeeding at a job interview is a process that begins long before you arrive and shake hands. Do advanced research, practice answering likely questions, and plan how you will dress and what you will bring to the interview.

- Know the law regarding employment interview questions and devise strategies in advance for how you might respond to them.

- Be specific but succinct in how you respond to interview questions. If allowed, create a brief media presentation that showcases your accomplishments relevant to the job.

  > Rehearse how you might respond to the following questions in a job interview: What are your greatest strengths and weaknesses? Why should we hire you? What have been your greatest challenges and victories? What do you want to be doing in five years?

  > Imagine you are being interviewed via videoconference for a job you really want. Rehearse how you might respond to the question, "Tell us more about yourself." Remember to pay attention to your posture, facial expressions, and where your gaze falls. Do you feel well prepared for a video interview? Why or why not?

OBJECTIVE 10.3   Explain how organizational communication patterns affect a work environment.

- Organizations have cultures of their own, and becoming acculturated to them is a process that may involve multiple strategies. Acculturation is always a two-way street to some extent. People adapt to cultures, and cultures adapt to them.

- Power is given, not taken, and it may be based on a variety of factors ranging from likability to the position someone holds.

  > Think of a time when you were in a new school, neighborhood, or job. Describe your acculturation strategy and experience.

  > Recall the passage "When Omar joined the staff, his great attitude and talent influenced everyone around him." What types of power does Omar seem to have?

OBJECTIVE 10.4   Describe ways to communicate effectively as a new employee in person and online.

- Some of the communication gaffes to avoid at work include belittling people, oversharing, being culturally insensitive, gossiping, and losing emotional control.

- People often overlook the powerful roles that followers play. Some followers tend to hang back, either because they are indifferent or because they prefer to watch and learn. Others take an active role—sometimes to the extent of putting their lives on the line for a cause.

- Online identity is important even after you land the job you want. To make the most of it, update your skills, build relationships online, dress and communicate in a professional manner, and be careful what you send and who you send it to.

  > Have you ever had an embarrassing moment at work or in public? If so, what happened? Are there ways you might avoid a similar experience in the future? If so, how?

  > Why do you think followership often gets less attention than leadership? Present an argument giving three compelling reasons to appreciate the important roles that followers play.

  > Analyze the level of professionalism in your online communication using the tips presented in this chapter.

KEY TERMS

acculturation p. 253

activists p. 259

all-channel network p. 254

bystanders p. 259

chain network p. 254

coercive power p. 255

connection power p. 255

diehards p. 259

expert power p. 254

gatekeeper p. 254

informational interview p. 244

isolates p. 258

legitimate power p. 254

networking p. 243

participants p. 259

power p. 254

referent power p. 255

reward power p. 255

selection interview p. 247

sociogram p. 253

wheel network p. 254

ACTIVITIES

1. Search for your name on Google and several other search engines. Do you feel that the photos and information revealed by the search would impress prospective employers? If not, how might you change your online image to be more professional? If your search leads to information about other people with the same name as you, how might you help prospective employers find the real you? If nothing much shows up about you online, how might you cultivate a greater presence?

2. Draft a cover letter and résumé for a job you would like to have. Share it with a mentor or representative at your university career center to get their feedback.

3. Conduct mock job interviews in which you take turns being the interviewer and the job candidate. Reflect on what you do well and how you might like to improve.

Leadership and Teamwork

11

CHAPTER OUTLINE

11.1 Communication Strategies for Leaders 265
Characteristics of Effective Leaders
Leadership Approaches
Trait Theories of Leadership
Situational Leadership
Transformational Leadership

11.2 Communicating in Groups and Teams 271
What Makes a Group a Team?
Motivational Factors
Rules in Small Groups
Individual Roles

11.3 Making the Most of Group Interaction 275
Enhance Cohesiveness
Managing Meetings Effectively
Using Discussion Formats Strategically

11.4 Group Problem Solving 280
Advantages of Group Problem Solving
Stages of Team Development
A Structured Problem-Solving Approach
Problem Solving in Virtual Groups

MAKING THE GRADE **288**
KEY TERMS **290**
ACTIVITIES **290**

LEARNING OBJECTIVES

11.1

Describe effective leadership skills based on the situation, goals, and team members' needs.

11.2

Identify factors that influence communication in small groups and either help or hinder their success.

11.3

Strategize ways to build cohesiveness and communicate effectively during group discussions.

11.4

Assess the advantages and stages of group problem solving, and practice applying Dewey's group problem-solving model.

Nelson Mandela

recognized the importance of encouraging team members by "leading from behind." Your experience probably demonstrates the importance of this principle in your own life.

Have you worked with a leader who appreciated and supported the contributions of team members? What kinds of communication conveyed the message that followers are important?

If you feel valued by a leader and other team members, how does that influence your contribution to the group effort?

ONE OF THE MOST INFLUENTIAL LEADERS in the modern world, Nelson Mandela spent his boyhood days tending cattle in the pastureland of South Africa. He quickly learned that it wasn't effective to stand in front of the herd and proclaim, "Follow me!"[1] It was more effective to walk behind and offer encouragement. Mandela explained:

> When you want to get the cattle to move in a certain direction, you stand at the back . . . and then you get a few of the cleverer cattle to go to the front and move . . . The rest of the cattle follow the few more-energetic cattle in the front, but you are really guiding them from the back.[2]

As Mandela grew up, he used his boyhood experiences to help him understand people. That's not to say he considered people to be as slow-witted as cows. But he understood that no one is likely to be inspired by leaders who simply issue orders. A more important factor is teamwork. Mandela was fond of saying, "When people are determined, they can overcome anything."[3]

No one could have predicted the degree to which Mandela's early leadership lessons would be put to the test. By the time he was a young man, South Africa was being torn apart by racist policies and poverty.[4] Under apartheid, the country's system of racial segregation and discrimination, white South Africans claimed almost 90% of the land and the vast majority of the country's wealth.

Mandela challenged the apartheid policies of the white South African government and spent 27 years as a political prisoner as a result. Yet he persevered, ultimately uniting the country and becoming one of the greatest civil rights leaders in history.[5] Throughout this chapter, consider Mandela's legendary approach to teamwork and leadership as a model to help you succeed in your own endeavors.

This chapter presents strategies for both leadership and teamwork. You will learn about:

- different approaches to leadership and which of them are most effective in various situations,
- the factors that make some groups difficult to work with and others enjoyable and productive,

- strategies to make the most of group interactions,
- when and how to engage in group problem solving, and
- how to resign from a job graciously if it's time for you to move on.

11.1 Communication Strategies for Leaders

Nelson Mandela began developing leadership skills long before he became a public figure. Whatever your career goals, starting early can be a boost for you as well. "If you want to become a leader, don't wait for the fancy title or the corner office," urges human resources expert Amy Gallo. "You can begin to act, think, and communicate like a leader long before that."[6]

The role you take now in leading a team project or helping solve a problem can contribute to your success as a professional. With that in mind, let's consider some essential points about leadership.

Characteristics of Effective Leaders

It may seem that the best leaders are showy and self-assertive, but the ones who achieve long-term success don't usually fit that description.[7,8] To the contrary, most leaders are remarkably humble. They're content to let others take the spotlight and quick to say that they are still learning and growing. Leaders as diverse as civil rights heroes Mandela, Mahatma Gandhi, and Rosa Parks, as well as corporate titans Bill Gates and Warren Buffett, have won hearts and minds through their ideas and actions, not their commanding personalities.

Think of a leaderless group to which you've belonged. Who did members look to for guidance and direction? **Emergent leaders** gain influence without being appointed by higher-ups. Those who prove to be respected and effective over time exhibit many of the characteristics common among successful leaders:[9]

- good listener
- open to innovation
- able to work well with teams
- good at facilitating change
- appreciative of diversity
- honest and ethical

As you might have noticed, these qualities are either accomplished by or conveyed through communication. The checklist in this section suggests some ways you might communicate to earn a reputation as a promising leader.

Leadership Approaches

Early scholars identified three basic leadership approaches. The first, **authoritarian leadership**, is based on a leader's position and ability to offer rewards or punishment. It's an approach in which relatively

---

CHECKLIST ✓

Demonstrating Your Leadership Potential

The following behaviors are effective ways to show others that you would be a good leader.

☐ **Stay engaged.** Getting involved won't guarantee that you'll be recognized as a leader, but failing to speak up will almost certainly knock you out of the running.

☐ **Demonstrate competence.** Make sure your comments identify you as someone who can help the team succeed. Talking for its own sake will only antagonize other members.

☐ **Be assertive, not aggressive.** It's fine to have a say, but don't try to overpower others. Treat every member's contributions respectfully, even if they differ from yours.

☐ **Provide solutions in a time of crisis.** How can the team obtain necessary resources? Resolve a disagreement? Meet a deadline? Members who problem-solve are likely to rise to positions of authority.

---

emergent leader A group member who assumes leadership without being appointed by higher-ups.

authoritarian leadership A style in which a designated leader uses coercive and reward power to dictate the group's actions.

democratic leadership A style in which a leader invites the group's participation in decision making.

laissez-faire leadership A style in which a leader takes a hands-off approach, imposing few rules on a team.

servant leadership A style in which a leader recruits outstanding team members and provides the support they need to do a good job.

trait theories of leadership A school of thought based on the belief that some people are born to be leaders and others are not.

situational leadership A theory arguing that the most effective leadership style varies according to the circumstances involved.

powerless team members are expected to obey an all-powerful leader. The second approach, **democratic leadership**, describes leaders who encourage others to share in decision making. It is a "power with" dynamic in which people work together. The third approach, **laissez-faire leadership**, reflects a leader's willingness to allow team members to function independently and to make decisions on their own. Some theorists now add a fourth approach—**servant leadership**, based on the perspective that a leader's job is mostly to recruit outstanding team members and provide the support they need to do a good job.[10] Unlike laissez-faire leaders, who tend to have a hands-off approach, servant leaders are often highly involved with team members and processes.

Each of these approaches is effective in some situations. As you might expect, morale tends to be higher in teams with servant leaders than in those with authoritarian leaders.[11] However, an authoritarian approach can sometimes produce faster results, which can be useful in a crisis. Satisfaction is typically high in teams led by democratic leaders, but inclusive decision making can be time consuming.[12] Highly experienced members may appreciate the hands-off approach of a laissez-faire leader, but for many teams, the ambiguity involved creates added stress.[13] A Gallup survey of millions of American workers revealed that nearly half of them don't have a clear sense of what their bosses expect of them.[14] Finally, servant leadership has been shown to enhance team members' satisfaction and lead them to feel more self-confident and optimistic than they would otherwise.[15]

Trait Theories of Leadership

More than 2,000 years ago, Aristotle proclaimed, "From the hour of their birth some are marked out for subjugation, and others for command."[16] This is a radical expression of **trait theories of leadership**, which are sometimes labeled the "great man" or "great woman" approach because they suggest that some people are born to be leaders while others are not. In fact, evidence shows that leaders of *any* personality type can be effective, and the leadership skills people acquire are typically more important than anything they are born with.[17]

Situational Leadership

Nelson Mandela wisely said, "The mark of great leaders is the ability to understand the context in which they are operating and act accordingly."[18] In contrast to trait theories of leadership, the principle of **situational leadership** holds that a leader's style should change with the circumstances.[19] Those who exercise situational leadership consider the nature of the task, including how prepared team members are to accomplish it. They also consider other aspects of the team involved, including their relationships with each other and with the leader.[20] The managerial grid developed by Robert Blake and Jane Mouton (Figure 11.1) portrays leadership on the basis of these two considerations: low to high emphasis on tasks and low to high emphasis on relationships.[21,22] Let's consider the management styles portrayed in the model.

Jack Ma, former executive director of the multibillion-dollar international online marketing firm Alibaba, achieved success by recruiting great team members and coaching them to be collaborative, have fun, and let their imaginations guide them.

Have you ever worked with a leader who acted more like a coach than a command-and-control authority figure? If so, how did that influence you as a team member?

Impoverished Leaders This type of leader has little interest in either tasks or relationships. If you've worked with a supervisor who overlooked poor performance and seemed oblivious to team members' needs, you know how frustrating it can be. There are a number of reasons leaders may take an impoverished approach. They may lack confidence and assertive communication skills. Or they may be experiencing burnout. Once-concerned leaders who feel depleted or discouraged may largely give up. And some leaders are so concerned with looking

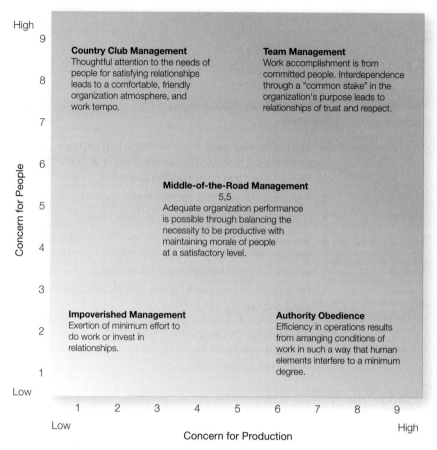

FIGURE 11.1 The Managerial Grid

good personally that they try do nearly everything themselves.[23] This gives others the impression that they don't care much about the team or how it functions. In turn, team members tend to feel resentful and unappreciated, a negative spiral that usually gets worse. Before you write this off as an approach you would never take, consider that inexperienced leaders often behave this way because they haven't learned to trust team members and they are eager to prove themselves to *their* bosses. To avoid the pitfalls of trying to do everything yourself, see the @Work box on delegation on page 268.

It's easy to see that impoverished leadership isn't ideal. Some people equate it with a laissez-faire approach. Indeed, some laissez-faire leaders truly don't care about task fulfillment or relationships. However, other laissez-faire leaders care a great deal about both, but they take a hands-off approach for another reason, such as that they trust the team, or they want them to develop confidence and independence. If that's their motivation, it's important that teams know it. Impoverished leaders take heed: Even when leaders take a hands-off approach, teams perform best when they perceive that leaders are invested in them personally and in their success.[24]

Country Club Leaders These leaders exhibit high regard for relationships but give little attention to task fulfillment. Country club leaders' primary goal is to keep members happy and maintain harmony. They tend to smile a lot, listen, and offer praise and encouragement. However, they may let misbehavior slide

ASK YOURSELF

Think of a group you have been part of. What type of leader or follower were you? What types of power did you use most effectively?

"I'll Do It Myself"—Or Should I?

It's a rookie mistake some leaders never outgrow—the tendency to do most tasks themselves instead of delegating. As a leader, the reasons for doing things yourself may seem appealing:

1. It's faster to do it myself.
2. Someone else might mess it up.
3. I don't want to ask others to do unpleasant tasks.
4. Other people are busy.
5. No one else knows how to do this.
6. It would take too long to train someone.
7. I like doing this task.
8. I wouldn't ask the team to do something I'm not willing to do myself.
9. If I'm not completing specific tasks, I might seem unnecessary.

The reasons to delegate aren't as numerous, but they are powerful:

1. As a leader, your most important job is to build high-performance, high-capacity teams,[25] and that doesn't happen when you do everything yourself.
2. Trying to do everything yourself slows down the process.
3. Leaders involved in tasks aren't available to lead.
4. Most team members, when asked, say they would gladly perform tasks on the boss's to-do list if that would free up time for the leader to listen to them more.
5. Leaders who try to do everything become tired, discouraged, and burned out.

The next time you're tempted to tackle the whole job yourself, consider whether delegating may be better for the team, yourself, and the task at hand.

ASK YOURSELF

Think about the last time you took on a new leadership role. Did you focus more on tasks or on relationships, or equally on both? How did this affect the way you communicated as a leader?

because they dread confrontations, and they typically have a hard time making difficult decisions. In many cases, people led by country club leaders like them personally but feel frustrated because their teams stagnate and underperform.[26] In many cases, country club management fosters an unproductive environment, as you know if you have been part of a project team in which meetings feel like social hours during which little actual work gets done.

Authority–Obedience Leaders   At the other extreme are managers who focus almost entirely on tasks and very little on relationships. This can be useful in small doses, as in emergencies and when inexperienced team members need task-related direction.[27] Authority–obedience leaders exhibit the traits of authoritarian leadership described earlier in that they tend to focus on commands and quotas rather than investing in people. As leadership coach Bob Weinstein puts it, such leaders often come off as "hard-nosed, tough, demanding perfectionist[s]" or "unyielding control freaks" who micromanage purely for the sense of being in control.[28] Clearly, few of us would want to work for long under those conditions. Another disadvantage is that change happens so quickly in today's environment that centralized decision making is often too slow to be effective.[29]

As frustrating as some of these leadership behaviors can be, don't leap to conclusions about your boss, or how you should respond, before you read the checklist on page 269 for tips on managing difficult bosses.

Middle-of-the-Road Leaders   In the center of Blake and Mouton's grid is an approach characterized by moderate interest in both tasks and relationships. If you've worked with a leader you consider not horrible or great, but simply "okay," you have probably experienced this approach firsthand.

Team Leaders These leaders are typically the most successful. They exhibit high regard for both tasks and relationships. This approach has a great deal in common with the model of transformational leadership, which is described next.

Transformational Leadership

Transformational leadership is defined by a leader's devotion to helping teams fulfill their potential.[30] Let's explore the central assumptions of transformational leadership.

- *People want to make a difference.* Transformational leaders believe that if the right people are on board, they will be motivated to accomplish important goals. Therefore, these leaders cultivate strong teams, actively listen to members, consider their feelings, and honor their contributions.

- *Empowerment is essential.* Transformational leaders aren't micromanagers who feel they have all the answers. Instead, they know that the best results come from well-prepared team members with the talent, training, and authority to make most decisions for themselves. The motto of transformational leaders could be, *"It's not about me. It's about the team and what we accomplish together."*

- *Mission is the driving force.* Transformational leaders expect 100% effort from everyone on the team because, otherwise, teams cannot live up to their full potential or accomplish their mission.

- *Transparency is key.* Although transformational leaders empower team members to make decisions for themselves as much as possible, when a tough decision is needed, these leaders aren't afraid to make it.[31] In those circumstances, they listen to diverse viewpoints, weigh all the factors, and when they announce a decision, they explain *why* they made it.[32]

Transformational leadership requires putting your ego aside and focusing on the team and the mission. Because they are willing to make tough calls, transformational leaders aren't popular with everyone all the time. But even in tough times, team members typically hold these leaders in high regard as being effective, trustworthy, and fair.[33]

Nelson Mandela exhibited the qualities of a transformational leader. For one, he was deeply committed to a noble vision. He willingly risked his life to advocate for a free and democratic South Africa. He was so committed to that ideal that, even after decades of atrocities against black South Africans and being imprisoned for his beliefs, Mandela did not call for revenge. Instead, his goal was even more transformational—forgiveness and harmony.[34] Second, Mandela believed in empowering and supporting others. The style of "leading from behind" that he learned as a boy stayed with him always. Third, despite Mandela's extraordinary achievements, he exhibited humility and gave credit to others. Upon his release after serving nearly three decades in prison, he told a crowd of supporters:

> I stand here before you not as a prophet but as a humble servant of you, the people. Your tireless and heroic sacrifices have made it

---

transformational leadership
Defined by a leader's devotion to help a team fulfill an important mission.

CHECKLIST ✓

Working with a Difficult Boss

☐ **Rise to the challenge.** Meeting your boss's expectations can make your life easier. If your boss is a micromanager, invite input.

☐ **Make up for the boss's shortcomings.** Proactively provide important details or information.

☐ **Seek advice.** Complaining about your boss is a bad idea. Instead, if other people in your organization have encountered the same problems, you might discover useful information by talking with them.

☐ **Have a heart-to-heart with your boss.** Rather than blaming the boss, use "I" language, such as "I feel confused when two managers give me different instructions." Solicit your boss's point of view, listen nondefensively, and use perception checking. Seek a win–win outcome.

☐ **Adjust your expectations.** You may not be able to change your boss's behavior, but you can control your attitude.

☐ **Maintain a professional demeanor.** You will gain nothing otherwise.

☐ **If necessary, make a gracious exit.** If you can't fix an intolerable situation, the smartest option may be to look for more rewarding employment. If so, leave on the most positive note you can.[35]

What's Your Leadership Style?

 Choose the item in each group that best characterizes your beliefs about leadership.

1. I believe a leader's most important job is to:

   a. Make sure people stay focused on the task at hand

   b. Take a hands-off approach so team members can figure things out on their own

   c. Make sure the workplace is a friendly environment

   d. Help team members build strong relationships so they can accomplish a lot together

2. When it comes to being an employee, I believe that people:

   a. Accomplish most when leaders set clear expectations

   b. Should do their work and let leaders do theirs

   c. Are most productive when they are enjoying themselves

   d. Have a natural inclination to work hard and do good work

3. As a leader, when a problem arises, I am most likely to:

   a. Announce a new policy or procedure to avoid the same problem in the future

   b. Ignore it; it will probably work itself out

   c. Try to smooth things over so no one feels upset about it

   d. Ask team members' input on how to solve it

4. If team members were asked to describe me in a few words, I hope they would say I am:

   a. Competent and results oriented

   b. Removed enough to make decisions without letting my emotions get in the way

   c. Pleasant and friendly

   d. Respectful and innovative

5. When I see team members talking in the hallway, I am likely to:

   a. Feel frustrated that they are goofing off

   b. Close my door so I can work without interruption

   c. Share my latest joke with them

   d. Feel encouraged that they get along so well

INTERPRETING YOUR RESPONSES

For insight about your leadership style, consider which of the following best describes your answers.

Authority-Obedience

If most of your answers were "a," you feel that people should stay focused on the job at hand, and you are frustrated by inefficiency and signs that people are "wasting time." The danger is that in your zeal to get the job done, you will overlook relationships. This can be counterproductive in the long run since teams often accomplish more than individuals working alone.

Impoverished

If the majority of your answers were "b," you tend to take a hands-off approach as a leader, investing neither in relationships nor in tasks. You may take pride that you aren't a micromanager, but you're probably going too far in the opposite direction. Team members often need guidance. And even those who work well without much supervision probably crave your attention and appreciation.

Country Club

If you selected "c" more than other options, your focus on strong relationships and a pleasant work environment is likely to make you likeable as a person. However, team members may be frustrated by less-than-optimal results. A focus on both relationships *and* tasks may ultimately be more rewarding for everyone involved.

Team or Transformational

If you chose "d" most often, you balance an emphasis on results with a respect for the people involved. Although your expectations are high, your support and empowerment are likely to bring out the best in people. Most people consider this to be the ideal leadership style.

possible for me to be here today. I therefore place the remaining years of my life in your hands.[36]

After white South African president F. W. de Klerk released Mandela from prison in 1990, the two worked together to help the nation heal from more than 50 years of bloodshed, cruelty, and injustice. They were jointly awarded the Nobel Peace

Prize in 1993. The following year, Mandela was elected as the first black president of South Africa. He won hundreds of leadership awards and has been lauded as one of the greatest leaders in history.

You don't have to be a world-famous leader to change the world around you. Take the *Understanding Your Communication* quiz to see which leadership style best describes you.

11.2 Communicating in Groups and Teams

Leaders get a lot of attention, but their influence wouldn't count for much without people working together to pursue the vision. One of Nelson Mandela's most important qualities was his ability to empower followers. Even while he was in prison, people continued to be inspired by his wisdom and vision, and the civil rights movement continued.[37] Closer to home, consider how well a workplace functions when a key leader is away.

This section focuses on the communication skills involved in group and teamwork. Group work sometimes gets a bad name. You may have been part of groups in which some members did more work than others, one or two people dominated group discussions, or members never trusted each other enough to become comfortable sharing their ideas. On the other hand, if you have ever been part of a great team, you know it can be one of life's most enjoyable experiences and can yield far greater results than working alone. Basketball great Michael Jordan once said, "Talent wins games, but teamwork and intelligence win championships."[38]

Being a team player is critical to career success. Employers rank teamwork skills among the 10 most desired traits of people they hire.[39] Employees give teamwork high priority as well. They typically say that effective teamwork makes them feel more powerful and empowered, more appreciated, more successful, closer to their colleagues, and more confident that colleagues will support and encourage them in the future.[40]

Away from work, groups also meet to solve problems. Volunteer groups plan fundraisers, athletic teams work to improve their collective performance, educators and parents work together to improve schools—the list goes on and on.

Most of the time, the difference between successful and frustrating group work involves communication. Let's consider the nature of small groups and the communication skills involved.

What Makes a Group a Team?

Groups may play a bigger role in your life than you realize. You are probably involved in them every day. Some groups are informal, such as friends and family. Others are part of work and school. Project groups, work teams, and study groups are common types. You may belong to some groups for fun (a band or athletic team) and others for profit (an investment group). Others center on personal growth (religious study, exercise) or advocacy (Save the Whales or Habitat for Humanity). You can probably think of more examples that illustrate how groups are a central part of life.

For our purposes, a **small group** consists of a limited number of people who interact with one another over time to reach shared goals. More precisely, small groups embody the following characteristics:

Melinda and Bill Gates are transformational leaders. They empower followers to achieve an inspiring and impactful goal—improving the lives of people in poverty.

Have you ever worked with a leader whose goal was to help make an important difference in the world? If so, how did their messages motivate others?

ON YOUR FEET

Think of a leader (either good or bad) who has influenced you in a powerful way. How would you describe that person's communication style and leadership philosophy? How have they influenced your own leadership approach? Prepare an oral presentation in which you share your answers.

small group A limited number of interdependent people who interact with one another over time to pursue shared goals.

- *Interaction.* Without interaction, a collection of people isn't a group. Students who passively listen to a lecture don't constitute a group until they begin actively to communicate with one another. This dynamic helps explain why some students feel isolated even though they spend a great deal of time on a crowded campus.

- *Interdependence.* In groups, members don't just interact; they are *interdependent.*[41] The behavior of one person affects all the others.[42] When one member behaves poorly, their actions shape the way the entire group functions. On the bright side, positive actions have ripple effects, too.

- *Time.* A collection of people who interact for a few minutes doesn't qualify as a group. True groups work together long enough to develop a sense of identity and history that shapes their ongoing effectiveness.

- *Size.* Our definition of *groups* includes the word *small.* Most experts in the field set the lower limit at three members.[43] There is less agreement about the maximum number of people.[44] As a rule of thumb, an effective group is small enough for members to know and react to every other member, and no larger than necessary to perform the task at hand effectively.[45] Small groups usually have between 3 and 20 members.

teams Groups of people who are highly successful at reaching clear, inspiring, and lofty goals.

The words *groups* and *teams* are sprinkled throughout this unit, but they are not synonymous. Teams share the same qualities as groups, but they take group work to a higher level. You probably know a team when you see it: Members are proud of their identity. They trust and value one another and cooperate. They seek, and often achieve, excellence. Teamwork doesn't come from *what* the group is doing, but *how* they do it. **Teams** tend to be unified by highly committed members and leaders who work well together to pursue clear, inspiring, and lofty goals.[46] High-profile examples are teams that win the World Cup, climb mountains, or collaborate to achieve medical or technological breakthroughs. But teamwork happens in everyday circumstances as well. Perhaps you have been part of a high-functioning team that united to help others, excel at a school assignment, or launch a campaign.

Teamwork can be especially rewarding, but not all groups need to function as teams. If the goal is fairly simple, routine, or done well alone, a group may accomplish it adequately. For example, you may be effective working alone to solve a math problem or write a press release. But when the job requires a great deal of thought, collaboration, and creativity, nothing beats teamwork. This is because collaboration operates on greater brainpower than any individual's alone and because people usually feel more confident tackling complex issues when they share the challenge as a team.[47]

Motivational Factors

Two underlying motives—group and individual goals—drive small-group communication. *Group goals* are the outcomes members collectively seek by joining together. Some group goals, such as meeting other people and having fun, are social. Others involve tasks or accomplishments, such as winning a contest, creating a product, or providing a service.

Individual goals are the personal motives of each member. Your individual goals might be to impress the boss, build your résumé, or develop a new skill. Sometimes individual goals can help the larger group. For example, a student seeking a top grade on a team project will probably help the team excel. However, problems arise when individual motives conflict with the group's goal. Consider a group member who monopolizes the discussion to get attention or one who engages in **social loafing**—lazy behavior some people use to avoid doing their share of the work. The checklist in this section presents some strategies for distributing the workload equitably.

Rules in Small Groups

All groups have guidelines in the form of rules and norms that govern members' behavior. Compare the way you act in class or at work with the way you behave with your friends.

Rules are official guidelines that govern what the group is supposed to do and how the members should behave. In a classroom, rules include how absences will be treated, whether late work will be accepted, and so on.

Norms are equally powerful, but they are typically conveyed by example rather than in words. There are three main types of small-group norms.

- *Social norms* govern how members interact with one another (e.g., what kinds of humor are/aren't appropriate, how much socializing is acceptable on the job).

- *Procedural norms* guide operations and decision making (*"We always start on time"* or *"When there's a disagreement, we try to reach consensus before forcing a vote"*).

- *Task norms* govern how members get the job done (*"Does the job have to be done perfectly, or is an adequate, if imperfect, solution good enough?"*).

It's important to realize two things about norms: First, norms don't always match what members might say is ideal behavior. Consider punctuality. The cultural expectation may be that meetings should begin at the scheduled time, yet the norm in some groups is to delay talking about real business until 10 or so minutes into the meeting. Second, group norms don't emerge immediately or automatically. When people first come together, it's common for them to feel unsure how to behave together. Even when groups have been together for a while, members' expectations may not match up perfectly. For example, the group norm may be for members to engage in rousing debates, but some members may wish meetings were calmer and quieter.

Individual Roles

The next time you see people working in small groups, observe how they behave. Some may seem stuck in silence with no one willing to speak up. In others, one or two members may dominate the discussion. In still others, all members may appear to be actively engaged—sharing ideas and enjoying the give and take. Small-group interaction is shaped by roles such as these that members play.

Whereas rules and norms establish expectations for how members behave overall, **roles** define patterns of behavior enacted by *particular* members. **Formal roles** are explicitly assigned by an organization or group. They usually come with a label, such as assistant coach, trea-

---

social loafing The tendency of some people to do less work as a group member than they would as an individual.

rules Explicit, official guidelines that govern group functions and member behavior.

norms Typically unspoken shared values, beliefs, behaviors, and processes that influence group members.

CHECKLIST ✓

Getting Slackers to Do Their Share

Experts offer the following suggestions to make sure everyone on your team does their fair share of the work.[48,49,50]

☐ Focus on the endgame: Motivation arises from working together toward an important goal.

☐ Match the goal to the group's size and talents.

☐ Establish clear goals and responsibilities.

☐ Provide training and tools to ensure that team members can accomplish the necessary tasks.

☐ Hold people accountable by asking team members to regularly share accomplishments.

☐ Focus on quality.

☐ If team members fall behind, ask why.

☐ Don't overlook poor performance. All team members need to pull their weight, or others may begin to slack off as well.

☐ Guard against burnout by paying attention to members' emotional states and workloads.

☐ Celebrate successes with praise and recognition.

role pattern of behavior enacted by particular members

formal role A role assigned to a person by group members or an organization, usually to establish order.

informal role A role usually not explicitly recognized by a group that describes functions of group members rather than their positions.

surer, or customer service representative. By contrast, **informal roles** (sometimes called "functional roles") are rarely acknowledged by the group in words.[51] Informal, functional group roles fall into two categories: task and social/maintenance (Table 11.1).

- *Task roles* help the group achieve particular outcomes, such as revising workplace policies or hosting an event. Task roles include information seeker, opinion giver, energizer, critic, and so on. For example, an initiator might ask, *"How about we take a different approach?"* and then present an idea or propose a solution.

- *Social roles* (also called "maintenance roles") help the relationships among group members run smoothly. For example, someone might encourage shy members to voice their opinions (an encourager), while another might offer compromises to help bring those with opposing viewpoints to a consensus (harmonizer). A gatekeeper might keep communication channels open by encouraging and facilitating interaction from quieter members (e.g., *"Susan, you haven't chimed in yet, but I know you've been studying the problem. What do you think about this idea?"*)

Not all informal roles are constructive. Some participants may bully others (attackers) or refuse to participate in offering opinions or ideas (withdrawers).

As you might expect, groups are most effective when people fulfill positive social roles and no one fulfills the dysfunctional ones.[52] Here are some strategies for using roles to benefit a group:

- *Make sure all helpful roles are filled.* When a group is experiencing problems, review the lists of functional roles in Table 11.1 to diagnose which important ones might be unfilled. Then you may be able to step into one of those roles yourself. For example, if key facts are missing, take the role of information seeker. If nobody is keeping track of the group's work, offer to be recordkeeper and take notes. You can also encourage others to fill key roles.

- *Avoid role fixation.* By always occupying the same role, you prevent others from having that experience. If you're always the one to suggest new ideas, the rest of the team may stop contributing. If you always support new ideas, others may be stuck playing devil's advocate to balance things out. Another danger is that you will fulfill a role out of habit, even when the situation doesn't require it. For example, you may be a world-class coordinator or critic, but these talents will only annoy others if you use them when they aren't needed.

- *Avoid dysfunctional roles.* It can be tempting to goof off or to sabotage annoying team members, but doing so may hurt the team and damage your reputation.

ASK YOURSELF

What functional (and possibly dysfunctional) roles do you usually play in small groups? How do those roles help (or hinder) the groups' effectiveness? Would you be a more productive member if you modified these roles?

TABLE 11.1

Informal, Functional Group Roles by Type

ROLE TYPE	EXAMPLES
Task	Initiator, contributor, information seeker/giver, opinion seeker/giver, clarifier, coordinator, diagnostician, orienter/summarizer, energizer, procedure developer, secretary, critic
Social/Maintenance	Supporter/encourager, harmonizer, tension reliever, conciliator (working with opposing sides to bring about agreement), gatekeeper, feeling expresser, follower

No group needs a blocker, aggressor, deserter, dominator, recognition seeker, incessant joker, or cynic. For example, a blocker would interfere with progress by saying something such as "That idea is absurd. You can talk all day, but my mind is made up." A cynic might say, "That idea probably won't solve the problem. Nothing we've tried so far has worked."

For tips on dealing with especially challenging team members, see the checklist in this section.

11.3 Making the Most of Group Interaction

Several years ago, five friends got together to brainstorm a market need they might fill. They conceived of an unlikely product—a mattress small enough (in its shipping package) to send directly to customers and to fit through tight doorways and stairwells. After a few years of research and market testing, the team launched Casper in 2014. The company's sales reached $1 million its first month and $100 million within two years.[53,54] What's important about this story, from our perspective, is the powerful role that teamwork played in it. Casper's success is a testament to effective teamwork, says management researcher Tricia Naddaff, who proposes that "there's nothing better than a new and enthusiastic team that . . . tries to do something that's never been done before."[55]

Even if you don't aspire to launch a start-up, the principles of effective teamwork can help you succeed. Let's consider some strategies for building a strong team.

Source: © Original Artist. Reproduction rights obtainable from www.CartoonStock.com

Dealing with Difficult Team Members

Every now and then you will run across a team member who consistently tests your patience. Perhaps they are whiny, bossy, aloof, aggressive, overly ingratiating, or a know-it-all. Here are some tips from the experts on coping effectively.[56,57]

☐ **Keep calm.** Some people thrive on creating drama. Don't play their game.

☐ **Look for underlying reasons.** Consider what factors might have led someone to feel ignored, hurt, or disrespected.

☐ **Surface the issue.** You might ask, "Do you feel that you didn't get a chance to explain your position?"

☐ **Lay ground rules.** Establishing specific expectations will make it easier to identify and address issues. For example, you might agree that there will be no yelling, interrupting, or maintaining side conversations.

☐ **Write down ideas.** People may be difficult because they don't feel they are being heard or respected. Writing down everyone's ideas on a board or flipchart captures what people are saying and prevents a potential source of frustration.

☐ **Make repercussions clear.** When you are coping with a difficult person whose behavior doesn't improve, it's important to be clear about the consequences. This may involve sharing the problem with a boss or instructor, or if you have the authority, making the repercussions clear yourself.

Enhancing Group Productivity

The following factors are associated with productivity.

- ☐ The group contains the smallest number of members necessary to accomplish its goals.

- ☐ Members care about and agree with the group's goals.

- ☐ Members are clear about and accept their roles, which match the abilities of each member.

- ☐ Group norms encourage high performance, quality, success, and innovation.

- ☐ Group members have sufficient time together to develop a mature working unit and accomplish shared goals.

- ☐ The group is highly cohesive and cooperative.

- ☐ The group spends time defining and discussing problems it must solve and decisions it must make.

- ☐ Periods of conflict may be frequent but are brief, and the group has effective strategies for dealing with conflict.

- ☐ The group has an open communication structure in which all members participate.

Source: Adapted from research summarized in S. A. Wheelan, D. Murphy, E. Tsumaura, & S. F. Kline. (1998). Member perceptions of internal group dynamics and productivity. *Small Group Research, 29,* 371–393.

cohesiveness The degree to which members feel part of a group and want to remain in it.

Enhance Cohesiveness

The degree to which members feel connected with and committed to a group is known as **cohesiveness**. You might think of cohesiveness as the glue that bonds individuals together, giving them a collective sense of identity. The Casper creators were already friends, but you can build strong relationships with people you don't know well. In highly cohesive groups, members spend more time interacting and express more positive feelings for one another than in those that lack cohesion. Members of cohesive groups typically report being highly satisfied and loyal. Cohesion keeps people coming back, even when the going is tough. With characteristics such as these, it's no surprise that highly cohesive groups have the potential to be productive. In fact, group cohesion is one of the strongest predictors of innovation, along with effective communication and encouragement.[58] The checklist in this section presents other factors correlated with group productivity.

Despite its advantages, cohesiveness doesn't guarantee success. Members may feel close to one another but not get the job done. In fact, in some cases, enjoying each other's company may diminish the group's efficiency. You've probably been part of study groups in which the members cared more about hanging out as friends than actually getting down to work. If so, your group was cohesive but not productive. Groups can maximize the positive aspects of cohesiveness in the following ways:

- *Focus on shared goals.* People draw closer when they share a similar aim or when their goals can be mutually satisfied.

- *Celebrate progress.* When a group is making progress, members tend to feel highly cohesive. When progress stops, cohesiveness often decreases.

- *Minimize competition.* Sometimes strife arises within groups. Perhaps there's a struggle over who will be the leader or decision maker. Whether the threat is real or imagined, the group must neutralize it or face the consequences of reduced cohesiveness.

- *Establish interdependence.* Groups become cohesive when their needs can be satisfied only with the help of other members.

- *Build relationships.* Groups often become close because the members genuinely like one another. It's important to devote time and energy to building camaraderie and friendship within the group.

Managing Meetings Effectively

If you've ever suffered through a meeting dominated by a few vocal or high-status members, you know that the quantity of speech doesn't equate with its quality. You can encourage the useful contributions of all members in a variety of ways.

Keep the Group Small In groups with three or four members, participation is roughly equal, but after the size increases to between five and eight, there is a dramatic gap between the contributions of members.[59]

Encourage Everyone to Participate   A team isn't likely to get optimal results without everyone's input. This can mean encouraging quiet members and moderating the input of overly talkative ones. Here are a few techniques you might use:

- *Thank normally quiet people for sharing.* Although it isn't necessary to gush about a quiet person's brilliant remark, saying thanks and acknowledging the value of their idea can increase the odds that they will speak up again in the future.

- *Assign tasks strategically.* You might ask quiet members to be in charge of specific tasks. The need to report on these tasks guarantees that they will speak up.

- *Ask outright for everyone's input.* If one member is talking too much, politely express a desire to hear from others.

- *Enlist the help of talkative members.* Speak privately with verbose team members and ask them to help you encourage quieter ones to be involved.

- *Change things up.* Group members who are hesitant to speak in large groups might be more comfortable writing down their ideas, talking in pairs, or brainstorming in small groups. (You'll read about discussion formats shortly.)

Sometimes it's difficult to know how much pressure to place on quiet members who might prefer not to be the center of attention. The *Ethical Challenge* feature presents some questions to consider in that regard.

Keep the Group on Topic   If the group is on a tangent and nothing else works, you might say something such as, "I'm sure last Saturday's party was awesome! But if we're going to meet the deadline, I think we'd better save those stories for happy hour."

ASK YOURSELF

What factors influence whether you feel highly committed to a team project? How does communication among team players influence your attitude?

ETHICAL CHALLENGE Balancing Overly Talkative and Quiet Group Members

Consider the perspectives of three real-life team members at different companies.

1) **Chris** is almost always silent in meetings on purpose. "I use my silence as a tool to actually listen to each word someone is saying," Chris says. He feels that most people in meetings talk too much and don't listen to (or learn from) each other enough.[60]

2) **Maria** would like to have a voice in meetings but feels belittled and ignored. "For example, at a meeting today, I made a point that was dismissed," she says, remembering that someone else advanced the same idea a few minutes later and everyone liked it. "I find myself clamming up at meetings."[61]

3) **Alex** has been accused of talking too much in meetings. "I'm definitely a talker," he acknowledges, but he finds that talking aloud helps him think through new information. Also, he feels that paraphrasing what others are saying helps several of his coworkers who "by their own admission [are] not very tech-savvy" better understand what is going on.[62]

Balancing participation in group discussions can involve reining in some members and urging others to speak up. What do you think?

- Are there instances in which you consider it helpful for meeting participants to talk a lot? To remain silent? If so, describe the circumstances.

- When is it a good idea to place quiet group members in the position of speaking up even when they would rather remain quiet? Are there situations in which it's unreasonable to urge quiet members to participate?

- Does discouraging talkative members ever violate the principles of free speech and tolerance for others' opinions? Describe when it is and is not appropriate to limit a member's contributions.

After developing your ethical guidelines, consider how you would feel if they were applied to you.

CHECKLIST

Coping with Information Overload

When you're overwhelmed by data and can't decide what's most important, consider the following tips offered by group expert J. Dan Rothwell:[71]

☐ **Specialize whenever possible.** Don't expect every member to explore every angle of a topic. Divide the research and share what you learn.

☐ **Be selective.** Take a quick look at each piece of information to see whether it has real value for your task. If it doesn't, move on to examine more promising material.

☐ **Limit your search.** Information specialists have discovered that there is often a curvilinear relationship between the amount of information a group possesses and the quality of its decision. After a certain point, gathering more material can slow you down without contributing to the quality of your group's decisions.

groupthink The tendency in some groups to support ideas without challenging them or providing alternatives.

brainstorming A group activity in which members share as many ideas about a topic as possible without stopping to judge them or rule anything out.

breakout groups A strategy used when the number of members is too large for effective discussion. Subgroups simultaneously address an issue and then report back to the group at large.

problem census A technique in which members write their ideas on cards, which are then posted and grouped by a leader to generate key ideas the group can then discuss.

round robin session in which members each address the group for specified amount of time

Avoid Information Underload and Overload   Make sure team members know the information and nuances that bear on a problem. At the same time, recognize that too much information makes it hard to sort out what's essential from what isn't. In such cases, experts suggest parceling out areas of responsibility.[63] Instead of expecting all members to explore everything about a topic, assign groups to explore particular aspects of it and share what they learn with the group at large. And if you find yourself drowning in information, use the checklist in this section for ideas.

Don't Pressure Members to Conform   The saying *"None of us is as dumb as all of us"* sometimes rings true. If you have ever gone along with a decision without thinking critically, you have engaged in **groupthink**, the tendency of some groups to support ideas without challenging them or providing alternatives.[64] Here are some of the reasons people might engage in groupthink:

- They wish to avoid a conflict.
- They want others to see them as "team players."
- They overestimate the group's good judgment or its privileged status.
- They fear that others won't like what they have to say.
- They just want to get the discussion over with.[65]

A range of factors may make groupthink seem appealing, but the results can range from disappointing to downright disastrous. On a small scale, your team might downplay customer complaints that could help you improve, maintaining instead that what you are already doing is *"the best it can be"* and *"you'll never please everyone."* In tragic examples, lives could even be at risk. After seven astronauts died in the *Challenger* explosion of 1986, an investigation revealed that engineers were discouraged from voicing their concerns before the launch.[66] More recently, in numerous high-profile cases, people didn't challenge or report child molesters because they were afraid of the repercussions.[67,68,69]

You can minimize the risk of groupthink by adopting the following practices:[70]

- *Recognize early warning signs.* It may seem like a good sign if agreement comes quickly and easily, but under the surface, the group may be avoiding the tough but necessary search for alternatives. Considering all the options now may save time and hardship later.

- *Minimize status differences.* Group members sometimes fall into the trap of agreeing with anything the leader says. To minimize this tendency, leaders should make it clear that they encourage open debate rather than blind obedience. They might also encourage members to conduct initial brainstorming sessions on their own, among peers.

- *Make respectful disagreement the norm.* After members recognize that questioning one another's positions doesn't signal personal animosity or disloyalty, a constructive exchange of ideas can lead to top-quality solutions. Sometimes it can be helpful to designate a person or subgroup as a "devil's advocate" who reminds the others about the dangers of groupthink and challenges them to consider potential disadvantages of popular ideas before reaching a decision.

Using Discussion Formats Strategically

Groups meet in a variety of settings and for a wide range of reasons. The formats they use are also varied. The following list is not exhaustive, but it provides a sense of how a group's structure can shape the type and quality of communication that members share.

Brainstorming During a **brainstorming** session, the goal is to think of as many ideas as possible about a specified topic without stopping to judge what is suggested or rule anything out. The checklist in this section presents tips for making the most of this strategy.

Breakout Groups When the number of members is too large for effective discussion, **breakout groups** can be used to maximize effective participation. In this approach, subgroups (usually consisting of five to seven members) simultaneously address an issue and then report back to the group at large. The best ideas of each breakout group are then discussed and synthesized.

Problem Census When some members are more vocal than others, **problem census** can help equalize participation. Working individually, members list their thoughts on cards, one idea per card. The leader then collects the cards, reads them to the group one by one, and posts them on a board visible to everyone. Because the name of the person who contributed each item isn't listed, issues are separated from personalities. As similar items are read, the leader posts and arranges them in clusters. After all cards are read and posted, members reflect on the ideas as a group.

Round Robins It can be useful to give everyone on the team a brief turn to talk. To initiate a **round robin** session, have members sit (preferably in a circle) where they can easily see and hear everyone else. Agree upon the time allotted to each member (perhaps 30 or 60 seconds) and the question everyone will address, such as *"What do you most hope we can achieve as a group?"* or *"Of the ideas we have discussed, which one are you most excited about and why?"* Since round robins aren't anonymous, they may not be effective in addressing highly sensitive issues. However, they allow members to get to know one another and appreciate the similarities and diversity of their ideas. Round robins are similar to problem census in that they give everyone a chance to share their thoughts, and like brainstorming, members in a round robin session are forbidden from critiquing or evaluating ideas as they are expressed. You might choose to have someone write all the ideas on a board to stimulate a follow-up discussion, or a round robin can be a good way to close out a meeting. You can also use round robins in virtual teams by asking each member to post a brief answer to the question on an online discussion board.

Dialogue Sometimes the best way to tackle a problem is to stop trying to find a solution and listen to one another. **Dialogue** is a process in which people let go of the notion that their ideas are superior to others'

CHECKLIST ✓

Making the Most of a Brainstorming Session

Brainstorming can be useful at any phase of problem solving. Here are some ways to make the most of this creative technique.

☐ **Include diverse participants.** Seeking a diversity of perspectives can reveal options not yet considered.[72]

☐ **Begin by brainstorming as individuals.** This way, a group can harness the diversity of individual thought *and* the synergy of group work.[73]

☐ **Choose a neutral facilitator.** A facilitator can keep the group focused.

☐ **Appoint a recorder or scribe.** It can be helpful to write all ideas on a board or flip chart where group members can see them.[74]

☐ **Forbid criticism.** Nothing stops the flow of ideas more quickly than negative evaluation. Criticism almost guarantees a defensive reaction, and it may inhibit people from thinking freely and sharing new ideas.

☐ **Share whatever comes to mind.** Even the most outlandish ideas sometimes prove workable or inspire a solution. The more ideas you generate, the more likely you are to come up with a good one.

☐ **Combine and build upon ideas.** Encourage members to consider modifying or combining ideas already suggested, potentially drawing diagrams to show how concepts relate to one another.[75]

---

dialogue A process in which people let go of the notion that their ideas are more correct or superior to those of others and instead seek to understand an issue from many different perspectives

---

@WORK

The Power of Constructive Dialogue

In a dialogue session, about 20 members of a large corporation explored ways to boost the effectiveness of their work team. One participant shared his belief that "people who come to work late are lazy and rude." Even members who thought the complaint was harsh listened respectfully as the speaker explained his ideas about what team members owe one another and the organization.

Another group member acknowledged that she had been arriving late recently and described a series of issues involving child care and transportation. She shared her frustration over the apparently impossible conflict between work obligations and family challenges, and her regret over letting down her colleagues. There was silence as the group pondered the dilemma she described.

Another member spoke out: "I don't have kids. You chose to have them. So your situation is not my problem." The emotional shockwaves in the room were evident on people's faces. The facilitator asked everyone to sit quietly for a few moments before responding.

One man in the group presented a different perspective: "Imagine we are a football team and we have just got new uniforms. We look great. We feel great. But the kicker says his feet hurt—his shoes are too small. Would we say, 'That's not our problem'?"

The session adjourned, as dialogues do, without any decisions being made, but leaving members with a good deal to think about. In the days that followed, many people approached the team member with child care problems to express their sympathy for her dilemma. Several said they had the opposite problem: They could come early, but they needed to leave a few minutes before the workday ended. They were able to make a time trade to suit the needs of everyone involved, without compromising the team's mission or inconveniencing the rest of the group.

In a follow-up dialogue, members of the team commented on how easily the situation was resolved once everyone understood one another. "Under different circumstances, we might have created a new policy we didn't need, which wouldn't have helped," said one team member. "Now we say, 'the kicker's feet hurt' any time we are tempted to jump to conclusions without listening first."

---

and instead try to understand an issue from many perspectives.[76] For example, if the problem is that some children in your community are not being immunized, you might invite a collection of diverse people together to talk about the issue. Perhaps you had assumed that parents were being irresponsible, but you learn by listening that some of them don't have transportation, can't afford the cost, or don't understand medical information very well. You will probably proceed very differently once you realize the complexities of the issue.

In a genuine dialogue, members acknowledge that everything they "know" and believe is an assumption based on their own, unavoidably limited experiences. People engage in curious and open-minded discussion about that assumption while guarding against either–or thinking. The goal is to understand one another better, not to reach a decision or debate an issue.

Through dialogue, observes theorist William Isaacs, problems are often not so much *solved* as *dissolved*.[77] The issue may cease to be a problem, or it may be so transformed that the solution is obvious or occurs without a formal decision. (For an example of an actual dialogue, see the *@Work* box in this section.)

11.4 Group Problem Solving

Nelson Mandela learned a lot about problem solving watching his guardian Jongintaba, who was a tribal leader. When people assembled at Jongintaba's home, they would sit in a circle and talk, sometimes engaging in heated debates.[78] During those meetings, Jongintaba usually sat quietly and listened. Only when everyone had been heard and all angles had been considered did he offer comment, in a calm,

ASK YOURSELF ?

Some meetings are brief and casual. At other times, formal structure can help a group tackle challenges in a systematic manner. When would each problem-solving format in this section be most helpful to you?

measured voice.[79] From this experience, Mandela learned to encourage open discussion and collaboration, even among people who might seem, on the surface, to be adversaries.

Groups may have the potential to solve problems effectively, but they don't always live up to this potential. When is group work a good option? What makes some groups succeed and others fail? Researchers have spent decades asking these questions. To discover their findings, read on.

Advantages of Group Problem Solving

Years of research show that, in most cases, groups can produce more and better solutions to a problem than can individuals working alone.[80] Groups have proven to be superior at everything from assembling jigsaw puzzles to solving complex reasoning problems. There are several reasons why groups are effective.[81]

In 1970 three U.S. astronauts radioed Mission Control from space with the famous words "Houston, We have a problem." The 1995 film *Apollo 13* dramatizes how the astronauts and the ground team used problem-solving skills to turn a potential catastrophe into a heroic outcome.

Think about a successful (or unsuccessful) group problem you attempted to solve. What skills did (didn't) the group use? How could the group have done a better job?

Resources   For many tasks, groups have access to a greater collection of resources than do individuals. Sometimes the resources involve physical effort. For example, three or four people can set up for an event or coordinate volunteers better than a lone person could. Resources may also include space, equipment, time, interpersonal connections, brain power, and more. Imagine trying to raise money for an important cause. Together, a group is likely to know far more potential contributors than any one person does. And if you have ever had the opportunity to take a test or quiz as a group, the odds are that you scored higher than any one member would have.

Accuracy   Another benefit of group work is the increased likelihood of catching errors. At one time or another, we all make stupid mistakes, like the man who built a boat in his basement and then wasn't able to get it out the door. Working in a group can help prevent foolish errors and oversights.

Commitment   Besides coming up with superior solutions, groups may also generate a higher commitment to carrying them out. Members are most likely to accept solutions they have helped create and to work harder to carry out those solutions. This fact has led to the principle of **participative decision making**, in which people contribute to the decisions that will affect them. Nearly any professor will tell you that students cooperate much more willingly when they help develop a policy or project for themselves than when it is imposed on them.

Diversity   Although we tend to think in terms of "lone geniuses" who make discoveries and solve the world's problems, most breakthroughs are actually the result of collective creativity—people working together to create options that no one would have thought of alone.[82]

Although diversity is a benefit of teamwork, it requires special effort, especially when members come from different cultural backgrounds. For example, in teams that consist of both Asian-born and American-born members, Americans tend to do most of the talking and are more likely than their Asian teammates to interrupt.[83] And women may be dismayed to find that members of some cultures, including their own, tend to dismiss their comments out of gender bias.[84] To make the most of multiculturalism and avoid some of the common pitfalls, see the *Understanding Diversity* feature in this section.

ASK YOURSELF

Think of a difficult task that could not have been accomplished by a person acting alone. Then reflect on the problems you face now, and decide which ones could be best tackled by a group.

participative decision making
A process in which people contribute to the decisions that will affect them.

UNDERSTANDING DIVERSITY

Maximizing the Effectiveness of Multicultural Teams

Evidence shows that multicultural teams are typically more creative than homogeneous groups or individuals.[85] But they also present unique challenges. Communication researchers offer the following tips to maximize the benefits and minimize the pitfalls of multicultural teams:[86]

- *Allow more time than usual.* When members have different backgrounds and perspectives, it can take patience and extra effort to understand and appreciate where each person is coming from.

- *Agree on clear guidelines for discussions, participation, and decision making.* If members come to the group with different expectations, it may be necessary to negotiate mutually acceptable ground rules.

- *Use a variety of communication formats.* People may be more or less comfortable speaking to the entire group, putting their thoughts in writing, speaking one on one, and so on. Variety will help everyone have a voice.

- *If possible, involve an even distribution of people from various cultures.* Research shows that being a "minority member" is especially challenging and not conducive to open communication.[87]

- *Educate team members about the cultures represented.* People are less likely to make unwarranted assumptions (that a person is lazy, disinterested, overbearing, or so on) if they understand the cultural patterns at play.

- *Open your mind to new possibilities.* It's easy to reject unfamiliar ideas without giving them proper consideration, but that shortcircuits the advantage of diverse perspectives. Make it a point to be curious and open-minded when a new idea is raised.

The results of multicultural teams are usually worth the effort. As one analyst puts it, "diversity makes us smarter."[88]

Despite the potential advantages of group work, it isn't always the quickest way to accomplish a task or make a decision. Following are some questions to consider when deciding whether a challenge is best addressed by a group or an individual. If you answer "yes" to the following conditions, collaborative problem solving will probably yield the best results.

The first ideas a group considers may not be the best ones. Brainstorming can boost the odds of making good decisions.

What can you do to encourage brainstorming in groups to which you belong?

- Is there more than one solution and no easy answer?
- Does the issue present implications for a large number of people?
- Is there the potential for disagreement?
- Will a good solution require more resources (e.g., time, money, supplies) than one person can provide?
- Would it be costly to make an error in judgment?
- Is buy-in by multiple people important to success?

By contrast, a problem with only one solution won't take full advantage of a group's talents. For example, phoning merchants to get price quotes and looking up a series of books in the library don't require much creative thinking. Jobs such as those can be handled by one or two people working alone. Of course, it may take a group meeting to decide how to divide the work to get the job done most efficiently.

Stages of Team Development

When it comes to solving problems in groups, the shortest distance to a solution isn't always a straight line. Communication scholar Aubrey Fisher analyzed tape recordings of problem-solving groups and discovered that many successful groups follow a four-stage process when arriving at decisions.[89] A useful way to remember these stages is with the rhyming words *forming*, *storming*, *norming*, and *performing*.[90] As you read about the stages, visualize how they have applied to problem-solving groups in your experience.

Orientation (Forming) Stage   In the **orientation (forming) stage**, members approach the problem and one another tentatively. Rather than stating their own positions outright, they test possible ideas cautiously and politely. This cautiousness doesn't mean that members agree with one another. Rather, they are assessing the situation before asserting themselves. There is little outward disagreement at this stage, but it can be viewed as a calm before the storm.

Conflict (Storming) Stage   After members understand the problem and become acquainted, a successful group enters the **conflict (storming) stage**. Members take strong positions and defend them. Coalitions are likely to form, and the discussion may become polarized. The conflict needn't be personal. It can focus on the issues at hand while preserving the members' respect for one another. Even when the climate does grow contentious, conflict seems to be a necessary stage in group development. The give and take of discussion tests the quality of ideas, and weaker ones may suffer a well-deserved death here.

Emergence (Norming) Stage   After a period of conflict, effective groups move to an **emergence (norming) stage**. One idea might emerge as the best one, or the group might combine the best parts of several plans into a new solution. As they approach consensus, members back off from their dogmatic positions. Statements become more tentative again: *"I guess that's a pretty good idea"* or *"I can see why you think that way."*

Reinforcement (Performing) Stage   Finally, an effective group reaches the **reinforcement (performing) stage**. At this point, not only do members accept the group's decision, they also endorse it. Even if members disagree with the outcome, they are not likely to voice their concerns. There is an unspoken drive toward consensus and harmony.

This isn't a one-time process. Ongoing groups can expect to move through these stages with each new issue, such that their interactions take on a cyclic pattern. They may begin discussion of a new issue tentatively, then experience conflict, emergent solutions, and reinforcement. In fact, a group that deals with several issues at once might find itself in a different stage for each problem.

Knowing that these phases are natural and predictable can be reassuring. It can help curb your impatience when the group is feeling its way through the orientation stage. It may also make you feel less threatened when inevitable and necessary conflicts take place. Understanding the nature of emergence and reinforcement can help you know when it is time to stop arguing and seek consensus.

A Structured Problem-Solving Approach

As early as 1910, John Dewey introduced his famous **reflective thinking method** as a systematic, multistep approach to solving problems.[91] Since then, other experts have suggested modifications, although it is still generally known as Dewey's reflective thinking method. Although no single approach is best for all situations, a structured procedure usually produces better results than "no pattern" discussions.[92]

---

orientation (forming) stage When group members become familiar with one another's positions and tentatively volunteer their own.

conflict (storming) stage When group members openly defend their positions and question those of others.

emergence (norming) stage When a group moves from conflict toward a single solution.

reinforcement (performing) stage When group members endorse the decision they have made.

reflective thinking method A systematic approach to solving problems that involves identifying a problem, analyzing it, developing and evaluating creative solutions, implementing a plan, and following up on the solution. Also known as Dewey's reflective thinking method.

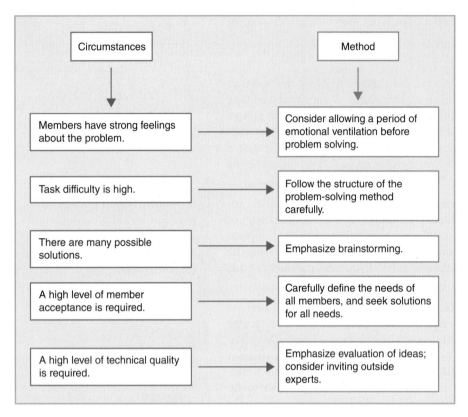

Circumstances		Method
Members have strong feelings about the problem.	→	Consider allowing a period of emotional ventilation before problem solving.
Task difficulty is high.	→	Follow the structure of the problem-solving method carefully.
There are many possible solutions.	→	Emphasize brainstorming.
A high level of member acceptance is required.	→	Carefully define the needs of all members, and seek solutions for all needs.
A high level of technical quality is required.	→	Emphasize evaluation of ideas; consider inviting outside experts.

FIGURE 11.2   Problem-Solving Options

The problem-solving model described here includes the elements common to most structured approaches developed in the last century. Although these steps provide a useful outline for solving problems, they are most valuable as a general set of guidelines and not as a precise formula. Certain parts of the model may need emphasis over others depending on the nature of a specific problem (Figure 11.2), but the general approach will give virtually any group a useful way to approach problem solving. This section applies Dewey's reflective thinking model to a hypothetical problem and then considers what happens when virtual groups engage in problem solving at a distance.

Identify the Problem   Sometimes a group's problem is easy to pinpoint. Many times, however, the challenge facing a group isn't so clear. For example, if a group is meeting to discuss a low-performing employee, it may be helpful to ask why that person is underperforming. It may be because this employee has personal problems, feels unappreciated by members, or hasn't been challenged. The best way to understand a problem is to look below the surface and identify the range of factors that may be involved. It may also be helpful at this stage to take stock of group goals and the individual goals of everyone on the team.

Analyze the Problem   After you have identified the general nature of the problem, examine it in more detail. Here are some ways to do that:

- *Word the problem as a broad, open question.* For example, if your group is trying to understand why employee turnover is high, you might ask yourselves, *"What can we do to make good hires and retain good employees?"* Open-ended questions like these encourage people to contribute ideas and to work cooperatively. They may also help the group identify the criteria for a successful solution.

- *Identify criteria for success.* Once you know what you're trying to achieve, you have a better chance of creating goal-oriented solutions and measuring your success. To continue the previous example, after analyzing the problem, you might set the goal of keeping employee turnover equal to or lower than what's typical for your business and community. Table 11.2 presents some ways to adapt the process to suit a team's needs.

- *Gather relevant information.* It's foolish to choose a solution before you know all the options and factors at play. In this stage, you might seek answers to questions such as: *How does our turnover compare to similar companies in this community? How can we measure employee satisfaction? What can we learn from current employees and those who have recently left? What can we learn from companies with less turnover?*

force field analysis A method of problem analysis that identifies the forces that work in favor of a desired outcome and those that will make it difficult to achieve.

nominal group technique A method for including the ideas of all group members in a problem-solving session, alternating between individual contributions and group discussion.

TABLE 11.2

Problem-Solving Methods for Special Circumstances

CIRCUMSTANCES	METHOD
Members have strong feelings about the problem.	Allow members to share their feelings at the outset.
Task difficulty is high.	Follow the structure of the problem-solving method carefully.
There are many possible solutions.	Emphasize brainstorming.
A high level of member acceptance is required.	Carefully define the needs of all members, and seek solutions that satisfy all needs.
A high level of technical quality is required.	Carefully evaluate ideas; consider inviting outside experts.

Source: Adapted from J. Brilhart & G. Galanes. (2001). Adapting problem-solving methods. In *Effective group discussion* (10th ed., p. 291). Copyright © 2001. Reprinted by permission of McGraw-Hill Companies, Inc.

- *Identify supporting and restraining forces.* A **force field analysis** involves listing the forces in favor of a desired outcome and those that will probably make it difficult to achieve.[93] For example, forces that can help you retain good employees might include offering higher pay and better training. Challenges might include a transient workforce and the inherent difficulty of the job.

Develop Creative Solutions   Considering more than one solution is important because the first solution, though the most obvious or most familiar, may not be the best one. During this development stage, creativity is essential.

Along with brainstorming, another method for including the ideas of all group members is the **nominal group technique**. It combines alternating cycles of individual work and group discussion.

- Each member works alone to develop a list of possible solutions.
- In round-robin fashion, each member offers one item from their list. The item is listed on a chart visible to everyone. Other members may ask questions to clarify an idea, but no evaluation is allowed during this step.
- Each member privately ranks the ideas that have been posted in order from least preferable (1 point) to most preferable (a point value equal to the number of ideas on the list). The rankings are collected, and the ideas with the most points are retained as the most promising solutions.
- A free discussion of the top ideas is held. The group weighs the merits of each idea until they select one, either by majority vote or by consensus.

Brainstorming and the nominal group technique can be useful means to think carefully about issues and to give everyone a voice in the process.

CHECKLIST ✓

Stages in Structured Problem Solving

The reflective thinking method inspired by John Dewey involves the following steps:[94]

☐ **Identify the problem.**
- Determine the group's goals.
- Determine individual members' goals.

☐ **Analyze the problem.**
- Word the problem as a broad, open question.
- Identify criteria for success.
- Gather relevant information.
- Identify supporting and restraining forces.

☐ **Develop creative solutions through brainstorming or the nominal group technique.**
- Avoid criticism at this stage.
- Encourage an unrestricted exchange of ideas.
- Develop a large number of ideas.
- Combine two or more individual ideas.

☐ **Evaluate the solutions by asking the following:**
- Which solution will best produce the desired changes?
- Which solution is most achievable?
- Which solution contains the fewest serious disadvantages?

☐ **Implement the plan.**
- Identify specific tasks.
- Determine necessary resources.
- Define individual responsibilities.
- Provide for emergencies.

☐ **Follow up on the solution.**
- Meet to evaluate progress.
- Revise the approach as necessary.

TABLE 11.3

Ways to Reach a Group Decision

APPROACH	ADVANTAGES	DISADVANTAGES
Reach consensus	Full participation can increase the quality of the decision as well as members' willingness to support it. This can be an effective approach when it's important to have everyone's support, the issue is critical or complex, and there is adequate time to involve everyone.	Consensus building can take a great deal of time, which makes it unsuitable for emergencies.
Let the majority decide	Majority rule may work well when it isn't necessary to have everyone's buy-in and when time is of the essence.	Decisions made under majority rule are often inferior to those hashed out by a group until the members reach consensus.[95]
Rely on the experts	Sometimes one group member will be considered an expert and will be given the power to make decisions. This can work well when that person's judgment is truly superior.	Some members may think they are the best qualified to make a decision even when that is not the case.
Let a few members decide	This approach works well with noncritical questions that would waste the whole group's time.	When an issue needs more support, it's best at least to have a subgroup report its findings for the approval of all members.
Honor authority rule	Though it sounds dictatorial, there are times when this approach has advantages. Sometimes there isn't time for a group to decide, or the matter is so routine that it doesn't require discussion.	Much of the time, group decisions are of higher quality and gain more support from members than those made by an individual.

Evaluate Potential Solutions A good way to evaluate solutions is to ask the following questions: *Which will best produce the desired changes? Which solution is most achievable?* and *Which solution contains the fewest serious disadvantages?* Table 11.3 outlines several means of group decision making and the pros and cons of each.

Implement the Plan Everyone who makes New Year's resolutions knows the difference between making a decision and carrying it out. There are several important steps in developing and implementing a plan of action:

- *Identify specific tasks to be accomplished.* What must be done? Even a relatively simple job usually involves several steps. Now is the time to anticipate all the tasks facing the group. This may help you avoid an oversight or last-minute rush later.

- *Determine necessary resources.* Identify the equipment, material, and other resources the group will need to get the job done.

- *Define individual responsibilities.* Who will do what? Do all the members know their jobs? The safest plan here is to put everyone's duties in writing with due dates. This might sound compulsive, but experience shows that it increases the chance of getting jobs done on time.

- *Plan ahead for emergencies.* Murphy's Law states that "whatever can go wrong will go wrong." Anyone experienced in group work knows the truth of this saying. People forget their obligations, get sick, or quit. The Internet goes down whenever it's most needed, and so on. Whenever possible, develop contingency plans to cover foreseeable problems. Probably the single best suggestion is to plan on having all work done well ahead of the deadline so you know that, even with last-minute problems, you can still finish on time.

Follow Up on the Solution   Even the best plans usually require some modifications after they're put into practice. You can improve the group's effectiveness by meeting periodically to evaluate progress and revise the approach as necessary.

Problem Solving in Virtual Groups

Imagine it's your first day on the job at Automattic, a billion-dollar web development company. What should you wear? Where will you sit? Where's the staff meeting? The answers are: Wear anything. Sit anywhere you like, as long as you have Wi-Fi. And the staff meeting? Online. Automattic employees work from different locations—at last count, more than 930 of them in 69 countries.[96] The company provides all the technology they need and several thousand dollars each for employees to create their own work environments. Teams work virtually most of the time, but a sizable travel budget allows them to schedule face-to-face meetings several times a year.

One advantage of virtual teamwork is autonomy. Automattic holds few formal meetings, and there are few rules and no set working hours. "I don't care if you spend the afternoon on the golf course and then work from 2 to 5 A.M.," CEO Matt Mullenweg says. His only question is: "What do you actually produce?"[97] Even the company-wide annual retreat is treated as an occasion for socializing and working on team projects rather than the typical sit-down-and-listen approach.

The lack of meetings doesn't mean communication is limited. Although Automatticians are spread all over the globe, communication among them flows more freely than it might if they shared the same roof. They communicate more or less continually in real time through a company-wide blog system. Anyone can review the posts, comment on them, and use them as an archive of information. "Everyone, from intern to CEO, can weigh in on anything,"[98] observes an industry analyst. The system supports transparency, frequent interaction, open debates, and shared learning—in short, dynamic communication 24 hours a day.

For either coworkers or students who are collaborating from a distance, online communication has three main benefits. It allows for greater diversity than might be possible in one geographic location; members who might have kept quiet in face-to-face sessions may be more comfortable speaking out online; and online meetings often generate a permanent record of the proceedings, which can be convenient.

But virtual interaction presents some unique challenges to problem solving:

- If members can't see one another clearly, it may be difficult to convey and understand one another's emotions and attitudes.

- It may take virtual teams longer to reach decisions than those who meet face to face.[99]

- Whereas high-tech videoconferencing may make it seem almost as if people are in the same room, other forms of interaction (such as email and instant messaging) can be laborious. Members may not bother to type messages online that they would have shared in person with less effort.

- The string of separate messages generated in a computerized medium can be hard to track, sort out, and synthesize in a meaningful way.

Research comparing the quality of decisions made by face-to-face and online groups is mixed. Overall, it suggests that virtual teams can be as effective as others, but only if the members have cultivated trusting relationships with one another.[100] Certain types of mediated communication generally work better than others. For example, asynchronous groups often make better decisions than those functioning

ASK YOURSELF

Have your team efforts been enhanced by the ability to communicate online? Have you encountered any difficulties or drawbacks? What lessons about virtual teamwork emerge from your experiences?

UNDERSTANDING COMMUNICATION TECHNOLOGY

Developing Trust Long Distance

Trust is a feeling. We may not be able to say clearly why one person strikes us as trustworthy and another triggers our defenses. But if we could observe trust building in slow motion, we'd find that it's based on a collection of signals, some of them very subtle.

Nonverbal cues are usually the first evidence we consider. Does this person smile, make eye contact, show evidence of being friendly and interested in what we have to say, exhibit appropriate use of space and touch? Over time, we take into account patterns of behavior that suggest the person's underlying character and commitment. Is this person dependable, consistent, committed to the team, honest, and so on?[101]

Some of these factors are difficult to judge at a distance. But emerging research suggests that, with a little effort, trust can flourish among virtual team members. Experts' tips include the following:[102]

- *Use video technology as often as possible.* This provides members with valuable visual and nonverbal cues about one another.

- *Pay particular attention to your nonverbal communication while you are on camera.* Although it is tempting to check your phone or leaf through papers, it may send the signal that you are uninterested.

- *Show enthusiasm for the group's mission and tasks.*

- *Encourage members to share information about themselves.*

in a "chat" mode, perhaps because members have time to digest and synthesize information.[103] Having a moderator also improves the effectiveness of many online groups.[104] (For more on trust building in virtual teams, see the *Understanding Communication Technology* feature in this section.)

It's fitting for this chapter to end where it began, with Nelson Mandela. Like all good team members, Mandela didn't simply do whatever leaders told him to do. He stood his ground when he believed their policies were unjust. Like all good leaders, he understood that true power lies in uniting and empowering people.

MAKING THE GRADE

At www.oup.com/he/adler-uhc14e, you will find a variety of resources to enhance your understanding, including video clips, animations, self-quizzes, additional activities, interactive self-assessments, and more.

OBJECTIVE 11.1 Describe effective leadership skills based on the situation, goals, and team members' needs.

- Most research suggests that people can learn the skills that contribute to effective leadership.

- No one leadership approach works well in all circumstances. Instead, leaders who understand the relative strengths of various styles are most likely to succeed.

- For the most part, leaders who focus on the overall mission, relationships, and task fulfillment accomplish more than those who are motivated by the desire to achieve personal glory or maintain harmony at all costs.

- Leaders often emerge through a process of elimination, which suggests that, whether or not they know it, they begin "auditioning" for leadership roles as soon as they join a group.

  > Think of leaders (people you know or public figures) who embody each of the following leadership styles: autocratic, democratic, laissez-faire, and servant leadership. In your opinion, which of these leaders has been most effective and why?

  > Imagine that the CEO of the retail clothing company where you work has challenged your team to increase sales by 50%. Describe how your team leader might

respond to this challenge differently using each of the five leadership approaches included in Blake and Mouton's Managerial Grid.

> Transformational leaders help people make a significant and valuable contribution in business, science, civil rights, or another arena. Describe a goal that is important to you, and explain how you might embody the qualities of transformational leadership to help people achieve it.

OBJECTIVE 11.2 Identify factors that influence communication in small groups and either help or hinder their success.

- Group work involves interaction and interdependence over time among a small number of participants with the purpose of achieving one or more goals.
- Some groups achieve the status of teams, which embody a high level of shared goals and identity, commitment to a common cause, and high ideals.
- Groups have their own goals, as do individual members.
- Members' goals fall into two main categories: task related and social.
- Social loafing is a common frustration in group work, but there are ways to help ensure that all group members feel accountable.
- Group norms suggest how members should interact with one another, how the group will do business, and who will carry out particular tasks.
- Group members play task roles (such as coordinator and diagnostician) and social roles (such as harmonizer and tension reliever). Some roles are helpful to the group, whereas others (such as dominator and deserter) can damage performance and member relationships.

> Describe the best and worst groups you have ever been part of, and then describe how communication differed in those groups.

> Think of a group you belong to, then make three lists: (1) your individual goals as a group member, (2) the individual goals of another group member, and (3) the group goals. How do the three lists compare? Are any of the individual goals you listed at odds with the team goals? If so, how?

> Think of a time you were tempted to do less than your share as a member of a group. What factors made that choice appealing? What factors motivated (or might have motivated) you to do as much as everyone else?

OBJECTIVE 11.3 Strategize ways to build cohesiveness and communicate effectively during group discussions.

- Effective groups strive to build a sense of cohesiveness among members.

- Groups use a variety of discussion formats when solving problems, and each format has advantages and disadvantages depending on the size of the group, the behavior of members, and the nature of the problem.
- Groups function best when they get the information they need without feeling overloaded.
- Members of effective teams make sure that they participate equally by encouraging the contributions of quiet members and by keeping more talkative people on topic.
- Effective team members guard against groupthink by minimizing pressure on members to conform for the sake of harmony or approval.
- Because face-to-face meetings can be time consuming and difficult to arrange, virtual teamwork is a good alternative for some group tasks. Mediated meetings provide a record of discussion, and they can make it easier for normally quiet members to participate. They can take more time, however, and they lack the nonverbal richness of face-to-face conversation.

> Imagine you have to lead a new committee responsible for redesigning your school's grading policy. What will you do to help ensure that the committee functions to its highest potential?

> Is a cohesive team always productive? Why or why not? Describe at least four methods of building team cohesiveness that can help a team get its job done.

> Think of a team meeting you have been part of. Assess the productivity of that meeting on a 1 (low) to 10 (high) scale. Using the tips for managing meetings effectively in this chapter, describe the reasons you think the team either was, or was not, very productive.

> Recall a group in which one or more members talked too much and others said very little. Describe an approach you might take to ensure more equal participation in the future.

> Teamwork involves inherent risks. Explain what strategies you would adopt in a new team to make sure that members get the right amount of information and the team avoids groupthink.

> Imagine that you and five other students have been asked to engage in a team project to benefit a nonprofit organization of the team's choice. What discussion formats would you use to decide on a project and host organization? If we expanded the project to include the entire class (let's say 30 people), would your choice of formats change? Why or why not? How might your communication strategies differ if members met virtually rather than in person?

OBJECTIVE 11.4 Assess the advantages and stages of group problem solving, and practice applying Dewey's group problem-solving model.

- Despite the sometimes-bad reputation of groups, research shows that they are often more effective than individuals in solving complex problems.

- Groups often have greater resources, both quantitatively and qualitatively, than do either individuals or collections of people working in isolation.

- Teamwork can result in greater accuracy than individual efforts, and people may feel more committed to solutions they have helped to produce.

- Strong teams move through the following stages as they solve a problem: orientation (forming), conflict (storming), emergence (norming), and reinforcement (performing).

- Problem-solving groups should begin by identifying the problem and recognizing the unexpressed needs of individual members. The next step is to analyze the problem, including forces that favor progress and those that may stand in the way. Only at this point should the group begin to develop possible solutions, taking care not to stifle creativity by evaluating any of them prematurely.

> Name six conditions in which it is usually more effective to address a problem as a group rather than as an individual.

> Describe a group experience in which you felt it would have been more productive to work alone. What factors caused you to feel that way?

> Think of a group you were part of that effectively dealt with a problem or opportunity. What factors helped the group succeed?

> Summarize the development of a team that is familiar to you, from orientation through reinforcement, giving an example of communication in each stage.

> Pretend that you are advising a newly formed team whose members will interact with each other online. What advice from this chapter would you share with them for developing trust at a distance?

KEY TERMS

authoritarian leadership p. 265

brainstorming p. 279

breakout groups p. 279

cohesiveness p. 276

conflict (storming) stage p. 283

democratic leadership p. 266

dialogue p. 279

emergence (norming) stage p. 283

emergent leader p. 265

force field analysis p. 285

formal role p. 273

groupthink p. 278

informal roles p. 274

laissez-faire leadership p. 266

nominal group technique p. 285

norms p. 273

orientation (forming) stage p. 283

participative decision making p. 281

problem census p. 279

reflective thinking method p. 283

reinforcement (performing) stage p. 283

roles p. 273

round robin p. 279

rules p. 273

servant leadership p. 266

situational leadership p. 266

small group p. 271

social loafing p. 273

teams p. 272

trait theories of leadership p. 266

transformational leadership p. 269

ACTIVITIES

1. Think about two groups to which you belong.

   a. What are your task-related goals in each?

   b. What are your social goals?

   c. Are your personal goals compatible or incompatible with those of other members?

   d. Are they compatible or incompatible with the group goals?

   e. What effect does the compatibility or incompatibility of goals have on the effectiveness of the group?

2. Describe the desirable norms and explicit rules you would like to see established in the following new groups, and describe the steps you could take to see that they are established.

   a. A group of classmates formed to develop and present a class research project

   b. A group of neighbors meeting for the first time to persuade the city to install a stop sign at a dangerous intersection

   c. A group of 8-year-olds you will coach in a team sport

   d. A group of fellow employees who will share new office space

3. Based on the information in Section 11.3 and your own experiences, give examples of groups that meet each of the following descriptions:

   a. A level of cohesiveness so low that it interferes with productivity

   b. An optimal level of cohesiveness

   c. A level of cohesiveness so high that it interferes with productivity

   d. For your answers to a and c, offer advice on how the level of cohesiveness could be adjusted to improve productivity.

   e. Are there ever situations in which maximizing cohesiveness is more important than maximizing productivity? Explain your answer, supporting it with examples.

4. Explain which of the following tasks would best be managed by a group:

   a. Collecting and editing a list of films illustrating communication principles

   b. Deciding what the group will eat for lunch at a one-day meeting

   c. Choosing the topic for a class project

   d. Finding which of six companies had the lowest auto insurance rates

   e. Designing a survey to measure community attitudes toward a subsidy for local artists

Preparing and Presenting Your Speech

12

CHAPTER OUTLINE

12.1 Getting Started 295
Choosing Your Topic
Defining Your Purpose
Writing a Purpose Statement
Stating Your Thesis

12.2 Analyzing the Speaking Situation 297
The Listeners
The Occasion

12.3 Gathering Information 301
Online Research
Library Research
Interviewing
Survey Research

12.4 Managing Communication Apprehension 304
Facilitative and Debilitative Communication Apprehension
Sources of Debilitative Communication Apprehension
Overcoming Debilitative Communication Apprehension

12.5 Presenting Your Speech 307
Choosing an Effective Type of Delivery
Practicing Your Speech

12.6 Guidelines for Delivery 308
Visual Aspects of Delivery
Auditory Aspects of Delivery

12.7 Sample Speech 312

MAKING THE GRADE **316**

KEY TERMS **316**

ACTIVITIES **317**

LEARNING OBJECTIVES

12.1

Describe the importance of topic, purpose, and thesis in effective speech preparation.

12.2

Analyze both the audience and occasion in any speaking situation.

12.3

Gather information on your chosen topic from a variety of sources.

12.4

Assess and manage debilitative speaking apprehension.

12.5

Make effective choices in the delivery of your speech.

Think about a topic you feel passionate about. Now think about these questions:

What would be your main objective in speaking about this topic?

How might interviewing help you prepare?

What information would you need to gather to support your ideas?

How could you deliver this speech most effectively?

ROCKY ROQUE DOESN'T LIKE to go into the details, but his mother was a victim of sexual harassment.

He will tell you that his mom was a receptionist, and the assistant manager who harassed her felt he could get away with it because Rocky's mother was an undocumented worker. She and several other victims kept quiet about the harassment for months, fearing deportation. She was one of the first employees to speak up, and she persuaded her coworkers to do the same. It wasn't an easy fight, but she won, and the assistant manager was fired.

In college, months after his mother was harassed, Rocky joined protest movements and became active in civil rights causes online. He also joined his college speech teams, first at the College of DuPage and then at Illinois State University. Over the course of 4 years, he presented a series of speeches advocating for immigrant rights, LGBTQ rights, gender equality, and other social issues.

Rocky's college speech experience inspired him to not only continue working as an activist but teach at the college level. "I want to teach students the art of rhetoric," he says. "I want them to be able to defend themselves, to use their narratives to give a different perspective to others. I want to create more educators."

In this chapter, we explore the process of creating and delivering an effective speech. Your classroom speech might not be as personal as Rocky Roque's, but speaking in front of an audience can still be nerve racking. You might be advocating for social change, making a job-related presentation, or speaking on a special occasion, such as at a wedding or funeral. You might find yourself speaking in favor of a civic-improvement project in your hometown or trying to persuade members of your club to work toward solving global problems such as poverty, religious conflicts, or environmental threats.

Despite the potential benefits of effective speeches, the prospect of standing before an audience terrifies many people. In fact, giving a speech seems to be one of the most anxiety-producing things we can do: When asked to list their common fears, research subjects mention public speaking more often than they do insects, heights, accidents, and even death.[1]

Even if you don't love the idea of giving speeches, we promise to give you the tools to speak in a way that is clear, interesting, and effective. And it's very likely that, as your skill grows, your confidence will too. This chapter covers major steps in that process, through careful speech planning.

12.1 Getting Started

Your first tasks are generally choosing a topic, determining your purpose, and writing a purpose statement.

Choosing Your Topic

The first question many student speakers face is, "What should I talk about?" When you need to choose a topic, you should try to pick one that is right for you, your audience, and the situation. Rocky Roque chose his topic because he had a close personal connection to it, but you can also pick a topic based on personal interests and curiosity. Decide on your topic as early as possible. Those who wait until the last possible moment usually find that they don't have enough time to research, outline, and practice their speech.

Defining Your Purpose

No one gives a speech—or expresses *any* kind of message—without having a reason to do so. Your first step in focusing your speech is to formulate a clear and precise statement of that purpose.

Writing a Purpose Statement

Your **purpose statement** should be expressed in the form of a complete sentence that describes your **specific purpose**—exactly what you want your speech to accomplish. It should stem from your **general purpose**, which might be to inform, persuade, or entertain. Beyond that, though, there are three criteria for an effective purpose statement:

1. **A purpose statement should be result oriented.** Having a *result orientation* means that your purpose is focused on the outcome you want to accomplish with your audience members. For example, if you were giving an informative talk on the high cost of college, this would be an inadequate purpose statement:

   > My purpose is to tell my audience about depression in college students.

   As that statement is worded, your purpose is "to tell" an audience something, which suggests that the speech could be successful even if no one listened. A result-oriented purpose statement should refer to the response you want from your audience: It should tell what the audience members will know or be able to do after listening to your speech: "After listening to my speech, audience members will know several factors contributing to depression in college students."

2. **A purpose statement should be specific.** To be effective, a purpose statement should be worded specifically, with enough details so that you would be able to measure or test your audience, after your speech, to see if you had achieved your purpose. In the example given earlier, simply knowing that college students get depressed is too vague; you need something more specific, such as:

   > After listening to my speech, my audience will be able to deal with depression when they experience it.

purpose statement A complete sentence that describes precisely what a speaker wants to accomplish.

specific purpose The precise effect that the speaker wants to have on an audience. It is expressed in the form of a purpose statement.

general purpose One of three basic ways a speaker seeks to affect an audience: to entertain, inform, or persuade.

ASK YOURSELF

Think of someone who is an effective public speaker. What makes this person so effective?

This is an improvement, but it can be made still better by applying a third criterion:

3. **A purpose statement should be realistic.** It's fine to be ambitious, but you need to design a purpose that has a reasonable chance of success. You can appreciate the importance of having a realistic goal by looking at some unrealistic ones, such as "My purpose is to convince my audience to make federal budget deficits illegal." Unless your audience happens to be a joint session of Congress, it won't have the power to change U.S. fiscal policy. But any audience can write its congressional representatives or sign a petition. In your speech on college students and depression, it would be a tall order to provide every audience member with the tools to handle their current or future bouts of depression. So a better purpose statement for this speech might sound something like this:

> After listening to my speech, my audience will be able to list four simple steps to deal with depression on a personal level.

Consider the following sets of purpose statements:

LESS EFFECTIVE	MORE EFFECTIVE
To get my audience thinking about the benefits of prison reform (not result oriented)	After listening to my speech, my audience members will vote yes to the referendum to set up a tutoring program for local prisoners.
To talk about professional wrestling (not receiver oriented)	After listening to my speech, my audience will understand that kids who imitate professional wrestlers can be seriously hurt.
To tell my audience about gun control (not specific)	After my speech, the audience will sign my petition calling for universal background checks.

You probably won't include your purpose statement word for word in your actual speech. Rather than being aimed at your listeners, a specific purpose statement usually is a tool to keep you focused on your goal as you plan your speech.

Stating Your Thesis

thesis statement A complete sentence describing the central idea of a speech.

After you have defined your purpose, you are ready to start planning what is arguably the most important sentence in your entire speech. The **thesis statement** tells your listeners the central idea of your speech. It is the one idea that you want your audience to remember after it has forgotten everything else you had to say. The thesis statement for a speech about your local recycling program might sound like this:

"A tour of our local MURF (Materials Recovery Facility) will demonstrate how recycling works."

Unlike your purpose statement, your thesis statement is almost always delivered directly to your audience. The thesis statement is usually formulated later in the speech-making process, after you have done some research on your topic. The progression from topic to purpose to thesis is, therefore, another focusing process, as you can see in the following example:

Topic: Organ donation

Specific Purpose: After listening to my speech, audience members will recognize the importance of organ donation and will sign an organ donor's card for themselves.

Thesis: Because not enough of us choose to become organ donors, thousands of us needlessly die every year. You can help prevent this needless dying.

12.2 Analyzing the Speaking Situation

There are two components to analyze in any speaking situation: the audience and the occasion. To be successful, every choice you make in putting together your speech—your purpose, topic, and all the material you use to develop your speech—must be appropriate to both of these components.

The Listeners

Audience analysis involves identifying and adapting your remarks to the most pertinent characteristics of your listeners.

Audience Purpose   Just as you have a purpose for speaking, audience members have a reason for gathering. Sometimes virtually all the members of your audience will have the same, obvious goal. Expectant parents at a natural childbirth class are all seeking a healthy delivery, and people attending an investment seminar are all looking for ways to increase their net worth.

There are other times, however, when audience purpose can't be so easily defined. In some instances, different listeners will have different goals, some of which might not be apparent to the speaker. Consider a church congregation, for example. Whereas most members might listen to a sermon with the hope of applying religious principles to their lives, a few might be interested in being entertained or in merely appearing pious. In the same way, the listeners in your speech class probably have a variety of motives for attending. Becoming aware of as many of these motives as possible will help you predict what will interest them. You can ask individual audience members about their point of view or simply listen carefully when they express themselves in class. As you do so, you can start to make some judgments about audience demographics.

Demographics   Your audience has a number of characteristics that you may be able to observe in advance. These factors, known as **demographics**, include cultural differences, age, gender, group membership, number of people, and so on. Demographic characteristics might affect your speech planning in a number of ways.[2] For example:

> **audience analysis** A consideration of characteristics, including the type, goals, demographics, beliefs, attitudes, and values of listeners.
>
> **demographics** Audience characteristics that can be analyzed statistically, such as age, gender, education, and group membership.

- **Cultural diversity.** Do audience members differ in terms of race, religion, or national origin? The guideline here might be, *Do not exclude or offend any portion of your audience on the basis of cultural differences.* If there is a dominant cultural group represented, you might decide to speak to it, but remember that the point is to analyze, not stereotype, your audience. If you talk down to any segment of your listeners, you have probably stereotyped them.

- **Gender.** Although masculine and feminine stereotypes are declining, it is still important to think about how gender can affect the way you choose and approach a topic. Every communication professor has a horror story about a student getting up in front of a class

To keep your audience engaged, it's essential to tailor your remarks to their demographic profile, interests, beliefs, and knowledge.

What demographic characteristics define the audiences you are likely to face? How can you adjust your remarks to reach them?

298

composed primarily, but not entirely, of men and speaking on a subject such as "Scoring with Tinder."

- **Age.** Our interests vary and change with our age. These differences may run relatively deep; our approach to literature, films, finance, health, and long-term success may change dramatically over just a few years, perhaps from graphic novels to serious literature, from punk to classical music, or from hip-hop to epic poetry.

- **Group membership.** Groups generally form around shared interests. By examining the groups to which your audience members belong, you can potentially surmise their political leanings, religious beliefs, or occupation. Group membership is often an important consideration in college classes. Consider the difference between a daytime class and a class that meets in the evening. At many colleges, the evening students are generally older and tend to belong to civic groups, church clubs, and the local chamber of commerce. Daytime students are more likely to belong to sororities and fraternities, sports clubs, and social action groups.[3]

- **Number of people.** Topic appropriateness varies with the size of an audience. With a small audience, you can be less formal and more intimate; you can, for example, talk more about your feelings and personal experiences. If you gave a speech before 5 people as impersonally as if they were a standing-room-only crowd in a lecture hall, they would probably find you stuffy. On the other hand, if you talked to 300 people about your unhappy childhood, you'd probably make them uncomfortable.

You have to decide which demographics of your audience are important for a particular speech. For example, when Britton Ody, a student at Berry College in Georgia, gave a speech on the lack of treatment for hepatitis C in state prisons, he knew he had to broaden the appeal of his topic beyond the prison demographic referred to in his speech. He adapted to his broader audience this way:

> Within prisons, hepatitis C is not being treated, allowing for the disease to attack new hosts uncontended. It's a virus feeding ground. And it continues to spread beyond prisons. According to the Bureau of Justice's statistics, 95% of inmates in the United States are being released back into the community, and hepatitis C with them.[4]

These five demographic characteristics are important examples, but the list goes on. Other demographic characteristics that might be important in a college classroom include the following:

- Educational level
- Economic status
- Hometown
- Year in school
- Major subject
- Ethnic background

A final factor to consider in audience analysis concerns members' attitudes, beliefs, and values.

Attitudes, Beliefs, and Values   Audience members' feelings about you, your subject, and your intentions are central issues in audience analysis. One way to approach these issues is through a consideration of attitudes, beliefs, and values.[5] Attitudes, beliefs, and values all deal with the way people think about different topics and how

---

ETHICAL CHALLENGE   If I Adapt, Do I Lose My Integrity?

A student asks:

Is it truly ethical for me to adapt to the audience, even if I believe something different? I hate it when politicians do that …

As a speaker, you have to decide how far to go when adapting to your audience. You should never take a position you do not sincerely hold. Audience adaptation is more a matter of recognizing where your listeners' beliefs differ from yours and then finding common ground. Former U.S. Secretary of Labor Robert Reich outlined several strategies toward this end, including the following:[6]

1) Don't avoid conversations with people who are likely to disagree with you. To the contrary, seek them out.

2) Instead of raising abstract points, begin by focusing on issues that affect your listeners directly.

3) Make it personal. Invite your audience to think about their own experiences and stories. Share yours. Point out common ground between their positions and yours.

4) Acknowledge your listeners' beliefs. Invite them to consider your thoughts and how your position might address their concerns.

5) Consider using humor, the great connector.

6) Remember, the point isn't to convince them you're right and they're wrong, which will only make them defensive.

FOR DISCUSSION: How might you construct a speech to win over others whose political views are different from your own?

---

they will respond to that topic. These three traits reside in human consciousness like layers of an onion (see Figure 12.1). **Attitudes**, which are closest to the surface, are most easily observed. Sometimes we can just look at a person and say, "He's got an attitude." That's because attitudes are directly reflected in things a person says and does. **Beliefs** lie a little deeper and deal with the truth of something. It takes a little longer for people to reveal the beliefs that lie beneath their attitudes. Often, beliefs are divulged only when someone is challenged on an attitude. **Values** are the most deeply rooted of all. Individuals might not be able to express or explain their values coherently, but those values still impact the way people see the world. Values tend to be part of an individual's identity. We often speak of an individual's belief system or value system, suggesting that these factors are part of a larger group's meaningful ideas.[7]

You can begin to appreciate the usefulness of these concepts by considering an example. Suppose you wanted to give a speech on the concept of political correctness. Consider how audience analysis would help you design the most promising approach:

Attitudes. How do your listeners feel about political correctness movements, such as Me Too and Black Lives Matter? Do they feel that such movements perform a useful and reasonable function, or do they sense that such movements are unnecessary and have gone too far? If they recognize the importance of such movements, you can proceed confidently, knowing they'll probably want to hear what you have to say. On the other hand, if they are vaguely disgusted by even thinking about the topic, you will need to dig deeper.

Beliefs. Does your audience accept the relationship between political correctness and deeper held beliefs, such as treating people equally and acting civilly in society? Or do you need to inform them about how these concepts are related?

Values. Which underlying values matter most to your listeners, and how does political correctness relate to them? For example, how would the tenets of their religion, or the lessons that their parents taught them about life, relate to the social justice movements that are part of political correctness?

attitude The predisposition to respond to an idea, person, or thing favorably or unfavorably.

belief An underlying conviction about the truth of an idea, often based on cultural training.

value A deeply rooted belief about a concept's inherent worth.

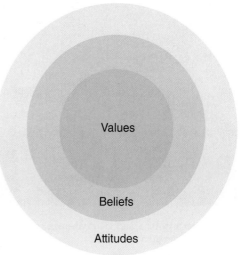

FIGURE 12.1 Attitudes, Beliefs, and Values

ASK YOURSELF ?

How might your own attitudes, beliefs, and values affect how you react to a speech?

Experts in audience analysis, such as professional speechwriters, often try to concentrate on values. As one team of researchers pointed out, "values have the advantage of being comparatively small in number, and owing to their abstract nature, are more likely to be shared by large numbers of people."[8] Stable American values include the ideas of good citizenship, a strong work ethic, tolerance of differing political views, individualism, and justice for all. Brianna Mahoney, a student at the University of Florida, appealed to her audience's values when she wanted to make the point that anti-homelessness laws were inhumane:

> Extreme poverty in the United States is a shockingly overlooked issue, making the homeless one of our most vulnerable populations. Recent legislation has worsened this by denying the homeless the ability to help themselves. People experiencing homelessness exist in every community; it's time to stop accusing them of becoming a burden when they are already struggling to carry their own.[9]

Mahoney pointed out that discriminating against homeless people was unfair and impractical; her analysis had suggested that the value of fairness would be important to this audience. She also surmised that they would be offended by unfairness combined with impracticality, as in this piece of evidence she used:

> Ninety-year-old Arnold Abbott made international headlines when he was arrested in Fort Lauderdale, Florida, and faced 60 days in prison for feeding hot meals to people who were homeless. When police arrived they grabbed trays of food and shoved them into the trash, while lines of hungry people were forced to just look on.[10]

You can often make an inference about audience members' attitudes by recognizing the beliefs and values they are likely to hold. In this example, Brianna knew that her audience, made up mostly of idealistic college students and professors, would dislike the idea of unfair and impractical discrimination.

Evaluation of hidden psychological states can be extremely helpful in audience analysis. For example, a religious group might hold the value of "obeying God's word." For some fundamentalists, this might lead to the belief, based on their religious training, that women are not meant to perform the same functions in society as men. This belief, in turn, might lead to the attitude that women ought not to pursue careers as firefighters, police officers, or construction workers.

You can also make a judgment about one attitude your audience members hold based on your knowledge of other attitudes they maintain. If your audience is made up of undergraduates who have a positive attitude toward liberation movements, it is a good bet they also have a positive attitude toward civil rights and ecology. If they have a negative attitude toward collegiate sports, they may also have a negative attitude toward fraternities and sororities. This should suggest not only some appropriate topics for each audience but also ways that those topics could be developed.

It can be fun to go for a quick laugh when making a speech. But keep in mind the impression that will linger.

Have you ever made remarks inappropriate for the occasion? What can you do to avoid this kind of mistake in the future?

The Occasion

The second phase in analyzing a speaking situation focuses on the occasion. The occasion of a speech is determined by the circumstances surrounding it. Three of these circumstances are time, place, and audience expectations.

Time   Your speech occupies an interval of time that is surrounded by other events. For example, other speeches might be presented before or after yours, or comments might be made that set a certain tone or mood. External events such as elections, the start of a new semester, or even the weather can color the occasion in one way or another. The date on which you give your speech might have some historical significance. If that historical significance relates in some way to your topic, you can use it to help build audience interest.

The time available for your speech is also an essential consideration. You should choose a topic that is broad enough to say something worthwhile but brief enough to fit your limits. "Wealth," for example, might be an inherently interesting topic to some college students, but it would be difficult to cover such a broad topic in a 10-minute speech and still say anything significant. However, a topic like "The problem of income inequality in America today" could conceivably be covered in 10 minutes in enough depth to be of some value. All speeches have limits, whether or not they are explicitly stated. If you are invited to say a few words, and you present a few volumes, you might not be invited back.

Place   Your speech also occupies a physical space. The beauty or squalor of your surroundings and the noise or stuffiness of the room should all be taken into consideration. These physical surroundings can be referred to in your speech if appropriate. If you were talking about world poverty, for example, you could compare your surroundings to those that might be found in a poor country.

Audience Expectations   Finally, your speech is surrounded by audience expectations. A speech presented in a college class, or a TED Talk such as the one that appears at the end of Chapter 14, is usually expected to reflect a high level of thought and intelligence. This doesn't necessarily mean that it has to be boring or humorless; wit and humor are, after all, indicative of intelligence. But it does mean that you have to put a little more effort into your presentation than if you were discussing the same subject with friends over coffee.

When you are considering the occasion of your speech, it pays to remember that every occasion is unique. Although there are obvious differences among the occasions of a college class, a church sermon, and a bachelor party "roast," there are also many subtle differences that will apply only to the circumstances of each unique event.

12.3 Gathering Information

This discussion about planning a speech purpose and analyzing the speech situation makes it apparent that it takes time, interest, and knowledge to develop a topic well. Setting aside a block of time to reflect on your own ideas is essential. However, you will also need to gather information from outside sources.

By this time you are probably familiar with both web searches and library research as forms of gathering information. Sometimes, however, speakers overlook interviewing, personal observation, and survey research as equally effective methods of gathering information. Let's review all these methods here and perhaps provide a new perspective on one or more of them.

ASK YOURSELF

Why is effective research so important to a college-level audience?

database A computerized collection of information that can be searched in a variety of ways to locate information that the user is seeking.

Online Research

The ease of using search engines like Google has made them the popular favorite for speech research. But students are sometimes so grateful to have found a website dealing with their topic that they forget to evaluate it. Like any other written sources you would use, websites should be accurate and rational. Beyond that, there are four specific criteria that you can use to evaluate the quality of a website: credibility, objectivity, currency, and functionality (see the checklist on this page).

In the case of some special search engines, like Google Scholar, the criteria of credibility, objectivity, and currency will be practically guaranteed. However, these guidelines are especially important when accessing information from Wikipedia, the popular online encyclopedia. Because anyone can edit a Wikipedia article at any time, many professors forbid the use of it as a primary resource. Others allow it to be used for general information and inspiration. Most will allow its use when articles have references to external sources (whether online or not) and the student reads the references and checks whether they really do support what the article says.

Library experts help you make sense of and determine the validity of the information you find, whether online or in print. And a library can be a great environment for concentration, a rare quiet place with minimal distractions.

Library Research

Libraries, like people, tend to be unique. Although many of your library's resources will be available online through your school's website, it can be extremely rewarding to get to know your library in person, to see what kind of special collections and services it offers, and just to find out where everything is. A few resources are common to most libraries, including the library catalog, reference works, periodicals, nonprint materials, and databases.

Databases, which can be particularly useful, are computerized collections of highly credible information from a wide variety of sources. One popular collection of databases is LexisNexis, which contains millions of articles from news services, magazines, scholarly journals, conference papers, books, law journals, and other sources. Other popular databases include ProQuest, Factiva, and Academic Search Premier, and there are dozens of specialized databases, such as Communication and Mass Media Complete. Database searches are slightly different from web searches; they generally don't respond well to long strings of terms or searches worded as questions. With databases it is best to use one or two key terms with a connector such as AND, OR, or NOT.[11] Once you learn this technique and a few other rules (perhaps with a librarian's help), you will be able to locate dozens of articles on your topic in just a few minutes.

Interviewing

An information-gathering interview allows you to view your topic from an expert's perspective and take advantage of that expert's experience, research, and thought. You can also use an interview to stimulate your own thinking. Often the interview will save you hours of Internet or library research and allow you to present ideas that you could not have uncovered any other way. And because an interview is an interaction with an expert, many ideas that otherwise might be unclear can become more understandable through questions and answers. Interviews can be conducted face to face, by telephone, or by email.

Allison McKibban, a student at Rice University, used highly effective excerpts from an interview in her speech on criminalizing rape survivors. She began that speech like this:

> As I sat down in the visitor's center of the Harris County jail, my hands shaking and heart pounding, Alycia bluntly asked, "Who are you?" It's never an easy question—but I started in with college and speech and when Alycia began to open up, our similarities were overwhelming. We grew up middle class, doting mothers, swim team, straight A's, bound for college—but it all changed at 17. Her senior year of high school, Alycia detailed in a personal interview from December 2nd of this year, was the first time her stepfather's friend sold her for sex.
>
> Two years later, after being trafficked through Texas against her will for money she never received, Alycia was freed from her abuser—and sentenced to 27 months in Texas prison for prostitution.[12]

Online and library research are essential to speech preparation. In addition, talking to credible sources can provide information that makes a speech more interesting and compelling.

Who can you interview to enrich the content of your speech?

Survey Research

One advantage of **survey research**—the distribution of questionnaires for people to respond to—is that it can give you up-to-date answers concerning "the way things are" for a specific audience. For example, if you handed out questionnaires a week or so before presenting a speech on the possible dangers of body piercing, you could present information like this in your speech:

> **survey research** Information gathering in which the responses of a population sample are collected to disclose information about the larger group.

> According to a survey I conducted last week, 90 percent of the students in this class believe that body piercing is basically safe. Only 10 percent are familiar with the scarring and injury that can result from this practice. Two of you, in fact, have experienced serious infections from body piercing: one from a pierced tongue and one from a simple pierced ear.

That statement would be of immediate interest to your audience members because *they* were the ones who were surveyed. Another advantage of conducting your own survey is that it is one of the best ways to find out about your audience: It is, in fact, *the* best way to collect the demographic data mentioned earlier. The one disadvantage of conducting your own survey is that, if it is used as evidence, it might not have as much credibility as published evidence found in the library. But the advantages seem to outweigh the disadvantages of survey research in public speaking.

No matter how you gather your information, remember that it is the *quality* rather than the quantity of the research that is most important. The key is to determine carefully what type of research will answer the questions you need to have answered. Sometimes only one type of research will be necessary; at other times every type mentioned here will have to be used. Generally, you will collect far more information than you'll use in your speech, but the winnowing process will ensure that the research you do use is of high quality.

Along with improving the quality of what you say, effective research will also minimize the anxiety of actually giving a speech. Let's take a close look at that form of anxiety.

12.4 Managing Communication Apprehension

The terror that strikes the hearts of so many beginning speakers is commonly known as *stage fright* or *speech anxiety,* or what communication scholars call *communication apprehension.*[13] Whatever term you choose, the important point to realize is that fear about speaking can be managed in a way that works for you rather than against you.

Facilitative and Debilitative Communication Apprehension

Although communication apprehension is a very real problem for many speakers, it can be overcome. The first step in feeling less apprehensive about speaking is to realize that a certain amount of nervousness is not only natural but also facilitative. That is, **facilitative communication apprehension** is a factor that can help improve your performance. Just as totally relaxed actors or musicians aren't likely to perform at the top of their potential, speakers think more rapidly and express themselves more energetically when their level of tension is moderate. In fact, experts suggest that you re-label communication apprehensive as something positive: "The crackle of excitement" or "a surge of extra energy."[14]

If you can't talk yourself out of it, **debilitative communication apprehension** tends to inhibit effective self-expression. Intense fear causes trouble in two ways. First, the strong emotion keeps you from thinking clearly.[15] This has been shown to be a problem even in the preparation process: Students who are highly anxious about giving a speech will find the preliminary steps, including research and organization, to be more difficult.[16] Second, intense fear leads to an urge to do something, anything, to make the problem go away. This urge to escape often causes a speaker to speed up delivery, which results in a rapid, almost machine-gun style. As you can imagine, this boost in speaking rate leads to even more mistakes, which only add to the speaker's anxiety. Thus, a relatively small amount of nervousness can begin to feed on itself until it grows into a serious problem.

facilitative communication apprehension A moderate level of anxiety about speaking before an audience that helps improve the speaker's performance

debilitative communication apprehension An intense level of anxiety about speaking before an audience, resulting in poor performance.

Sources of Debilitative Communication Apprehension

Before we describe how to manage debilitative communication apprehension, let's consider why people are afflicted with the problem in the first place.[17]

Previous Negative Experience   People often feel apprehensive about speech giving because of unpleasant past experiences. Most of us are uncomfortable doing *anything* in public, especially if it is a type of performance in which our talents and abilities are being evaluated. An unpleasant experience in one type of performance can cause you to expect that a future similar situation will also be unpleasant.[18] These expectations can be realized through the self-fulfilling prophecies discussed in Chapter 3. A traumatic failure at an earlier speech and low self-esteem from critical parents during childhood are common examples of experiences that can cause later communication apprehension.

You might object to the idea that past experiences cause communication apprehension. After all, not everyone who has bungled a speech or had critical parents is debilitated in the future. To understand why past experiences affect some people more strongly than others, we need to consider another cause of communication apprehension.

Jennifer Lawrence, whose films have earned over $6 billion, is often in the public eye. In spite of her cool demeanor, she admits she suffers from communication apprehension. Her tip for reducing anxiety: "I just try to acknowledge that this scrutiny is stressful, and that anyone would find it stressful."

How can you apply the information in this chapter to manage—and benefit from—your natural apprehension?

Irrational Thinking   Cognitive psychologists argue that it is not events that cause people to feel nervous but rather the beliefs they have about those events. Certain irrational beliefs leave people feeling unnecessarily apprehensive. Psychologist Albert Ellis lists several such beliefs, or examples of **irrational thinking**, which we will call "fallacies" because of their illogical nature.[19]

- **Catastrophic failure**. People who succumb to the **fallacy of catastrophic failure** operate on the assumption that if something bad can happen, it probably will. Their thoughts before a speech resemble these:

  "As soon as I stand up to speak, I'll forget everything I wanted to say."

  "Everyone will think my ideas are stupid."

  "Somebody will probably laugh at me."

  Although it is naive to imagine that all your speeches will be totally successful, it is equally naive to assume they will all fail miserably. One way to escape the fallacy of catastrophic failure is to take a more realistic look at the situation. Would your audience members really hoot you off the stage? Will they really think your ideas are stupid? Even if you did forget your remarks for a moment, would the results be a genuine disaster? It helps to remember that nervousness is more apparent to the speaker than to the audience.[20] Beginning public speakers, when congratulated for their poise during a speech, are apt to say, "Are you kidding? I was dying up there."

- **Perfection**. Speakers who succumb to the **fallacy of perfection** expect themselves to behave flawlessly. Whereas such a standard of perfection might serve as a target and a source of inspiration (like the desire to make a hole in one while golfing), it is totally unrealistic to expect that you will write and deliver a perfect speech, especially as a beginner. It helps to remember that audiences don't expect you to be perfect.

- **Approval**. The mistaken belief called the **fallacy of approval** is based on the idea that it is vital—not just desirable—to gain the approval of everyone in the audience. It is rare that even the best speakers please everyone, especially on topics that are at all controversial. To paraphrase Abraham Lincoln, you can't please all the people all the time, and it is irrational to expect you will.

- **Overgeneralization**. The **fallacy of overgeneralization** might also be labeled the fallacy of exaggeration because it occurs when a person blows one experience out of proportion. Consider these examples:

  "I'm so stupid! I mispronounced that word."

  "I completely blew it—I forgot one of my supporting points."

  "My hands were shaking. The audience must have thought I was crazy."

  A second type of exaggeration occurs when a speaker treats occasional lapses as if they were the rule rather than the exception. This sort of mistake usually involves extreme labels, such as "always" or "never."

  "I always forget what I want to say."

  "I can never come up with a good topic."

  "I can't do anything right."

irrational thinking Beliefs that have no basis in reality or logic; one source of debilitative communication apprehension.

fallacy of catastrophic failure The irrational belief that the worst possible outcome will probably occur.

fallacy of perfection The irrational belief that a worthwhile communicator should be able to handle every situation with complete confidence and skill.

fallacy of approval The irrational belief that it is vital to win the approval of virtually every person a communicator deals with.

fallacy of overgeneralization Irrational beliefs in which (1) conclusions (usually negative) are based on limited evidence or (2) communicators exaggerate their shortcomings.

Overcoming Debilitative Communication Apprehension

There are five strategies that can help you manage debilitative communication apprehension:

- **Use nervousness to your advantage.** Paralyzing fear is obviously a problem, but a little nervousness can actually help you deliver a successful speech. Being completely calm can take away the passion that is one element of a good speech. Control your anxiety, but don't try to completely eliminate it.

- **Understand the difference between rational and irrational fears.** Some fears about speaking are rational. For example, you ought to be worried if you haven't properly prepared for your speech. But fears based on the fallacies you just read about aren't constructive. It's not realistic to expect that you'll deliver a perfect speech, and it's not rational to indulge in catastrophic fantasies about what might go wrong.

- **Maintain a receiver orientation.** Paying too much attention to your own feelings—even when you're feeling good about yourself—will take energy away from communicating with your listeners. Concentrate on your audience members rather than on yourself. Focus your energy on keeping them interested and on making sure they understand you.

- **Keep a positive attitude.** Build and maintain a positive attitude toward your audience, your speech, and yourself as a speaker. Some communication consultants suggest that public speakers should concentrate on three statements immediately before speaking. The three statements are as follows:

  I'm glad I have the chance to talk about this topic.

  I know what I'm talking about.

  I care about my audience.

  Repeating these statements (until you believe them) can help you maintain a positive attitude.

  Another technique for building a positive attitude is known as **visualization,**[21] an approach that has been used successfully with athletes. It requires you to use your imagination to visualize the successful completion of your speech. Visualization can help make the self-fulfilling prophecy discussed in Chapter 3 work in your favor.

- **Be prepared!** Preparation is the most important key to controlling communication apprehension. You can feel confident if you know from practice that your remarks are well organized and supported and that your delivery is smooth. Researchers have determined that the highest level of communication apprehension occurs just before speaking; the second highest level at the time the assignment is announced and explained; and the lowest level during the time you spend preparing your speech.[22] You should take advantage of this relatively low-stress time to work through the problems that would tend to make you nervous during the actual speech. For example, if, on one hand, your anxiety is based on a fear of forgetting what you are going to say, make sure that your note cards are complete and effective, and that you have practiced your speech thoroughly (we'll go into speech practice in more detail in a moment). If, on the other hand, your great fear is "sounding stupid," then getting started early with lots of research and advance thinking is the key to relieving your communication apprehension.

visualization A technique for rehearsal using a mental visualization of the successful completion of a speech.

Speech Anxiety Symptoms

To what degree do you experience the following anxiety symptoms while speaking?

1. Sweating
   a. Nonexistent
   b. Moderate
   c. Severe

2. Rapid breathing
   a. Nonexistent
   b. Moderate
   c. Severe

3. Difficulty catching your breath
   a. Nonexistent
   b. Moderate
   c. Severe

4. Rapid heartbeat
   a. Nonexistent
   b. Moderate
   c. Severe

5. Restless energy
   a. Nonexistent
   b. Moderate
   c. Severe

6. Forgetting what you wanted to say
   a. Nonexistent
   b. Moderate
   c. Severe

EVALUATING YOUR RESPONSES

Give yourself one point for every "a," two points for every "b," and three points for every "c." If your score is:

6 to 9 You have nerves of steel. You're probably a natural public speaker, but you can always improve.

10 to 13 You are the typical public speaker. Practice the strategies discussed in this chapter to improve your skills.

14 to 18 You tend to have significant apprehension about public speaking. You need to consider each strategy in this chapter carefully. Although you will benefit from the tips provided, you should keep in mind that some of the greatest speakers of all time have considered themselves highly anxious.

12.5 Presenting Your Speech

Once you have done all the planning and analysis that precedes speechmaking, you can prepare for your actual presentation. Your tasks at this point include choosing an effective type of delivery, formulating a plan for practicing your speech, and thinking carefully about the visual and auditory choices you will make.

Choosing an Effective Type of Delivery

There are four basic types of delivery: extemporaneous, impromptu, manuscript, and memorized. Each type creates a different impression and is appropriate under different conditions. Any speech may incorporate more than one of these types of delivery. For purposes of discussion, however, it is best to consider them separately.

1. An **extemporaneous speech** is planned in advance but presented in a direct, spontaneous manner. Extemporaneous speeches are conversational in tone, which means that they give the audience members the impression that you are talking to them, directly and honestly. Extemporaneous speaking is the most common type of delivery in both the classroom and the "outside" world. In fact, extemporaneous speaking is one of the factors that researchers believe led to Donald Trump's election win in 2016. Trump's adherents found his off-the-cuff remarks compelling compared

> **extemporaneous speech** A speech that is planned in advance but presented in a direct, conversational manner.

The best type of speech delivery depends on the situation.

What styles are most appropriate in the situations where you might deliver remarks? Why?

impromptu speech A speech given "off the top of one's head," without preparation.

manuscript speech A speech that is read word for word from a prepared text.

memorized speech A speech learned and delivered by rote without a written text.

with the carefully prepared speeches of more typical politicians.[23]

2. An **impromptu speech** is given off the top of one's head, without preparation. This type of speech is spontaneous by definition, but it is a delivery style that is necessary for informal talks, group discussions, and comments on others' speeches. It is also a highly effective training aid that teaches you to think on your feet and to organize your thoughts quickly.

3. **Manuscript speeches** are read word for word from a prepared text. They are necessary when you are speaking for the record, such as at legal proceedings or when presenting scientific findings. The greatest disadvantage of a manuscript speech is, of course, the lack of spontaneity.

4. **Memorized speeches**—those learned by heart—are the most difficult and often the least effective. They often seem excessively formal. However, like manuscript speeches, they may be necessary on special occasions. They are used in oratory contests and as training devices for memory.

One guideline holds for each type of speech: Practice.

Practicing Your Speech

A smooth and natural delivery is the result of extensive practice. Get to know your material until you feel comfortable with your presentation. One way to do that is to go through some or all of the steps listed in the checklist on page 309. In each of these steps, critique your speech according to the guidelines that follow.

🔘 12.6 Guidelines for Delivery

Let's examine some nonverbal aspects of presenting a speech. As you read in Chapter 7, nonverbal behavior can change, or even contradict, the meaning of the words a speaker utters. If audience members want to interpret how you feel about something, they are likely to trust your nonverbal communication more than the words you speak. If you tell them, "It's great to be here today," but you stand before them slouched over with your hands in your pockets and an expression on your face like you're about to be shot, they are likely to discount what you say. This might cause your audience members to react negatively to your speech, and their negative reaction might make you even more nervous. This cycle of speaker and audience reinforcing each other's feelings can work for you, though, if you approach a subject with genuine enthusiasm. Enthusiasm is shown through both the visual and auditory aspects of your delivery.

Visual Aspects of Delivery

Visual aspects of delivery include appearance, movement, posture, facial expression, and eye contact.

Appearance   This is not a presentation variable as much as a preparation variable. Some communication consultants suggest new clothes, new glasses, and new hairstyles for their clients. In case you consider any of these grooming aids, be forewarned that you should be attractive to your audience but not flashy. Research suggests that audiences like speakers who are similar to them, but they prefer the similarity to be shown conservatively.[24] Speakers, it seems, are perceived to be more credible when they look businesslike. Part of looking businesslike, of course, is looking like you took care in the preparation of your wardrobe and appearance.

Movement   The way you walk to the front of your audience will express your confidence and enthusiasm. And after you begin speaking, nervous energy can cause your body to shake and twitch, and that can be distressing both to you and to your audience. One way to control involuntary movement is to move voluntarily when you feel the need to move. Don't feel that you have to stand in one spot or that all your gestures need to be carefully planned. Simply get involved in your message, and let your involvement create the motivation for your movement. That way, when you move, you will emphasize what you are saying in the same way you would emphasize it if you were talking to a group of friends.

Movement can also help you maintain contact with all members of your audience. Those closest to you will feel the greatest contact. This creates what is known as the "action zone" in the typical classroom, within the area of the front and center of the room. Movement enables you to extend this action zone, to include in it people who would otherwise remain uninvolved. Without overdoing it, you should feel free to move toward, away from, or from side to side in front of your audience.

Remember: Move with the understanding that it will add to the meaning of the words you use. It is difficult to bang your fist on a podium or take a step without conveying emphasis. Make the emphasis natural by allowing your message to create your motivation to move.

Posture   Generally speaking, good posture means standing with your spine relatively straight, your shoulders relatively squared off, and your feet angled out to keep your body from falling over sideways. In other words, rather than standing at military attention, you should be comfortably erect.

Good posture can help you control nervousness by allowing your breathing apparatus to work properly; when your brain receives enough oxygen, it's easier for you to think clearly. Good posture also increases your audience contact because the audience members will feel that you are interested enough in them to stand formally, yet relaxed enough to be at ease with them.

Facial Expression   The expression on your face can be more meaningful to an audience than the words you say. Try it yourself with a mirror. Say, "You're a terrific audience," for example, with a smirk, with a warm smile, with a deadpan expression, and then with a scowl. It just doesn't mean the same thing. But don't try to fake it. Like your movement, your facial expressions will reflect your genuine involvement with your message.

Eye Contact   Eye contact is perhaps the most important nonverbal facet of delivery.[25] Eye contact increases your connection with your audience, and at the same time it helps you control your nervousness.

CHECKLIST ✓

Practicing Your Presentation

Be sure to give yourself plenty of time to practice. Here are a few suggestions.

☐ First, present the speech to yourself. Talk through the entire speech, including your examples and forms of support. Don't skip parts by using placeholders. Make sure you have a clear plan for presenting your statistics and explanations.

☐ Record the speech on your phone, and listen to it. Because we are sometimes surprised at what we sound like and how we appear, video recording has been shown to be an especially effective tool for rehearsals.[26]

☐ Present the speech in front of a small group of friends or relatives.[27]

☐ Present the speech to at least one listener in the room where you will present the final speech (or, if that room is not available, a similar room).

ETHICAL CHALLENGE Speaking Sincerely to Distasteful Audiences

At some point, you are likely to be speaking to an audience composed of listeners who you don't like personally or whose positions are distasteful to you. Speakers are most effective when their delivery is perceived as being sincere. The challenge: To explore the ethical dimensions of this situation, image what kind of listeners you would find offensive. How can you deliver your remarks in a way that masks your feelings without being insincere?

Consider discussing your approach with other class members. Collectively, develop a list of strategies to manage the auditory and visual approach that would be both ethical and effective.

Direct eye contact is a form of reality testing. The most frightening aspect of speaking is the unknown. How will the audience react? Direct eye contact allows you to test your perception of your audience as you speak. Usually, especially in a college class, you will find your audience is more "with" you than you think. By deliberately establishing eye contact with audience members, you might engage them and find they are more interested than they previously appeared.

To maintain eye contact, you could try to meet the eyes of each member of your audience squarely at least once during any given presentation. After you have made definite eye contact, move on to another audience member. You can learn to do this quickly, so you can visually latch on to every member of a good-sized class in a relatively short time.

The characteristics of appearance, movement, posture, facial expression, and eye contact are visual, nonverbal facets of delivery. Now consider the auditory nonverbal messages that you might send during a presentation.

Auditory Aspects of Delivery

As you read in Chapter 6, your paralanguage—the way you use your voice—says a good deal about you, especially about your sincerity and enthusiasm. Like eye contact, using your voice well can help you control your nervousness. It's another cycle: Controlling your vocal characteristics will decrease your nervousness, which will enable you to control your voice even more. But this cycle can also work in the opposite direction. If your voice is out of control, your nerves will probably be in the same state. Controlling your voice is mostly a matter of recognizing and using appropriate volume, rate, pitch, and articulation.

Volume   The loudness of your voice is determined by the amount of air you push past the vocal folds in your throat. The key to controlling volume, then, is controlling the amount of air you use. The key to determining the right volume is audience contact. Your delivery should be loud enough so that your audience members can hear everything you say but not so loud that they feel you are talking to someone in the next room. Too much volume is seldom the problem for beginning speakers. Usually, they either are not loud enough or tend to fade off at the end of a thought. Sometimes, when they lose faith in an idea in midsentence, they compromise by mumbling the end of the sentence so that it isn't quite coherent.

rate The speed at which a speaker utters words.

Rate   There is a range of personal differences in speaking speed, or **rate**. Daniel Webster, for example, is said to have spoken at around 90 words per minute, whereas one actor who is known for his fast-talking commercials speaks at about 250. Normal speaking speed, however, is between 120 and 150 words per minute. If you talk much more slowly than that, you may tend to lull your audience to sleep. Faster speaking rates are stereotypically associated with speaker competence,[28] but if you speak too rapidly, you will tend to be unintelligible. Once again, your involvement in your message is the key to achieving an effective rate.

Pitch   The highness or lowness of your voice—**pitch**—is controlled by the frequency at which your vocal folds vibrate as you push air through them. Because taut vocal folds vibrate at a greater frequency, pitch is influenced by muscular tension. This explains why nervous speakers have a tendency occasionally to "squeak," whereas relaxed speakers seem to be in more control. Pitch will tend to follow rate and volume. As you speed up or become louder, your pitch will tend to rise. If your range in pitch is too narrow, your voice will have a singsong quality; if it is too wide, you may sound overly dramatic. You should control your pitch so that your listeners believe you are talking with them rather than performing in front of them. Once again, your involvement in your message should take care of this problem naturally for you.

When considering volume, rate, and pitch, keep emphasis in mind. Remember that a change in volume, pitch, or rate will result in emphasis. If you pause or speed up, your rate will suggest emphasis. Words you whisper or scream will be emphasized by their volume.

Articulation   The final auditory nonverbal behavior, articulation, is perhaps the most important. For our purposes here, **articulation** means pronouncing all the parts of all the necessary words and nothing else.

It is not our purpose to condemn regional or ethnic dialects within this discussion. It is true that a considerable amount of research suggests that regional dialects can cause negative impressions,[29] but our purpose here is to suggest careful, not standardized, articulation. Incorrect articulation is usually nothing more than careless articulation. It is caused by (1) leaving off parts of words (deletion), (2) replacing parts of words (substitution), (3) adding parts to words (addition), or (4) overlapping two or more words (slurring).

Deletion   The most common mistake in articulation is **deletion**, or leaving off part of a word. As you are thinking the complete word, it is often difficult to recognize that you are saying only part of it. The most common deletions occur at the ends of words, especially -*ing* words. *Going, doing,* and *stopping* become *goin', doin',* and *stoppin'*. Parts of words can be left off in the middle, too, as in *terr'iss* for *terrorist, Innernet* for *Internet,* and *asst* for *asked.*

Substitution   **Substitution** takes place when you replace part of a word with an incorrect sound. The ending -*th* is often replaced at the end of a word with a single *t,* as when *with* becomes *wit.* The *th-* sound is also a problem at the beginning of words, as *this, that,* and *those* tend to become *dis, dat,* and *dose.* (This tendency is especially prevalent in many parts of the northeastern United States.)

Addition   The articulation problem of **addition** is caused by adding extra parts to words, such as *incentative* instead of *incentive, athalete* instead of *athlete,* and *orientated* instead of *oriented.* Sometimes this type of addition is caused by incorrect word choice, as when *irregardless* is used for *regardless.*

Another type of addition is the use of "tag questions," such as *you know?* or *you see?* or *right?* at the end of sentences. To have every other sentence punctuated with one of these barely audible superfluous phrases can be annoying.

Probably the worst type of addition, or at least the most common, is the use of *uh* and *anda* between words. *Anda* is often stuck between two words when *and* isn't even needed. If you find yourself doing that, you might just want to pause or swallow instead.[30]

pitch The highness or lowness of one's voice.

articulation The process of pronouncing all the necessary parts of a word.

deletion An articulation error that involves leaving off parts of words.

substitution The articulation error that involves replacing part of a word with an incorrect sound.

addition The articulation error that involves adding extra parts to words.

slurring The articulation error that involves overlapping the end of one word with the beginning of the next.

Slurring   This error is caused by trying to say two or more words at once—or at least overlapping the end of one word with the beginning of the next. Word pairs ending with *of* are the worst offenders in this category. *Sort of* becomes *sorta*, *kind of* becomes *kinda*, and *because of* becomes *becausa*. Word combinations ending with *to* are often slurred, as when *want to* becomes *wanna*. Sometimes even more than two words are blended together, as when *that is the way* becomes *thatsaway*. Careful articulation means using your lips, teeth, tongue, and jaw to bite off your words, cleanly and separately, one at a time.

12.7 Sample Speech

Rocky Roque, whose profile began this chapter, presented the following speech as a member of the speech team at Illinois State University in 2018.

As he explained in the vignette that started this chapter, Rocky chose a speech topic that was important to him. "The current political climate was a major influence when I chose my topic. Immigration had been on my mind since the 2016 election. I feared that at any given moment, I was going to receive the call that one of my family members was going to be detained and sent back."

Rocky thought about a speech topic by asking himself, "What makes me angry?" He then listed his answers on a sheet of paper, circled the ones that upset him the most, and chose the one that made him angriest.

When determining his purpose, Rocky says, "I felt that it was my responsibility as a Mexican American to use my privilege as a voice to help others that cannot speak for themselves." His purpose statement, therefore, became:

> After listening to my speech, my audience members will consider ways to help protect undocumented immigrant workers.

After working on his speech for a while, his thesis statement became evident to him:

> We must stop employers who threaten their undocumented workers with deportation.

In analyzing his audience, Rocky decided that some of his audience members would be immigrants and children of immigrants, but many others would not be directly affected by the issue he wanted to discuss. They were, however, interested in social justice, so Rocky decided to focus his speech in that direction.

Rocky spent many hours, over a number of weeks, researching his topic. "I read many articles and selected the ones that were the most emotional, credible, and logical. I used these articles to build my argument."

Unlike most classroom speeches, Rocky began this one with a manuscript, and practiced it until it was memorized. A similar speech, however, could be presented extemporaneously from notes.

Rocky won a "Top 2" award at the 2018 Illinois Intercollegiate Forensics Association State Tournament with this speech, and placed 4th at the Interstate Oratorical Association Tournament, the oldest public speaking tournament in the country.

An outline for this speech can be found on page 321. Rocky's bibliography follows the speech.

SAMPLE SPEECH Rocky Roque

Working to Document the Common Neighbor

1   After working in Oklahoma as a nanny for more than 10 years, Alicia Morales, an undocumented immigrant, understood the hardships of taking care of her children and her employer's children. From making beds to changing diapers, life with the new family was going well until she requested a raise that she had been promised the day she started working there.

> Rocky used this interesting anecdote as an attention-getter.

2   Weeks went by with no raise, and when her hours were extended as well, Morales confronted her employers. In response, she was greeted with a threat: She would be reported to ICE if she asked again. Caught between deportation and providing for her family, Morales decided to stay.

3   Her story is not new, as the Los Angeles Times reported on January 2, 2018. U.S. employers regularly threaten to report their undocumented workers to the U.S. Immigration and Customs Enforcement, or ICE, when the workers petition for better working conditions, pay raises, and protection from sexual violence. In fact, the British online newspaper *The Independent* found that the number of reported cases has multiplied five times since 2012 and doubled within the 1st year of Trump's presidency. Because of existing loopholes, this act has been undermined by many people, to the point where we do not have a concrete number of how many immigrants are being affected.

> Rocky used several of these formal citations to build the credibility of his arguments.

4   We must stop employers who threaten their undocumented workers with deportation. Today we will look at problems, causes and solutions to what some experts are calling a form of modern-day slavery.

> Thesis Statement, followed by a preview of main points.

5   In 2011, Josue Diaz told the *Los Angeles Times* that his employer was forcing him to collect animal carcasses found in contaminated water with his bare hands. When he reported the treatment to his boss, his paycheck was cut in half. This is an example of two common problems: Abusive Conditions and No Way Out.

> Here Rocky introduces two subpoints.

6   **First: Abusive Conditions.** A November 2017 report from the University of Illinois–Chicago stated, and I quote, "In Houston alone, 26% of undocumented workers in construction, factories, and sewage plants reported not getting paid after completing their work." Undocumented workers are exploited to work hazardous jobs because they lack protection from the Fair Labor Standards Act. For example, an employee from Florida was fired after cutting his hand at work and then was arrested for not having proper identification. Because of his citizenship, his employer had the legal right to not give him money for medical care after being detained. When these workers are injured, they become disposable.

7   **Second Problem: No Way Out**. Because immigrants have to pay their bills and feed their families, many are desperate to keep their jobs and will tolerate mistreatment from their employers. Dangerous working conditions that would be fined by the law remain unexposed because immigrants fear that their status will be exposed.

8   In 1995, an immigrant by the name of Fernando fought a 20-year legal battle after he was threatened with deportation when he found out that his employer didn't complete his I-9 forms, which would have established his work eligibility. The employer's power is rooted in two causes: Dispensable Workers and New Discriminatory Policies.

> Here he introduces his next two subpoints.

9 **First: Dispensable Workers**. The *Daily Press* newspaper in California has reported that 1.4 million immigrants are working with fake social security numbers. There is no proof of their existence. When employers become aware of workers in this situation, they can withhold the workers' I-9 forms. I-9 forms are rarely checked, and even if they are caught, the fines are minimal for employers—$216 for uncompleted forms, which is less than the cost of printing paper and ink cartridges for a month. As reported by Pew Research, 8 million immigrants are working or looking for a job. That is 8 million people that employers are free to pick, choose, and punish at will.

10 **Second Cause: New Discrimination Policies**. While the Immigration and Nationality Act prohibits citizenship status discrimination, Congress recently passed the Criminal Alien Gang Member Removal Act. One guideline of the act defines a "Criminal Gang" member as an illegal eligible for deportation who has committed crimes of violence.

11 This guideline enables an employer to manipulate reported cases of self-defense from abuse in the workplace. This vague language gives ICE unlimited discretion to determine if an immigrant is eligible for deportation. Under the same act, a person can also be investigated for looking like a gang member, allowing employers to racially profile their employees. ICE has posted on their page that they will investigate these employers, but if found guilty the employer's punishment will be a strongly worded letter to stop.

Here he introduces two more subpoints.

12 When 46-year-old Olivia Solis was threatened with deportation if she didn't comply with her assistant manager's advances, she had enough. She gathered her female coworkers who were also undocumented to band together and stop the abuse. Olivia's retaliation is a step in the right direction and a foundation for two solutions: Request Protection Acts and Know Your Rights.

13 **First: Request Protection Acts**. The California Immigrant Worker Protection Act went into effect on January 1, 2018. Under the rules of this state law, employers in California cannot allow inspectors in the workplace without a proper warrant. Similarly, the Illinois Trust Act was signed on August of 2017 to provide protection for immigrants in the workplace. This bill was signed after a large movement of Latinx individuals who campaigned for employment equality.

14 If you're not a California or Illinois resident, call your local representatives to campaign for these bills. Progress starts with a movement. I've attached my email on these cards. If you want more information on what your state is doing to enforce protection laws, send me a message and I will send you information. We, as a group of advocates for what is right, have a moral obligation to support these bills.

15 **Second Solution: Know Your Rights**. The Equal Employment Opportunity Commission, the EEOC, entitles undocumented workers to the same protections as citizens. The EEOC has a tab on their website titled "National Origin Discrimination," which explains that it is illegal for employers to treat employees unfavorably because they are from a particular country. If you are undocumented and have been harassed, you should take the steps to file a complaint. I've attached the link to filing a harassment claim on these cards.

16 **[Rocky hands out cards]**

Rocky's personal connection to this topic gives the speech added impact.

17 Take a card, and give it to someone who might need it, because that's what 46-year-old Olivia Solis did. A courageous single mother from Chicago—that was my mom. When she was threatened with deportation, she filed a harassment claim against her assistant manager and won. She taught me that there is still hope and that undocumented workers are willing to fight for their survival. Because no matter who you are or where you are from . . . we're glad you're our neighbor.

18 In his 2018 State of the Union address, President Trump said that we must summon our unity as "One team, one people, and one American family." However, we are not one American family if there are people who still objectify my mother and millions of immigrants and try to have them deported.

19 After addressing this problem, and addressing its causes and solutions, we saw that undocumented workers deserve to be able to work without fearing those who run a professional workplace.

BIBLIOGRAPHY

Austin, T. (2017, December 1). Undocumented workers in Houston face hazardous conditions and unpaid wages. Retrieved from https://www.wsws.org/en/articles/2017/12/01/undo-d01.html.

Associated Press. (2017, December 23). Fla. Companies dump undocumented workers after gaining profits. *Herald-Tribune*. Retrieved from https://www.heraldtribune.com/news/20171223/fla-companies-dump-undocumented-workers-after-gaining-profits

Average salary information for US workers. *The Balance Careers*. Retrieved from https://www.thebalancecareers.com/average-salary-information-for-us-workers-2060808

Gomez, A. (2017, March 22). These undocumented immigrants thought they could stay. Trump says deport them. *USA Today*. Retrieved from https://www.usatoday.com/story/news/world/2017/03/22/faces-of-deportations-under-president-trump/99455428/

Guzzardi, J. (2017, June 29). 1.4 million Illegal immigrants using fake Social Security numbers. *Daily Press*. Retrieved from https://www.vvdailypress.com/news/20170629/14-million-illegal-immigrants-using-fake-social-security-numbers

Justice Department. (2017, February 3). Civil monetary penalties inflation adjustment for 2017. *Federal Register*. Retrieved from https://www.federalregister.gov/documents/2017/02/03/2017-01306/civil-monetary-penalties-inflation-adjustment-for-2017

Khouri, A. (2018, January 2). More workers say their bosses are threatening to have them deported. *Los Angeles Times*. Retrieved from http://www.latimes.com/business/la-fi-immigration-retaliation-20180102-story.html

Low wages. *National Farm Worker Ministry*. Retrieved from http://nfwm.org/resources/low-wages/

Meyerson, H. (2011, June 24). Protecting undocumented workers. *Los Angeles Times*. Retrieved from http://articles.latimes.com/2011/jun/24/opinion/la-oe-meyerson-undocumented-abuses-20110624

Moores, C. (2017, October 19). California Immigrant Worker Protection Act: New obligations for employers. *Cook Brown*. Retrieved from https://www.cookbrown.com/california-immigrant-workers-protection-act/

National origin discrimination. *US Equal Employment Opportunity Commission*. Retrieved from https://www.eeoc.gov/laws/types/nationalorigin.cfm(2017, November 3). Size of US unauthorized immigrant workforce stable after the great recession. *Pew Research Center*. Retrieved from http://www.pewhispanic.org/2016/11/03/size-of-u-s-unauthorized-immigrant-workforce-stable-after-the-great-recession/

White, J. (2018, January 4). Number of employers said to threaten immigrant workers with deportation Spikes in California since Trump's win. *Independent*. Retrieved from https://www.independent.co.uk/news/world/americas/california-immigrant-workers-deportation-threat-employers-trump-a8140676.html

MAKING THE GRADE

At www.oup.com/he/adler-uhc14e, you will find a variety of resources to enhance your understanding, including video clips, animations, self-quizzes, additional activities, audio and video summaries, interactive self-assessments, and more.

OBJECTIVE 12.1   Describe the importance of topic, purpose, and thesis in effective speech preparation.

- Choose a topic that is right for you, your audience, and the situation.
- Formulating a clear purpose statement serves to keep you focused while preparing your speech.
- A straightforward thesis statement helps the audience understand your intent.
  > Why is a carefully worded purpose statement essential for speech success?
  > Why do you believe your topic has the potential to be effective for you as a speaker?
  > How would you word your thesis statement for your next speech?

OBJECTIVE 12.2   Analyze both the audience and occasion in any speaking situation.

- When analyzing your audience, you should consider the audience's purpose, demographics, attitudes, beliefs, and values.
- When analyzing the occasion, you should consider the time (and date) when your speech will take place, the location, and audience expectations given the occasion.
  > What are the most important aspects of analyzing your audience?
  > What are the most important aspects of analyzing the occasion for your next speech?
  > How will you adapt your next speech to both your audience and occasion?

OBJECTIVE 12.3   Gather information on your chosen topic from a variety of sources.

- When researching a speech, students usually think first of online searches.
- Also consider searching the collections or databases at a library, interviewing, making personal observations, and conducting survey research.
  > Why are multiple forms of research important in speech preparation?

  > How do you analyze the reliability of information you find online?
  > What is the most important piece of research you need for your next speech?

OBJECTIVE 12.4   Assess and manage debilitative speaking apprehension.

- Sources of debilitative communication apprehension often include irrational thinking, such as a belief in one or more fallacies.
- To help overcome communication apprehension, remember that nervousness is natural, and use it to your advantage.
- Other methods of overcoming communication apprehension involve being rational, receiver oriented, positive, and prepared.
  > What is the relationship between self-talk and communication apprehension?
  > Which types of self-defeating thoughts create the greatest challenge for you?
  > What forms of rational self-talk can you use to overcome self-defeating thoughts?

OBJECTIVE 12.5   Make effective choices in the delivery of your speech.

- Choose an effective type of delivery, or combinations of extemporaneous, impromptu, manuscript, and memorized speeches.
- Practice your speech thoroughly.
- Consider both visual and auditory aspects of delivery.
  > What are the primary differences among the main types of speeches?
  > In your last speech, why did you make the delivery choices that you did? Were they as effective as they could have been?
  > What visual and auditory aspects of delivery will be most important in the next speech you present.

KEY TERMS

addition p. 311

articulation p. 311

attitude p. 299

audience analysis p. 297

belief p. 299

database p. 302

debilitative communication apprehension p. 304

deletion p. 311

demographics p. 297

extemporaneous speech p. 307

facilitative communication apprehension p. 304

fallacy of approval p. 305

fallacy of catastrophic failure p. 305

fallacy of overgeneralization p. 305

fallacy of perfection p. 305

general purpose p. 295

impromptu speech p. 308

irrational thinking p. 305

manuscript speech p. 308

memorized speech p. 308

pitch p. 311

purpose statement p. 295

rate p. 310

slurring p. 311

specific purpose p. 295

substitution p. 311

survey research p. 303

thesis statement p. 296

value p. 299

visualization p. 306

ACTIVITIES

1. **Formulating Purpose Statements** Write a specific purpose statement for each of the following speeches:

   a. An after-dinner speech at an awards banquet in which you will honor a team that has a winning, but not championship, record. (You pick the team. For example: "After listening to my speech, my audience members will appreciate the individual sacrifices made by the members of the chess team.")

   b. A classroom speech in which you explain how to do something. (Again, you choose the topic: "After

listening to my speech, my audience members will know at least three ways to maximize their comfort and convenience on an economy class flight.")

   c. A campaign speech in which you support the candidate of your choice. (For example: "After listening to my speech, my audience members will consider voting for Alexandra Rodman in order to clean up student government.")

   d. Answer the following questions about each of the purpose statements you make up: Is it result oriented? Is it precise? Is it attainable?

2. **Formulating Thesis Statements** Turn each of the following purpose statements into a statement that expresses a possible thesis. For example, if you had a purpose statement such as this:

   After listening to my speech, my audience will recognize the primary advantages and disadvantages of home teeth bleaching.

   you might turn it into a thesis statement such as this:

   Home bleaching your teeth can significantly improve your appearance, but watch out for injury to the gums and teeth.

   a. At the end of my speech, the audience members will be willing to sign my petition supporting the local needle exchange program for drug addicts.

   b. After listening to my speech, the audience members will be able to list five disadvantages of tattoos.

   c. During my speech on the trials and tribulations of writing a research paper, the audience members will show their interest by paying attention and their amusement by occasionally laughing.

3. **Communication Apprehension: A Personal Analysis** To analyze your own reaction to communication apprehension, think back to your last public speech, and rate yourself on how rational, receiver oriented, positive, and prepared you were. How did these attributes affect your anxiety level?

Speech Organization and Support

CHAPTER OUTLINE

13.1 Structuring Your Speech 320
Your Working Outline
Your Formal Outline
Your Speaking Notes

13.2 Principles of Outlining 324
Standard Symbols
Standard Format
The Rule of Division
The Rule of Parallel Wording

13.3 Organizing Your Outline into a Logical Pattern 326
Time Patterns
Space Patterns
Topic Patterns
Problem-Solution Patterns
Cause-Effect Patterns
Monroe's Motivated Sequence

13.4 Beginnings, Endings, and Transitions 330
The Introduction
The Conclusion
Transitions

13.5 Supporting Material 334
Functions of Supporting Material
Types of Supporting Material
Styles of Support: Narration and Citation

13.6 Sample Speech 340
Speech Outline
Bibliography

MAKING THE GRADE **344**

KEY TERMS **345**

ACTIVITIES **345**

LEARNING OBJECTIVES

13.1

Describe different types of speech outlines and their functions.

13.2

Construct an effective speech outline using the organizing principles described in this chapter.

13.3

Choose an appropriate organizational pattern for your speech.

13.4

Develop a compelling introduction and conclusion, and use transitions at key points in your speech.

13.5

Choose supporting material that will help make your points clear, interesting, memorable, and convincing.

Public speaking skills helped Hunter Barclay succeed in college and communicate ideas he cares about. Consider the following questions as you read this chapter:

In what ways could public speaking skills potentially improve your life?

What speakers have you encountered who present meaningful information in a clear and interesting way? What lessons can you learn from them?

What personal experiences might form the basis for an informative speech? How might you organize and support your presentation in a compelling way?

basic speech structure The division of a speech into introduction, body, and conclusion.

HUNTER BARCLAY IS AN ORGANIZED person. As a student at Marshall University in West Virginia, he earned top grades while preparing for medical school and serving as the student body president. Summers were just as productive: he participated in the Fulbright Summer Institute as well as Oxford University's History, Politics, and Society Summer School. As a member of Marshall's Speech and Debate Team, he won multiple championships at the state and international levels.

Hunter was an International Affairs major, but his lifelong ambition was to become the first physician in his family. While applying to medical schools, he discovered that medical students and young physicians had an extremely high rate of suicide. Further research showed that many of those who had committed suicide were similar to him in their ambition and drive.

To highlight the issue, Hunter chose physician suicide as the topic for the persuasive speech he would present at tournaments with the Marshall speech team. He gave special attention to organizing and supporting his points. You can read the speech itself at the end of this chapter. With it, Hunter won the 2018 West Virginia Persuasive Speaking State Championship. In 2019, Hunter graduated summa cum laude and began medical school at West Virginia University.

As Hunter Barclay learned in his study of public speaking, it pays to be organized and it pays to back up what you have to say. In the following pages, you will learn methods of organizing and supporting your thoughts effectively.

13.1 Structuring Your Speech

As discussed in Chapter 3, people tend to arrange their perceptions in some meaningful way in order to make sense of the world. Being clear to your audience, however, isn't the only benefit of good organization; structuring a message effectively will help you refine your own ideas and construct more persuasive messages.

A good speech is like a good building: Both grow from a careful plan. Chapter 12 showed you how to begin this planning by formulating a purpose, analyzing your audience, and conducting research. You apply that information to the structure of the speech through outlining. Like any other plan, a speech outline is the framework on which your message is built. It contains your main ideas and shows how they relate to one another and to your thesis. Virtually every speech outline ought to follow the basic structure outlined in Figure 13.1.

This **basic speech structure** demonstrates the old aphorism for speakers: "Tell what you're going to say, say it, and then tell what you said." Although this structure sounds redundant, the research on listening cited in Chapter 5 demonstrates that receivers forget much of what they hear. The clear, repetitive nature of the basic speech structure reduces the potential for memory loss because audiences tend to listen more carefully during the beginning and ending of a speech.[1] Your outline will reflect this basic speech structure.

Outlines come in all shapes and sizes, but the three types that are most important to us here are working outlines, formal outlines, and speaking notes.

Your Working Outline

A **working outline** is a construction tool used to map out your speech. The working outline will probably follow the basic speech structure but only in rough form. It is for your eyes only, and you'll probably create several drafts as you refine your ideas. As your ideas solidify, your outline will change accordingly, becoming more polished as you go along. Figure 13.1 is a working outline for the sample speech at the end of Chapter 12 (pages 313–316.)

working outline A constantly changing organizational aid used in planning a speech.

Working Outline

"Working to Document the Common Neighbor"

Rocky Roque

I. Introduction

   A. Attention Getter: Alicia Morales's story on how she was forced to nanny children.

   B. Thesis: We must stop employers who threaten their undocumented workers with deportation.

   C. Preview of Main Points: By looking at problems and causes, we will identify some solutions to what some call a form of modern-day slavery.

II. Body: Problems

Transition:

Josue Diaz's story in the *LA Times*.

   A. Sub Point 1: Abusive conditions

   B. Sub Point 2: No way out.

III. Main Body: Causes

Transition: In 1995, an immigrant by the name Fernando held a 20-year legal battle after he was threatened with deportation when he found out that his employer didn't complete his I-9 forms.

   A. Sub Point 1: Dispensable workers

   B. Sub Point 2: New discrimination policies

IV. Main Body: Solutions

Transition:

When 46-year-old Olivia Solis retaliated after being threatened with deportation.

   A. Sub Point 1: Request protection laws

   B. Sub Point 2: Know your rights

   C. Final impact/statement: Olivia Solis's retaliation and success

V. Conclusion

   A. Review

   B. Final Statement

FIGURE 13.1 Working Outline

formal outline A consistent format and set of symbols used to identify the structure of ideas.

Your Formal Outline

A **formal outline** uses a consistent format and set of symbols to identify the structure of ideas.

A formal outline serves several purposes. In simplified form, it can be displayed as a visual aid or distributed as a handout. It can also serve as a record of a speech that was delivered; many organizations send outlines to members who miss meetings at which presentations were given. Finally, in speech classes, instructors often use speech outlines to analyze student speeches. When one is used for that purpose, it is usually a full-sentence outline and includes the purpose, thesis, and topic or title. Most instructors also require a bibliography of sources at the end of the outline. The bibliography should include full research citations, the correct form for which can be found in any style guide, such as *The Craft of Research* by Wayne Booth et al.[2] There are at least six standard bibliographic styles. Whichever style you use, you should be consistent in form and remember the two primary functions of a bibliographic citation: to demonstrate the credibility of your source and to enable the readers—in this case, your professor or your fellow students—to find the source if they want to check its accuracy or explore your topic in more detail.

Another person should be able to understand the basic ideas included in your speech by reading the formal outline. In fact, that's one test of the effectiveness of your outline. Figure 13.2 is a brief formal outline for the sample speech at the end of this chapter. An expanded version of this outline appears on pages 340–341.

Introduction

I. Attention-Getter

II. Preview of Main Points

Body

I.   Physician suicide has two main causes.

   A. Hazing Culture

   B. Assembly-Line Medicine

II. Physician suicide affects us all.

   A. Disguised Mental Anguish

   B. Physician Burnout

III. Physician suicide can be solved.

   A. Public Action

      1. We can educate the public.

      2. We can use the documentary, *Do No Harm*.

   B. Individual Action

      1. We must care for our doctors.

      2. We must thank them when they help us.

Conclusion

I. Review

II. Final Remarks

FIGURE 13-2 Formal Outline

Figure 13.3 is a detailed analysis of all the elements of a full-sentence outline. It specifies the critical elements of a speech, including the title, general purpose, specific purpose, thesis statement, introduction, main points, subpoints, transitions, visual aids, citations, conclusion, and bibliography.

Speech Rubric

General Purpose: To inform (or persuade, etc.)

Specific Purpose: After listening to my speech, my audience will . . .

Practice Time Average: (Average all your timed practices—cut if long, add if short)

Title

I. Introduction

    A. The introduction of a speech should contain the opening statements of a speech.

    B. The introduction also includes the thesis of the speech—a kind of preview.

    C. Typically, the introduction will comprise 20% of delivery time of the speech.

    D. It is also best to use "attention-getters" in this portion of the speech.

***At this point there should be a transition from the introduction to the body.

II. Body

    A. The body of the speech can contain three to four main points about the central idea being presented.

        1. For each main idea, subpoints may be included to help discuss the topic further.

        2. You may have citations to lend credibility or quotes to draw in your audience.

Transition

    B. The bulk of the speech is usually centered in the body of the speech.

        1. Thus, typically comprising 70% of the delivery time for the speech.

        2. You should note what visual aids you plan to use and when you will incorporate them into the speech.

Transition

    C. This would be the third main point; remember to add visuals to keep the audience engaged.

        1. Cue visuals

        2. Check time

***At this point, there should be a transition from the body to the conclusion.

III. Conclusion

    A. The closing statements of a speech will sum up the three main points of the speech.

    B. Typically, this aspect of the speech will comprise 10% of delivery time.

IV. Bibliography (This can be an additional outline number, an additional page, or an additional paragraph with the title *Bibliography* centered at the top.)

    A. In this portion of the outline, a list of all sources used in the preparation of the speech will be included.

    B. APA format

FIGURE 13.3 Speech Rubric

Source: Adapted from "Outlining the Speech," Speech Program, University of Colorado at Colorado Springs. Available at https://www.uccs.edu/Documents/eberhardt/Outlining%20a%20Speech.doc

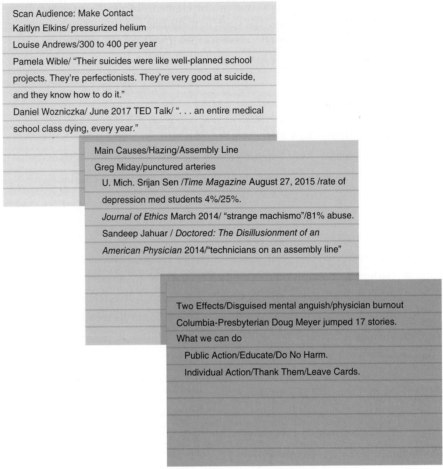

Scan Audience: Make Contact
Kaitlyn Elkins/ pressurized helium
Louise Andrews/300 to 400 per year
Pamela Wible/ "Their suicides were like well-planned school projects. They're perfectionists. They're very good at suicide, and they know how to do it."
Daniel Wozniczka/ June 2017 TED Talk/ ". . . an entire medical school class dying, every year."

Main Causes/Hazing/Assembly Line
Greg Miday/punctured arteries
U. Mich. Srijan Sen /*Time Magazine* August 27, 2015 /rate of depression med students 4%/25%.
Journal of Ethics March 2014/ "strange machismo"/81% abuse.
Sandeep Jahuar / *Doctored: The Disillusionment of an American Physician* 2014/"technicians on an assembly line"

Two Effects/Disguised mental anguish/physician burnout
Columbia-Presbyterian Doug Meyer jumped 17 stories.
What we can do
Public Action/Educate/Do No Harm.
Individual Action/Thank Them/Leave Cards.

FIGURE 13.4 Speaking Notes These note cards are based on the Physician Suicide speech outlined above and on pages 340–341. Speaking notes are unique to the speaker; yours could be completely different for the same speech.

Your Speaking Notes

Like your working outline, your speaking notes are for your use only, so the format is up to you. Many teachers suggest that speaking notes should be in the form of a brief keyword outline, with just enough information listed to jog your memory but not enough to get lost in.

Many teachers suggest that you fit your notes on one side of one 3- by 5-inch note card. Other teachers recommend that you also have your introduction and conclusion on note cards, and still others recommend that your longer quotations be written out on note cards.

13.2 Principles of Outlining

Over the years, a series of rules or principles for the construction of outlines has evolved. These rules are based on the use of the standard symbols and format discussed next.

Standard Symbols

A speech outline generally uses the following symbols:

I. Main point (roman numeral)

   A. Subpoint (capital letter)

      1. Sub-subpoint (standard number)

        a. Sub-subsubpoint (lowercase letter)

In the examples in this chapter, the major divisions of the speech—introduction, body, and conclusion—are not given symbols. They are listed by name, and the roman numerals for their main points begin anew in each division. An alternative form is to list these major divisions with roman numerals, main points with capital letters, and so on.

Standard Format

In the sample outlines in this chapter, notice that each symbol is indented a number of spaces from the symbol above it. Besides keeping the outline neat, the indentation of different-order ideas is actually the key to the technique of outlining; it enables you to coordinate and order ideas in the form in which they are most comprehensible to the human mind. If the standard format is used in your working outline, it will help you create a well-organized speech. If it is used in speaking notes, it will help you remember everything you want to say.

Proper outline form is based on a few rules and guidelines, the first of which is the rule of division.

The Rule of Division

In formal outlines, main points and subpoints always represent a division of a whole. Because it is impossible to divide something into fewer than two parts, you always have at least two main points for every topic. Then, if your main points are divided, you will always have at least two subpoints, and so on. Thus, the rule for formal outlines is as follows: Never a "I" without a "II," never an "A" without a "B," and so on.

Three to five is considered to be the ideal number of main points. It is also considered best to divide those main points into three to five subpoints, when necessary and possible.

The Rule of Parallel Wording

Your main points should be worded in a similar, or "parallel," manner. For example, if you are developing a speech against capital punishment, your main points might look like this:

  I. Capital punishment is not effective: it is not a deterrent to crime.

 II. Capital punishment is not constitutional: it does not comply with the Eighth Amendment.

III. Capital punishment is not civilized: it does not allow for a reverence for life.

Whenever possible, subpoints should also be worded in a parallel manner. For your points to be worded in a parallel manner, they should each contain one, and only one, idea. (After all, they can't really be parallel if one is longer or contains more ideas than the others.) This will enable you to completely develop one idea before moving on to another one in your speech. If you were discussing cures for indigestion, your topic might be divided incorrectly if your main points looked like this:

  I. "Preventive cures" help you before eating.

 II. "Participation cures" help you during and after eating.

You might actually have three ideas there and thus three main points:

  I. Prevention cures (before eating)

 II. Participation cures (during eating)

III. Postparticipation cures (after eating)

Main Points and Subpoints

 To get an idea of your ability to distinguish main points from subpoints, set the "timer" function on your mobile phone and see how long it takes you to fit the following concepts for a speech entitled "The College Application Process" into outline form:

CONCEPTS	RECOMMENDED OUTLINE FORM
Participation in extracurricular activities	I.
Visit and evaluate college websites	A.
Prepare application materials	B.
Career ambitions	II.
Choose desired college	A.
Letters of recommendation	B.
Write personal statement	C.
Visit and evaluate college campuses	III.
Choose interesting topic	A.
Test scores	B.
Include important personal details	1.
Volunteer work	2.
Transcripts	3.

You can score yourself as follows:

A minute or less: Congratulations, organization comes naturally to you.

61–90 seconds: You have typical skills in this area.

More than 90 seconds: Give yourself extra time while building your speech outline.

13.3 Organizing Your Outline into a Logical Pattern

An outline should reflect a logical order for your points. You might arrange them from newest to oldest, largest to smallest, best to worst, or in a number of other ways that follow. The organizing pattern you choose ought to be the one that best develops your thesis.

Time Patterns

time pattern An organizing plan for a speech based on chronology.

Arrangement according to **time patterns**, or chronology, is one of the most common patterns of organization. The period of time could be anything from centuries to seconds. In a speech on airline food, a time pattern might look like this:

    I. Early airline food: a gourmet treat

    II. The middle period: institutional food at 30,000 feet

    III. Today's airline food: the passenger starves

Arranging points according to the steps that make up a process is another form of time patterning. The topic "Recording a Hit Song" might use this type of patterning:

I. Record the demo.
II. Post a YouTube video.
III. Get a recording company to listen and view.

Time patterns are also the basis of **climax patterns**, which are used to create suspense. For example, if you wanted to create suspense in a speech about military intervention, you could chronologically trace the steps that eventually led us into Afghanistan or Iraq in such a way that you build up your audience's curiosity. If you detail these steps through the eyes of a soldier who entered military service right before one of those wars, you will be building suspense as your audience wonders what will become of that soldier.

The climax pattern can also be reversed. When it is reversed, it is called *anti-climactic* organization. If you started your military intervention speech by telling the audience that you were going to explain why a specific soldier was killed in a specific war, and then you went on to explain the things that caused that soldier to become involved in that war, you would be using anticlimactic organization. This pattern is helpful when you have an essentially uninterested audience and you need to build interest early in your speech to get the audience to listen to the rest of it.

climax pattern An organizing plan for a speech that builds ideas to the point of maximum interest or tension.

space pattern An organizing plan in a speech that arranges points according to their physical location.

topic pattern An organizing plan for a speech that arranges points according to logical types or categories.

Space Patterns

Speech organization by physical area is known as a **space pattern**. The area could be stated in terms of continents or centimeters or anything in between. If you were discussing the Great Lakes, for example, you could arrange them from west to east:

I. Superior
II. Michigan
III. Huron
IV. Erie
V. Ontario

Topic Patterns

A topical arrangement or **topic pattern** is based on types or categories. These categories could be either well known or original; both have their advantages. For example, a division of college students according to well-known categories might look like this:

I. Freshmen
II. Sophomores
III. Juniors
IV. Seniors

Well-known categories are advantageous because audiences quickly understand them. But familiarity also has its disadvantages. One disadvantage is the "Oh, this again" syndrome. If the members of an audience feel they have nothing new to learn about the components of your topic, they might not listen to you. To

Any outline you show your audience should be much briefer than the one in your speaking notes or the formal one you prepare.

How would your audience react if you displayed your formal outline during your speech?

avoid this, you could invent original categories that freshen up your topic by suggesting an original analysis. For example, original categories for "college students" might look like this:

I. Grinds: Students who go to every class and read every assignment before it is due.
II. Renaissance students: Students who find a satisfying balance of scholarly and social pursuits.
III. Burnouts: Students who have a difficult time finding the classroom, let alone doing the work.

Sometimes topics are arranged in the order that will be easiest for your audience to remember. To return to our Great Lakes example, the names of the lakes could be arranged so that their first letters spell the word "HOMES." Words used in this way are known as *mnemonics*. Carol Koehler, a professor of communication and medicine, uses the mnemonic "CARE" to describe the characteristics of a caring doctor:

C stands for *concentrate*. Physicians should pay attention with their eyes and ears . . .

A stands for *acknowledge*. Show them that you are listening . . .

R stands for *response*. Clarify issues by asking questions, providing periodic recaps . . .

E stands for *exercise emotional control*. When your "hot buttons" are pushed . . .[3]

Problem-Solution Patterns

Describing what's wrong and proposing a way to make things better is known as a **problem-solution pattern**. It is usually (but not always) divisible into two distinct parts, as in this example:

I. The Problem: Addiction (which could then be broken down into addiction to cigarettes, alcohol, prescribed drugs, and street drugs)
II. The Solution: A national addiction institute (which would study the root causes of addiction in the same way that the National Cancer Institute studies the root causes of cancer)

We will discuss this pattern in more detail in Chapter 15.

Cause-Effect Patterns

Cause-effect patterns are similar to problem-solution patterns in that they are basically two-part patterns: First you discuss something that happened, and then you discuss its effects.

A variation of this pattern reverses the order and presents the effects first and then the causes. Persuasive speeches often have effect-cause or cause-effect as the first two main points. Elizabeth Hallum, a student at Arizona State University, organized the first two points of a speech on "workplace revenge"[4] like this:

I. The effects of the problem
   A. Lost productivity
   B. Costs of sabotage

II. The causes of the problem
   A. Employees feeling alienated
   B. Employers' light treatment of incidents of revenge

problem-solution pattern An organizing pattern for a speech that describes an unsatisfactory state of affairs and then proposes a plan to remedy the problem.

cause-effect pattern An organizing plan for a speech that demonstrates how one or more events result in another event or events.

Nontraditional Patterns of Organization

In addition to the traditional patterns usually taught in public speaking classes, researchers are looking at the use of less linear forms.

One of these is the wave pattern, in which the speaker uses repetitions and variations of themes and ideas. The major points of the speech come at the crest of the wave. The speaker follows these with a variety of examples leading up to another crest, where she repeats the theme or makes another major point.

Perhaps the most famous speech that illustrates this pattern is the Reverend Dr. Martin Luther King Jr.'s "I Have a Dream." King used this memorable line as the crest of a wave that he followed with examples of what he saw in his dream; then he repeated the line. He ended with a "peak" conclusion that emerged from the final wave in the speech—repetition and variation on the phrase "Let freedom ring."

An excerpt from Sojourner Truth's "Ain't I a Woman?" speech also illustrates this pattern:

That man over there says that women need to be helped into carriages, and lifted over ditches, and to have the best place everywhere. Nobody ever helps me into carriages, or over mud-puddles, or gives me any best place!

And ain't I a woman?

Look at me! Look at my arm! I have ploughed and planted, and gathered into barns, and no man could head me!

And ain't I a woman?

I could work as much and eat as much as a man—when I could get it—and bear the lash as well!

And ain't I a woman?

I have borne thirteen children and seen them most all sold off to slavery, and when I cried out with my mother's grief, none but Jesus heard me!

And ain't I a woman?

Jaffe, C. (2007). *Public speaking: Concepts and skills for a diverse society* (5th ed.). Boston: Wadsworth, © 2007. Reprinted with permission of Wadsworth, an imprint of the Wadsworth Group, a division of Thomson Learning.

Why is diversity important in the messages you send and receive?

Why is message organization such an important factor?

The third main point in this type of persuasive speech is often "solutions," and the fourth main point is often "the desired audience behavior." Hallum's final points were as follows:

III. Solutions: Support the National Employee Rights Institute.
IV. Desired Audience Response: Log on to www.disgruntled.com.

Cause-effect and problem-solution patterns are often combined in various arrangements. One extension of the problem-solution organizational pattern is Monroe's Motivated Sequence.

Monroe's Motivated Sequence

The Motivated Sequence was proposed by a scholar named Alan Monroe in the 1930s.[5] In this persuasive pattern, the problem is broken down into an attention step and a need step, and the solution is broken down into a satisfaction step, a visualization step, and an action step. In a speech on "random acts of kindness,"[6] the Motivated Sequence might break down like this:

I. The attention step draws attention to your subject. ("Just the other day Ron saved George's life with a small, random, seemingly unimportant act of kindness.")

II. The need step establishes the problem. ("Millions of Americans suffer from depression, a life-threatening disease.")

III. The satisfaction step proposes a solution. ("One random act of kindness can lift a person from depression.")

IV. The visualization step describes the results of the solution. ("Imagine yourself having that kind of effect on another person.")

V. The action step is a direct appeal for the audience to do something. ("Try a random act of kindness today!")

Chapter 15 has more to say about the organization of persuasive speeches.

13.4 Beginnings, Endings, and Transitions

The introduction and conclusion of a speech are vitally important, although they usually will occupy less than 20% of your speaking time. Listeners form their impression of a speaker early, and they remember what they hear last; it is, therefore, vital to make those few moments at the beginning and end of your speech work to your advantage. It is also essential that you connect sections within your speech using effective transitions.

The Introduction

introduction (of a speech) The first structural unit of a speech, in which the speaker captures the audience's attention and previews the main points to be covered.

The **introduction** of a speech has four functions: to capture the audience's attention, preview the main points, set the mood and tone of the speech, and demonstrate the importance of the topic.

Capturing Attention   There are several ways to capture an audience's attention, including the following:

- **Refer to the audience**. This technique is especially effective if it is complimentary: "Zoe's speech about how animals communicate was so interesting that I decided to explore a related topic."

- **Refer to the occasion**. "Given our assignment, it seems appropriate to talk about something you might have wondered."

- **Refer to the relationship between the audience and the subject**. "It's fair to say that all of us here believe it's important to care for our environment."

- **Refer to something familiar to the audience**. "Most of us have cared for a plant at some point in our lives."

- **Cite a startling fact or opinion**. "New scientific evidence suggests that plants appreciate human company, kind words, and classical music."

- **Ask a question**. "Have you ever wondered why some people have a green thumb, whereas others couldn't make a weed grow?"

- **Tell an anecdote**. "The other day, while taking a walk near campus, I saw a man talking quite animatedly to a sunflower."

- **Use a quotation**. "The naturalist Max Thornton recently said, 'Psychobiology has shown that plants can communicate.'"

- **Tell an (appropriate) joke**. "We once worried about people who talked to plants, but that's no longer the case. Now we only worry if the plants talk back."

Previewing Main Points   After you capture the attention of the audience, an effective introduction will almost always state the speaker's thesis and give the listeners an idea of the upcoming main points. Katharine Graham, the former publisher of the *Washington Post*, addressed a group of businessmen and their wives in this way:

> I am delighted to be here. It is a privilege to address you. And I am especially glad the rules have been bent for tonight, allowing so many of you to bring along your husbands. I think it's nice for them to get out once in a while and see how the other half lives. Gentlemen, we welcome you.

> Actually, I have other reasons for appreciating this chance to talk with you tonight. It gives me an opportunity to address some current questions about the press and its responsibilities—whom we are responsible to, what we are responsible for, and generally how responsible our performance has been.[7]

TED Talk speakers know how to capture their audience's attention by using the strategies listed on page 330.

Think of at least three strategies you can use to capture the attention of your listeners.

Thus, Graham previewed her main points:

1. To explain who the press is responsible to
2. To explain what the press is responsible for
3. To explain how responsible the press has been

Sometimes your preview of main points will be even more straightforward:

> "I have three points to discuss: They are _____, _____, and _____."

Sometimes you will not want to refer directly to your main points in your introduction. Your reasons for not doing so might be based on a plan calling for suspense, humorous effect, or stalling for time to win over a hostile audience. In that case, you might preview only your thesis:

> "I am going to say a few words about _____."

> "Did you ever wonder about _____?"

> "_____ is one of the most important issues facing us today."

Setting the Mood and Tone of Your Speech   Notice, in the example just given, how Katharine Graham began her speech by joking with her audience. She was a powerful woman speaking before an all-male organization; the only women in the audience were the members' wives. That is why Ms. Graham felt it necessary to put her audience members at ease by joking with them about women's traditional role in society. By beginning in this manner, she assured the men that she would not berate them for the sexist bylaws of their organization. She also showed them that she was going to approach her topic with wit and intelligence. Thus, she set the mood and tone for her entire speech. Imagine how different that mood and tone would have been if she had begun this way:

> Before I start today, I would just like to say that I would never have accepted your invitation to speak here had I known that your organization does not accept

women as members. Just where do you Cro-Magnons get off, excluding more than half the human race from your little club?

Demonstrating the Importance of Your Topic to Your Audience Your audience members will listen to you more carefully if your speech relates to them as individuals. Based on your audience analysis, you should state directly *why* your topic is of importance to your audience members. This importance should be related as closely as possible to their specific needs at that specific time.

For example, when Anna Claire Tucker, a student at Berry College, presented a speech about medical waste, she began her introduction with a striking example of how medical practices result in overcharging patients.

> When you need milk, a walk down the dairy aisle provides you with a variety of options: pint, quart, gallon. With so many choices, you wouldn't buy a gallon if you only needed a glass. But when it comes to thousand-dollar chemotherapy treatments, patients aren't given the same choice. Shaun Recchi didn't get to choose between small, medium, or large vials of rituximab. Manufacturers made that decision for him when they produced it in one size. Their greed cost Shaun $13,000 for a drug he only needed half of, and it only gets worse. While an open jug of milk could be saved, Shaun's medication couldn't, so the remaining $6,500 worth was simply thrown in the trash.[8]

That example caught her audience's attention, and Anna followed it up immediately with a reminder that her problem was a problem for everyone in the audience:

> From overprescribed pills to unused surgical tools, medical waste is a $765 billion problem. According to the National Academy of Medicine, it accounts for one quarter of all healthcare spending. Manufacturers and hospitals perpetuate the problem, but as Senator Amy Klobuchar decries, "It's the American taxpayers footing the bill."[9]

Thus, she reminded the audience that they all, as potential patients and definitely as taxpayers, were affected by the topic of her speech.

Establishing Credibility One final consideration for your introduction is to establish your credibility to speak on your topic. One way to do this is to be *well prepared*. Another is to *appear confident* as soon as you face your audience. A third technique is to *tell your audience about your personal experience* with the topic, in order to establish why it is important to you. We will discuss credibility, and how to establish it, in greater detail in Chapter 15.

The Conclusion

Like the introduction, the **conclusion** is an especially important part of your speech. It has three essential functions: to restate the thesis, to review your main points, and to provide a memorable final remark.

You can review your thesis either by repeating it or by paraphrasing it. Or you might devise a striking summary statement for your conclusion to help your audience remember your thesis. Grant Anderson, a student at Minnesota State University, gave a speech against the policy of rejecting blood donations from homosexuals. He ended his conclusion with this statement: "The gay community still has a whole host of issues to contend with, but together all of us can all take a step forward by recognizing this unjust and discriminatory measure. So stand up and raise whatever arm they poke you in to draw blood and say 'Blood is Blood' no matter who you are."[10]

ASK YOURSELF (?)

Think of an idea you could present to your class. Which pattern of organization would be most effective for that presentation?

conclusion (of a speech) The final part of a speech, in which the main points are reviewed and final remarks are made to motivate the audience or help listeners remember key ideas.

Grant's statement was concise but memorable.

Your main points can also be reviewed artistically. For example, first look back at that example introduction by Katharine Graham, and then read her conclusion to that speech:

> So instead of seeking flat and absolute answers to the kinds of problems I have discussed tonight, what we should be trying to foster is respect for one another's conception of where duty lies, and understanding of the real worlds in which we try to do our best. And we should be hoping for the energy and sense to keep on arguing and questioning, because there is no better sign that our society is healthy and strong.

Let's take a closer look at how and why this conclusion was effective. Graham posed three questions in her introduction. She dealt with those questions in her speech and reminded her audience, in her conclusion, that she had answered the questions.

PREVIEW (FROM INTRODUCTION OF SPEECH)	REVIEW (FROM CONCLUSION)
1. To whom is the press responsible?	1. To its own conception of where its duty lies
2. What is the press responsible for?	2. For doing its best in the "real world"
3. How responsible has the press been?	3. It has done its best

Transitions

To keep your message moving forward, **transitions** perform the following functions:

They tell how the introduction relates to the body of the speech.

They tell how one main point relates to the next main point.

They tell how your subpoints relate to the points they are part of.

They tell how your supporting points relate to the points they support.

To be effective, transitions should refer to the previous point and to the upcoming point, showing how they relate to each other and to the thesis. They usually sound something like this:

"Like [previous point], another important consideration in [topic] is [upcoming point]."

"But _____ isn't the only thing we have to worry about. _____ is even more potentially dangerous."

"Yes, the problem is obvious. But what are the solutions? Well, one possible solution is . . ."

Sometimes a transition includes an internal review (a restatement of preceding points), an internal preview (a look ahead to upcoming points), or both:

CHECKLIST ✓

Effective Conclusions

You can make your final remarks most effective by avoiding the following mistakes:

☐ **Don't end abruptly.** Make sure your conclusion accomplishes everything it is supposed to accomplish. Develop it fully. You might want to use signposts such as, "Finally . . . ," "In conclusion . . . ," or "To sum up what I've been talking about here . . ." to let your audience know that you have reached the conclusion of the speech.

☐ **Don't ramble either.** Prepare a definite conclusion, and never, ever end by mumbling something like, "Well, I guess that's about all I wanted to say . . . ."

☐ **Don't introduce new points.** The worst kind of rambling is, "Oh, yes, and something I forgot to mention is . . . ."

☐ **Don't apologize.** Don't make statements such as "I'm sorry I didn't have more time to research this subject." You will only highlight the possible weaknesses of your speech, which may have been far more apparent to you than to your audience. It's best to end strong. You can use any of the attention-getters suggested for the introduction to make the conclusion memorable, or you can revisit your attention-getting introduction.

---

transition A phrase that connects ideas in a speech by showing how one relates to the other.

"So far we've discussed _____, _____, and _____. Our next points are _____, _____, and _____."

It isn't always necessary to provide a transition between every set of points. You have to choose when one is necessary for your given audience to follow the progression of your ideas. You can find several examples of transitions in the sample speech at the end of this chapter.

📍 13.5 Supporting Material

It is important to organize ideas clearly and logically. But clarity and logic by themselves won't guarantee that you'll interest, enlighten, or persuade others; these results call for the use of supporting materials. These materials—the facts and information that back up and prove your ideas and opinions—are the flesh that fills out the skeleton of your speech.

Functions of Supporting Material

Supporting material has four functions.

To Clarify   As explained in Chapter 4, people of different backgrounds tend to attach different meanings to words. Supporting material can help you overcome this potential source of confusion by helping you to clarify key terms and ideas. For example, when the biologist Prosanta Chakrabarty talks about the confusion inherent in the idea of evolutionary theory, he explains it like this:

> For instance, we're taught to say "the theory of evolution." There are actually many theories, and just like the process itself, the ones that best fit the data are the ones that survive to this day. The one we know best is Darwinian natural selection. That's the process by which organisms that best fit an environment survive and get to reproduce, while those that are less fit slowly die off. And that's it. Evolution is as simple as that, and it's a fact.

> Evolution is a fact as much as the "theory of gravity." You can prove it just as easily. You just need to look at your bellybutton that you share with other placental mammals, or your backbone that you share with other vertebrates, or your DNA that you share with all other life on earth. Those traits didn't pop up in humans. They were passed down from different ancestors to all their descendants, not just us.[11]

To Prove   A second function of support is to be used as evidence, to prove the truth of what you are saying. If you were giving a speech on how college fraternities are changing today, you might want to quote from a researcher who has been looking into that topic.

> Americans demonize fraternities as bastions of toxic masculinity where young men go to indulge their worst impulses. Universities have cracked down: Since November 2017, more than a dozen have suspended all fraternity events. But Alexandra Robbins, a researcher who writes about the lives of young Americans, spent more than two years interviewing fraternity members nationwide for a book about what college students think it means to "be a man," and what she learned was often heartening. Contrary to negative headlines and popular opinion, many fraternities are encouraging brothers to defy stereotypical hypermasculine standards and to simply be good people.[12]

Organizing Business Presentations

When top business executives plan an important speech, they often call in a communication consultant to help organize their remarks. Even though they are experts, executives are so close to the topic of their message that they may have difficulty arranging their ideas so others will understand or be motivated by them.

Consultants stress how important organization and message structure are in giving presentations. Seminar leader and corporate trainer T. Stephen Eggleston sums up the basic approach: "Any presentation . . . regardless of complexity . . . should consist of the same four basic parts: an opening, body, summary and closing."[13]

Ethel Cook, a Massachusetts consultant, is very specific about how much time should be spent on each section of a speech. "In timing your presentation," she says, "an ideal breakdown would be:

Opening—10 to 20%

Body—65 to 75%

Closing—10 to 20%"[14]

Business coach Vadim Kotelnikov gives his clients a step-by-step procedure to organize their ideas within the body of a presentation. "List all the points you plan to cover," he advises. "Group them in sections and put your list of sections in the order that best achieves your objectives. Begin with the most important topics."[15]

Toastmasters International, an organization that runs training programs for business professionals, suggests alternative organizational patterns:

> To organize your ideas into an effective proposal, use an approach developed in the field of journalism—the "inverted pyramid." In the "inverted pyramid" format, the most important information is given in the first few paragraphs. As you present the pitch, the information becomes less and less crucial. This way, your presentation can be cut short, yet remain effective.[16]

Imagine a business presentation you might have to make in your future career. Why would organization be important in such a presentation?

To Make Interesting A third function of support is to make an idea interesting or to catch your audience's attention. For example, when the marketing executive Aparna Mehta got up to speak about the problem online retailers have with returns, audience members who were not in her field might have been tempted to run out for a snack. But before they could, Mehta began telling the following anecdote:

> Hi. My name is Aparna. I am a shopaholic—and I'm addicted to online returns.
>
> Well, at least I was. At one time, I had two or three packages of clothing delivered to me every other day. I would intentionally buy the same item in a couple different sizes and many colors, because I did not know what I really wanted. So I overordered, I tried things on, and then I sent what didn't work back. Once my daughter was watching me return some of those packages, and she said, "Mom, I think you have a problem."
>
> I didn't think so. I mean, it's free shipping and free returns, right? I didn't even think twice about it, until I heard a statistic at work that shocked me.
>
> You see, I'm a global solutions director for top-tier retail, and we were in a meeting with one of my largest customers, discussing how to streamline costs. One of their biggest concerns was managing returns. Just this past holiday season alone, they had 7.5 million pieces of clothing returned to them.
>
> I could not stop thinking about it. What happens to all these returned clothes? So I came home and researched. *And I learned that every year, four billion pounds of returned clothing ends up in the landfill. That's like every resident in the US did a load of laundry last night and decided to throw it in the trash today.*[17]

To Make Memorable A final function of supporting materials, related to the preceding one, is to make a point memorable. We have already mentioned the importance of "memorable" statements in a speech conclusion; use of supporting material in the introduction and body of the speech provides another way to help your audience retain important information. For example, when Nathan Hill, a student at the University of Akron, spoke about how certain local ordinances discriminated against victims of abuse, he started with this example:

> Overjoyed. That's the word single mother Lakesha Briggs used to describe how she felt when she finally found a safe home for herself and her two daughters. That is, until Lakesha's boyfriend, a guest in the home, began to abuse her daily . . . . When the authorities responded, the family was startled when the police had a stark warning . . . for Ms. Briggs: "Strike One." The officers told her that if the police were called again to the property, they would make sure her family's landlord prepared an eviction. The police were referring to a local "Nuisance Property Ordinance," which labels a property as a nuisance when it is the site of a designated number of calls to police . . . making no distinction between perpetrators and victims of crimes.[18]

Types of Supporting Material

As you may have noted, each function of support could be fulfilled by several different types of material. Let's take a look at these different types of supporting material.

Definitions It's a good idea to give your audience members definitions of your key terms, especially if those terms are unfamiliar to them or are being used in an unusual way. A good definition is simple and concise, and might even encapsulate the main points of your speech. For example, when Reagan Williams, a student at Arkansas State University, gave a speech on Rape Culture, she defined her key term as follows:

> This term refers to a society or culture that normalizes sexual violence through patriarchal myths about the rape, the nature of a reaction to the crime, and rigid gender constructs that normalize violence.[19]

Examples An example is a specific case that is used to demonstrate a general idea. Examples can be either factual or hypothetical, personal or borrowed. In Reagan Williams's speech, she used the following **factual example** to demonstrate how rapes are often perpetrated by someone the victim knows:

> On January 8, 2012, Daisy Coleman's mother found her lying on the front lawn, hair frozen to the ground. Daisy was fourteen. Her mother drove her to the hospital, where she underwent a rape kit. When the result came back, their worst fears were confirmed.
>
> The evening before, the senior quarterback asked Daisy and her friend to join him at his home. When they arrived, they were encouraged to drink vodka from the "bitch cup." They were inexperienced with alcohol, and as the night went on, everything got blurry. They both blacked out. Less than an hour later, five senior boys left a fourteen-year-old girl lying on the ground in below freezing temperature. Years later, Daisy would learn that her 13-year-old friend had been raped in the room right next door.[20]

Hypothetical examples, which ask audience members to imagine something, can often be more powerful than factual examples because they create active participation in the thought. Stephanie Wideman of the University of West Florida used a hypothetical example to start off her speech on oil prices:

> The year is 2025. One day you are asked not to come into work, not because of a holiday, but instead because there is not enough energy available to power your office. You see, it is not that the power is out, but that they are out of power.[21]

factual example A true, specific case that is used to demonstrate a general idea.

hypothetical example An example that asks an audience to imagine an object or event.

UNDERSTANDING COMMUNICATION TECHNOLOGY

Plagiarism in a Digital Age

Some experts believe that the social media are redefining how students understand the concept of authorship and originality. After all, the Internet is the home of file sharing that allows us to download music, movies, and TV programs without payment. Also, Google and Wikipedia are our main portals to random free information. It all seems to belong to us, residing on our computer as it does. Information wants to be free.

According to one expert on the topic, "Now we have a whole generation of students who've grown up with information that just seems to be hanging out there in cyberspace and doesn't seem to have an author. It's possible to believe this information is just out there for anyone to take."[22] Other experts beg to differ. They say students are fully aware of what plagiarism is, online or off, and they know it's cheating. It's just that it's so easy to copy and paste online material, and students like to save time wherever they can.

Public speaking instructors are on the front lines of those fighting plagiarism because it's so important for successful student speakers to speak from the heart, in their own words and with their own voice. In addition, citing research enhances credibility. Plagiarism in public speaking isn't just cheating, it's ineffective.

The general rule for the digital age is as follows: Thou shalt not cut and paste into a final draft—not for a paper and not for a speech. Cutting and pasting is fine for research, but everything that's cut and pasted should be placed in a separate "research" file, complete with a full citation for the website in which you found it. Then switch to your "draft" file to put everything in your own words, and go back to the research file to find the attribution information when you need to cite facts and ideas that you got from those sources.

When would it be easy for you to plagiarize information? What would be the consequences if you were caught?

Statistics   Data organized to show that a fact or principle is true for a large percentage of cases are known as statistics. These are actually collections of examples, which is why they are often more effective as proof than are isolated examples. Here's the way a newspaper columnist used statistics to demonstrate a point about gun violence:

> I had coffee the other day with Marian Wright Edelman, president of the Children's Defense Fund, and she mentioned that since the murders of Robert Kennedy and the Rev. Martin Luther King Jr. in 1968, well over a million Americans have been killed by firearms in the United States. That's more than the combined U.S. combat deaths in all the wars in all of American history. "We're losing eight children and teenagers a day to gun violence," she said.[23]

Because statistics can be powerful support, you have to follow certain rules when using them. You should make sure that the statistics make sense and that they come from a credible source. You should also cite the source of the statistic when you use it. A final rule is based on effectiveness rather than ethics. You should reduce the statistic to a concrete image if possible. For example, $1 billion in $100 bills. Using concrete images such as this will make your statistics more than "just numbers" when you use them. For example, one observer expressed the idea of Bill Gates's wealth this way:

> Examine Bill Gates' wealth compared to yours: Consider the average American of reasonable but modest wealth. Perhaps he has a net worth of $100,000. Mr. Gates' worth is 400,000 times larger. Which means that if something costs $100,000 to him, to Bill it's as though it costs 25 cents. So for example, you might think a new Lamborghini Diablo would cost $250,000, but in Bill Gates dollars that's 63 cents. [24]

statistic Numbers arranged or organized to show how a fact or principle is true for a large percentage of cases.

ASK YOURSELF

Recall a recent occasion in which you tried to change someone's mind. What were your arguments? Which forms of support did you use to back them up?

analogies An extended comparison that can be used as supporting material in a speech.

anecdote A brief, personal story used to illustrate or support a point in a speech.

Analogies/Comparison-Contrast We use **analogies**, or comparisons, all the time, often in the form of figures of speech, such as similes and metaphors. A simile is a direct comparison that usually uses *like* or *as*, whereas a metaphor is an implied comparison that does not use *like* or *as*. So if you said that the rush of refugees from a war-torn country was "like a tidal wave," you would be using a simile. If you used the expression "a tidal wave of refugees," you would be using a metaphor.

Analogies are extended metaphors. They can be used to compare or contrast an unknown concept with a known one. For example, here's how one writer made her point against separate Academy Awards for men and women:

> Many hours into the Academy Awards ceremony this Sunday, the Oscar for best actor will go to Morgan Freeman, Jeff Bridges, George Clooney, Colin Firth, or Jeremy Renner. Suppose, however, that the Academy of Motion Picture Arts and Sciences presented separate honors for best white actor and best non-white actor, and that Mr. Freeman was prohibited from competing against the likes of Mr. Clooney and Mr. Bridges. Surely, the Academy would be derided as intolerant and out of touch; public outcry would swiftly ensure that Oscar nominations never again fell along racial lines.
>
> Why, then, is it considered acceptable to segregate nominations by sex, offering different Oscars for best actor and best actress?[25]

Anecdotes An **anecdote** is a brief story with a point, often (but not always) based on personal experience. (The word *anecdote* comes from the Greek, meaning "unpublished item.") Alyssa Gieseck, a student at the University of Akron, used the following anecdote in her speech about the problems some Deaf people encounter with police:

> Jonathan Meister, a deaf man, was retrieving his personal belongings from a friend's home when police arrived, responding to a report of suspicious behavior. Trying to sign to the police officers, Meister was seen as a threat, which is when the officers decided to handcuff him. Handcuffing a deaf person is equivalent to putting duct tape over a hearing person's mouth. Meister initially pulled away to sign that he was deaf, but one police officer pushed him up against a fence, kneed him twice in the abdomen, put him in a choke hold, and tasered him. A second officer repeatedly punched him in the face, tasering him again. But this wasn't the end of the assault. He was then shoved to the ground, kicked, elbowed, tasered for a third time, and put in a second chokehold, which left him unconscious.[26]

Quotations/Testimonies Using a familiar, artistically stated quotation will enable you to take advantage of someone else's memorable wording. For example, if you were giving a speech on personal integrity, you might quote Mark Twain, who said, "Always do right. This will gratify some people, and astonish the rest." A quotation like that fits Alexander Pope's definition of "true wit": "What was often thought, but ne'er so well expressed."

You can also use quotations as **testimony** to prove a point by using the support of someone who is more authoritative or experienced on the subject than you are. When Julia Boyle, a student at Northern Illinois University, wanted to prove that spyware stalking was a serious problem, she used testimony this way:

> Michella Cash, advocate for the Women's Service network, noted, "Spyware technology is the new form of domestic violence abuse that enables perpetrators to exert round the clock control over their victims."[27]

Sometimes testimony can be paraphrased. For example, when one business executive was talking on the subject of diversity, he used a conversation he had with Jesse Jackson Sr., an African American political activist, as testimony:

> At one point in our conversation, Jesse talked about the stages of advancement toward a society where diversity is fully valued. He said the first stage was emancipation—the end of slavery. The second stage was the right to vote and the third stage was the political power to actively participate in government—to be part of city hall, the governor's office and Capitol Hill. Jesse was clearly focused, though, on the fourth stage—which he described as the ability to participate fully in the prosperity that this nation enjoys. In other words, economic power.[28]

Styles of Support: Narration and Citation

Most of the forms of support discussed in the preceding section could be presented in either of two ways: through narration or through citation. **Narration** involves telling a story with your information. You put it in the form of a small drama, with a beginning, middle, and end. For example, Evan McCarley of the University of Mississippi narrated the following example in his speech on the importance of drug courts:

> Oakland contractor Josef Corbin has a lot to be proud of. Last year his firm, Corbin Building Inc., posted revenue of over 3 million dollars after funding dozens of urban restoration projects. His company was ranked as one of the 800 fastest-growing companies in the country, all due to what his friends call his motivation for success. Unfortunately, Corbin used this motivation to rob and steal on the streets of San Francisco to support a heroin and cocaine habit. But when he was charged with possession, Josef was given the option to participate in a state drug court, a program targeted at those recently charged with drug use, possession, or distribution. The drug court offers offenders free drug treatment, therapy, employment, education, and weekly meetings with a judge, parole officer and other accused drug offenders.[29]

Citation, unlike narration, is a simple statement of the facts. Citation is shorter and more precise than narration, in the sense that the source is carefully stated. Citation will always include such phrases as, "According to the July 25, 2016, edition of *Time* magazine," or, "As Mr. Smith made clear in an interview last April 24." Evan McCarley cited statistics later in his speech on drug courts:

> Fortunately, Corbin's story, as reported in the May 30th *San Francisco Chronicle*, is not unique, since there are currently over 300 drug courts operating in 21 states, turning first-time and repeat offenders into successful citizens with a 70% success rate.[30]

Some forms of support, such as anecdotes, are inherently more likely to be expressed as narration. Statistics, on the other hand, are nearly always cited rather than narrated. However, when you are using examples, quotation/testimony, definitions, and analogies, you often have a choice.

testimony Supporting material that proves or illustrates a point by citing an authoritative source.

narration The presentation of speech supporting material as a story with a beginning, middle, and end.

citation A brief statement of supporting material in a speech.

13.6 Sample Speech

The following speech was given by Hunter Barclay, whose profile opened this chapter. Hunter is an organized person, a quality that showed up in his speech preparation. He explains it this way:

> When I organized this speech, I wanted to follow a logical pathway that would maximize my audience members' comprehension of why physician suicide is happening, the effects of this epidemic, and finally how my audience could evoke positive change.
>
> I chose to begin with an introduction that capitalizes on pathos because I could share many statistics, but numbers are impersonal. Thus, I began my speech with an anecdote and continued to share heart-wrenching stories throughout the remainder of my speech because physician suicide ultimately affects people.
>
> I personally prefer the cause-effect-solution organizational structure because my first question is always, "Why is this problem happening?" After answering this question, I explain the effects of this problem, which reminds the audience that this problem has implications greater than what one may initially expect. The effects section serves to show my audience how this problem will impact their lives.[31]

A full outline for Barclay's speech might appear like the one below. This is an expansion of the outline shown on page 322 and the speaking notes on page 324. Numbers in parentheses correspond to the numbered paragraphs of the speech.

Speech Outline

Physician Suicide: America's Silent Epidemic

Introduction

 I. Attention-getter: 23-year-old medical student Kaitlyn Elkins purposefully ingested pressurized helium, resulting in immediate brain death. [1–2]

 II. Thesis Statement: physician suicide is an epidemic that threatens both practicing and training physicians. [3]

 III. Preview Main Points: let's examine the causes, effects, and solutions to a medical emergency that equates to an entire medical school class dying, every year. [3]

Body

 I. Physician suicide has two main causes (4–10)
   A. Hazing culture plagues our medical institutions. (5–7)
   B. Assembly-line medicine is a parasite on physicians' lives. (8–10)

 II. Physician suicide affects us all. (11–15)
   A. The medical field's unrelenting demands lead to disguised mental anguish. (11–13)
   B. Increasing expectations lead to physician burnout. (14–15)

 III. Physician suicide can be solved. (16–18)
   A. We can take public action. (16)
     1. We must educate the public. (17)
     2. One way to do that is to show the documentary *Do No Harm*. (17–18)

B. We can take individual action. (19)

    1. We must care for our doctors. (19)

    2. We must thank them when they help us. (19–20)

Conclusion

  I. Final Remarks (21)

  II. Review (22)

Regarding his supporting material, Hunter says:

> I focused on the studies of leading scholars who have researched physician suicide. I also desired to incorporate diverse forms of supporting evidence: TEDx Talks, journal articles, books, and blog posts. I also attempted to include recent sources, which reinforces the relevance of this issue.

The thoroughness of Hunter's research is reflected in his bibliography:

Bibliography

Andrew, L., & Brenner, B. (2017). Physician Suicide. *Medscape.*

Beloff, R. (2016). A Call to Arms: On the Need for Clinical Leadership. *Des Moines University Magazine.*

Bright, R. (2011). Depression and Suicide Among Professions. *Current Psychiatry.*

In Memory of My Daughter, Kaitlyn Nicole Elkins. (2013). *Journeys through Grief.* Retrieved December 23, 2017, from https://journeysthrugrief.wordpress.com/2013/09/27/in-memory-of-my-daughter-elkins

Jauhar, S. (2014). *Doctored: The Disillusionment of an American Physician.* New York: Farrar, Straus and Giroux.

Kishore, S. (2016). Breaking the Culture of Silence on Physician Suicide. *National Academy of Medicine.*

Major, A. (2014). The Problem of Mistreatment in Medical Education. *AMA Journal of Ethics.*

Oaklander, M. (2015). Doctors on Life Support. *Time.* Retrieved December 23, 2017, from http://time.com/4012840/doctors-on-life-support

Simon, R. (2018). *Do No Harm.* Retrieved March 15, 2018, from http://donoharmfilm.com

Suicide in Physicians. (2009). *Academy of Medicine.*

Wible, P. (2018). *What I've Learned from 757 Doctor Suicides.* Retrieved April 14, 2018, from http://www.idealmedicalcare.org/ive-learned-547-doctor-suicides

Wible, P. (2015). *Why Doctors Kill Themselves. TEDMED.* Retrieved December 21, 2017, from https://www.tedmed.com/talks/show?id=528918

Winakur, J. (2016). *In America, the Art of Doctoring Is Dying. Washington Post.* Retrieved December 26, 2017, from https://www.washingtonpost.com/opinions/the-dying-art-of-doctoring/2016/02/12/bb08a16a-cdd0-11e5-88cd-753e80cd29ad_story.html?noredirect=on&utm_term=.e88b69d3cdd0

Wozniczka, D. (2017). *Millennials in Medicine: Doctors of the Future. TEDx.* Retrieved December 23, 2017, from https://www.youtube.com/watch?v=Kykj3k2wBXg

Physician Suicide: America's Silent Epidemic

1   23-year-old medical student Kaitlyn Elkins was the definition of a star student: valedictorian, graduated from college in 2.5 years, and aced her medical school exams. But as a third-year medical student at Wake Forest University, Kaitlyn purposefully ingested pressurized helium, resulting in immediate brain death.

He begins his speech with a striking, attention-getting example.

2   Dr. Louise Andrews explains in a June 12, 2017, MedScape article that Kaitlyn's death is not an anomaly because approximately 300 to 400 physicians and medical students kill themselves each year, meaning we are losing a doctor per day to suicide. Family physician Dr. Pamela Wible further states, "Their suicides were like well-planned school projects. They're perfectionists. They're very good at suicide, and they know how to do it."

He follows up with recent, credible statistics about his problem.

3   Clearly, physician suicide is an epidemic that threatens both practicing and training physicians. Therefore, let's examine the causes, effects, and solutions to a medical emergency that Dr. Daniel Wozniczka equates in a June 2017 TED Talk, to an entire medical school class dying, every year.

Statement of thesis and review of main points.

4   Skillfully puncturing the arteries to his hands and feet, Dr. Greg Miday succumbed to the pressures of his oncology fellowship. Dr. Miday's suicide exemplifies two causes of the physician suicide epidemic: hazing culture and assembly-line medicine.

Introduces his first main point with another striking example.

5   First, hazing culture plagues our medical institutions. University of Michigan psychiatrist Dr. Srijan Sen elaborates in *Time Magazine* of August 27, 2015, that the rate of depression among medical students is approximately 4% following the first two years of medical school—a rate on par with the national average. That number, however, skyrockets to 25% as the students leave the classroom and begin their hospital internships.

First subpoint, backed up with recent credible statistics.

6   What prompts this drastic increase? Dr. Sen explains that medical interns are paid very little, yelled at a lot, and become scapegoats when things go wrong. A March 2014 *Journal of Ethics* publication elucidates that a strange machismo pervades medicine where doctors feel pressure to project intellectual, emotional, and physical prowess beyond their individual capabilities.

Subpoint explained with expert testimony.

7   The publication further states that 81% of medical students reported physical or emotional abuse throughout medical school—a rate of abuse greater than those of America's most violent prisons. This barbaric hazing culture pushes our nation's healers into a dangerous downward spiral.

Support in the form of an analogy.

8   Second, assembly-line medicine is a parasite on physician's lives. Dr. Sandeep Jahuar explains in his 2014 book *Doctored: The Disillusionment of an American Physician* that today's physicians see themselves not as the "pillars of any community" but as "technicians on an assembly line" who are pressured to see many patients in a small amount of time . . . because time is money.

Second subpoint introduced with more expert testimony.

9   In the April 2016 edition of *Des Moines University Magazine*, Professor Richard Belloff exposes that in 1985, 85% of physicians had no doubt about their career choice. Conversely, in 2008, that number plummeted to 6% of doctors. Retired family physician Dr. Jerald Winakur concludes in a February 12, 2016, *Washington Post* article that assembly-line medicine depersonalizes patients and removes doctors from the intimate doctor–patient relationship.

Further development with statistics and testimony.

10   Assembly-line medicine makes doctors slaves to profit-seeking medical executives who prize profits over patients. Perhaps these business-minded executives should be reminded that the customer, in this case the patient, always comes first.

Summing up this subpoint as a transition.

11   Once voted the most compassionate physician at Columbia-Presbyterian Hospital, New York Dr. Doug Meyer shocked his coworkers when he jumped 17 stories to his death, revealing two effects of the medical field's unyielding demands on doctors: disguised mental anguish and physician burnout.

Second main point—effects—introduced with another striking example.

12   First, the medical field's unrelenting demands lead to disguised mental anguish. A June 2009 Academy of Medicine publication reveals that physicians are unlikely to seek professional help because approximately 70% of medical boards require doctors to disclose any mental diagnosis or treatment in order for them to renew their licenses.

First subpoint introduced with a well-chosen statistic.

13   Instead of providing struggling physicians with confidential access to mental health care, these medical boards choose to revoke their licenses. Mayo Clinic Dr. Robert Bright further states in an April 2011 *Current Psychiatry* publication that depressed physicians are six times more likely to make a medication error than their peers are. Thus, concealed depression extends to unsuspecting patients, like you and me. When medical boards prize physician silence over treatment, they endanger both doctors' and patients' lives, turning the Hippocratic Oath into the Hypocritic Hoax.

Further developing this point with a statistic, Hunter relates his topic to his audience's well-being.

14   Second, increasing expectations lead to physician burnout. In her October 28, 2017 article titled "What I've Learned from 757 Doctor Suicides," Dr. Pamela Wible highlights that pressure from insurance companies and government mandates crush many doctors who cite inhumane working conditions in their suicide notes. She explains that this epidemic's primary victims are female doctors who are 300 percent more likely to kill themselves than a female in the general population.

Second subpoint introduced with testimony and statistics.

15   A June 2016 National Academy of Medicine report reveals that sleep deprivation, extensive paperwork, looming malpractice suits, and life and death decisions lead to one in three physicians experiencing burnout at any given time. Knowing that one in three of your doctors is currently burned out, would you really place your complete confidence in their medical decisions?

A further analysis reinforces the importance of this topic to his audience.

16   Although we usually rely on our doctors to be the lifesavers, it is time to turn the tables and save our doctors from a disease that does not appear on any lab report. Only public and individual action can prompt necessary change.

Transition to his final main point—Solutions.

17   Initially, we must creatively educate the public about this epidemic. Two-time Emmy award-winning director Robyn Symon has directed *Do No Harm*—a documentary exploring why physicians have the highest suicide rate among any profession. The documentary is in its final stages of production and will be released next month. Please pick up a film poster and scan the QR code to view the film's trailer.

First subpoint.

18   The trailer includes disturbing footage, but now is not the time to turn a blind eye. When this film becomes available, I urge you to follow my example and to organize a screening for your college or community. If you are interested in hosting a public screening, the QR code also includes links to promotional materials that you can share with your family, friends, and medical providers.

19   Individually, we must care for our doctors. Healing our healers begins with us—the patients. During your next doctor visit, I encourage you to leave one of these cards in the examination room or waiting room. Each card includes a biography about a doctor or medical student who chose to take their lives. Their stories put names and faces to this epidemic, showing our doctors they are not alone in their struggles.

Second subpoint

20   Moreover, each card includes a phone number to Dr. Pamela Wible's suicide hot-line. This hotline is specifically for doctors who want to talk to someone who can empathize with their unique concerns. Finally, each card has a few blank lines for you to write an encouraging message to your doctor. Your message could be simple "thank you" or a heartfelt compliment. Breast cancer surgeon Dr. Andrew Higgins states that gratitude notes from his patients are motivational wonders. Please, take a card and let your doctors know that their lives matter.

Here he shows the cards to his audience.

21   I am a pre-med student, and physician suicide scares me more than any other illness. When I look at my fellow classmates, I fear some of us could become the next Kaitlyn Elkins, the next Dr. Greg Miday, the next Dr. Doug Meyer, the next victim of this silent epidemic.

Final remarks

22   Today, we have shattered this silence by examining the causes, effects, and solutions to America's physician suicide epidemic. I want to become a doctor to save lives, but aspiring doctors, like myself, should not have to fear becoming a side effect of our profession.

Review of main points.

MAKING THE GRADE

At www.oup.com/he/adler-uhc14e, you will find a variety of resources to enhance your understanding, including video clips, animations, self-quizzes, additional activities, audio and video summaries, interactive self-assessments, and more.

OBJECTIVE 13.1   Describe different types of speech outlines and their functions.

- A working outline is used to map out the structure of your speech, in rough form.
- A formal outline uses a standard format and a consistent set of symbols.
- Speaking notes are used to jog your memory while giving a speech. Like a working outline, they are for your eyes only.
  > Explain the primary differences among working outlines, formal outlines, and speaking notes.
  > Which type of outline would be most important for your next speech?
  > What style of speaking notes would you use for your next speech?

OBJECTIVE 13.2   Construct an effective speech outline using the organizing principles described in this chapter.

- Principles for the effective construction of outlines are based on the use of standard symbols and a standard format.
- The rule of division requires at least two divisions of every point or subpoint.

- The rule of parallel wording requires that points at each level of division be worded in a similar manner whenever possible.
  > What are the standard symbols used in a formal outline?
  > Which principle of outlining is least intuitive for you? Which one is most intuitive?
  > How would you divide your next speech into main points and subpoints?

OBJECTIVE 11.3   Choose an appropriate organizational pattern for your speech.

- The organization of ideas should follow a pattern, such as time, space, topic, problem-solution, cause-effect, or Monroe's motivated sequence.
- Each pattern of organization has its own advantages in helping your audience understand and remember what you have to say.
  > Why is it important to organize your ideas according to a pattern?
  > When should you use each of the six patterns of organization discussed in this chapter?
  > Which pattern of organization would be best for your next speech?

OBJECTIVE 13.4   Develop a compelling introduction and conclusion, and use effective transitions at key points in your speech.

- The main idea of the speech is established in the introduction, developed in the body, and reviewed in the conclusion.

- The introduction will also gain the audience's attention, preview the main points, set the mood and tone of the speech, and demonstrate the importance of the topic to the audience.
- The conclusion will review your main points and supply the audience with a memory aid in the form of compelling final remarks.
- Effective transitions keep your message moving forward and demonstrate how your ideas are related.
  - > Why are the beginning and end of a speech so important? Why are transitions?
  - > What would be the most important function of the introduction of your next speech?
  - > What idea would you like to leave your audience with in your next speech?

OBJECTIVE 13.5 Choose supporting material that will help make your points clear, interesting, memorable, and convincing.

- Supporting materials are the facts and information you use to back up what you say.
- Types of support include *definitions, examples, statistics, analogies, anecdotes, quotations,* and *testimony.*
- Support may be narrated or cited.
  - > Give examples of five forms of support that could be used for a speech on financing a college education.
  - > Which form of support do you find to be most effective for most speeches?
  - > What are the main forms of support that you would use for your next speech?

KEY TERMS

analogy p. 338

anecdote p. 338

basic speech structure p. 320

cause-effect pattern p. 328

citation p. 339

climax pattern p. 327

conclusion (of a speech) p. 332

factual example p. 336

formal outline p. 322

hypothetical example p. 336

introduction (of a speech) p. 330

narration p. 339

problem-solution pattern p. 328

space pattern p. 327

statistic p. 337

testimony p. 339

time pattern p. 326

topic pattern p. 327

transition p. 333

working outline p. 321

ACTIVITIES

1. **Dividing Ideas** For practice in the principle of division, divide each of the following into three to five subcategories:
   a. Clothing
   b. Academic studies
   c. Crime
   d. Health care
   e. Fun
   f. Charities

2. **Organizational Effectiveness** Take any written statement at least three paragraphs long that you consider effective. This statement might be an editorial in your local newspaper, a short magazine article, or even a section of one of your textbooks. Outline this statement according to the rules discussed here. Was the statement well organized? Did its organization contribute to its effectiveness?

3. **The Functions of Support** For practice in recognizing the functions of support, identify three instances of support in each of the speeches at the end of Chapters 13 and 14. Explain the function of each instance of support. (Keep in mind that any instance of support *could* perform more than one function.)

Informative Speaking

CHAPTER OUTLINE

14.1 Types of Informative Speaking 349
By Content
By Purpose

14.2 Informative Versus Persuasive Topics 351
Type of Topic
Speech Purpose

14.3 Techniques of Informative Speaking 351
Define a Specific Informative Purpose
Create Information Hunger
Make It Easy to Listen
Use Clear, Simple Language
Use a Clear Organization and Structure

14.4 Using Supporting Material Effectively 355
Emphasizing Important Points
Generating Audience Involvement
Using Visual Aids
Using Presentation Software
Alternative Media for Presenting Graphics
Rules for Using Visual Aids

14.5 Sample Speech 363

MAKING THE GRADE 367
KEY TERMS 368
ACTIVITIES 368

LEARNING OBJECTIVES

14.1

Distinguish among the main types of informative speaking.

14.2

Describe the differences between informative and persuasive speaking.

14.3

Outline techniques that increase the effectiveness of informative speeches.

14.4

Explain how you might use visual aids appropriately and effectively.

Lera Boroditsky

demonstrates that highly technical information can be presented in a clear, interesting manner to listeners unfamiliar with a topic.

What can you learn from successful informative speakers you have seen?

What knowledge do you possess that could form the basis for an informative speech? What additional research might you need to do?

How could you develop such a speech by applying the information in this chapter?

LERA BORODITSKY IS BRILLIANT, and she gives a great informative speech. Those traits don't always go together. You've probably heard speakers who are accomplished in their field but whose explanations are confusing and sometimes downright incomprehensible to nonexperts.

Boroditsky is an associate professor of cognitive science at the University of California, San Diego. She was born in Belarus and grew up fluent in Russian, Belarusian, and Polish. When Boroditsky was 12 years old, her family immigrated to the United States, where she mastered English as her fourth language.

As a teenager, Boroditsky began thinking about the degree to which language differences could shape an argument and exaggerate the differences between people. After college and graduate school in cognitive psychology at Stanford University, Boroditsky became one of the best-known researchers in the field of linguistic relativity. Her research provides new insights into whether and how the language we speak might shape the way we think.

In addition to her scholarly work, Boroditsky gives lectures to the general public. One of her speeches is the sample you'll find at the end of this chapter. You'll see for yourself that Boroditsky's lecture is interesting, easy to follow, and even funny. Perhaps more importantly, it demonstrates how highly technical information can be made clear to a general audience. That's the essence of informative speaking. Even if you're not an expert, you'll face occasions when it's necessary to translate information into terms that make sense to your listeners.

There is a huge amount of information competing for our attention these days. And that amount is increasing exponentially. One expert estimates that every two days now we create as much information as we did from the dawn of civilization up until 2003.[1]

Social scientists describe a form of psychological stress known as information overload: the difficulty of sorting through all the information available to them.[2] Some experts use the term **information anxiety** for the state of being overwhelmed by information. To check how information overload affects you personally, try the self-assessment in this introduction.

information anxiety The psychological stress of dealing with too much information.

The informative speaker's responsibility is not just to provide new information. When it's done effectively, informative speaking relieves information overload by turning information into knowledge for an audience. Information is the raw material, the sometimes contradictory statements and competing claims you encounter. Knowledge results from being able to make sense of that raw material. Effective public speakers filter, organize, and illustrate information to reach audiences with tailored messages, in an environment where they can see if the audience is understanding. If they aren't, the speaker can adjust as needed.

UNDERSTANDING YOUR COMMUNICATION

Are You Overloaded?

Problems in informative speaking are often the result of information overload—on both the speaker's and the audience's parts. For each statement to the right, select "often," "sometimes," or "seldom."

1. I forget information I need to know.

    OFTEN   SOMETIMES   SELDOM

2. I have difficulty concentrating on important tasks.

    OFTEN   SOMETIMES   SELDOM

3. When I go online, I feel anxious about the work I don't have time to do.

    OFTEN   SOMETIMES   SELDOM

4. I have email messages sitting in my inbox that are more than 2 weeks old.

    OFTEN   SOMETIMES   SELDOM

5. I constantly check my online services because I am afraid that if I don't, I will never catch up.

    OFTEN   SOMETIMES   SELDOM

6. I find myself easily distracted by things that allow me to avoid work I need to do.

    OFTEN   SOMETIMES   SELDOM

7. I feel fatigued by the amount of information I encounter.

    OFTEN   SOMETIMES   SELDOM

8. I delay making decisions because of too many choices.

    OFTEN   SOMETIMES   SELDOM

9. I make wrong decisions because of too many choices.

    OFTEN   SOMETIMES   SELDOM

10. I spend too much time seeking information that is *nice to know* rather than information I *need to know*.

    OFTEN   SOMETIMES   SELDOM

Scoring: Give yourself 3 points for each "often," 2 points for each "sometimes," and 1 point for each "seldom." If your score is:

10–15: Information overload is not a big problem for you. However, it's probably still a significant problem for at least some members of your audience, so try to follow the guidelines for informative speaking outlined in this chapter.

16–24: You have a normal level of information overload. The guidelines in this chapter will help you be a more effective speaker.

25–30: You have a high level of information overload. Along with observing the guidelines in this chapter, you might also want to search online for guidelines to help you overcome this problem.

Informative speaking goes on all around you: in your professors' lectures or in a mechanic's explanation of how to keep your car from breaking down. You engage in this type of speaking frequently whether you realize it or not. Sometimes it is formal, as when you give a report in class. At other times, it is more casual, as when you tell a friend how to prepare your favorite dish. The main objective of this chapter is to give you the skills you need to enhance all of your informative speaking.

14.1 Types of Informative Speaking

There are several types of informative speaking. The primary types have to do with the content and purpose of the speech.

By Content

Informative speeches are generally categorized according to their content and include the following types.

Speeches About Objects   This type of informative speech is about anything that is tangible (that is, capable of being seen or touched). Speeches about objects might include an appreciation of the Grand Canyon (or any other natural wonder) or a demonstration of the newest smartphone (or any other product).

Speeches About Processes   A process is any series of actions that leads to a specific result. If you spoke on the process of aging, the process of learning to juggle, or the process of breaking into a social networking business, you would be giving this type of speech.

Speeches About Events   You would be giving this type of informative speech if your topic dealt with anything notable that happened, was happening, or might happen: an upcoming protest against hydraulic fracturing ("fracking"), for example, or the prospects of your favorite baseball team winning the national championship.

Speeches About Concepts   Concepts include intangible ideas, such as beliefs, theories, ideas, and principles. The sample speech at the end of this chapter deals with such a concept. If you gave an informative speech about postmodernism, vegetarianism, or any other "ism," you would be giving this type of speech. Other topics would include everything from New Age religions to theories about extraterrestrial life to rules for making millions of dollars.

By Purpose

We also distinguish among types of informative speeches depending on the speaker's purpose. We ask, "Does the speaker seek to describe, explain, or instruct?"

description A type of speech that uses details to create a "word picture" of something's essential factors.

explanations Speeches or presentations that clarify ideas and concepts already known but not understood by an audience.

instructions Remarks that teach something to an audience in a logical, step-by-step manner.

Descriptions   A speech of **description** is the most straightforward type of informative speech. You might introduce a new product like a wearable computer to a group of customers, or you might describe what a career in nursing would be like. Whatever its topic, a descriptive speech uses details to create a "word picture" of the essential factors that make that thing what it is.

Explanations   **Explanations** clarify ideas and concepts that are already known but not understood by an audience. For example, your audience members might already know that a U.S. national debt exists, but they might be baffled by the reasons why it has become so large. Explanations often deal with the question of *why* or *how*. Why do we have to wait until the age of 21 to drink legally? How did China evolve from an impoverished economy to a world power in a single generation? Why did tuition need to be increased this semester?

Instructions   **Instructions** teach something to the audience in a logical, step-by-step manner. They are the basis of training programs and orientations. They often deal with the question of *how to*. This type of speech sometimes features a demonstration or a visual aid. Thus, if you were giving instructions on "how to promote your career via social networking sites," you might demonstrate by showing the social media profile of successful people. For instructions on "how to perform CPR," you could use a volunteer or a dummy.

These types of informative speeches aren't mutually exclusive. As you'll see in the sample speech at the end of this chapter, there is considerable overlap, as when you give a speech about a complex concept that has the purpose of explaining it. Still, even this imperfect categorization demonstrates how wide a range of informative topics is available. One final distinction we need to make, however, is the difference between an informative and a persuasive speech topic.

🔵 14.2 Informative Versus Persuasive Topics

There are many similarities between an informative and a persuasive speech. In an informative speech, for example, you are constantly trying to "persuade" your audience to listen, understand, and remember. In a persuasive speech, you "inform" your audience about your arguments, your evidence, and so on. Nonetheless, two basic characteristics differentiate an informative topic from a persuasive topic.

Type of Topic

In an informative speech, you generally do not present information that your audience is likely to disagree with. Again, this is a matter of degree. For example, you might want to give a purely informative talk on the differences between hospital births and home-based midwife births by simply describing what the practitioners of each method believe and do. By contrast, a talk either boosting or criticizing one method over the other would clearly be persuasive.

The noncontroversial nature of informative speaking doesn't mean that your speech topic should be uninteresting to your audience; rather, it means that your approach to it should not engender conflict. You could speak about the animal rights movement, for example, by explaining the points of view of both sides in an interesting but objective manner.

Speech Purpose

The informative speaker does seek a response (such as attention and interest) from the listener and does try to make the topic important to the audience. But the speaker's primary intent isn't to change attitudes or to make the audience members *feel* differently about the topic. For example, an informative speaker might explain how facial recognition software works but will not try to change attitudes about whether this technology is a boon or a threat.

The speaker's intent is best expressed in a specific informative purpose statement, which brings us to the first of our techniques of informative speaking.

🔵 14.3 Techniques of Informative Speaking

The techniques of informative speaking are based on a number of principles covered in earlier chapters. The most important principles to apply to informative speaking include those that help an audience understand and care about your speech. Let's look at how these principles apply to specific techniques.

Define a Specific Informative Purpose

As Chapter 12 explained, any good speech must be based on a purpose statement that is audience oriented, precise, and attainable. When you are preparing an informative speech, it is especially important to define in advance, for yourself, a clear informative purpose. An **informative purpose statement** will generally be worded to stress audience knowledge, ability, or both:

> After listening to my speech, my audience will be able to recall the three most important questions to ask when shopping for a smartphone.

informative purpose statement
A complete statement of the objective of a speech, worded to stress audience knowledge and/or ability.

How Culture Affects Information

Cultural background is always a part of informative speaking, although it's not always easy to spot. Sometimes this is because of *ethnocentrism*, the belief in the inherent superiority of one's own ethnic group or culture. According to communication scholars Larry Samovar and Richard Porter, ethnocentrism is exemplified by how subjects are taught in schools.

Each culture, whether consciously or unconsciously, tends to glorify its historical, scientific, and artistic accomplishments while frequently minimizing the accomplishments of other cultures. In this way, schools in all cultures, whether or

not they intend to, teach ethnocentrism. For instance, the next time you look at a world map published in the USA, notice that the United States is prominently located in the center. Would this be so if you were looking at a Chinese or Russian map? Similarly, many students in the United States, if asked to identify the great books of the world, would likely produce a list of books by mainly Western authors.

From Samovar, L., & Porter, R. (2010). *Communication Between Cultures* (5th ed.). Boston: Wadsworth. © 2010. Reprinted with permission of Wadsworth, an imprint of the Wadsworth Group, a division of Cengage Learning.

After listening to my speech, my audience will be able to identify the four reasons that online memes go viral.

After listening to my speech, my audience will be able to discuss the pros and cons of using drones in warfare.

Notice that in each of these purpose statements a specific verb such as *to recall*, *to identify*, or *to discuss* points out what the audience will be able to do after hearing the speech. Other key verbs for informative purpose statements include these:

Accomplish	Contrast	Integrate	Perform
Analyze	Describe	List	Review
Apply	Explain	Name	Summarize
Choose	Identify	Operate	

A clear purpose statement will lead to a clear thesis statement. As you remember from Chapter 12, a thesis statement presents the central idea of your speech. Sometimes your thesis statement for an informative speech will just preview the central idea:

Today's smartphones have so many features that it is difficult for the uninformed consumer to make a choice.

Understanding how memes go viral could make you very wealthy someday.

Soldiers and civilians have different views on the morality of drones.

At other times, the thesis statement for an informative speech will delineate the main points of that speech:

When shopping for a smartphone, the informed consumer seeks to balance price, dependability, and user friendliness.

The four basic principles of aerodynamics—lift, thrust, drag, and gravity—can explain why memes go viral.

Drones can save warrior lives but cost the lives of civilians.

Setting a clear informative purpose will help keep you focused as you prepare and present your speech.

Create Information Hunger

An effective informative speech creates **information hunger**: a reason for your audience members to want to listen to and learn from your speech. To do so, you can use the analysis of communication functions discussed in Chapter 1 as a guide. You read there that communication of all types helps us meet our physical needs, identity needs, social needs, and practical needs. In informative speaking, you could tap into your audience members' physical needs by relating your topic to their survival or to the improvement of their living conditions. If you gave a speech on food (eating it, cooking it, or shopping for it), you would be dealing with that basic physical need. In the same way, you could appeal to identity needs by showing your audience members how to be respected—or simply by showing them that you respect them. You could relate to social needs by showing them how your topic could help them be well liked. Finally, you can relate your topic to practical audience needs by telling your audience members how to succeed in their courses, their job search, or their quest for the perfect outfit.

Make It Easy to Listen

Keep in mind the complex nature of listening, discussed in Chapter 5, and make it easy for your audience members to hear, pay attention, understand, and remember. This means first that you should speak clearly and with enough volume to be heard by all your listeners. It also means that as you put your speech together, you should take into consideration techniques that recognize the way human beings process information.

Limit the Amount of Information You Present   Remember that you probably won't have enough time to transmit all your research to your audience in one sitting. It's better to make careful choices about the three to five main ideas you want to get across and then develop those ideas fully. Remember, too much information leads to overload, anxiety, and lack of attention on the part of your audience.

Use Familiar Information to Increase Understanding of the Unfamiliar
Move your audience members from familiar information (on the basis of your audience analysis) to your newer information. For example, if you are giving a speech about how the stock market works, you could compare the daily activity of a broker with that of a salesperson in a retail store, or you could compare the idea of capital growth (a new concept to some listeners) with interest earned in a savings account (a more familiar concept).

Use Simple Information to Build Up Understanding of Complex Information   Just as you move your audience members from the familiar to the unfamiliar, you can move them from the simple to the complex. An average college audience, for example, can understand the complexities of genetic modification if you begin with the concept of inherited characteristics.

Use Clear, Simple Language

Another technique for effective informative speaking is to use clear language, which means using precise, simple wording and avoiding jargon. As you plan your speech, consult online dictionaries such as Dictionary.com to make sure you are selecting precise vocabulary.

> **information hunger** Audience desire, created by a speaker, to learn information.

CHECKLIST ✓

Techniques of Informative Speaking

☐ **Define a specific informative purpose.**

☐ **Create information hunger by relating to audience needs.**

☐ **Make it easy for audience members to listen.**
- Limit the amount of information presented.
- Use familiar information to introduce unfamiliar information.
- Start with simple information before moving to more complex ideas.

☐ **Use clear, simple language.**

☐ **Use clear organization and structure.**

☐ **Support and illustrate your points.**
- Provide interesting, relevant facts and examples, citing your sources.
- Use visual aids that help make your points clear, interesting, and memorable.

☐ **Emphasize important points.**
- Repeat key information in more than one way.
- Use signposts: words or phrases that highlight what you are about to say.

☐ **Generate audience involvement.**
- Personalize the speech.
- Use audience participation.
- Use volunteers.
- Have a question-and-answer period at the end.

Remember that picking the right word seldom means using a word that is unfamiliar to your audience; in fact, just the opposite is true. Important ideas do not have to sound complicated. Along with simple, precise vocabulary, you should also strive for direct, short sentence structure. For example, when Warren Buffett, one of the world's most successful investors, wanted to explain the impact of taxes on investing, he didn't use unusual vocabulary or complicated sentences. He explained it like this:

> Suppose that an investor you admire and trust comes to you with an investment idea. "This is a good one," he says enthusiastically. "I'm in it, and I think you should be, too." Would your reply possibly be this? "Well, it all depends on what my tax rate will be on the gain you're saying we're going to make. If the taxes are too high, I would rather leave the money in my savings account, earning a quarter of 1 percent." So let's forget about the rich and ultrarich going on strike and stuffing their ample funds under their mattresses if—gasp—capital gains rates and ordinary income rates are increased. The ultrarich, including me, will forever pursue investment opportunities.[3]

Each idea within that explanation is stated directly, using simple, clear language.

Use a Clear Organization and Structure

Because of the way humans process information (that is, in a limited number of chunks at any one time),[4] organization is extremely important in an informative speech. Rules for structure may be mere suggestions for other types of speeches, but for informative speeches they are ironclad.

Chapter 12 discusses some of these rules:

- Limit your speech to three to five main points.
- Divide, coordinate, and order those main points.
- Use a strong introduction that previews your ideas.
- Use a conclusion that reviews your ideas and makes them memorable.
- Use transitions, internal summaries, and internal previews.

The repetition that is inherent in strong organization will help your audience members understand and remember those points. This will be especially true if you use a well-organized introduction, body, and conclusion.

The Introduction   The following principles of organization from Chapter 12 become especially important in the introduction of an informative speech:

1. Establish the importance of your topic to your audience.
2. Preview the thesis, the one central idea you want your audience to remember.
3. Preview your main points.

For example, Kevin Allocca, the trends manager at YouTube, began his TED Talk "Why Videos Go Viral" with the following introduction:

> I professionally watch YouTube videos. It's true. So we're going to talk a little bit today about how videos go viral and then why that even matters. Web video has made it so that any of us or any of the creative things that we do can become completely famous in a part of our world's culture. Any one of you could be famous on the Internet by next Saturday. But there are over 48 hours of video uploaded to YouTube every minute. And of that, only a tiny percentage ever goes viral and gets tons of views and becomes a cultural moment.

ASK YOURSELF

Do you organize speeches in the order ideas occur to you, or do you plan the organization more strategically? How could you organize your speeches more effectively?

So how does it happen? Three things: tastemakers, communities of participation, and unexpectedness.[5]

The Body   In the body of an informative speech, the following organizational principles take on special importance:

1. Limit your division of main points to three to five subpoints.

2. Use transitions, internal summaries, and internal previews.

3. Order your points in the way that they will be easiest to understand and remember.

Kevin Allocca followed these principles for organizing his speech on why some videos go viral and some do not. He developed his speech with the following three main points:

1. Tastemakers: Tastemakers like Jimmy Kimmel introduce us to new and interesting things and bring them to a larger audience.

2. Communities of participation: A community of people who share this big inside joke start talking about it and doing things with it.

3. Unexpectedness: In a world where more than two days of video get uploaded every minute, only those that are truly unique can go viral.

The Conclusion   Organizational principles are also important in the conclusion of an informative speech:

1. Review your main points.

2. Remind your audience members of the importance of your topic to them.

3. Provide your audience with a memory aid.

For example, this is how Kevin Allocca concluded his speech on viral videos:

> Tastemakers, creative participating communities, complete unexpectedness, these are characteristics of a new kind of media and a new kind of culture where anyone has access and the audience defines the popularity. One of the biggest stars in the world right now, Justin Bieber, got his start on YouTube. No one has to green-light your idea. And we all now feel some ownership in our own pop culture. And these are not characteristics of old media, and they're barely true of the media of today, but they will define the entertainment of the future.

🎧14.4 Using Supporting Material Effectively

Another technique for effective informative speaking has to do with the supporting material discussed in Chapter 12. All of the purposes of support (to clarify, to prove, to make interesting, to make memorable) are essential to informative speaking. Therefore, you should be careful to support your thesis in every way possible. Notice the way Lera Boroditsky uses solid supporting material in the sample speech at the end of this chapter. In particular, notice her use of striking photographs, which can grab your audience members' attention and keep them attuned to your topic throughout your speech.

You should also try to include **vocal citations**, or brief explanations of where your supporting material came from. These citations build the credibility of your

vocal citation A concise, spoken statement of the source of your evidence.

explanations and increase audience trust in the accuracy of what you are saying. For example, when Brian Davis of Orange Coast College gave a speech on the importance of black male teachers, he used the following vocal citation:

> Not only do black students continue to statistically underperform, they are also antagonized. Slate on June 9, 2016, posits that there is "a tendency of white teachers to grade black and Latino students more harshly." This explains the 22 percent achievement gap between white students and students of color, showing the consequences of having a faculty with little to no diversity.[6]

By telling concisely and simply where his information came from, Brian reassured his audience that his facts were credible.

Emphasizing Important Points

One specific principle of informative speaking is to stress the important points in your speech through repetition and the use of signposts.

Repetition   Repetition is one of the age-old rules of learning. Human beings are more likely to comprehend information that is stated more than once. This is especially true in a speaking situation, because, unlike a written paper, your audience members cannot go back to reread something they have missed. If their minds have wandered the first time you say something, they just might pick it up the second time.

Of course, simply repeating something in the same words might bore the audience members who actually are paying attention, so effective speakers learn to say the same thing in more than one way, sometimes by adding an additional example. Shayla Cabalan, a student at the University of Indianapolis, used this technique in her speech on child marriage:

> Parents who marry off their minor children are often motivated by the need to control sexuality and unwanted behavior, prevent unsuitable relationships, protect family honor and perceived religious or cultural ideals, or achieve financial gain. The *Washington Post* of February 10, 2017, details the story of 15-year-old Sara Siddiqui of Nevada, whose father married her off to a man 13 years her senior, simply because she was seeing someone of a different cultural background.[7]

Redundancy can be effective when you use it to emphasize important points.[8] It is ineffective only when (1) you are redundant with obvious, trivial, or boring points or (2) you run an important point into the ground. There is no sure rule for making certain you have not overemphasized a point. You just have to use your best judgment to make sure that you have stated the point enough that your audience members get it without repeating it so often that they want to give it back.

signpost A phrase that emphasizes the importance of upcoming material in a speech.

audience involvement The level of commitment and attention that listeners devote to a speech.

Signposts   Another way to emphasize important material is by using **signposts**: words or phrases that emphasize the importance of what you are about to say. You can state, simply enough, "What I'm about to say is important," or you can use some variation of that statement: "But listen to this . . . ," or "The most important thing to remember is . . . ," or "The three keys to this situation are . . . ," and so on.

Generating Audience Involvement

The final technique for effective informative speaking is to get your audience involved in your speech. **Audience involvement** is the level of commitment and attention that listeners devote to a speech. Educational psychologists have long known that the best way to teach people something is to have them do it; social psychologists have added to this rule by proving, in many studies, that involvement in a message increases audience comprehension of, and agreement with, that message.

There are many ways to encourage audience involvement in your speech. One way is to follow the rules for good delivery by maintaining enthusiasm, energy, eye contact, and so on. Other ways include personalizing your speech, using audience participation, using volunteers, and having a question-and-answer period.

Personalize Your Speech   One way to encourage audience involvement is to give audience members a human being to connect to. In other words, don't be afraid to be yourself and to inject a little of your own personality into the speech. If you happen to be good at storytelling, make a narration part of your speech. If humor is a personal strength, be funny. If you feel passion about your topic, show it. Certainly, if you have any experience that relates to your topic, use it.

Kathryn Schulz, author of *Being Wrong* and a self-proclaimed "wrongologist," personalized her TED speech, "Being Wrong," this way:

> I'm in college, and a friend and I go on a road trip from Providence, Rhode Island, to Portland, Oregon. And you know, we're young and unemployed, so we do the whole thing on back roads through state parks and national forests— basically the longest route we can possibly take. And somewhere in the middle of South Dakota, I turn to my friend and I ask her a question that's been bothering me for 2,000 miles. "What's up with the Chinese character I keep seeing by the side of the road?"

> My friend looks at me totally blankly. There's actually a gentleman in the front row who's doing a perfect imitation of her look. (Laughter) And I'm like, "You know, all the signs we keep seeing with the Chinese character on them." She just stares at me for a few moments, and then she cracks up, because she figures out what I'm talking about. And what I'm talking about is this:

> Right: The Famous Chinese Character for Picnic Area.[9]

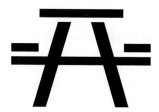

Another way to personalize your speech is to link it to the experience of audience members . . . maybe even naming one or more.

Use Audience Participation   Having your listeners actually do something during your speech—**audience participation**—is another way to increase their involvement in your message. For example, if you were giving a demonstration on isometric exercises (which don't require too much room for movement), you could have the entire audience stand up and do one or two sample exercises. If you were explaining how to fill out a federal income tax form, you could give each class member a sample form to fill out as you explain it. Outlines and checklists can be used in a similar manner for just about any speech.

Lera Boroditsky, the researcher whose sample speech appears at the end of this chapter, habitually uses audience participation in her lectures on the importance of language:

> I often start my undergraduate lectures by asking students the following question: which cognitive faculty would you most hate to lose? Most of them pick the sense of sight; a few pick hearing. Once in a while, a wisecracking student

audience participation Listener activity during a speech; a technique to increase audience involvement.

Rubes®   By Leigh Rubin

"In order to adequately demonstrate just how many ways there are to skin a cat, I'll need a volunteer from the audience."

Source: By permission of Leigh Rubin and Creators Syndicate, Inc.

might pick her sense of humor or her fashion sense. Almost never do any of them spontaneously say that the faculty they'd most hate to lose is language. Yet if you lose (or are born without) your sight or hearing, you can still have a wonderfully rich social existence. You can have friends, you can get an education, you can hold a job, you can start a family. But what would your life be like if you had never learned a language?[10]

Use Volunteers Some points or actions are more easily demonstrated with one or two volunteers. Selecting volunteers from the audience will increase the psychological involvement of all audience members because they will tend to identify with the volunteers.

Kathryn Schulz, in her speech on being wrong, subtly enlisted volunteers when she wanted to impress an important point on her audience. She began by addressing a rhetorical question to her entire audience but then directed it to a few individuals in the front row:

> So let me ask you guys something—or actually, let me ask you guys something, because you're right here: How does it feel—emotionally—how does it feel to be wrong?

Schulz then listened to the responses and repeated them for the rest of the audience:

> Dreadful. Thumbs down. Embarrassing. . . . Thank you, these are great answers, but they're answers to a different question. You guys are answering the question: How does it feel to *realize* you're wrong? When we're wrong about something—not when we realize it, but before that—*it feels like being right*.

Have a Question-and-Answer Period One way to increase audience involvement that is nearly always appropriate if time allows is to answer questions at the end of your speech. You should encourage your audience to ask questions. Solicit questions and be patient waiting for the first one. Often no one wants to ask the first question. When the questions do start coming, the following suggestions might increase your effectiveness in answering them:

1. Listen to the substance of the question. Don't **zero in on** irrelevant details; listen for the big picture, the basic, overall question that is being asked. If you are not really sure what the substance of a question is, ask the questioner to paraphrase it. Don't be afraid to let the questioners do their share of the work.

2. Paraphrase confusing or quietly asked questions. Use the active listening skills described in Chapter 5. You can paraphrase the question in just a few words: "If I understand your question, you are asking _____. Is that right?"

3. Avoid defensive reactions to questions. Even if the questioner seems to be calling you a liar or stupid or biased, try to listen to the substance of the question and not to the possible personal attack.

4. Answer the question briefly. Then check the questioner's comprehension of your answer by observing his or her nonverbal response or by asking, "Does that answer your question?"

cultural idiom

zero in on: focus directly on

Using Visual Aids

Visual aids are graphic devices used in a speech to illustrate or support ideas. Although they can be used in any type of speech, they are especially important in informative speeches. For example, they can be extremely useful when you want to show how things look (photos of your trek to Nepal or the effects of malnutrition) or how things work (a demonstration of a new ski binding or a diagram of how seawater is made drinkable). Visual aids can also show how things relate to one another (a graph showing the relationships among gender, education, and income).

There is a wide variety of types of visual aids. The most common types include the following.

Objects and Models   Sometimes the most effective visual aid is the actual thing you are talking about. This is true when the thing you are talking about is easily portable and simple enough to use during a demonstration before an audience (e.g., a lacrosse racket). **Models** are scaled representations of the object you are discussing and are used when that object is too large (the new campus arts complex) or too small (a DNA molecule) or simply doesn't exist anymore (a *Tyrannosaurus rex*).

Photos, Videos, and Audio Files   Speaking venues, including college classrooms, are often set up with digital projectors connected to online computers, DVD players, and outlets for personal laptops. In these cases, a wealth of photos, video clips, and audio files are available to enhance your speech. If you can project photos, you can use them to help explain geological formations, for example, or underwater habitats. Videos can be used to play a visual **sound bite** (a brief recorded excerpt) from an authoritative source, or to show a process such as plant growth. Using audio files can help you compare musical styles or demonstrate the differences in the sounds of gas and diesel engines.

Photos can stay up as long as you are referring to them, but videos and audio files should be used sparingly because they allow audience members to receive information passively. You want your audience to actively participate in the presentation. The general rule when using videos and sound files is *Don't let them get in the way of the direct, person-to-person contact that is the primary advantage of public speaking.*

Videos and sound files should be carefully introduced, controlled, and summarized at the end. They should also be very brief, and they should include what media professionals call **wraparounds** or **intros** and **outros:** careful introductions and closing summaries of each clip. Using these devices enables the speakers to remain in control of the message.

Other types of visual aids are charts and graphs for representing facts and data. They include the following.

Diagrams   A **diagram** is any kind of line drawing that shows the most important properties of an object. Diagrams do not try to show everything but just those parts of a thing that the audience most needs to be aware of and understand. Blueprints and architectural plans are common types of diagrams, as are maps and organizational charts. A diagram is most appropriate when you need to simplify a complex object or phenomenon and make it more understandable to the audience. Figure 14.1 shows a humorous depiction of one student's perception of what was covered on a final exam, in the form of a Venn diagram.

Word and Number Charts   **Word charts** and **number charts** are visual depictions of key facts or statistics. Your audience will understand and remember these facts and numbers better if you show them than if you just talk about them. Many speakers arrange the main points of their speech, often in outline form, as a word chart. Other speakers list their main statistics. An important guideline for word and number

visual aids Graphic devices used in a speech to illustrate or support ideas.

model (in speeches and presentations) A replica of an object being discussed. It is usually used when it would be difficult or impossible to use the actual object.

sound bite A brief recorded excerpt from a longer statement.

wraparound A brief introduction before a visual aid is presented, accompanied by a brief conclusion afterward.

intro A brief explanation or comment before a visual aid is used.

outro A brief summary or conclusion after a visual aid has been used.

diagram A line drawing that shows the most important components of an object.

word chart A visual aid that lists words or terms in tabular form in order to clarify information.

number chart A visual aid that lists numbers in tabular form to clarify information.

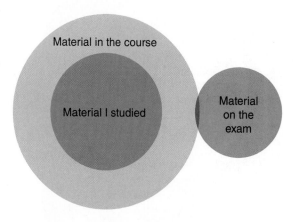

FIGURE 14.1 Venn Diagram

pie chart A visual aid that divides a circle into wedges, representing percentages of the whole.

pictogram A visual aid that conveys its meaning through an image of an actual object.

bar chart A visual aid that compares two or more values by showing them as elongated horizontal rectangles.

column chart A visual aid that compares two or more values by showing them as elongated vertical rectangles.

charts is, *Don't read them to your audience; use them to enhance what you are saying and help your audience remember key points.*

Pie Charts Circular graphs cut into with wedges are known as **pie charts**. They are used to show divisions of any whole: where your tax dollars go, the percentage of the population involved in various occupations, and so on. Pie charts are often made up of percentages that add up to 100. Usually, the wedges of the pie are organized from largest to smallest. The pie chart in Figure 14.2 represents one person's perception of "how princesses spend their time," and Figure 14.3 shows how the U.S. government adapted a pie chart for a new nutrition diagram. Coincidentally, Figure 14.3 is also a **pictogram**, which is a visual aid that conveys its meaning through images of an actual object.

Bar and Column Charts Figure 14.4 is a **bar chart**, a type of chart that compares two or more values by stretching them out in the form of horizontal rectangles. **Column charts**, such as the one shown in Figure 14.5, perform the same function as bar charts but use vertical rectangles.

Line Charts A **line chart** maps out the direction of a moving point; it is ideally suited for showing changes over time. The time element is usually placed on the horizontal axis so that the line visually represents the trend over time. Figure 14.6 is a line chart.

Flow Charts A **flow chart** is a diagram that depicts the steps in a process with shapes and arrows. Often, it branches according to yes/no decisions (if this, then this). Figure 14.7 represents one speaker's perception of how you identify "mansplaining."

Using Presentation Software

Several specialized programs exist just to produce visual aids. Among the most popular of these programs are Microsoft PowerPoint, Apple's Keynote, and Prezi.

In its simplest form, presentation software lets you build an effective slide show out of your basic outline. You can choose color-coordinated backgrounds and consistent

How Princesses Spend Their Time

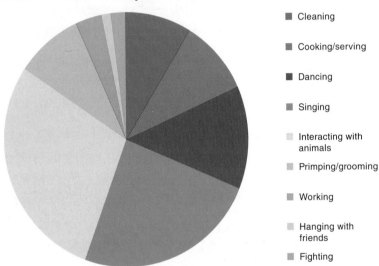

- Cleaning
- Cooking/serving
- Dancing
- Singing
- Interacting with animals
- Primping/grooming
- Working
- Hanging with friends
- Fighting

FIGURE 14.2 Pie Chart

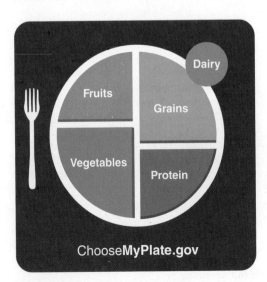

FIGURE 14.3 Adaptation of a Pie Chart

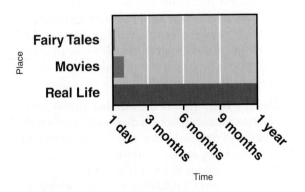

FIGURE 14.4 Bar Chart: Time It Takes to Fall in Love

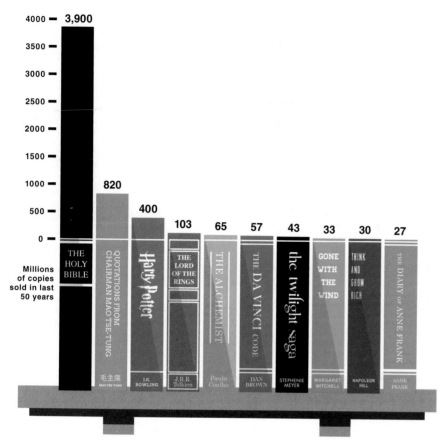

FIGURE 14.5 Column Chart: Most Read Books in the World

formatting that match the tone and purpose of your presentation. Most presentation software programs contain a clip art library that allows you to choose images to accompany your words. They also allow you to import images from outside sources and to build your own charts.

If you would like to learn more about using PowerPoint, Keynote, and Prezi, you can easily find several web-based tutorial programs by typing the name of your preferred program into your favorite search engine.

Alternative Media for Presenting Graphics

When a digital projector and/or screen are unavailable, you can use other methods to present visual aids.

Chalkboards, Whiteboards, and Polymer Marking Surfaces The

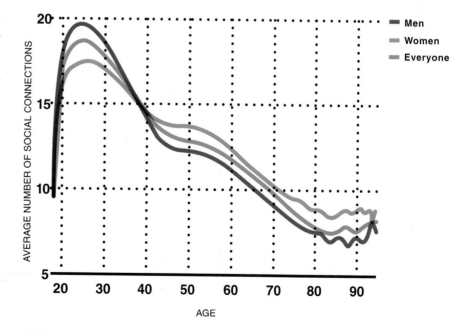

FIGURE 14.6 Line Chart: The Age When People Are the Most Popular

major advantage of these write-as-you-go media is their spontaneity. With them you can create your visual aid as you speak, including items generated from audience responses. Along with the odor of whiteboard markers and the squeaking of

line chart A visual aid consisting of a grid that maps out the direction of a trend by plotting a series of points.

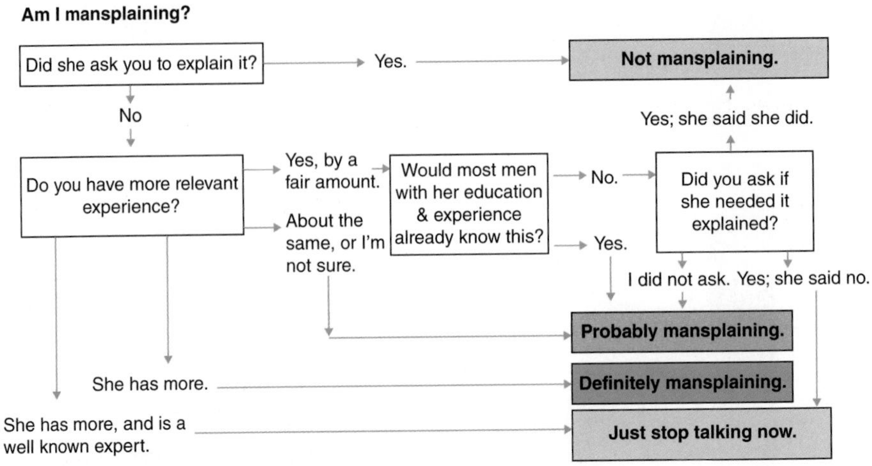

Am I mansplaining?

FIGURE 14.7 Flow Chart: Am I Mansplaining?

Source: Kim Goodwin Twitter post July 19, 2018 at https://twitter.com/kimgoodwin/status/102002957226643865 7?lang=en

chalk, a major disadvantage of these media is the difficulty of preparing visual aids on them in advance, especially if several speeches are scheduled in the same room at the same hour.

Flip Pads and Poster Board   Flip pads are like oversized writing tablets attached to a portable easel. Flip pads enable you to combine the spontaneity of the chalkboard (you can write on them as you go) with portability, which enables you to prepare them in advance. If you plan to use your visuals more than once, you can prepare them in advance on rigid poster board and display them on the same type of easel.

Despite their advantages, flip pads and poster boards are bulky, and preparing professional-looking exhibits on them requires a fair amount of artistic ability.

Handouts   The major advantage of handouts is that audience members can take away the information they contain after your speech. For this reason, handouts are excellent memory and reference aids. The major disadvantage is that they are distracting when handed out during a speech: First, there is the distraction of passing them out, and second, there is the distraction of having them in front of the audience members while you have gone on to something else. It's best, therefore, to pass them out at the end of the speech, so that audience members can use them as take-aways.

Rules for Using Visual Aids

It's easy to see that each type of visual aid and each medium for its presentation have their own advantages and disadvantages. No matter which type you use, however, there are a few rules to follow.

Simplicity   Keep your visual aids simple. Your goal is to clarify, not confuse. Use only key words or phrases, not sentences. The "rule of seven" states that each exhibit you use should contain no more than seven lines of text, each with no more than seven words. Keep all printing horizontal. Omit all nonessential details.

Size   Visual aids should be large enough for your entire audience to see them at one time but portable enough for you to get them out of the way when they no longer pertain to the point you are making.

Attractiveness   Visual aids should be visually interesting and as neat as possible. If you don't have the necessary artistic or computer skills, try to get help from a friend or at the computer or audiovisual center on your campus.

The Pros and Cons of Presentation Software

PowerPoint is by far the most popular form of work presentation today.[11] As with any software, however, it is not without its drawbacks.

The Pros Proponents say that PowerPoint slides can focus the attention of audience members on important information at the appropriate time. In addition, the slides can help listeners appreciate the relationship between different pieces of information. The software may also help speakers organize their thoughts in advance. But its primary benefit may be in providing a second, visual channel of information. One psychology professor puts it this way: "If you zone out for 30 seconds—and who doesn't?—it is nice to be able to glance up on the screen and see what you missed."[12]

The Cons For all its popularity, PowerPoint has received some bad press.[13,14] One particularly strong criticism came from Edward R. Tufte, an influential author of several books on the effective design of visual aids.[15] According to Tufte, the use of low-content PowerPoint slides trivializes important information. It encourages oversimplification by asking the presenter to summarize key concepts in as few words as possible—the ever-present bullet points.

Tufte also insists that PowerPoint makes it easier for a speaker to hide lies and logical fallacies.

Perhaps most seriously, opponents of PowerPoint say it is an enemy of interaction. One expert argued, "Instead of human contact, we are given human display."[16]

The Middle Ground? PowerPoint proponents say that it is just a tool, one that can be used effectively or ineffectively. They are the first to admit that a poorly done PowerPoint presentation can be boring and ineffective. In the infamous "triple delivery," the same text is seen on the screen, spoken aloud, and printed on the handout in front of you. Effective speakers know that PowerPoint should not be allowed to overpower a presentation—it should be just one element, not the whole thing.

After reviewing the pros and cons, would you say that PowerPoint is overall a benefit or detriment to effective public speaking? Why?

Appropriateness   Visuals must be appropriate to all the components of the speaking situation—you, your audience, and your topic—and they must emphasize the point you are trying to make. Don't make the mistake of using a visual aid that looks good but has only a weak link to the point you want to make—such as showing a map of a city transit system while talking about the condition of the individual cars.

Reliability   You must be in control of your visual aid at all times. Test all electronic media (projectors, computers, and so on) in advance, preferably in the room where you will speak. Just to be safe, have nonelectronic backups ready in case of disaster. Be conservative when you choose demonstrations: Wild animals, chemical reactions, and gimmicks meant to shock a crowd can often backfire.

When it comes time for you to use the visual aid, remember one more point: Talk to your audience, not to your visual aid. Some speakers become so wrapped up in their props that they turn their backs on their audience and sacrifice all their eye contact.

🔊 14.5 Sample Speech

At TEDWomen 2017 in New Orleans, Lera Boroditsky presented a speech arguing that the language we speak can affect the way we view the world.[17] Boroditsky develops well-chosen examples with clear language and two striking photos as visual aids. One of those photos is from her field research, and the other is a humorous example of the point she is making.

Her informative purpose statement could be worded:

After listening to my speech, my audience will be able to identify and cite examples of how language shapes the way we think.

Her thesis statement could be worded:

The language we speak shapes the way we think.

This is a complicated concept that has produced volumes of scholarly research over the years. Boroditsky makes her explanations and descriptions interesting, enjoyable, and fully comprehensible to an audience of nonexperts. Toward this end, she also organizes and structures her speech carefully. Her outline would look like this:

Introduction (1–4)

    I.   Language is a magical ability that humans have. (1)

    II.   Language is a physical process. (1)

    III. Language can put bizarre new ideas in your mind. (2–4)

        Example: Waltzing Jellyfish

Body

    I.   Languages are diverse.

        A.   There are about 7,000 languages spoken around the world. (5)

        B.   They are all different. (5)

    II.   Languages shape the way we think. (5–6)

        A.   Some languages increase our ability to navigate in the world. (8)

            1. People who speak languages like this stay oriented well. (10)

               Example: Tribesmen

            2. People who don't speak languages like this don't stay oriented well. (11–14)

               Example: All of us here.

        B.   Languages can influence our sense of cause-effect and personal responsibility. (15)

           Example: Breaking a vase.

            1. It has implications for what we pay attention to. (17)

            2. It has implications for eyewitness testimony. (18)

            3. It has implications for blame and punishment. (18)

    III. Languages are important.

        A.   Linguistic diversity reveals how ingenious the human mind is. (21)

        B.   Linguistic diversity is rapidly being lost. (22)

            1. We're losing about one language a week.

            2. We will lose about half of the world's languages in the next hundred years.

            3. We are losing a global research perspective. (23)

Conclusion (24)

    I.   Review of main points

    II.   Idea to remember: Linguistic diversity is about you.

        A.   Why do you think the way you do?

        B.   How could you think differently?

        C.   What thoughts do you wish to create?

Throughout her speech, Boroditsky makes it easy for her audience to listen by limiting the number of examples she presents and by choosing examples the audience will find interesting. Notice also her use of clear, simple language, and the way she generates audience involvement by personalizing her speech with humor.

How Language Shapes the Way We Think

1 I'll be speaking to you using language . . . because I can. This is one of these magical abilities that we humans have. We can transmit really complicated thoughts to one another. So what I'm doing right now is, I'm making sounds with my mouth as I'm exhaling. I'm making tones and hisses and puffs, and those are creating vibrations in the air. Those air vibrations are traveling to you, they're hitting your eardrums, and then your brain takes those vibrations from your eardrums and transforms them into thoughts. I hope.

2 I hope that's happening. So because of this ability, we humans are able to transmit our ideas across vast reaches of space and time. We're able to transmit knowledge across minds. I can put a bizarre new idea in your mind right now. I could say, "Imagine a jellyfish waltzing in a library while thinking about quantum mechanics."

3 Now, if everything has gone relatively well in your life so far, you probably haven't had that thought before.

4 But now I've just made you think it, through language.

5 Now, of course, there isn't just one language in the world, there are about 7,000 languages spoken around the world. And all the languages differ from one another in all kinds of ways. Some languages have different sounds, they have different vocabularies, and they also have different structures. That begs the question: Does the language we speak shape the way we think?

6 Until recently, there hasn't been any data to help us decide either way. Recently, in my lab and other labs around the world, we've started doing research, and now we have actual scientific data to weigh in on this question.

7 So let me tell you about some of my favorite examples. I'll start with an example from an Aboriginal community in Australia that I had the chance to work with. These are the Kuuk Thaayorre people.

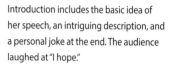

8 They live in Pormpuraaw at the very west edge of Cape York. What's cool about Kuuk Thaayorre is that, in Kuuk Thaayorre, they don't use words like "left" and "right," and instead, everything is in cardinal directions: north, south, east, and west. And when I say everything, I really mean everything. You would say something like, "Oh, there's an ant on your southwest leg." Or "Move your cup to the north-northeast a little bit. In fact, the way that you say "hello" in Kuuk Thaayorre is you say, "Which way are you going?" And the answer should be, "North-northeast in the far distance. How about you?"

9 So imagine as you're walking around your day, every person you greet, you have to report your heading direction.

Introduction includes the basic idea of her speech, an intriguing description, and a personal joke at the end. The audience laughed at "I hope."

A striking example that calls for audience participation and imagination, and establishes the importance of this topic to her audience.

Another joke personalizes the speech, and the audience laughs in appreciation.

An explanation that drives home the effectiveness of her example.

Thesis statement, in question form.

Establishing credibility to speak as an expert, without being overly technical about it.

This example is accompanied by a photograph as visual aid.

Simple, clear language.

Laughter here.

10   But that would actually get you oriented pretty fast, right? Because you literally couldn't get past "hello," if you didn't know which way you were going. In fact, people who speak languages like this stay oriented really well. They stay oriented better than we used to think humans could. We used to think that humans were worse than other creatures because of some biological excuse: "Oh, we don't have magnets in our beaks or in our scales." No; if your language and your culture trains you to do it, actually, you can do it. There are humans around the world who stay oriented really well.

Using familiar information to increase understanding of the unfamiliar.

11   And just to get us in agreement about how different this is from the way we do it, I want you all to close your eyes for a second and point southeast.

12   Keep your eyes closed. Point. OK, so you can open your eyes. I see you guys pointing there, there, there, there, there . . . I don't know which way it is myself—

13   You have not been a lot of help.

14   So let's just say the accuracy in this room was not very high. This is a big difference in cognitive ability across languages, right? Where one group—very distinguished group like you guys—doesn't know which way is which, but in another group, I could ask a five-year-old and they would know.

Audience participation, this time provoking audience laughter as they realize how difficult this is for them.

More laughter, as she shows the audience that she has the same language limitations they do.

Here she uses this simple demonstration to build understanding of a complex concept. More laughter from the audience.

Visual aid, photo used as example.

15   Languages also differ in how they describe events, right? You take an event like this, an accident.

16   In English, it's fine to say, "He broke the vase." In a language like Spanish, you might be more likely to say, "The vase broke," or "The vase broke itself." If it's an accident, you wouldn't say that someone did it. In English, quite weirdly, we can even say things like, "I broke my arm." Now, in lots of languages, you couldn't use that construction unless you are a lunatic and you went out looking to break your arm—and you succeeded. If it was an accident, you would use a different construction.

Laughter at this example.

17   Now, this has consequences. So, people who speak different languages will pay attention to different things, depending on what their language usually requires them to do. So we show the same accident to English speakers and Spanish speakers. English speakers will remember who did it because English requires you to say, "He did it; he broke the vase." Whereas Spanish speakers might be less likely to remember who did it if it's an accident, but they're more likely to remember that it was an accident. They're more likely to remember the intention.

18 So, two people watch the same event, witness the same crime, but end up remembering different things about that event. This has implications, of course, for eyewitness testimony. It also has implications for blame and punishment. So if you take English speakers and I just show you someone breaking a vase, and I say, "He broke the vase," as opposed to "The vase broke," even though you can witness it yourself, you can watch the video, you can watch the crime against the vase, you will punish someone more, you will blame someone more if I just said, "He broke it," as opposed to, "It broke." The language guides our reasoning about events.

19 I've given you two examples of how language can profoundly shape the way we think. So language can have big effects, like we saw with space, where people can lay out space in completely different coordinate frames from each other.

Internal review keeps us oriented.

20 I also gave you an example of how language can shape things that have personal weight to us—ideas like blame and punishment or eyewitness memory. These are important things in our daily lives.

21 Now, the beauty of linguistic diversity is that it reveals to us just how ingenious and how flexible the human mind is. Human minds have invented not one cognitive universe, but 7,000—there are 7,000 languages spoken around the world.

Using repetition to emphasize an important idea.

22 And we can create many more—languages, of course, are living things, things that we can hone and change to suit our needs. The tragic thing is that we're losing so much of this linguistic diversity all the time. We're losing about one language a week, and by some estimates, half of the world's languages will be gone in the next hundred years.

Again, repetition.

23 And the even worse news is that right now, almost everything we know about the human mind and human brain is based on studies of usually American English-speaking undergraduates at universities. That excludes almost all humans. Right? So what we know about the human mind is actually incredibly narrow and biased, and our science has to do better.

24 I want to leave you with this final thought. I've told you about how speakers of different languages think differently, but of course, that's not about how people elsewhere think. It's about how you think. It's how the language that you speak shapes the way that you think. And that gives you the opportunity to ask, "Why do I think the way that I do?" "How could I think differently?" And also, "What thoughts do I wish to create?"

Her conclusion reviews her main points and reminds her audience members about the importance of this topic to them.

25 Thank you very much.

MAKING THE GRADE

At www.oup.com/he/adler-uhc14e, you will find a variety of resources to enhance your understanding, including video clips, animations, self-quizzes, additional activities, audio and video summaries, interactive self-assessments, and more.

OBJECTIVE 14.1 Distinguish among the main types of informative speaking.

- Informative speeches can be classified by the type of content, including speeches about objects, processes, events, and concepts.
- Informative speeches can also be classified according to their purpose, including descriptions, explanations, and instructions.

> Explain the primary differences among types of informative speeches.

> Which type of informative speech is most important in your own life?

> Which type of speech will you use for your next informative speech?

OBJECTIVE 14.2   Describe the differences between informative and persuasive speaking.

● Two basic characteristics differentiate an informative topic from a persuasive topic.

● In an informative speech, you generally do not present information that your audience is likely to disagree with.

● The speaker's primary intent is not to change attitudes or to make the audience members *feel* differently about the topic.

> Explain the primary difference between informative and persuasive speaking.

> How are informative and persuasive speeches similar?

> Why is it important to distinguish between informative and persuasive speaking?

OBJECTIVE 14.3   Outline techniques that increase the effectiveness of informative speeches.

● Decide on a specific purpose statement.

● Create information hunger by tapping into audience needs.

● Make it easy to listen by limiting the amount of information and by using familiar information, straightforward organization, clear language, and effective supporting material.

● Involve your audience through audience participation, the use of volunteers, and a question-and-answer period.

> List at least three techniques for increasing informative effectiveness.

> Which is the most important technique in informative speaking?

> Which technique do you need to work on for your next speech?

OBJECTIVE 14.4   Explain how you might use visual aids appropriately and effectively.

● Visual aids include objects and models, photos, videos, audio files, charts, and graphs.

● Media for the presentation of visual aids include digital projectors with presentation software, chalkboards, whiteboards, flip pads, and handouts.

● Each type of visual aid has its own advantages and disadvantages.

● Keep visual aids simple, large enough for the audience to see, and visually interesting.

> List at least three types of visual aids, giving examples of when to use them.

> Describe the advantages and disadvantages of the visual aids on your list.

> Which types of visual aids will you use for your next informative speech?

KEY TERMS

audience involvement p. 356

audience participation p. 357

bar chart p. 360

column chart p. 360

description p. 350

diagram p. 359

explanations p. 350

flow chart p. 360

information anxiety p. 348

information hunger p. 353

informative purpose statement p. 351

instructions p. 350

intro p. 359

line chart p. 360

model (in speeches and presentations) p. 359

number chart p. 359

outro p. 359

pictogram p. 360

pie chart p. 360

signpost p. 356

sound bite p. 359

visual aids p. 359

vocal citation p. 355

word chart p. 359

wraparound p. 359

ACTIVITIES

1. **Informative Purpose Statements** For practice in defining informative speech purposes, reword the following statements so that they specifically point out what the audience will be able to do after hearing the speech.

a. My talk today is about building a wood deck.

b. My purpose is to tell you about vintage car restoration.

c. I am going to talk about toilet training.

d. I'd like to talk to you today about sexist language.

e. There are six basic types of machines.

f. The two sides of the brain have different functions.

g. Do you realize that many of you are sleep deprived?

2. **Effective Repetition** Create a list of three statements, or use the three that follow. Restate each of these ideas in three different ways.

a. The magazine *Modern Maturity* has a circulation of more than 20 million readers.

b. Before buying a used car, you should have it checked out by an independent mechanic.

c. One hundred thousand pounds of dandelions are imported into the United States annually for medical purposes.

3. **Using Clear Language** For practice in using clear language, select an article from any issue of a professional journal in your major field. Using the suggestions in this chapter, rewrite a paragraph from the article so that it will be clear and interesting to a layperson.

4. **Inventing Visual Aids** Take any sample speech. Analyze it for where visual aids might be effective. Describe the visual aids that you think will work best. Compare the visuals you devise with those of your classmates.

15

Persuasive Speaking

CHAPTER OUTLINE

15.1 Characteristics of Persuasion 373
- Persuasion Is Not Coercive
- Persuasion Is Usually Incremental
- Persuasion Is Interactive
- Persuasion Can Be Ethical

15.2 Categorizing Persuasive Attempts 376
- By Type of Proposition
- By Desired Outcome
- By Directness of Approach
- By Type of Appeal: Aristotle's Ethos, Pathos, and Logos

15.3 Creating a Persuasive Message 379
- Set a Clear, Persuasive Purpose
- Structure the Message Carefully
- Use Solid Evidence
- Avoid Fallacies

15.4 Adapting to the Audience 385
- Establish Common Ground
- Organize According to the Expected Response
- Neutralize Potential Hostility

15.5 Building Credibility as a Speaker 386
- Competence
- Character
- Charisma

15.6 Sample Speech 389

MAKING THE GRADE **394**

KEY TERMS **395**

ACTIVITIES **395**

LEARNING OBJECTIVES

15.1

Identify the primary characteristics of persuasion.

15.2

Compare and contrast different types of persuasion.

15.3

Apply the guidelines for persuasive speaking to a speech you will prepare.

15.4

Explain how to best adapt a specific speech to a specific audience.

15.5

Improve your credibility in your next persuasive speech.

Shayla Cabalan's story shows how public speaking can create change through persuasion.

What persuasive messages have changed the ways you think and act?

What audiences would you hope to reach through your own public speaking? How would you like to change the way they think and act?

SHAYLA CABALAN'S FIRST EXPERIENCE in public speaking was as a sixth grader who suffered from social anxiety. A teacher suggested that joining the school's speech team would help her become more self-assured.

That teacher was right. Public speaking turned out to be a big confidence-builder for Shayla. As she became more comfortable in front of audiences, her desire grew to speak out against injustice. She developed a passionate pursuit of advocacy, especially when it came to the rights and protection of children.

Shayla brought that interest to the University of Indianapolis. She became the opinion editor for the school newspaper, news director at the campus radio station, and a member of the school's speech and debate team. In all these positions, Shayla spoke out against things she saw as wrong and sought solutions through various kinds of persuasion. One topic she selected for a speech was child marriage. She explained, "As someone who works with and has an affinity for kids, I wanted to help expose this ongoing atrocity in any way I could."[1]

This speech wasn't just an assignment or a chance to practice her speaking skills. Shayla truly wanted to change her audience:

> I wanted to persuade my audience of fellow competitors, observers, and judges to support organizations attempting to combat child marriage, as well as advocate for legal action against child marriage loophole laws.[2]

Audience members approached Shayla after each round of her competitions. They told her that because of her speech, they would take the actions she called for. Shayla felt she had found her voice and hoped to extend her passion into a lifelong career helping others. You'll find her speech at the end of this chapter.

You probably aim to persuade others all the time—in personal relationships, at work, and perhaps even in impersonal settings. You know that persuasion can help you succeed in your career. And you probably feel strongly about both personal and social issues. Learning the art and science of persuasion can help you become more effective in bringing about results in all the areas that matter to you.

🔘 15.1 Characteristics of Persuasion

Persuasion is the process of motivating others through communication to change a particular belief, attitude, or behavior. Implicit in this definition are several characteristics of persuasion.

Persuasion Is Not Coercive

Persuasion is not the same thing as coercion. If you put someone in a headlock and said, "Do this, or I'll choke you," you would be acting coercively. Besides being illegal, this approach would be ineffective. As soon as the authorities took you away, the person would stop following your demands.

The failure of coercion to achieve lasting results is also apparent in less dramatic circumstances. Children whose parents are coercive often rebel as soon as they can; students who perform from fear of an instructor's threats rarely appreciate the subject matter; and employees who work for abusive and demanding employers are often unproductive and eager to switch jobs as soon as possible. Persuasion, by contrast, makes a listener *want* to think or act differently.

Persuasion Is Usually Incremental

Attitudes do not normally change instantly or dramatically. Persuasion is a process. When it is successful, it generally succeeds over time, usually in increments. So, despite your passion, the best way to change hearts and minds is to have modest but achievable goals.

Social judgment theory explains how and why a gradualist approach can be most effective.[3] This theory tells us that when members of an audience hear a persuasive appeal, they compare it to opinions they already hold. The preexisting opinion is called an **anchor**, but around this anchor there exist what are called **latitudes of acceptance**, **latitudes of rejection**, and **latitudes of noncommitment**. A diagram of any opinion, therefore, might look something like Figure 15.1.

People who care very strongly about a point of view will have a narrow latitude of noncommitment, and those who care less strongly will have a wider latitude of noncommitment. Research suggests that audience members respond most favorably to appeals that fall within their latitude of rejection. This means that persuasion in the real world takes place in a series of small movements. One persuasive speech may be but a single step in an overall persuasive campaign. The best example of this is the various communications that take place during the months of a political campaign. Candidates watch the opinion polls carefully, adjusting their appeals to the latitudes of acceptance and noncommitment of the uncommitted voters.

Public speakers who heed the principle of social judgment theory tend to seek realistic, if modest, goals in their speeches. For example, if you want to change audience views on the pro-life/pro-choice question, social judgment theory suggests that the first step would be to consider a range of arguments such as this:

Abortion is a sin.

Abortion should be absolutely illegal.

Abortion should be allowed only in cases of rape and incest.

persuasion The act of motivating a listener, through communication, to change a particular belief, attitude, value, or behavior.

social judgment theory The theory that opinions will change only in small increments and only when the target opinions lie within the receiver's latitudes of acceptance and noncommitment.

anchor The position supported by audience members before a persuasion attempt.

latitude of acceptance In social judgment theory, statements that a receiver would not reject.

latitude of rejection In social judgment theory, statements that a receiver could not possibly accept.

latitude of noncommitment In social judgment theory, statements that a receiver would not care strongly about one way or another.

Strongly disagree	Don't care	Agree	Strongly agree
(Latitude of rejection)	(Latitude of noncommitment)	(Latitude of acceptance)	(Anchor)

FIGURE 15.1 Latitudes of Acceptance, Rejection, and Noncommitment

A woman should be required to have her husband's permission to have an abortion.

A girl under the age of 18 should be required to have a parent's permission before she has an abortion.

Abortion should be allowed during the first 3 months of pregnancy.

A girl under the age of 18 should not be required to have a parent's permission before she has an abortion.

A woman should not be required to have her husband's permission to have an abortion.

Abortion is a woman's personal decision.

Abortion should be discouraged but legal.

Abortion should be available anytime to anyone.

Abortion should be considered simply a form of birth control.

You could then arrange these positions on a continuum and estimate how listeners would react to each one. The statement that best represented the listeners' point of view would be their anchor. Other items that might also seem reasonable to them would make up their latitude of acceptance, and opinions they would reject would make up their latitude of rejection. The remaining statements would be the listeners' latitude of noncommitment.

Social judgment theory suggests that the best chance of changing audience attitudes would come by presenting an argument that falls somewhere within the listeners' latitude of noncommitment—even if this wasn't the position that you ultimately want them to accept. If you push too hard by arguing a position in your audience's latitude of rejection, your appeals would probably backfire, making your audience *more* opposed to you than before.

Persuasion Is Interactive

It's easy to think of public speaking as a one-way event in which you talk and the audience listens. But the transactional model of communication described in Chapter 1 is a reminder that persuasion isn't something you do *to* audience members but rather something you do *with* them. Pay attention to your listeners' nonverbal behavior during the speech. If you notice they seem confused, you might decide to elaborate on a point. If they seem annoyed, you might recognize the need to address their objections. If they're enthusiastic, you might decide to make your goals more ambitious. If time permits after you've spoken, responding to listeners' questions can be a way to address their concerns.

Persuasion Can Be Ethical

Even when you understand the difference between persuasion and coercion, you might still feel uncomfortable with the idea of persuasive speaking. You might see it as the work of high-pressure hucksters: salespeople with their feet stuck in the door, unscrupulous politicians taking advantage of beleaguered taxpayers, and so on. Indeed, many of the principles we are about to discuss have been used by unethical speakers for unethical purposes, but that's not what all—or even most—persuasion is about.

Persuasion can influence others' lives in worthwhile ways. Saying "I don't want to influence other people," really means, "I don't want to get involved with other people." Look at the good you can accomplish through persuasion: You can convince people to live healthier lives; you can get members of your community to conserve energy or to join together to refurbish a park; or you can persuade an employer to hire you for a job in which your own talents, interests, and abilities will be put to their best use.

Persuasion is considered ethical if it conforms to accepted standards. But what are the standards today? If your plan is selfish and not in the best interest of your audience members, but you are honest about your motives—is that ethical? If your plan is in the best interest of your audience members, yet you lie to them to get them to accept the plan—is that ethical? Philosophers and rhetoricians have argued for centuries over questions like these.

There are many ways to define **ethical persuasion**.[4] For our purpose, we will consider it as *communication in the best interest of the audience that does not depend on false or misleading information to change an audience's attitude or behavior*. The best way to appreciate the value of this simple definition is to consider the many strategies listed in Table 15.1 that do not fit it. For example, faking enthusiasm about a speech topic, plagiarizing material from another source, and passing it off as your own and making up statistics to support your case are clearly unethical.

Besides being wrong on moral grounds, unethical attempts at persuasion have a major practical disadvantage: If your deception is uncovered, your credibility will suffer. If, for example, prospective buyers uncover your attempt to withhold a structural flaw in the condominium you are trying to sell, they will probably suspect that the property has other hidden problems. Likewise, if your speech instructor suspects that you are lifting material from other sources without giving credit, your entire presentation will be suspect. One unethical act can cast doubt on future truthful statements. Thus, for pragmatic as well as moral reasons, honesty really is the best policy.

ASK YOURSELF

Present examples of both ethical and unethical persuasion. What distinguishes them from one another?

ethical persuasion Persuasion in an audience's best interest that does not depend on false or misleading information to induce change in that audience.

TABLE 15.1

Unethical Communication Behaviors

1. Committing Plagiarism
   a. Claiming someone else's ideas as your own
   b. Quoting without citing the source

2. Relaying False Information
   a. Deliberate lying
   b. Ignorant misstatement
   c. Deliberate distortion and suppression of material
   d. Fallacious reasoning to misrepresent truth

3. Withholding Information; Suppression
   a. About self (speaker); not disclosing private motives or special interests
   b. About speech purpose
   c. About sources (not revealing sources; plagiarism)
   d. About evidence; omission of certain evidence (card stacking)
   e. About opposing arguments; presenting only one side

4. Appearing to Be What One Is Not; Insincerity
   a. In words, saying what one does not mean or believe
   b. In delivery (for example, feigning enthusiasm)

5. Using Emotional Appeals to Hinder Truth
   a. Using emotional appeals as a substitute or cover-up for lack of sound reasoning and valid evidence
   b. Failing to use balanced appeals

Source: Adapted from Andersen, M. K. (1979). *An analysis of the treatment of ethos in selected speech communication textbooks* (Unpublished dissertation). University of Michigan, Ann Arbor, pp. 244–247.

ETHICAL CHALLENGE
You Versus the Experts

Read Table 15.1 carefully. The behaviors listed there are presented in what some (but certainly not all) communication experts would describe as "most serious to least serious" ethical faults. Do you agree or disagree with the order of this list? Explain your answer and whether you would change the order of any of these behaviors. Are there any other behaviors that you would add to this list?

🌐 15.2 Categorizing Persuasive Attempts

There are several ways to categorize the persuasive attempts you will make as a speaker. What kinds of subjects will you focus on? What results will you be seeking? How will you go about getting those results? In this section, we look at each of these questions.

By Type of Proposition

proposition of fact A claim bearing on issue in which there are two or more sides of conflicting factual evidence.

proposition of value A claim bearing on an issue involving the worth of some idea, person, or object.

proposition of policy A claim bearing on an issue that involves adopting or rejecting a specific course of action.

Persuasive topics fall into one of three categories, depending on the type of thesis statement (referred to as a "proposition" in persuasion) you are advancing. The three categories are propositions of fact, propositions of value, and propositions of policy.

Propositions of Fact Some persuasive messages focus on **propositions of fact**: issues in which there are two or more sides about conflicting information, in which listeners are required to choose the truth for themselves. For example:

> Global climate change has already had serious economic consequences.
>
> Early childhood vaccination is essential for public health.
>
> Ride-hailing services have contributed to traffic gridlock in many cities.

These examples show that many questions of fact require careful examination and interpretation of evidence, usually collected from a variety of sources. That's why it is possible to debate questions of fact, and that also explains why these propositions form the basis of persuasive speeches and not informative ones.

Propositions of Value Beyond issues of truth or falsity, **propositions of value** explore the worth of some idea, person, or object. For example:

> We are/aren't obliged to obey all laws, even those that seem foolish or morally wrong.
>
> The United States is/is not justified in attacking countries that harbor terrorist organizations.
>
> The use of laboratory animals for scientific experiments is/is not cruel and immoral.

In order to deal with most propositions of value, you will have to explore certain propositions of fact. For example, you won't be able to debate whether the experimental use of animals in research is immoral—a proposition of value—until you have dealt with propositions of fact such as how many animals are used in experiments and whether experts believe they suffer.

Propositions of Policy A step beyond questions of fact or value are **propositions of policy**, which recommend a specific course of action (a "policy"). For example:

> The World Bank should/should not create a program of microloans for citizens of impoverished nations.
>
> The Electoral College should/should not be abolished.
>
> Genetic engineering of plants and livestock is/is not an appropriate way to increase the food supply.

Looking at persuasion according to the type of proposition is a convenient way to generate topics for a persuasive speech because each type of proposition suggests different topics. Selected topics could also be handled differently depending on how they are approached. For example, a campaign speech could be

Engraved on the Statue of Liberty's pedestal are the words "Give me your tired, your poor, your huddled masses yearning to breathe free." Deciding whether those sentiments should apply to today's immigrants involves questions of values and policy.

What values guide your beliefs about immigration policy? How could you use the principles in this chapter to persuade others?

approached as a proposition of fact ("Candidate X has done more for this community than the opponent"), a proposition of value ("Candidate X is a better person than the opponent"), or a proposition of policy ("We should get out and vote for Candidate X"). Remember, however, that a fully developed persuasive speech is likely to contain all three types of propositions. If you were preparing a speech advocating that college athletes should be paid in cash for their talents (a proposition of policy), you might want to first prove that the practice is already widespread (a proposition of fact) and that it is unfair to athletes from other schools (a proposition of value).

By Desired Outcome

We can also categorize persuasion according to two major outcomes: convincing and actuating.

Convincing   When you set out to **convince** an audience, you want to change the way its members think. When we say that convincing an audience changes the way its members think, we do not mean that you have to swing them from one belief or attitude to a completely different one. Sometimes audience members will already think the way you want them to, but they will not be firmly enough committed to that way of thinking. When that is the case, you reinforce, or strengthen, their opinions. For example, if your audience already believed that the federal budget should be balanced but did not consider the idea important, your job would be to reinforce members' current beliefs. Reinforcing is still a type of change, however, because you are causing an audience to adhere more strongly to a belief or attitude. In other cases, a speech to convince will begin to shift attitudes without bringing about a total change of thinking. For example, an effective speech to convince might get a group of skeptics to consider the possibility that bilingual education is/isn't a good idea.

Actuating   When you set out to **actuate** an audience, you want to move its members to a specific behavior. Whereas a speech to convince might move an audience to action, it won't be any specific action that you have recommended. In a speech to actuate, you do recommend that specific action.

You can ask for two types of action—adoption or discontinuance. Adoption asks an audience to engage in a new behavior, whereas discontinuance asks an audience to stop behaving in an established way. If you gave a speech for a political candidate and then asked for contributions to that candidate's campaign, you would be asking your audience to adopt a new behavior. If you gave a speech against smoking and then asked your audience members to sign a pledge to quit, you would be asking them to discontinue an established behavior.

By Directness of Approach

We can also categorize persuasion according to the speaker's directness of approach.

Direct Persuasion   Using **direct persuasion**, speakers make their purpose clear, usually by stating it outright early in the speech. This is the best strategy to use with a friendly audience, especially when you are asking for a response the audience is likely to give you. Direct persuasion is the kind we hear in most academic situations.

By Indirect Persuasion   In using **indirect persuasion**, the speaker's purpose is disguised or deemphasized in some way. The question, "Is a season ticket to the symphony worth the money?" (when you intend to prove it is), is based on indirect persuasion, as is any strategy that does not express the speaker's purpose at the outset.

Indirect persuasion is sometimes easy to spot. A television commercial that shows us attractive young men and women romping in the surf on a beautiful day

convincing   A speech goal that aims at changing audience members' beliefs, values, or attitudes.

actuate   To move members of an audience toward a specific behavior.

direct persuasion   Persuasion that does not try to hide or disguise the speaker's persuasive purpose.

indirect persuasion   Persuasion that disguises or deemphasizes the speaker's persuasive goal.

and then flashes the product name on the screen is pretty indisputably indirect persuasion. Political oratory also is sometimes indirect persuasion, and it can be more difficult to identify as such. A political hopeful ostensibly might be speaking on some great social issue when the real persuasive message is, "Please remember my name, and vote for me in the next election."

In public speaking, indirect persuasion is usually disguised as informative speaking, but this approach isn't necessarily unethical. In fact, it is probably the best approach to use when your audience is hostile to either you or your topic. It is also often necessary to use the indirect approach to get a hearing from listeners who would tune you out if you took a more direct approach. Under such circumstances, you might want to ease into your speech slowly.[5] You might take some time to make your audience feel good about you or the social action you are advocating. If you are speaking in favor of your candidacy for city council, but you are in favor of a tax increase and your audience is not, you might talk for a while about the benefits that a well-financed city council could provide to the community. You might even want to change your desired audience response. Rather than trying to get audience members to rush out to vote for you, you might want them simply to read a policy statement that you have written or become more informed on a particular issue. The one thing you cannot do in this instance is to begin by saying, "My appearance here today has nothing to do with my candidacy for city council." That would be a false statement. It is more than indirect; it is untrue and therefore unethical.

To test the ethics of an indirect approach, ask whether you would express your persuasive purpose directly if you were asked to do so. In other words, if someone in the audience stopped you and asked, "Don't you want us to vote for you for city council?," you would admit to it rather than denying your true purpose, if you were ethical.

By Type of Appeal: Aristotle's Ethos, Pathos, and Logos

Over 2,000 years ago Aristotle created the first comprehensive theory of persuasion. His *Rhetoric* is generally regarded as "the most important single work on persuasion ever written."[6] He proposed three approaches to influencing others: ethos (speaker credibility), pathos (emotional appeals), and logos (logical reasoning). Aristotle advised a balance of these three approaches, called the **Rhetorical Triad** (Figure 15.2). This chapter will examine each type of appeal.

Ethos   From the Greek word for *character*, **ethos**-based proofs rely on the audience's faith in a speaker. This faith is based on judgments of the speaker's competence, character, and charisma. Later in this chapter you'll learn how to boost your persuasiveness by demonstrating each of these qualities.

Ethos: Appeals based on the credibility of the speaker.

Pathos: Appeals based on emotion.

Logos: Appeals based on logical reasoning.

Pathos   In modern terms, **pathos** involves appealing to emotion. Most people claim to be rational, but a large body of research shows that emotions often shape reasoning, rather than the other way around. Social scientists have suggested that emotions are like elephants, while reason is like an elephant rider.

Perched atop the Elephant, the Rider holds the reins and seems to be the leader. But the Rider's control is precarious because the Rider is so small

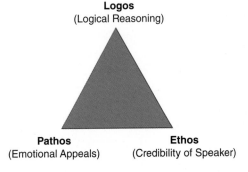

FIGURE 15.2 Aristotle's Rhetorical Triad: Analysis of persuasion as a combination of appeals based on credibility, emotion, and logic

relative to the Elephant. Anytime the Elephant and the Rider disagree about which direction to go, the Rider is going to lose.[7]

As psychologist Jonathan Haidt puts it, "If you want to change someone's mind about a moral or political issue, *talk to the elephant first*."[8]

Effective speakers appeal to emotions in a variety of ways, including vivid descriptions. emotional stories, and powerful testimony. You can see examples of emotional appeals in Shayla Cabalan's speech at the end of this chapter.

Logos   Later in this chapter, you'll read about boosting the persuasiveness of your message by using solid evidence and reasoning. But even the best evidence-based reasoning won't work if you antagonize your audience, even if you don't mean to. Social psychologist Jonatan Haidt uses the elephant-rider metaphor to emphasize that an argument has to be delivered with respect in order to be heard:

> When discussions are hostile, the odds of change are slight. The elephant [the audience's emotional reaction] leans away from the opponent [speaker], and the rider [the audience's logic] works frantically to rebut the opponent's charges. But if there is affection, admiration, or a desire to please the other person, then the elephant leans toward that person and the rider tries to find truth in the other person's arguments.[9]

🎤 15.3 Creating a Persuasive Message

Persuasive speaking has been defined as "reason-giving discourse." Its principal technique, therefore, involves proposing claims and then backing those claims up with reasons that are true. Preparing an effective persuasive speech isn't easy, but it can be made easier by observing a few simple rules.

Set a Clear, Persuasive Purpose

Remember that your objective in a persuasive speech is to move the audience to a specific, attainable attitude or behavior. In a speech to convince, the purpose statement will probably stress an attitude:

> After listening to my speech, my audience members will agree that steps should be taken to save whales from extinction.

In a speech to actuate, the purpose statement will stress behavior in the form of a desired audience response. That desired audience response should be as straightforward and clear-cut as possible.

As Chapter 12 explained, your purpose statement should always be specific, attainable, and worded from the audience's point of view. "The purpose of my speech is to save the whales" is not a purpose statement that has been carefully thought out. Your audience members wouldn't be able to jump into the ocean and save the whales, even if your speech motivated them into a frenzy. They might, however, be able to support a specific piece of legislation.

A clear, specific purpose statement will help you stay on track throughout all the stages of preparation of your persuasive speech. Because the main purpose of your speech is to have an effect on your audience, you have a continual test that you can use for every idea, every piece of evidence, and every organizational structure that you think of using. The question you ask is: "Will this help me to get the audience members to think/feel/behave in the manner I have described in my purpose statement?" If the answer is "yes," you forge ahead.

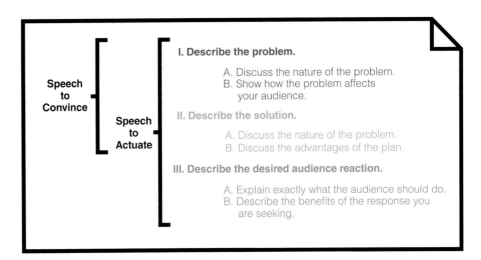

FIGURE 15.3 Sample Structure for a Persuasive Speech

Structure the Message Carefully

A sample structure of the body of a persuasive speech is outlined in Figure 15.3. With this structure, if your objective is to convince, you concentrate on the first two components: establishing the problem and describing the solution. If your objective is to actuate, you add the third component, describing the desired audience reaction.

Of course, other structures can be used for persuasive speeches. This one can be used as a basic model, however, because it's easily applied to most persuasive topics.

Describe the Problem   To convince an audience that something needs to be changed, you have to show members that a problem exists. After all, if your listeners don't recognize the problem, they won't find your arguments for a solution very important.

Rebecca Yocum, a student at West Chester University of Pennsylvania, presented a speech on problems with the 911 emergency response system. She described the problem this way:

> On May 15th of last year, a fire broke out at Michelle Dzoch's home in Wilkes-Barre, Pennsylvania. One of Michelle's neighbors called 911 only three minutes after the start of the incident. 911 operators quickly dispatched three fire and rescue units—to an incorrect address fifteen miles away from the actual fire. Meanwhile, Michelle Dzoch lay dying in the second-floor bathroom of her burning home.

An effective description of the problem will answer two questions, either directly or indirectly.

Show How the Problem Affects Your Audience   It's not enough to prove that a problem exists. You have to show your listeners that the problem affects them in some way. This is relatively easy in some cases: the high cost of tuition, the lack of convenient parking near campus, the quality of food in the student center. In other cases, you will need to spell out the impact to your listeners more clearly. Rebecca Yocum did this when she explained the extent of the problem with the 911 emergency system:

The National Emergency Number Association website estimates that 240 million calls are made to 911 in the United States every year, with the potential for millions of cases of inaccurate dispatches. In order to mitigate this risk for each of us, we need a better system for locating those in need so that our emergency response units can do their jobs with efficiency.[10]

The problem section of a persuasive speech is often broken up into segments discussing the cause and effect of the problem. (The sample speech at the end of this chapter is an example of this type of organization.)

Describe the Solution   Your next step in persuading your audience members is to convince them that there is an answer to the problem you have just introduced. To describe your solution, you should show that the solution will work and then you should explain the advantages of this solution.

Show That the Solution Will Work   A skeptical audience might agree with the desirability of your solution but still not believe that it has a chance of succeeding. In the homeless speech discussed previously, you would need to prove that establishing a shelter can help unlucky families get back on their feet—especially if your audience analysis shows that some listeners might view such a shelter as a way of coddling people who are too lazy to work.

Explain What Advantages Will Result from Your Solution   You need to describe in specific terms how your solution will lead to the desired changes. In this step, you will paint a vivid picture of the benefits of your proposal. In a speech proposing a shelter for homeless families, the benefits you describe would probably include the following:

1. Families will have a safe place to stay, free of the danger of living on the street.

2. Parents will have the resources that will help them find jobs: an address, telephone, clothes washers, and showers.

3. The police won't have to apply antivagrancy laws (such as prohibitions against sleeping in cars) to people who aren't the intended target of those laws.

4. The community (including your listeners) won't need to feel guilty about ignoring the plight of unfortunate citizens.

Describe the Desired Audience Response   When you want to go beyond simply convincing your audience members and impel them to follow your solution, you need to describe exactly what you want them to do. This action step, like the previous ones, should answer two questions.

Explain How the Audience Can Put Your Solution into Action   Make the behavior you are asking your audience members to adopt as clear and simple as possible for them. If you want them to vote in a referendum, tell them when and where to go to vote and how to go about registering, if necessary (some activists even provide transportation). If you're asking them to support a legislative change, don't expect them to write their congressional representative. *You* write the letter or draft a petition and ask them to sign it. If you're asking for a donation, pass the hat at the conclusion of your speech, or give audience members a stamped, addressed envelope and simple forms that they can return easily.

Demonstrate the Direct Rewards   Your solution might be important to society, but your audience members will be more likely to adopt it if you can show that they will get a personal payoff. Show that supporting legislation to reduce acid rain will produce a wide range of benefits, from reduced lung damage to healthier forests to longer life for their car's paint. Explain that saying "no" to a second drink before driving will not only save lives but also help your listeners avoid expensive court costs, keep their insurance rates low, and prevent personal humiliation. Show how helping to establish and staff a homeless shelter can lead to personal feelings of satisfaction and provide an impressive demonstration of community service on a job-seeking résumé.

Use Solid Evidence

All the forms of support discussed in Chapter 12 can be used to back up your persuasive arguments.[11] Your objective here is not just to find supporting material that clarifies your ideas, but rather to find the perfect example, statistic, definition, analogy, anecdote, or testimony to establish the truth of your claim in the mind of this specific audience.

Whatever type of evidence you use, cite your sources carefully. It is important that your audience know that your sources are credible, unbiased, and current. If you are quoting the source of an interview, give a full statement of the source's credentials:

According to Sean Wilentz, Dayton–Stockton Professor of History, Director of American Studies at Princeton University, and the author of several books on this topic . . .

If the currency of the interview is important, you might add, "I spoke to Professor Wilentz just last week. . . ." If you're quoting an article, give a quick statement of the author's credentials and the full date and title of the magazine:

According to Professor Sean Wilentz of Princeton University, in an article in the April 21, 2016, *Rolling Stone Magazine* . . .

You don't need to give the title of the article (although you may, if it helps in any way) or the page number. If you're quoting from a book, include a quick statement of the author's credentials:

According to Professor Sean Wilentz of Princeton University, in his book *The Rise of American Democracy* . . .

Avoid Fallacies

fallacy An error in logic.

ad hominem fallacy A fallacious argument that attacks the integrity of a person to weaken the person's position.

A **fallacy** (from the Latin word meaning *false*) is an error in logic. Although the original meaning of the term implied purposeful deception, most logical fallacies are not recognized as such by those who use them. Scholars have devoted lives and volumes to the description of various types of logical fallacies.[12] Here are some of the most common types to keep in mind when building your persuasive argument:[13]

Attack on the Person Instead of the Argument (*Ad Hominem*) In an ***ad hominem* fallacy** the speaker attacks the integrity of a person in order to weaken the argument. At its crudest level, an *ad hominem* argument is easy to detect. "How can you believe that fat slob?" is hardly persuasive. It takes critical thinking to catch more subtle *ad hominem* arguments, however. Consider this argument: "All this talk about 'family values' is hypocritical. Take Senator _____, who made a speech about the 'sanctity of marriage' last year. Now it turns out he was having an affair with a staffer, and his wife is suing him for divorce." Although the senator certainly does seem to be a hypocrite, his behavior doesn't necessarily weaken the merits of family values.

Reduction to the Absurd (*Reductio ad Absurdum*)   A ***reductio ad absurdum fallacy*** unfairly attacks an argument by extending it to such extreme lengths that it looks ridiculous. "If we allow developers to build homes in one section of this area, soon we will have no open spaces left. Fresh air and wildlife will be a thing of the past." "If we allow the administration to raise tuition this year, soon they will be raising it every year, and before we know it only the wealthiest students will be able to go to school here." This extension of reasoning doesn't make any sense: Developing one area doesn't necessarily mean that other areas have to be developed, and one tuition increase doesn't mean that others will occur. Any of these policies might be unwise or unfair, but the *ad absurdum* reasoning doesn't prove it.

Either–Or   An **either–or fallacy** sets up false alternatives, suggesting that if the inferior one must be rejected, then the other must be accepted. An angry citizen used either–or thinking to support a proposed city ordinance: "Either we outlaw alcohol in city parks, or there will be no way to get rid of drunks." This reasoning overlooks the possibility that there may be other ways to control public drunkenness besides banning all alcoholic beverages. The old saying "America, love it or leave it" provides another example of either–or reasoning. For instance, when an Asian-born college professor pointed out examples of lingering discrimination in the United States, some suggested that if she didn't like her adopted country, she should return to her native home—ignoring the fact that it is possible to admire a country and still envision ways to make it a better place.

False Cause (*Post Hoc Ergo Propter Hoc*)   A ***post hoc* fallacy** mistakenly assumes that one event causes another because they occur sequentially. An old (and not especially funny) joke illustrates the *post hoc* fallacy. Mac approaches Jack and asks,

> ***reductio ad absurdum* fallacy** Fallacious reasoning that unfairly attacks an argument by extending it to such extreme lengths that it looks ridiculous.
>
> **either–or fallacy** Fallacious reasoning that sets up false alternatives, suggesting that if the inferior one must be rejected, then the other must be accepted.
>
> ***post hoc* fallacy** Fallacious reasoning that mistakenly assumes that one event causes another because they occur sequentially.

UNDERSTANDING DIVERSITY

Cultural Differences in Persuasion

Different individuals have a tendency to view persuasion differently, and often these differences are based on cultural background. Even the ability to recognize logical argument is, to a certain extent, culturally determined. Not all cultures use logic in the same way that the European American culture does. The influence of the dominant culture is seen even in the way we talk about argumentation. When we talk about "defending" ideas and "attacking our opponent's position," we are using male-oriented militaristic/aggressive terms. Logic is also based on a trust in objective reality, on information that is verifiable through our senses. As one researcher points out, such a perspective can be culturally influenced:

> Western culture assumes a reality that is materialist and limited to comprehension via the five senses. African

culture assumes a reality that is both material and spiritual viewed as one and the same.[14]

The way logic is viewed also differs between Eastern and Western Hemisphere cultures. As Larry A. Samovar and Richard E. Porter point out:

> Westerners discover truth by active searching and the application of Aristotelian modes of reasoning. On the contrary, many Easterners wait patiently, and if truth is to be known it will make itself apparent.[15]

It is because of cultural differences such as these that speech experts have always recommended a blend of logical and emotional evidence.

TABLE 15.2

Other Common Fallacies

FALLACY	DEFINITION	EXAMPLE
Straw Man	Setting up an argument that was not proposed and then attacking it as if it were the original argument.	"You say we should support animal rights, but many animal rights activists have supported the destruction of research facilities." (The speaker then goes on to argue that the destruction of research facilities is wrong.)
Red Herring	Shifting the focus to a tangential subject, similar to dragging a fish across a trail to distract a bloodhound.	"Bill says that buying a term paper is immoral. But what is morality, anyway?" (The speaker then goes on to discuss this philosophical question.)
Begging the Question	Repeating an argument but never providing support for a point of view.	"I can't believe people eat dog. That's just plain gross. Why? Because it's a dog, of course. How could someone eat a dog?"
Faulty Analogy	Using a comparison suggesting that two things are more alike than they really are.	"If we legalize gay marriage, next we'll legalize marriage between people and their pets."
Hasty Generalization	Reaching an unjustifiable conclusion after making assumptions or misunderstanding statistics.	"You are likely to be shot if you visit New York City." (In fact, fewer people are murdered, per capita, in New York City than in most rural American small towns.)

"Hey, why are you snapping your fingers?" Jack replies, "To keep the elephants away." Mac is incredulous: "What are you talking about? There aren't any elephants within a thousand miles of here." Jack smiles and keeps on snapping: "I know. Works pretty well, doesn't it?"

In real life, *post hoc* fallacies aren't always so easy to detect. For example, one critic of education pointed out that the increase in sexual promiscuity among adolescents began at about the same time that the courts prohibited prayer in public schools. A causal link in this case may exist: Decreased emphasis on spirituality could contribute to promiscuity. But it would take evidence to establish a *definite* connection between the two phenomena.

Appeal to Authority (*Argumentum ad Verecundiam*) An ***argumentum ad verecundiam* fallacy** involves relying on the testimony of someone who is not an authority in the case being argued. Relying on experts is not a fallacy, of course. A movie star might be just the right person to offer advice on how to seem more glamorous, and a professional athlete could be the best person to comment on what it takes to succeed in organized sports. But an *ad verecundiam* fallacy occurs when the movie star promotes a political candidate or the athlete tells us why we should buy a certain kind of automobile. When considering endorsements and claims, it's smart to ask yourself whether the source is qualified to make them.

Bandwagon Appeal (*Argumentum ad Populum*) An ***argumentum ad populum* fallacy** is based on the often dubious notion that, just because many people favor an idea, you should, too. Sometimes, of course, the mass appeal of an idea can be a sign of its merit. If most of your friends have enjoyed a film or a new book, there is probably a good chance that you will, too. But in other cases widespread acceptance of an idea is no guarantee of its validity. In the face of almost universal belief to the contrary, Galileo reasoned accurately that the Earth is not the center of the universe,

***argumentum ad verecundiam* fallacy** Fallacious reasoning that tries to support a belief by relying on the testimony of someone who is not an authority on the issue being argued.

***argumentum ad populum* fallacy** Fallacious reasoning based on the dubious notion that because many people favor an idea, you should, too.

and he suffered for his convictions. The lesson here is simple to comprehend but often difficult to follow: When faced with an idea, don't just follow the crowd. Consider the facts carefully and make up your own mind.

Other Common Fallacies   There is a wide range of other common fallacies, as shown in Table 15.2. Often, dogmatic speakers don't even realize they are using faulty logic; other times, it is purposeful manipulation. How many of these do you recognize from advertising, politics, or everyday arguments? How many other fallacies can you name?

🎯 15.4 Adapting to the Audience

When making a persuasive speech, it is important to know as much as possible about your audience. For one thing, you should appeal to the values of your audience whenever possible, even if they are not *your* strongest values. This advice does not mean you should pretend to believe in something. According to our definition of *ethical persuasion*, pretense is against the rules. It does mean, however, that you have to stress those values that the members of your audience feel most forcefully.[16]

In addition, you should analyze your audience carefully to predict the type of response you will get. Sometimes you have to pick out one part of your audience—a **target audience**, the subgroup you must persuade to reach your goal—and aim your speech mostly at those members. Some of your audience members might be so opposed to what you are advocating that you have no hope of reaching them. Still others might already agree with you, so they do not need to be persuaded. A middle portion of your audience members might be undecided or uncommitted, and they would be the most productive target for your appeals.

> **target audience** That part of an audience that must be influenced to achieve a persuasive goal.

Of course, you need not ignore that portion of your audience that does not fit your target. For example, if you were giving a speech against smoking, your target might be the smokers in your class. Your main purpose would be to get them to quit, but at the same time, you could convince the nonsmokers not to start and to use their influence to help their smoking friends quit.

All of the methods of audience analysis described in Chapter 12—surveys, observation, interviews, and research—are valuable in collecting information about your audience for a persuasive speech.

Establish Common Ground

It helps to stress as many similarities as possible between yourself and your audience members. This technique helps prove that you understand them—if not, why should they listen to you? Also, if you share a lot of common ground, it shows you agree on many things. Therefore, it should be easy to settle one disagreement: the one related to the attitude or behavior you would like them to change.

The manager of public affairs for *Playboy* magazine gave a good demonstration of establishing common ground when he reminded a group of Southern Baptists that they shared some important values with him:

> I am sure we are all aware of the seeming incongruity of a representative of *Playboy* magazine speaking to an assemblage of representatives of the Southern Baptist convention. I was intrigued by the invitation when it came last fall, though I was not surprised. I am grateful for your genuine and warm hospitality, and I am flattered (although again not surprised) by the implication that I would have something to say that could have meaning to you people.

Both *Playboy* and the Baptists have indeed been considering many of the same issues and ethical problems; and even if we have not arrived at the same conclusions, I am impressed and gratified by your openness and willingness to listen to our views.[17]

Organize According to the Expected Response

It is much easier to get an audience to agree with you if the members have already agreed with you on a previous point. Therefore, you should arrange your points in a persuasive speech so that you develop a "yes" response. In effect, you get your audience into the habit of agreeing with you. For example, if you were giving a speech on organ donation, you might begin by asking the members of the audience if they would like to be able to get a kidney if they needed one. Then you might ask them if they would like to have a major role in curbing tragic and needless dying. The presumed response to both questions is "yes." It is only when you have built a pattern of "yes" responses that you would ask the audience to sign organ donor cards.

An example of a speaker who was careful to organize material according to expected audience response is the late Robert Kennedy. Kennedy, when speaking on civil rights before a group of South Africans who believed in racial discrimination, arranged his ideas so that he spoke first on values that he and his audience shared—values like independence and freedom.[18]

If audience members are already basically in agreement with you, you can organize your material to reinforce their attitudes quickly and then spend most of your time convincing them to take a specific course of action. If, on the other hand, they are hostile to your ideas, you have to spend more time getting the first "yes" out of them.

Neutralize Potential Hostility

One of the trickier problems in audience adaptation occurs when you face an audience hostile to you or your ideas. Hostile audiences are those who have a significant number of members who feel adversely toward you, your topic, or the speech situation. Members of a hostile audience could range from unfriendly to violent. Two guidelines for handling this type of audience are to (1) show that you understand their point of view, and (2) if possible, use appropriate humor.

ASK YOURSELF

Imagine an audience that is as different from you as possible. Assign this audience backgrounds, beliefs, and values that are the opposite of yours. How would you adapt your speech accordingly?

credibility The believability of a speaker or other source of information.

🔍 15.5 Building Credibility as a Speaker

Credibility refers to the believability of a speaker. Credibility isn't an objective quality; rather, it is a perception in the minds of the audience. In a class such as the one you're taking now, students often wonder how they can build their credibility. After all, the members of the class tend to know one another well by the time the speech assignments roll around. This familiarity illustrates why it's important to earn a good reputation before you speak, through your class comments and the general attitude you've shown.

It is also possible for credibility to change during a speaking event. In fact, researchers speak in terms of initial credibility (what you have when you first get up to speak), derived credibility (what you acquire while speaking), and terminal credibility (what you have after you finish speaking). It is not uncommon for a student with low initial credibility to earn increased credibility while speaking and to finish with much higher terminal credibility.

UNDERSTANDING YOUR COMMUNICATION

Persuasive Speech

Use this self-assessment for a persuasive speech you have presented or plan to present.

1. Have you set a clear, persuasive purpose?

    I'VE DONE MY BEST.   I'VE GOT WORK TO DO.   I'VE BARELY STARTED.

2. Is your purpose in the best interest of the audience?

    I'VE DONE MY BEST.   I'VE GOT WORK TO DO.   I'VE BARELY STARTED.

3. Have you structured the message to achieve a "yes" response?

    I'VE DONE MY BEST.   I'VE GOT WORK TO DO.   I'VE BARELY STARTED.

4. Have you used solid evidence for each point?

    I'VE DONE MY BEST.   I'VE GOT WORK TO DO.   I'VE BARELY STARTED.

5. Have you used solid reasoning for each point?

    I'VE DONE MY BEST.   I'VE GOT WORK TO DO.   I'VE BARELY STARTED.

6. Have you adapted to your audience?

    I'VE DONE MY BEST.   I'VE GOT WORK TO DO.   I'VE BARELY STARTED.

7. Have you built your own credibility?

    I'VE DONE MY BEST.   I'VE GOT WORK TO DO.   I'VE BARELY STARTED.

8. Is your information true to the best of your knowledge?

    I'VE DONE MY BEST.   I'VE GOT WORK TO DO.   I'VE BARELY STARTED.

Scoring on this assessment is self-evident.

Without credibility, you won't be able to convince your listeners that your ideas are worth accepting, even if your material is outstanding. On the other hand, if you can develop a high degree of credibility in the eyes of your listeners, they will likely open up to ideas they wouldn't otherwise accept. Members of an audience form judgments about the credibility of a speaker based on their perception of many characteristics, the most important of which might be called the "three Cs" of credibility: competence, character, and charisma.[19]

Competence

Competence refers to the speaker's expertise on the topic. Sometimes this competence can come from personal experience that will lead your audience to regard you as an authority on the topic you are discussing. If everyone in the audience knows you've earned big profits in the stock market, they will probably take your investment advice seriously. If you say that you lost 25 pounds from a diet-and-exercise program, most audience members will be likely to respect your opinions on weight loss.

The other way to be seen as competent is to be well prepared for speaking. A speech that is well researched, organized, and presented will greatly increase the audience's perception of the speaker's competence. Your personal credibility will

Persuasion in the World of Sales

Persuasive skills have a range of applications in the work-place. Business consultant George Rodriguez makes it clear that developing a successful sales plan is very much like the planning involved in building a persuasive speech.

"A sales plan is basically your strategic and tactical plan for achieving your marketing objectives," Rodriguez explains. "It is a step-by-step and detailed process that will show how you will acquire new business; and how you will gain more business from your existing customer base."[20]

The process of audience analysis is as important in sales-plan development as it is in persuasive speaking. "The first step is to clearly identify your target markets," Rodriguez says. "Who are more likely to buy your product? The more defined your target market, the better. Your target market can be defined as high-income men ages 30–60 who love to buy the latest electronic gadgets; or mothers with babies 0–12 months old living in urban areas."

And don't forget that persuasion is interactive. "Prospects are more likely to purchase if you can talk to them about

solving their problems," Rodriguez points out. He is far from alone in pointing out the importance of thinking in terms of problems and solutions. Business consultant Barbara Sanfilippo advises her clients to "prepare, prepare, and plan your calls. Today's customers and prospects have very little time to waste. They want solutions. A sales consultant who demonstrates a keen understanding of customers' needs and shows up prepared will earn the business."[21] Sanfilippo suggests reviewing the customer's website and interviewing key people in advance of the meeting.

Sanfilippo also points out the importance of building cred-ibility: "How can you stand out from the pack of sales pro-fessionals and consultants all offering similar services?" she asks rhetorically. "Establish Credibility and Differentiate!" But George Rodriguez probably has the last word on the value of persuasive speaking to the sales professional. Before you make that first sales call, he says, "You may want to take courses on how to improve your confidence and presenta-tion skills."

therefore be enhanced by the credibility of your evidence, including the sources you cite, the examples you choose, the way you present statistics, the quality of your visual aids, and the precision of your language.

Character

Competence is the first component of being believed by an audience. The second is being trusted, which is a matter of character. *Character* involves the audience's perception of at least two ingredients: honesty and impartiality. You should try to find ways to talk about yourself (without boasting, of course) that demon-strate your integrity. You might describe how much time you spent researching the subject or demonstrate your open-mindedness by telling your audience that you changed your mind after your investigation. For example, if you were giving a speech arguing against a proposed tax cut in your community, you might begin this way:

> You might say I'm an expert on the municipal services of this town. As a lifelong resident, I owe a debt to its schools and recreation programs. I've been protected by its police and firefighters and served by its hospitals, roads, and sanitation crews.
>
> I'm also a taxpayer who's on a tight budget. When I first heard about the tax cut that's been proposed, I liked the idea. But then I did some in-depth investigation into the possible effects, not just to my tax bill but to the qual-ity of life of our entire community. I looked into our municipal expenses and

into the expenses of similar communities where tax cuts have been mandated by law.

Charisma

Charisma is spoken about in the popular press as an almost indefinable, mystical quality. Even the dictionary defines it as "a special quality of leadership that captures the popular imagination and inspires unswerving allegiance and devotion." Luckily, communication scholars favor a more down-to-earth definition. For them, charisma is the audience's perception of two factors: the speaker's enthusiasm and likability. Whatever the definition, history and research have both shown us that audiences are more likely to be persuaded by a charismatic speaker than by a less charismatic one who delivers the same information.

Communication scholars sometimes call enthusiasm "dynamism." Your enthusiasm will mostly be perceived from the way you deliver your remarks, not from what you say. The nonverbal parts of your speech, far better than your words, will show that you believe in what you are saying. Is your voice animated and sincere? Do your gestures reflect your enthusiasm? Do your facial expression and eye contact show you care about your audience?

You can boost your likability by showing that you like and respect your audience. Insincere flattery will probably boomerang, but if you can find a way to give your listeners a genuine compliment, they'll be more receptive to your ideas.

Building your personal credibility through recognition of the roles of competence, character, and charisma is an important component of your persuasive strategy. When combined with careful consideration of audience adaptation and persuasive structure and purpose, it will enable you to formulate the most effective strategy possible.

Aristotle warned of the strength of human emotions. Extending the point, he warned that political leaders who used only emotional appeals rather than reasoning and ethical standards were dangerous. In a phenomenon researchers call **confirmation bias**, people have an emotional tendency to interpret new information as confirmation of their existing beliefs. For example, if workers already feel that regulations may cost them their jobs, they will believe this even more strongly when presented with evidence to the contrary. If people want to believe something because of the emotions involved, they will, even if it is both untrue and ultimately not in their best interests.

> **confirmation bias** The emotional tendency to interpret new information as confirmation of one's existing beliefs.

🔊 15.6 Sample Speech

The sample speech for this chapter was presented by Shayla Cabalan, the student from the University of Indianapolis whose profile began this chapter.

Shayla presents a proposition of policy (child marriage should be ended) that she felt strongly about. Although Shayla presented this speech competitively, she had a broader purpose than winning tournaments. She sought to actuate her audience toward a specific action:

> Via handouts and word of mouth, I directed them to numerous resources they could use in order to get in touch with their local lawmakers or organizations fighting for a shift in the struggle against child marriage.[22]

She bolsters her credibility throughout by showing the depth of her research and analysis into the problem she seeks to solve, while carefully balancing her emotional and logical evidence.

Shayla's outline, with the paragraphs of the speech noted in parentheses, might look like this:

Speech Outline

INTRODUCTION

  I. What's marriage? (1)

  II. Thesis statement: This happens here. (2–3)

  III. Transition/Preview main points: (4)

BODY

I. Causes (5–10)

  A. First, antiquated customs, (6–7)

  B. And second, resistant state lawmakers. (8–10)

    1. While nearly every state "prohibits" people younger than 18 from marrying, in reality, there are always loopholes and exceptions to the law. (9)

    2. Bills to change these laws in Virginia, New York, and Maryland have all died, while many states haven't even acted at all. (10)

Transition: who have these laws been working for? (11–12)

  II. Effects (13–17)

  A. Child marriage can negatively impact a child's mental and physical health. (14–16)

    1. Married children are more susceptible to disease, due to stress and lack of education. (14)

    2. Married children are more susceptible to psychiatric disorders. (15)

  B. Child marriage takes away a child's future, priming her for abuse. (16)

    1. Married children are more likely to drop out of high school. (16)

    2. Married children are four times less likely to graduate from college. (16)

    3. Married children are more likely to live in poverty. (16)

  C. Child marriage primes girls for abuse. (17)

    1. Married children are more likely to be physically and emotionally abused by their spouses. (17)

    2. Married children do not have access to domestic violence shelters. (17)

    3. Married children do not have access to legal representation. (17)

Transition: Johnson's fight led her to the Statehouse for a solution. (18)

  III. Solutions (19–24)

  A. We should support organizations that work to ban child marriage. (20–21)

  B. We should encourage lawmakers to change laws regarding marriage. (22–24)

Transition: Review of main points. (25)

CONCLUSION

  I. Restatement of thesis (26)

  II. Memorable statement "It was supposed to be the happiest day of her life." (27)

Shayla's bibliography appears after the speech.

SAMPLE SPEECH   **Shayla Cabalan**

Eleven and Engaged: On America's Unseen Child Marriage Crisis

1   The *New York Times* of May 26, 2017, details the story of an 11-year-old girl who was brutally raped by a 20-year-old member of her parish. When child welfare authorities started questioning the girl's family and members of her church, the child's mother asked her if she wanted to get married in order to avoid the whole sticky situation. Her response: "I don't know. What's marriage?" The girl was told she would find out after the ceremony.

An anecdote designed to engage audience emotions by appealing to their values.

2   Unfortunately, this girl's story is not uncommon, and neither is it foreign. Subconsciously, you may be assigning blame for this abuse to far-off people in far off places, such as Tanzania, Bangladesh, or Pakistan—but in reality, this little girl's name is Sherry Johnson, she grew up in Tampa, Florida, and at the age of 11, she was forced to marry her 20-year-old rapist.

3   According to the same *New York Times* article, the practice of child marriage is a startling reality right here in the United States, right now, in 2018. Unchained at Last, an organization attempting to ban child marriage throughout the United States, found that in our country, there were nearly a quarter of a million child marriages between 2000 and 2010, marriages that were fraught with abuse, neglect, and trauma.

Logical appeal with factual evidence.

4   In order to solve this immediate problem, we must analyze the causes, effects, and solutions of this human rights abuse—an abuse that Johnson, in *The Guardian* of February 6, 2018, states "put a definite end to her childhood." Sherry Johnson graphically describes her forced marriage by stating, "Instead of putting the handcuffs on him and sending him to prison, they put the handcuffs on me and imprisoned me in a marriage."

Main points, introduced with more emotional engagement.

5   The causes of our child marriage crisis are twofold: first, antiquated customs, and second, resistant state lawmakers.

First two subpoints.

6   First, antiquated customs. In *Shattering the Silence Surrounding Forced and Early Marriages in the US*, Dr. Julia Alanen states that parents who marry off their minor children are often motivated by the need to "control sexuality and unwanted behavior, [prevent] unsuitable relationships, [protect] family honor and perceived religious or cultural ideals, or achieve financial gain."

Logical support . . .

7   The *Washington Post* of February 10, 2017, details the story of 15-year-old Sara Siddiqui of Nevada, whose father married her off to a man 13 years her senior, simply because she was seeing someone of a different cultural background. Siddiqui notes, "I couldn't even drive yet when I was handed over to this man. I wasn't ready to take care of myself, and yet, I was thrown into taking care of a husband and being a mother." And this is how the epidemic spreads, perpetuated by the helplessness of a child who has no real resources.

. . . backed up with emotional evidence.

8   Second, resistant state lawmakers. According to the *Pew Research Center* of November 2016, child marriage is "legal in almost every state."

Development of second subpoint begins . . .

9   While nearly every state "prohibits" people younger than 18 from marrying, in reality, there are always loopholes and exceptions to the law. For example, in 36 states, minors can be married with judicial consent; in 34 states, 16- and 17-year-olds can marry with parents' permission; and shockingly enough, in New Hampshire and Massachusetts, 13-year-old girls and 14-year-old boys can be married with either judicial consent or parents' permission.

. . . with logical evidence . . .

… again, backed up with emotional evidence.

10 Bills to change these laws in Virginia, New York, and Maryland have all died, while many states haven't even acted at all. The previously mentioned *New York Times* article details the story of 13-year-old Girl Scout, Cassandra Levesque of New Hampshire, who attempted to change the underage marriage law in her state—but was met with ridicule. "We're asking the legislature to repeal a law that's been on the books for over a century, which has been working, without difficulty, on the basis of a request from a minor doing a Girl Scout project," scoffed state representative David Bates.

Transition to next set of subpoints.

11 The question at hand, however, is exactly who have these laws been working for? Because it's certainly not the children.

Development of these subpoints begins.

12 The previously mentioned *Guardian* notes that reaching out to local politicians about her forced marriage "was part of [Johnson's] healing process to tell [her] story." Unfortunately, her plight often fell on deaf ears.

Transition to third main point.

13 This social abuse brings some very detrimental effects. Child marriage can negatively impact a child's mental and physical health and takes away their future, priming them for abuse.

First subpoint developed with a logic argument and backed up with emotional evidence.

14 First, their mental and physical health. Pediatric doctors Matthew Dupre and Sarah Meadow explain that women who marry at 18 or younger face a 23% higher risk of heart attack, diabetes, cancer, and stroke than do women who marry between ages 19 and 25, largely because early marriage leads to increased stress and a lack of education.

15 The same doctors found that child marriage increases the risk of lifetime and current psychiatric disorders. Truly, according to Dupre and Meadow, it seems that children entering marriage are set to face "a lifetime … of mental disorders," in comparison to adults entering into marriage.

Second subpoint developed.

16 Second, child marriage takes away a child's future. The previously-mentioned Sherry Johnson recalls spending countless days changing diapers and taking care of her family instead of going to school. Dr. Gordon B. Dahl found that American girls who get married prior to the age of 19 are 50% more likely than their unmarried counterparts to drop out of high school, as well as four times less likely to graduate from college. Shockingly, Dahl discovered that they are also "31 percentage points" more likely to live in poverty when they are older.

Third subpoint.

17 Furthermore, child marriage primes children for future abuse. The *World Policy Analysis Center* in 2015 found that women who marry before the age of 18 are three times more likely to be physically and emotionally abused by their spouses. Moreover, the previously mentioned *Washington Post* notes that while adults fleeing marriage have access to domestic violence shelters and legal representation, children fleeing marriage absolutely do not. Domestic violence shelters do not take in children, and not only can children not afford to pay attorney's fees, they also cannot file legal action in their own names. Essentially, it is extremely easy for a child to fall into a life of abuse permanently.

18 Johnson's fight to change Florida's child marriage laws eventually led her to the State House, where she faced immense scrutiny from both male and female politicians. One female politician even asked, "Won't you increase abortion rates if you end child marriage?" Johnson was speechless.

Beginning of final main point, Solutions.

19 We must take action by first, supporting organizations trying to ban child marriage, and second, encouraging lawmakers to change state law regarding marriage.

Her call to action is made as easy as possible for the audience to follow.

20 First, support organizations and groups advocating for an end to underage marriage. One particular organization that you can easily support and donate to is the previously mentioned nonprofit, Unchained at Last. Founded by Fraidy Reiss, a forced underage marriage survivor, Unchained at Last aims to help any woman or girl in the United States who has been wrongfully forced into marriage.

21 While there are many organizations around the world attempting to end underage marriage, Unchained at Last is the only organization attempting to end it in the United States specifically. Their website provides resources for advocates, information about current legislation, and even provides an automated email system to help visitors contact their elected officials. You can find more information and take action at www.unchainedatlast.org.

22 Second, demand lawmakers change laws regarding underage marriage in the United States. Just three months ago, Sherry's Johnson's fight to change Florida state underage marriage laws found success. The *Miami Herald* of February 1, 2018, notes that the Florida Senate voted to unanimously pass a bill that would end child marriage in the state. A similar House bill now moves to a committee vote before passing to the entire House for a full vote.

23 While this undoubtedly sounds like a victory, there is still a lot of work that needs to be done. In my home state of Indiana, the minimum legal marriage age is 17 with parental consent and 18 without. However, there are those legal loopholes that need to be addressed, where a child as young as 15 years old can still be married in cases of pregnancy, birth of child, and with approval of judge of superior or county court.

24 These legal loopholes must be addressed. Johnson is now 58 years old. She was forced into marriage at 10 years old. Forty-eight years is too long to wait for redemption.

25 Today, we discussed the causes, effects, and solutions to underage marriage in the United States. *Transition to conclusion.*

26 To this day, according to the previously cited *Washington Post,* laws in 27 states still do not specify a minimum marriage age below which a child cannot marry, and more states have not acted to raise the age to 18 while also taking into account loopholes. *Restatement of thesis ...*

27 Despite this, Sherry Johnson's fight, for now, seems to have a positive outlook. She contends, "To know that there's something that's being done about it ... I smile from within to know that children will not have to face what I've been through." Still, she recalls someone telling her not to cry on her wedding day, as it was supposed to be the happiest day of her life. Perhaps it could have been if it didn't happen in her childhood. *... and final memorable statement.*

Bibliography

Alanen, J. (2012). Shattering the silence surrounding forced and early marriage in the United States. *SSRN Electronic Journal.* doi:10.2139/ssrn.2143910

Conflict of laws. Marriage. Polygamous marriage. (1923). *Columbia Law Review, 23*(5), 489. doi:10.2307/1112343

Dahl, G. (2005). Early teen marriage and future poverty. *Demography.* doi:10.3386/w11328

Dupre M. E., & Meadows, S. O. (2007). Disaggregating the effects of marital trajectories on health. *Journal of Family Issues, 28*(5), 623–652. doi:10.1177/0192513x06296296

Girls Not Brides. (2015, July 22). World policy analysis center assessing national action. Retrieved from https://www.girlsnotbrides.org/resource-centre/assessing-national-action-on-protection-from-child-marriage/world-policy-analysis-center-assessing-national-action

Koh, E. (2018, February 1). Raped at 8 and forced to wed at 11, this woman tries to end child marriage. Retrieved from http://www.miamiherald.com/news/politics-government/state-politics/article197899194.html

Kristof, N. (2017, May 26). 11 years old, a mom, and pushed to marry her rapist in Florida. Retrieved from https://www.nytimes.com/2017/05/26/opinion/sunday/it-was-forced-on-me-child-marriage-in-the-us.html

McClendon D., & Sandstrom, A. (2016, November 01). Child marriage is rare in the U.S., though this varies by state. Retrieved from http://www.pewresearch.org/fact-tank/2016/11/01/child-marriage-is-rare-in-the-u-s-though-this-varies-by-state

Reiss, F. (2017, February 10). Perspective|Why can 12-year-olds still get married in the United States? Retrieved from https://www.washingtonpost.com/posteverything/wp/2017/02/10/why-does-the-united-states-still-let-12-year-old-girls-get-married/?utm_term=.c1a354fc1126

Zee, R. V. (2018, February 06). "It put an end to my childhood": The hidden scandal of US child marriage. Retrieved from https://www.theguardian.com/inequality/2018/feb/06/it-put-an-end-to-my-childhood-the-hidden-scandal-of-us-child-marriage

MAKING THE GRADE

At www.oup.com/he/adler-uhc14e, you will find a variety of resources to enhance your understanding, including video clips, animations, self-quizzes, additional activities, audio and video summaries, interactive self-assessments, and more.

OBJECTIVE 15.1   Identify the primary characteristics of persuasion.

- Persuasion is the act of moving someone, through communication, toward a belief, attitude, or behavior. Despite a sometimes bad reputation, persuasion can be both worthwhile and ethical.
- Ethical persuasion requires that the speaker be sincere and honest and avoid such behaviors as plagiarism.
- Ethical persuasion must also serve the best interest of the audience, as perceived by the speaker.
  - > What are some examples of persuasive speaking in your everyday life?
  - > Using the examples you identified earlier, describe the difference between ethical and unethical persuasion.
  - > Describe an ethical approach to a persuasive presentation you could deliver.

OBJECTIVE 15.2   Compare and contrast different types of persuasion.

- Persuasion can be categorized according to the type of proposition (fact, value, or policy).
- Persuasion can be categorized according to the desired outcome (convincing or actuating).
- Persuasion can be categorized according to the type of approach (direct or indirect).
  - > For persuasive speeches you could deliver, describe propositions of fact, value, and policy.
  - > In your next persuasive speech, do you intend to convince or to actuate your audience? Why have you chosen this goal?
  - > In your next persuasive speech, are you planning to use a direct or indirect persuasive approach? Why?

OBJECTIVE 15.3   Apply the guidelines for persuasive speaking to a speech you will prepare.

A persuasive strategy is put into effect through the use of several strategies. These include
- setting a specific, clear persuasive purpose,
- structuring the message carefully,
- using solid evidence (including emotional evidence),
- using careful reasoning,
- adapting to the audience, and
- building credibility as a speaker.
  - > For a speech from your personal experience or one that you have watched online (e.g., a TED Talk), identify its purpose, message structure, use of evidence and reasoning, audience adaptation, and enhancement of speaker credibility.
  - > Apply the preceding guidelines to a speech you are developing.

OBJECTIVE 15.4 Explain how to best adapt a specific speech to a specific audience.

In adapting to your audience, you should
- establish common ground,
- organize your speech in such a way that you can expect a "yes" response along each step of your persuasive plan, and
- take special care with a hostile audience.

> Give examples from speeches you have observed (either in person or online) that illustrate these three strategies. Explain why each strategy contributes to the success of the speech.

> Apply these strategies to a speech you are developing.

OBJECTIVE 15.5 Improve your credibility in your next persuasive speech.

- In building credibility, you should keep in mind the audience's perception of your competence, character, and charisma.

> For the most effective persuasive speech you can recall, describe how the speaker enhanced his or her competence, character, and charisma.

> How can you enhance your perceived competence, character, and charisma in a speech you are developing? Suggest specific improvements.

KEY TERMS

actuate p. 377
ad hominem fallacy p. 382
anchor p. 373
argumentum ad populum fallacy p. 384
argumentum ad verecundiam fallacy p. 383
confirmation bias p. 389
convincing p. 377
credibility p. 386
direct persuasion p. 377
either–or fallacy p. 382
ethical persuasion p. 375
ethos p. 378
fallacy p. 382
indirect persuasion p. 377
latitude of acceptance p. 373
latitude of noncommitment p. 373
latitude of rejection p. 373
logos p. 378
pathos p. 379
persuasion p. 373
post hoc fallacy p. 383

proposition of fact p. 376
proposition of policy p. 376
proposition of value p. 376
reductio ad absurdum fallacy p. 382
social judgment theory p. 373
target audience p. 385

ACTIVITIES

1. **Audience Latitudes of Acceptance** To better understand the concept of latitudes of acceptance, rejection, and noncommitment, formulate a list of perspectives on a topic of your choice. This list should contain 8 to 10 statements that represent a variety of attitudes, such as the list pertaining to the pro-life/pro-choice issue on pages 373–374. Arrange this list from your own point of view, from most acceptable to least acceptable. Then circle the single statement that best represents your own point of view. This will be your "anchor." Underline those items that also seem reasonable. These make up your latitude of acceptance on this issue. Then cross out the numbers in front of any items that express opinions that you cannot accept. These make up your latitude of rejection. Those statements that are left would be your latitude of noncommitment. Do you agree that someone seeking to persuade you on this issue would do best by advancing propositions that fall within this latitude of noncommitment?

2. **Personal Persuasion** When was the last time you changed your attitude about something after discussing it with someone? In your opinion, was this persuasion interactive? Not coercive? Incremental? Ethical? Explain your answer.

3. **Propositions of Fact, Value, and Policy** Which of the following are propositions of fact, propositions of value, and propositions of policy?

   a. "Three Strikes" laws that put felons away for life after their third conviction are/are not fair.
   b. Elder care should/should not be the responsibility of the government.
   c. The mercury in dental fillings is/is not healthy for the dental patient.
   d. Congressional pay raises should/should not be delayed until an election has intervened.
   e. Third-party candidates strengthen/weaken American democracy.
   f. National medical insurance should/should not be provided to all citizens of the United States.
   g. Elderly people who are wealthy do/do not receive too many social security benefits.

h. Tobacco advertising should/should not be banned from all media.

i. Domestic violence is/is not on the rise.

j. Pit bulls are/are not dangerous animals.

4. Structuring Persuasive Speeches For practice in structuring persuasive speeches, choose one of the following topics, and provide a full-sentence outline that conforms to the outline in Figure 15.3.

a. It should/should not be more difficult to purchase a handgun.

b. Public relations messages that appear in news reports should/should not be labeled as advertising.

c. Newspaper recycling is/is not important for the environment.

d. Police should/should not be required to carry nonlethal weapons only.

e. Parole should/should not be abolished.

f. The capital of the United States should/should not be moved to a more central location.

g. We should/should not ban capital punishment.

h. Bilingual education should/should not be offered in all schools in which students speak English as a second language.

5. Find the Fallacy Test your ability to detect shaky reasoning by identifying which fallacy is exhibited in each of the following statements.

> *Ad hominem*

> *Ad absurdum*

> Either–or

> *Post hoc*

> *Ad verecundiam*

> *Ad populum*

a. Some companies claim to be in favor of protecting the environment, but you can't trust them. Businesses exist to make a profit, and the cost of saving Earth is just another expense to be cut.

b. Take it from me, imported cars are much better than domestic cars. I used to buy only American, but the cars made here are all junk.

c. Rap music ought to be boycotted. After all, the number of assaults on police officers went up right after rap became popular.

d. Carpooling to cut down on the parking problem is a stupid idea. Look around—nobody carpools!

e. I know that staying in the sun can cause cancer, but if I start worrying about every environmental risk I'll have to stay inside a bomb shelter breathing filtered air, never drive a car or ride my bike, and I won't be able to eat anything.

f. The biblical account of creation is just another fairy tale. You can't seriously consider the arguments of those Bible-thumping, know-nothing fundamentalists, can you?

6. The Credibility of Persuaders Identify someone who tries to persuade you via public speaking or mass communication. This person might be a politician, a teacher, a member of the clergy, a coach, a boss, or anyone else. Analyze this person's credibility in terms of the three dimensions discussed in the chapter. Which dimension is most important in terms of this person's effectiveness?

Notes

FRONT MATTER

1. Research supporting these claims is cited in Burgoon, J. K., & Hoobler, G. D. (2002). Nonverbal signals. In M. L. Knapp & J. A. Daly (Eds.), *Handbook of interpersonal communication* (3rd ed., pp. 240–299). Thousand Oaks, CA: Sage.
2. Employers: Verbal communication most important candidate skill. (2016, February 24). National Association of Colleges and Employers Center for Career Development and Talent Acquisition. Retrieved from http://www.naceweb.org/career-readiness/competencies/employers-verbal-communication-most-important-candidate-skill.
3. Mikkelson, A. C., York, J. A., & Arritola, J. (2015). Communication competence, leadership behaviors, and employee outcomes in supervisor-employee relationships. *Business and Professional Communication Quarterly, 78*(3), 336–354.
4. Degges-White, S. (2015, March 23). The 13 essential traits of good friends. *Psychology Today*. Retrieved from https://www.psychologytoday.com/blog/lifetime-connections/201503/the-13-essential-traits-good-friends.
5. Tartakovsky, M. (2014). 5 things that make a good partner. *Psych Central*. Retrieved from https://psychcentral.com/blog/archives/2014/06/01/5-things-that-make-a-good-partner.

CHAPTER 1

1. Lee, J. (2017, August 15). Route 7 is one of Metro Transit's most challenging bus lines, and driver Nathan Vass loves it. *The Seattle Times*. Retrieved from https://www.seattletimes.com/seattle-news/transportation/route-7-is-one-of-metro-transits-most-challenging-bus-lines-and-driver-nathan-vass-loves-it.
2. 2017 Wall of Fame honorees. (n.d.). Nathan Vass, transit operator. King County. Retrieved from https://kingcounty.gov/depts/transportation/metro/employees/2017-wall-of-fame.aspx.
3. Ishisaka, N. (2018, November). Seattle's most influential people 2018: The next generation. *The Seattle Times*. Retrieved from https://www.seattlemag.com/news-and-features/seattles-most-influential-people-2018-next-generation.
4. This bus driver is making the most of his route. (2018, November 21). K5News. Retrieved from https://www.king5.com/article/entertainment/television/programs/new-day-northwest/this-bus-driver-is-making-the-most-of-his-route/281-616547321. Quote is at 4:29 mins.
5. Average hours per day spent on socializing and communicating by the U.S. population from 2009 to 2014. (2015). *Statistica*. Retrieved from http://www.statista.com/statistics/189527/daily-time-spent-on-socializing-and-communicating-in-the-us-since-2009.

6. Jaremka, L. M., Andridge, R. R., Fagundes, C. P., Alfano, C. M., Povoski, S. P., Lipari, A. M., . . . Kiecolt-Glaser, J. K. (2014). Pain, depression, and fatigue: Loneliness as a longitudinal risk factor. *Health Psychology, 33*(9), 948–957.
7. Newall, N. G., Chipperfield, J. G., Bailis, D. S., & Stewart, T. L. (2013). Consequences of loneliness on physical activity and mortality in older adults and the power of positive emotions. *Health Psychology, 32*(8), 921–924.
8. Cacioppo, S., Capitanio, J. P., & Cacioppo, J. T. (2014). Toward a neurology of loneliness. *Psychological Bulletin, 140*(6), 1464–1504.
9. Arroyo, A., & Segrin, C. (2011). The relationship between self- and other-perceptions of communication competence and friendship quality. *Communication Studies, 62*(5), 547–562.
10. Määttä, K., & Uusiautti, S. (2013). Silence is not golden: Review of studies of couple interaction. *Communication Studies, 64*(1), 33–48.
11. National Association of Colleges and Employers. (2014). *Job outlook 2015*. Bethlehem, PA: Author. Retrieved from https://www.umuc.edu/upload/NACE-Job-Outlook-2015.pdf.
12. Kramer, M. W., Lee, S. K., & Guo, Y. (2019). Using communication technology to manage uncertainty during organizational assimilation: Information-seeking and information-giving. *Western Journal of Communication, 83*, 304–325.
13. Rubin, R. B., Perse, E. M., & Barbato, C. A. (1988). Conceptualization and measurement of interpersonal communication motives. *Human Communication Research, 14*, 602–628.
14. Gergen, K. (1991). *The saturated self: Dilemmas of identity in contemporary life*. New York: Basic Books, p. 158.
15. Vass, N. (2018). *The lines that make us: Stories from Nathan's bus*. Seattle, WA: Tome Press. Quote appears on p. 79.
16. Chokshi, N. (2014, July 29). The handicap symbol gets an update—at least in New York State. *The Washington Post*. Retrieved from https://www.washingtonpost.com/blogs/govbeat/wp/2014/07/29/the-handicap-symbol-gets-an-update-at-least-in-new-york-state/?utm_term=.0201faa2cd9b. Quoted words appear in para. 3.
17. Tusler, A. (2014). The great blue man debate: Two views. *New Mobility*. Retrieved from http://www.newmobility.com/2014/12/accessibility-symbol/. Quote appears in para. 4.
18. Watts-Jones, D. (2017). Confronting the language of subtle racism. *Psychotherapy Networker*. Retrieved from https://www.psychotherapynetworker.org/blog/details/1318/confronting-the-language-of-subtle-racism. Quote appears in para. 1.
19. Chow, K. (2017, December 11). Ask Code Switch: Who can call themselves "brown"? National Public Radio. Retrieved from https://www.npr.org/2017/12/11/569983724/ask-code-switch-who-can-call-themselves-brown. (Question is paraphrased.)

20. Ailin, H. (2018, March 9). Why are whites in the USA generally not referred to as "European Americans"? *Quora*. Retrieved from https://www.quora.com/Why-are-whites-in-the-USA-generally-not-referred-to-as-European-Americans. Quote appears in paras. 2 and 7.

21. Shoneye, T. (2018, April 22). As a black woman, I hate the term "people of colour." *The Independent*. Retrieved from https://www.independent.co.uk/voices/black-women-people-of-colour-racism-beyonce-coachella-black-lives-matter-a8316561.html. Quote appears in para. 4.

22. Momenian, D. (2018, January 24). Middle Easterners are not Caucasian, should not have to identify as such. *Collegiate Times*. Retrieved from http://www.collegiatetimes.com/opinions/middle-easterners-are-not-caucasian-should-not-have-to-identify/article_30ad42ba-0180-11e8-be25-5b5758984e58.html.

23. Teitel, E. (2017, April 26). Critics of new "dynamic" disability symbol not just anti-PC cranks. *The Star*. Retrieved from https://www.thestar.com/news/gta/2017/04/26/critics-of-new-dynamic-disability-symbol-not-just-anti-pc-cranks-teitel.html. Quote appears in para. 9.

24. Shannon, C. E., & Weaver, W. (1949). *The mathematical theory of communication*. Urbana: University of Illinois Press.

25. Boaz, K., Epley, N., Carter, T. & Swanson, A. (2011). *Journal of Experimental Social Psychology, 47*, 269–273.

26. See, for example, Dunne, M., & Ng, S. H. (1994). Simultaneous speech in small group conversation: All-together-now and one-at-a-time? *Journal of Language and Social Psychology, 13*, 45–71.

27. Wang, A. B. (2017, April 5). Nivea's "white is purity" ad campaign didn't end well. *The Washington Post*. Retrieved from https://www.washingtonpost.com/news/business/wp/2017/04/05/niveas-white-is-purity-ad-campaign-didnt-end-well/?utm_term=.0a42ac8f8f87.

28. The issue of intentionality has been a matter of debate by communication theorists. For a sample of the arguments on both sides, see Greene, J. O. (Ed.). (1997). *Message production: Advances in communication theory*. Mahwah, NJ: Erlbaum; Motley, M. T. (1990). On whether one can(not) communicate: An examination via traditional communication postulates. *Western Journal of Speech Communication, 54*, 1–20; Bavelas, J. B. (1990). Behaving and communicating: A reply to Motley. *Western Journal of Speech Communication, 54*, 593–602; and Stewart, J. (1991). A postmodern look at traditional communication postulates. *Western Journal of Speech Communication, 55*, 354–379.

29. For an in-depth look at this topic, see Cunningham, S. B. (2012). Intrapersonal communication: A review and critique. In S. Deetz (Ed.), *Communication Yearbook* 15 (pp. 597–620). Newbury Park, CA: Sage.

30. Goo, S. K. (2015, February 19). *The skills Americans say kids need to succeed in life*. Pew Research Center. Retrieved from http://www.pewresearch.org/fact-tank/2015/02/19/skills-for-success/?utm_source=Pew+Research+Center&utm_campaign=ad0be41c05-Internet_newsletter_022015&utm_medium=email&utm_term=0_3e953b9b70-ad0be41c05-399444569.

31. National Association of Colleges and Employers. (2012, October 24). The skills and qualities employers want in their class of 2013 recruits. Retrieved from http://www.naceweb.org/s10242012/skills-abilities-qualities-new-hires.

32. Anderson, C., & Gantz, J. F. (2013). Skills requirements for tomorrow's best jobs: Helping educators provide students with skills and tools they need. Framingham, MA: International Data Corporation. Retrieved from http://news.microsoft.com/download/presskits/education/docs/IDC_101513.pdf.

33. Chui, M., Manyika, J., Bughin, J., Dobbs, R., Roxburgh, C., Sarrazin, H., . . . Westergren, M. (2012). *The social economy: Unlocking value and productivity through social technologies*. New York: McKinsey Global Institute. Retrieved from http://www.mckinsey.com/insights/high_tech_telecoms_internet/the_social_economy; Project Management Institute. (2013). The high cost of low performance: The essential role of communications. Retrieved from http://www.pmi.org/~/media/PDF/Business-Solutions/The-High-Cost-Low-Performance-The-Essential-Role-of-Communications.ashx; Silverman, R. E. (2012, February 14). Where's the boss? Trapped in a meeting. *Wall Street Journal*. Retrieved from http://www.wsj.com/articles/SB10001424052970204642604577215013504567548.

34. Gill, B. (2013, June). E-mail: Not dead, evolving. *Harvard Business Review, 91*(6), 32–33.

35. Lucas, K., & Rawlins, J. D. (2015). The competency pivot: Introducing a revised approach to the business communication curriculum. *Business and Professional Communication Quarterly, 78*(2), 167–193.

36. Samuels, E. (2018, August 2). Grocery store workers lets autistic teen stock shelves, causing a "miracle in action." *The Washington Post*. Retrieved from https://www.washingtonpost.com/news/inspired-life/wp/2018/08/02/grocery-store-worker-lets-an-autistic-teen-stock-shelves-causing-a-miracle-in-action/?noredirect=on&utm_term=.e2be8ac01da9. Quote appears in para. 9.

37. Tweeting yourself out of a job—"Cisco Fatty" style. *Career Goods*. Retrieved from https://www.careergoods.com/how-to-lose-a-job-via-twitter-the-cisco-fatty-story/. Quotes appear in paras. 5 and 8.

38. Tran, K. M. (2018, August 18). I won't be marginalized by online harassment. *New York Times*. Retrieved from https://www.nytimes.com/2018/08/21/movies/kelly-marie-tran.html. Quote appears in para 2.

39. Griffith, E. (2015, January 22). How to quit social media (and why you should). *PC Magazine*. Retrieved from http://www.pcmag.com/article2/0,2817,2475453,00.asp. Quote appears in para. 5.

40. Roberts, J. A., Luc Honore Petnji, Y., & Manolis, C. (2014). The invisible addiction: Cell-phone activities and addiction among male and female college students. *Journal of Behavioral Addictions, 3*(4), 254–265.

41. U.S. Bureau of Labor Statistics. (2015). American time use survey: Time use on an average weekday of full-time university and college students. Retrieved from http://www.bls.gov/tus/charts/students.htm.

42. Mullen, C. (n.d.). 2 reasons you should limit your social media time. Elevate. Design your best life [blog post]. Retrieved from http://chrismullen.org/2-reasons-you-should-limit-your-social-media-time-2/. Quotes appear in para. 3.

43. Moose, A. (2015, May 28). 7 reasons why you need to take a break from social media. *Elite Daily*. Retrieved from http://elitedaily.com/life/limit-time-on-social-media/1036660/. Quote appears in para. 3.

44. Grumstrup, C. (2015, June 3). 7 reasons to limit your social media that will lead to a happier you. *DOSE* [blog post]. Retrieved from http://www.dose.com/theworld/21439/7-Reasons-To-Limit-Your-Social-Media-That-Will-Lead-To-A-Happier-You. Quote appears in para. 3.

45. Moose A. (2015, May 28). 7 reasons why you need to take a break from social media. *Elite Daily*. Retrieved from http://elitedaily.com/life/limit-time-on-social-media/1036660. Quote appears in para. 5.

46. Griffith, E. (2015, January 22.) How to quit social media (and why you should). *PC Magazine*. Retrieved from http://www.pcmag.com/article2/0,2817,2475453,00.asp. Quote appears in para. 2.

47. Jenks, I. (2009). *Living on the future edge.* Presentation handout, 21st Century Fluency Project. Kelowna, BC, Canada: The Info Savvy Group.

48. Allen, T. B. (2019, February). The future is calling. *National Geographic.* Retrieved from https://www.nationalgeographic.com/science/space/universe/future-is-calling.

49. United Nations Cyberschoolbus. (n.d.). Retrieved from http://www.un.org/Pubs/CyberSchoolBus/aboutus.html.

50. Aristotle. (1991). *On rhetoric: A theory of civic discourse* (George A. Kennedy, Trans.). New York: Oxford University Press.

51. Paulsson, S. (2011, February 17). A view of the holocaust. British Broadcasting Corporation. Retrieved from http://www.bbc.co.uk/history/worldwars/genocide/holocaust_overview_01.shtml.

52. Heath, R. L., & Bryant, J. (2000). *Human communication theory and research.* Mahwah, NJ: Erlbaum.

53. Heath & Bryant, *Human communication theory and research.*

54. See Wiemann, J. M., Takai, J., Ota, H., & Wiemann, M. (1997). A relational model of communication competence. In B. Kovacic (Ed.), *Emerging theories of human communication.* Albany, NY: SUNY Press. These goals, and the strategies used to achieve them, needn't be conscious. See Fitzsimons, G. M., & Bargh, J. A. (2003). Thinking of you: Nonconscious pursuit of interpersonal goals associated with relationship partners. *Journal of Personality and Social Psychology, 84,* 148–164.

55. Light, J., & Mcnaughton, D. (2014). Communicative competence for individuals who require augmentative and alternative communication: A new definition for a new era of communication? *Augmentative and Alternative Communication, 30*(1), 1–18.

56. Rubin, R. B., Graham, E. E., & Mignerey, J. T. (1990). A longitudinal study of college students' communication competence. *Communication Education, 39,* 1–14.

57. Rubin, Graham, & Mignerey, A longitudinal study of college students' communication competence.

58. Mathrick, R., Meagher, T., & Norbury, C. F. (2017). Evaluation of an interview skills training package for adolescents with speech, language and communication needs. *International Journal of Language and Communication Disorders,* nonpaginated epublication.

59. Bruno, G., & Gareth, R. (2014, July). Do we notice when communication goes awry? An investigation of people's sensitivity to coherence in spontaneous conversation. *Plos ONE, 9*(7), E103182.

60. Wang, S., Hu, Q., & Dong, B. (2015). Managing personal networks: An examination of how high self-monitors achieve better job performance. *Journal of Vocational Behavior, 91,* 180–188.

61. Jones, K. (2014). At-risk students and communication skill deficiencies: A preliminary study. *Journal of Education and Human Development, 3,* 1–8.

62. Desta, Y. (2015, March 13). Obsessing over the perfect social media post is ruining your life, study says. *Mashable.* Retrieved from http://mashable.com/2015/03/13/social-media-ruining-your-life/#qbjK5m5pKaq3. Quote appears in para. 14.

63. Confusing sentences that actually make sense. (2017, April 7). Grammarly. Retrieved from https://www.grammarly.com/blog/confusing-sentences-actually-make-sense. Quote appears in item 3.

64. Vass, *The lines that make us,* p. 97.

65. Koenig Kellas, J., Horstman, H. K., Willer, E. K., & Carr, K. (2015). The benefits and risks of telling and listening to stories of difficulty over time: Experimentally testing the expressive writing paradigm in the context of interpersonal communication between friends. *Health Communication, 30*(9), 843.

66. Smith, J. L., Ickes, W., & Hodges, S. (Eds.). (2010). *Managing interpersonal sensitivity: Knowing when—and when not—to understand others.* Hauppauge, NY: Nova Science.

67. Goldsmith, D. J., & Domann-Scholz, K. (2013). The meanings of "open communication" among couples coping with a cardiac event. *Journal of Communication, 63,* 266–286.

68. Nathan Vass Leroy speech at Fresh Ground Stories. (2014, December 27). YouTube. Retrieved from https://www.youtube.com/watch?v=aQecGi50WGA. Quote is at 5:22 mins.

69. Lee, J. (2017, August 15). Route 7 is one of Metro Transit's most challenging bus lines, and driver Nathan Vass loves it. *The Seattle Times.* Retrieved from https://www.seattletimes.com/seattle-news/transportation/route-7-is-one-of-metro-transits-most-challenging-bus-lines-and-driver-nathan-vass-loves-it/. Quote appears in para. 4.

70. Vass, *The lines that make us,* p. 208.

CHAPTER 2

1. All quotes are from Stuart-Ulin, C. R. (2018, April 26). You could be flirting on dating apps with paid impersonators. Quartz. Retrieved from https://qz.com/1247382/online-dating-is-so-awful-that-people-are-paying-virtual-dating-assistants-to-impersonate-them/?utm_source=atlfb.

2. Feezell, J. T. (2017). Agenda setting through social media: The importance of incidental news exposure and social filtering in the digital era. *Political Research Quarterly, 71,* 482–494.

3. Bayer, J. B., Ellison, N. B., Schoenbeck, S. Y., & Falk, E. B. (2016). Sharing the small moments: Ephemeral social interaction on Snapchat. *Information, Communication and Society 19*(7), 956–977.

4. Severin, W. J., & Tankard, J. W. (1997). *Communication theories: Origins, methods, and uses in the mass media* (4th ed.). New York: Longman, pp. 197–214.

5. Ruggiero, T. E. (2000). Uses and gratifications theory in the 21st century. *Mass Communication and Society, 3,* 3–37. For a somewhat different categorization of uses and gratifications, see Joinson, A. N. (2008, April 5–10). "Looking at," "looking up" or "keeping up with" people? Motives and uses of Facebook. In *Proceedings of the 26th annual SIGCHI Conference on Human Factors in Computing Systems* (Florence, Italy) (pp. 1027–1036). New York: ACM. See also Flanagan, A. J. (2005). IM online: Instant messaging use among college students. *Communication Research Reports, 22,* 173–187.

6. O'Sullivan, P. B., & Carr, C. T. (2018). Masspersonal communication: A model bridging the mass-interpersonal divide. *New Media and Society, 20,* 1161–1180.

7. The Strive Series https://www.youtube.com/channel/UCSQkQjPhnZw12Hj-SfsbX8w

8. Fagan, K. (August 26, 2018), Meet the people making a living live-streaming their niche hobbies, travel adventures, and everyday lives on Twitch. *Business Insider.* Retrieved from https://www.businessinsider.com/what-is-irl-streaming-on-twitch-2018-8.

9. Otondo, R. F., Van Scotter, J. R., Allen, D. G., & Palvia, P. (2008). The complexity of richness: Media, message, and communication outcomes. Information and Management, 45, 21–30.

10. Surinder, K. S., & Cooper, R. B. (2003). Exploring the core concepts of media richness theory: The impact of cue multiplicity and feedback immediacy on decision quality. *Journal of Management Information Systems, 20,* 263–299.

11. Walther, J. B. (2007). Selective self-presentation in computer-mediated communication: Hyperpersonal dimensions of technology, language, and cognition. *Computers in Human Behavior.* 23 (5): 2538–2557. doi:10.1016/j.chb.2006.05.002.

12. Jiang, C. & Bazarova, N. & Hancock, J.. (2011). Jiang, C. L., The disclosure-intimacy link in computer-mediated communication: An attributional extension of the hyperpersonal model. *Human Communication Research, 37,* 58-77..doi: 10.1111/j.1468-2958.2010.01393

13. McEwan, B., & Zanolla, D. (2013). When online meets offline: A field investigation of modality switching. *Computers in Human Behavior, 29*, 1565-1571.

14. Rains, S. A., & Tsetsi, E. (2017). Social support and digital inequality: Does Internet use magnify or mitigate traditional inequities in support availability? *Communication Monographs, 84*, 54–74.

15. Miller, D., and Madianou, M. (2012). *Migration and new media: Transnational families and polymedia.* London: Routledge.

16. Kalinov, K. (2017). "Transmedia narratives: definition and social transformations in the consumption of media content in the globalized world" (PDF). *Postmodernism Problems, 7*, 60–68. Retrieved from http://ppm.swu.bg/media/45765/kalinov_k_%20transmedia_narratives.pdf.

17. Ozkul, D., & Humphreys, L. (2015). Record and remember: Memory and meaning-making practices through mobile media. *Mobile Media and Communication, 3*, 351–365.

18. Macaulay, T. (2017, September 14). What is the right to be forgotten and where did it come from? *Techworld*. https://www.techworld.com/data/could-right-be-forgotten-put-people-back-in-control-of-their-data-3663849.

19. Walk, H. (2017, December 23). Should regulators force Facebook to ship a "Start Over" Button for users? Hunter Walk Podcast, https://hunterwalk.com/2017/12/23/should-regulators-force-facebook-to-ship-a-start-over-button-for-users.

20. Piwek, L., & Joinson, A. (2016). What do they Snapchat about? Patterns of use in time- limited instant messaging service. *Computers in Human Behavior, 54*, 358–367.

21. Bayer, J. B., Ellison, N. B., Schoenebeck, S. Y., & Falk, E. B. (2016). Sharing the small moments: Ephemeral social interaction on Snapchat. *Information, Communication and Society, 19*, 956–977.

22. Utz, S., Muscanell, N., & Khalid, C. (2015). Snapchat elicits more jealousy than Facebook: A comparison of Snapchat and Facebook use. *Cyberpsychology, Behavior, and Social Networking, 18*, 141–146.

23. Kirkpatrick, D. (1992, March 23). Here comes the payoff from PCs. *Fortune.*

24. Anderson, M., & Jiang, J. (2018). Teens' social media habits and experiences. Pew Research Center. http://www.pewinternet.org/2018/11/28/teens-social-media-habits-and-experiences.

25. Anderson, J. Q., & Rainie, L. (2010, July 2). *The future of social relations.* Pew Internet and American Life Project.

26. Lenhart, A. (2016, August 6). Chapter 4: Social media and friendships. In *Teens, technology and friendships.* Pew Research Center Report. Retrieved from http://www.pewinternet.org/2015/08/06/chapter-4-social-media-and-friendships.

27. Rosenfeld, M. J. (2017). Marriage, choice, and couplehood in the age of the Internet. *Sociological Science, 4*, 490–510.

28. Cacioppo, J. T., Cacioppo, S., Gonzaga, G. C., Ogburn, E. L., & VanderWeele, T. J. (2013). Marital satisfaction and break-ups differ across on-line and off-line meeting venues. *PNAS, 110*, 10135–10140.

29. Cacioppo, J. T., Cacioppo, S., Gonzaga, G. C., Ogburn, E. L., & VanderWeele, T. J. (2013). Marital satisfaction and break-ups differ across on-line and off-line meeting venues. *Proceedings of the National Academy of Sciences, 110*(25), 10135–10140.

30. Smith, A., & Duggan, M. (October 21, 2013). *Online dating and relationships.* Pew Internet and American Life Project.

31. Porter, C. E. (2006). A typology of virtual communities: A multi-disciplinary foundation for future research. *Journal of Computer-Mediated Communication, 10*, Article 3; Schwammlein, E., & Wodzicki, K. (2012). What to tell about me? Self-presentation in online communities. *Journal of Computer-Mediated Communication, 17*, 387–407.

32. Orr, E. S., Sisic, M., Ross, C., Simmering, M. G., Arseneault, J. M., & Orr, R. R. (2009). The influence of shyness on the use of Facebook in an undergraduate sample. *Cyberpsychology Behavior, 12*, 337–340.

33. Baker, L. R., & Oswald, D. L. (2010). Shyness and online social networking services. *Journal of Social and Personal Relationships, 27*, 873–889.

34. Cotten, S. R., Anderson, W. A., & McCullough, B. M. (2013). Impact of Internet use on loneliness and contact with others among older adults: Cross-sectional analysis. *Journal of Medical Internet Research, 15*, e39.

35. Lee, K., Noh, M., & Koo, D. (2013). Lonely people are no longer lonely on social networking sites: The mediating role of self-disclosure and social support. *Cyberpsychology, Behavior, and Social Networking, 16*, 413–418.

36. Tong, S. T., & Walther, J. B. (2011). Relational maintenance and computer-mediated communication. In K. B. Wright & L. M. Webb (Eds.), *Computer mediated communication and personal relationships* (pp. 98–118). New York: Peter Lang; Ledbetter, A. M. (2010). Assessing the measurement invariance of relational maintenance behavior when face-to-face and online. *Communication Research Reports, 27*, 30–37.

37. Ezumah, B. A. (2013). College students' use of social media: Site preferences, uses and gratifications theory revisited. *International Journal of Business and Social Science, 4* (5). http://www.ijbssnet.com/journals/Vol_4_No_5_May_2013/3.pdf.

38. Craig, E., & Wright, B. (2012). Computer-mediated relational development and maintenance on Facebook. *Communication Research Reports, 29*, 119–129; Dainton, M. (2013). Relationship maintenance on Facebook: Development of a measure, relationship to general maintenance, and relationship satisfaction. *College Student Journal, 47*, 112–121.

39. Tong, S. T., & Walther, J. B. (2011.) Relational maintenance and computer-mediated communication. In K. Wright & L. Webb (Eds.) *Computer-mediated communication and personal relationships* (pp. 98–118), Cresskill, NJ: Hampton Press.

40. Bergen, K. M., Kirby, E. & McBride, M. C. (2007). "'How do you get two houses cleaned?'": Accomplishing family caregiving in commuter marriages. *Journal of Family Communication, 7*, 287–307.

41. Stafford, L. (2005). *Maintaining long-distance and cross-residential relationships.* Mahwah, NJ: Erlbaum.

42. Jiang, C., & Hancock, J. T. (2013). Absence makes the communication grow fonder: Geographic separation, interpersonal media, and intimacy in dating relationships. *Journal of Communication, 63*, 566–577.

43. Pearson, C. (August 12, 2013). Long distance relationship benefits include greater intimacy, study says. *Huffington Post.* Paragraph 8.

44. Vitak, J. (2014). Facebook makes the heart grow fonder: Relationship maintenance strategies among geographically dispersed and communication-restricted connections. In Proceedings of the 17th ACM Conference on Computer Supported Cooperative Work and Social Computing. New York: ACM; Walther, J. B., & Ramirez, A., Jr. (2010). New technologies and new directions in online relating. In S. W. Smith & S. R. Wilson (Eds.), *New directions in interpersonal communication research* (pp. 264–284). Thousand Oaks, CA: Sage.

45. Valenzuela, S., Halpern, D., & Katz, J. E. (2014). Social network sites, marriage well-being and divorce: Survey and state-level evidence from the United States. *Computers in Human Behavior, 36.* 94–101.

46. McClure, E. A., Acquavita, S. P., Dunn, K. E., Stoller, K. B., & Sitzer, M. L. (2014). Characterizing smoking, cessation services,

and quit interest across outpatient substance abuse treatment modalities. *Journal of Substance Abuse Treatment, 46,* 194–201.

47. Luxton, D. D., June, J. D. & Kinn, J. T. (2011). Technology-based suicide prevention: Current applications and future directions. *Telemedicine and e-Health,* 17, 50–54.

48. Hawdon, J., & Ryan, R. (2012). Well-being after the Virginia Tech mass murder: The relative effectiveness of face-to-face and virtual interactions in providing support to survivors. *Traumatology,* 18, 3–12.

49. K., Brad. (2017). Social networking, survival, and healing. In Adler, R. B., & Proctor, R. F. *Looking out/looking in.* (15th ed.) Boston: Cengage, p. 46.

50. Fox, S. (2011, June). *Peer-to-peer healthcare.* Pew Internet and American Life Project; Rains, S. A., & Keating, D. M. (2011). The social dimension of blogging about health: Health blogging, social support, and well-being. *Communication Monographs, 78,* 511–553.

51. Sanford, A. A. (2010). "I can air my feelings instead of eating them": Blogging as social support for the morbidly obese. *Communication Studies, 61,* 567–584.

52. Anderson, M., Toor, S., Rainie, L. & Smith, A. (2018). Public attitudes toward political engagement on social media. Pew Research Center. http://www.pewinternet.org/2018/07/11/public-attitudes-toward-political-engagement-on-social-media.

53. Anderson, M., et al. (2018). Public attitudes toward political engagement on social media. Pew Research Center. http://www.pewinternet.org/2018/07/11/public-attitudes-toward-political-engagement-on-social-media.

54. Flanagin, A. J. (2005). IM online: Instant messaging use among college students. *Communication Research Reports, 22,* 175–187.

55. Boase, J., Horrigan, J. B., Wellman, B., & Rainie, L. (2006, January 25). *The strength of Internet ties.* Pew Internet and American Life Project.

56. DeAndrea, D. C., Tong, S. T., & Walther, J. B. (2010). Dark sides of computer-mediated communication. In W. R. Cupach & B. H. Spitzberg (Eds.), *The dark side of close relationships II* (pp. 95–118). New York: Routledge.

57. Dunbar, R. (2010). *How many friends does one person need? Dunbar's number and other evolutionary quirks.* Cambridge, MA: Harvard University Press.

58. Bryant, E. M., & Marmo, J. (2012). The rules of Facebook friendship: A two-stage examination of interaction rules in close, casual, and acquaintance friendships. *Journal of Social and Personal Relationships, 29,* 1013–1035.

59. Parks, M. R. (2007). *Personal networks and personal relationships.* Mahwah, NJ: Lawrence Erlbaum.

60. Dunbar, R. (2012). Social cognition on the Internet: Testing constraints on social network size. *Philosophical Transactions of the Royal Society, 367,* 2192–2201.

61. Loveys, K. (January 24, 2010). 5,000 friends on Facebook? Scientists prove 150 is the most we can cope with. *Mail Online.*

62. Tong, S. T., Van Der Heide, B., Langwell, L., & Walther, J. B. (2008). Too much of a good thing? The relationship between number of friends and interpersonal impressions on Facebook. *Journal of Computer-Mediated Communication, 13,* 531–549.

63. Lee, J. R., Moore, D. C., Park, E., & Park, S. G. (2012). Who wants to be "friend rich"? Social compensatory friending on Facebook and the moderating role of public self-consciousness. *Computers in Human Behavior, 28,* 1036–1043; Kim, J., & Lee, J. R. (2011). The Facebook paths to happiness: Effects of the number of Facebook friends and self-presentation on subjective well-being. *Cyberpsychology, Behavior, and Social Networking, 14,* 359–364.

64. Turkle, S. (2014, April 22.) The flight from conversation. *New York Times.* Retrieved from http://www.nytimes.com/2012/04/22/opinion/sunday/the-flight-from-conversation.html?pagewanted=all&_r=0.

65. Caplan, S. E. (2003). Preference for online social interaction: A theory of problematic Internet use and psychosocial well-being. *Communication Research, 30,* 625–648.

66. Phu, B., & Gow, A. (2019). Facebook use and its association with subjective happiness and loneliness. *Computers in Human Behavior, 92,* 151–159.

67. Lundy, B. L., & Drouin, M. (2016). From social anxiety to interpersonal connectedness: Relationship building within face-to-face, phone and instant messaging mediums. *Computers in Human Behavior, 54,* 271–277.

68. Walther, J. B., Van Der Heide, B., Hamel, L., & Shulman, H. (2009). Self-generated versus other-generated statements and impressions in computer-mediated communication: A test of warranting theory using Facebook. *Communication Research, 36,* 229–253.

69. Caplan, S. E. (2005). A social skill account of problematic Internet use. *Journal of Communication, 55,* 721–736.

70. Hand, M. M., Thomas, D. B., Walter, C., Deemer, E. D., & Buyanjargal, M. (2013). Facebook and romantic relationships: Intimacy and couple satisfaction associated with online social network use. *Cyberpsychology, Behavior, and Social Networking, 16,* 8–13.

71. Mirsa, S., Cheng, L., Genevie., J., & Yuan, M. (2014). The iPhone effect: The quality of in-person social interactions in the presence of mobile devices. *Environment and Behavior;* Przybylski, A. K., & Weinstein, N. (2013). Can you connect with me now? How the presence of mobile communication technology influences face-to-face conversation quality. *Journal of Social and Personal Relationships, 30,* 237–246.

72. Clayton, R. B., Nagumey, A., & Smith, J. R. (2013). Cheating, breakup, and divorce: Is Facebook to blame? *CyberPsychology, Behavior and Social Networking, 16,* 717–720.

73. Cravens, J. D., Leckie, K. R., & Whiting, J. B. (2013). Facebook infidelity: When poking becomes problematic. *Contemporary Family Therapy, 35,* 74–90; Schneider, J. P., Weiss, R., & Samenow, C. (2012). Is it really cheating? Understanding the emotional reactions and clinical treatment of spouses and partners affected by cybersex infidelity. *Sexual Addiction and Compulsivity, 19,* 123–139.

74. Valenzuela, S., Halpern, D., & Katz, J. E. (2014). Social network sites, marriage well-being and divorce: Survey and state-level evidence from the United States. *Computers in Human Behavior, 36,* 94–101.

75. Woods, H. C., and Scott, H. (2–16). #Sleepyteens: Social media use in adolescence is associated with poor sleep quality, anxiety, depression and low self-esteem. *Journal of Adolescence 51,* 41–49.

76. Caplan, S. E. (2005). A social skill account of problematic Internet use. *Journal of Communication, 55,* 721–736; Schiffrin, H., Edelman, A., Falkenstein, M., & Stewart, C. (2010). Associations among computer-mediated communication, relationships, and well-being. *Cyberpsychology, Behavior, and Social Networking, 13,* 1–14; Morrison, C. M., & Gore, H. (2010). The relationship between excessive Internet use and depression: A questionnaire-based study of 1,319 young people and adults. *Psychopathology, 43,* 121–126.

77. Vogel, E., Rose, J. Roberts, L. & Eckles, K. (2014). Social comparison, social media, and self-esteem. *Psychology of Popular Media Culture, 3,* 208–232.

78. Verduyn, P., Ybarra, O., Résibois, M., Jonides, J., & Kross, E. (2017). Do social network sites enhance or undermine subjective well-being?: A critical review. *Social Issues and Policy Review, 11,* 274–302.

79. Toma, C. L., Hancock, J. T., & Ellison, N. B. (2008). Separating fact from fiction: An examination of deceptive self-presentation in online dating profiles. *Personality and Social Psychology Bulletin, 34*, 1023–1036.

80. DeAndrea, D. C., & Walther, J. B. (2011). Attributions for inconsistencies between online and offline self-presentations. *Communication Research, 38*, 805–825.

81. Drouin, M., Miller, D. Wehle, S., & Hernandez, E. (2016). Why do people lie online? Because everyone lies on the Internet. *Computers in Human Behavior, 64*, 134–142.

82. Korsgaard, C. (1986.) The right to lie: Kant on dealing with evil. *Philosophy and Public Affairs 15*(4), 325–349.

83. Bok, S. (2011). *Lying: Moral choice in public and private life.* New York: Vintage.

84. Lyndon, A., Bonds-Raacke, J., & Cratty, A. D. (2011). College students' Facebook stalking of ex-partners. *Cyberpsychology, Behavior, and Social Networking, 14*, 711–716.

85. Reyns, B. W., Henson, B., & Fisher, B. S. (2012). Stalking in the twilight zone: Extent of cyberstalking victimization and offending among college students. *Deviant Behavior, 33*, 1–25.

86. DreBing, H., Bailer, J., Anders, A., Wagner, H., & Gallas, C. (2014). Cyberstalking in a large sample of social network users: Prevalence, characteristics, and impact upon victims. *Cyberpsychology, Behavior, and Social Networking, 17*, 61–67.

87. Shahani, A. (September 15, 2014). *Smartphones are used to stalk, control domestic abuse victims.* All Tech Considered.

88. Smith, A., & Duggan, M. (2018). Crossing the line: What counts as online harassment? Pew Research Center. Retrieved from http://www.pewinternet.org/2018/01/04/crossing-the-line-what-counts-as-online-harassment.

89. Bauman, S. (2011). *Cyberbullying: What counselors need to know.* Alexandria, VA: American Counseling Association; Holfeld, B., & Grabe, M. (2012). An examination of the history, prevalence, characteristics, and reporting of cyberbullying in the United States. In Q. Li, D. Cross, & P. K. Smith (Eds.), *Cyberbullying in the global playground: Research from international perspectives* (pp. 117–142). San Francisco, CA: Wiley-Blackwell.

90. Cassidy, W., Faucher, C., & Jackson, M. (2013). Cyberbullying among youth: A comprehensive review of current international research and its implications and application to policy and practice. *School Psychology International, 34*, 575–612; Roberto, A. J., Eden, J., Savage, M. W., Ramos-Salazar, L., & Deiss, D. M. (2014). Prevalence and predictors of cyberbullying perpetration by high school seniors. *Communication Quarterly, 62*, 97–114.

91. Anderson, M. (2018). A majority of teens have experienced some form of cyberbullying. Pew Research Center. Retrieved from http://www.pewinternet.org/2018/09/27/a-majority-of-teens-have-experienced-some-form-of-cyberbullying.

92. Bauman, S., Toomey, R. B., & Walker, J. L. (2013). Associations among bullying, cyberbullying, and suicide in high school students. *Journal of Adolescence, 36*, 341–350; Huang, Y.-Y., & Chou, C. (2010). An analysis of multiple factors of cyberbullying among junior high school students in Taiwan. *Computers in Human Behavior, 26*, 1581–1590.

93. National Crime Prevention Council (2007, February 28). *Teens and cyberbullying.* National Crime Prevention Council.

94. Hossenei, M., & Tammimy, Z. (2016). Recognizing users' gender in social media using linguistic features. *Computers in Human Behavior, 56*, 192–197.

95. Pennebaker, J. W. (2011). *The secret lives of pronouns: What our words say about us.* New York: Bloomsbury.

96. Schwartz, H.A., Eichstaedt, J. C., Kern, M. L., Dziurzynski, L., Ramones, S. M., Agrawal, M., . . . Ungar, L. H. (2013).

97. Palomares, N. A., & Lee, E. (2010). Virtual gender identity: The linguistic assimilation to gendered avatars in computer-mediated communication. *Journal of Language and Social Psychology, 29*, 5–23.

98. Kapidzic, S., & Herring, S. C. (2011). Gender, communication, and self-presentation in teen chatrooms revisited: Have patterns changed? *Journal of Computer-Mediated Communication, 17*, 39–59.

99. Prensky, M. (2001). Digital natives, digital immigrants. *On the Horizon, 9*, 1–6; Rainie, L. (October 27, 2006). *Digital natives: How today's youth are different from their "digital immigrant" elders and what that means for libraries.* Pew Research Internet Project.

100. Hyman, I. (January 26, 2014). Cell phones are changing social interaction. *Psychology Today.* Retrieved from https://www.psychologytoday.com/us/blog/mental-mishaps/201401/cell-phones-are-changing-social-interaction

101. Smith, A. (September 19, 2011). *Americans and text messaging.* Pew Internet and American Life Project.

102. Kluger, J. (August 16, 2012). We never talk anymore: The problem with text messaging. *Time.* http://techland.time.com/2012/08/16/we-never-talk-anymore-the-problem-with-text-messaging/

103. Brenner, J., & Smith, A. (August 5, 2013). *72% of online adults are social networking site users.* Pew Internet and American Life Project.

104. Smith, A. (April 3, 2014). *Older adults and technology use.* Pew Research Internet Project.

105. Schwartz, H. A., Eichstaedt, J. C., Kern, M. L., Dziurzynski, L., Ramones, S. M., Agrawal, M., . . . Ungar, L. H. (2013). Personality, gender, and age in the language of social media: The open-vocabulary approach. *PLoS ONE, 8*, e73791.

106. https://link.springer.com/article/10.1007/s40692-018-0100-6.

107. http://users.clas.ufl.edu/burt/Bibliomania!/WhenYourPunctuationSaysIt.pdf

108. Bauerlein, M. (2009, September 4). Why Gen-Y Johnny can't read nonverbal cues. *Wall Street Journal.* Retrieved from http://www.wsj.com/articles/SB10001424052970203863204574348493483201758.

109. Watts, S. A. (2007). Evaluative feedback: Perspectives on media effects. *Journal of Computer-Mediated Communication, 12*(2), 384–411. Retrieved from http://jcmc.indiana.edu/vol12/issue2/watts.html. See also Turnage, A. K. (2007). Email flaming behaviors and organizational conflict. *Journal of Computer-Mediated Communication, 13*(1), 43–59. Retrieved from http://jcmc.indiana.edu/vol13/issue1/turnage.html.

110. Adapted from Radford, B. (2018, August 3). Post-mortem on a misunderstood post: Online misunderstandings and how to avoid them. *Center for Inquiry.* https://centerforinquiry.org/blog/post-mortem-on-a-misunderstood-post-online-misunderstandings-and-how-to-avoid-them.

111. Nye, W. (2015) Commencement address, Rutgers University. https://commencement.rutgers.edu/2015-university-commencement-videos.

112. Roberts, J. J. (2017, June 5). Harvard Yanks 10 acceptance letters over offensive Facebook posts. *Fortune.* Retrieved from http://fortune.com/2017/06/05/harvard-acceptance-rescinded/?utm_campaign=time&utm_source=facebook.com&utm_medium=social&xid=time_socialflow_facebook, Paragraph 6.

113. Lenhart, A. (2009, December 15). Teens and sexting. *Pew Research Center: Internet, Science and Tech.* Retrieved from http://www.pewinternet.org/Reports/2009/Teens-and-Sexting.aspx.

114. *A thin line: MTV-AP digital abuse study* (executive summary). (2009). Retrieved from http://www.athinline.org/MTV-AP_Digital_Abuse_Study_Executive_Summary.pdf.

115. Meyer, E. J. (2009, December 16). "Sexting" and suicide. *Psychology Today Online*. Retrieved from http://www.psychologytoday.com/blog/gender-and-schooling/200912/sexting-and-suicide.

116. Stieger, S., & Lewetz, D. (2018). A week without using social media: Results from an ecological momentary intervention study using smartphones. *Cyberpsychology, Behavior, and Social Networking, 21*, 618–624.

117. See, for example, Lenhart, A., Madden, M., Smith, A., & MacGill, A. (2007, December 19). Teens creating content. *Pew Internet and American Life Project*. Retrieved from http://www.pewinternet.org/Reports/2007/Teens-and-Social-Media/3-Teens-creating-content/18-Videos-are-not-restricted-as-often-as-photos.aspx?r=1.

118. Strayer, D. L., Drews, F. A., Crouch, D. J., & Johnston, W. A. (2005). Why do cell phone conversations interfere with driving? In W. R. Walker & D. Herrmann (Eds.), *Cognitive technology: Transforming thought and society* (pp. 51–68). Jefferson, NC: McFarland.

119. Distracted Driving. U.S. Department of Transportation, National Highway Safety Administration. https://www.nhtsa.gov/risky-driving/distracted-driving. Retrieved February 8, 2018 from http://www-nrd.nhtsa.dot.gov/Pubs/811379.pdf.

120. Strayer, D. L., & Drew, F. A. (2004, Winter). Profiles in driver distraction: Effects of cell phone conversations on younger and older drivers. *Human Factors, 46*, 640–649.

CHAPTER 3

1. Birkeland, M. S., Breivik, K., & Wold, B. (2014, January). Peer acceptance protects global self-esteem from negative effects of low closeness to parents during adolescence and early adulthdood. *Journal of Youth and Adolescence, 43*, 70–80.

2. Back, M., Mund, M., Finn, C., Hagemeyer, B., Zimmermann, J., & Neyer, F. J. (2015). The dynamics of self-esteem in partner relationships. *European Journal of Personality, 2*, 235–249.

3. Zhang, L., Zhang, S., Yang, Y., & Li, C. (2017). Attachment orientations and dispositional gratitude: The mediating roles of perceived social support and self-esteem. *Personality & Individual Differences, 114*, 193–197.

4. Luerssen, A., Jhita, G. J., & Ayduk, O. (2017). Putting yourself on the line: Self-esteem and expressing affection in romantic relationships. *Personality and Social Psychology Bulletin, 7*, 940–956.

5. Kille, D. R., Eibach, R. P., Wood, J. V., & Holmes, J. G. (2017). Who can't take a compliment? The role of construal level and self-esteem in accepting positive feedback from close others. *Journal of Experimental Social Psychology, 68*, 40–49.

6. Dredge, R., Gleeson, J. M., & de la Piedad Garcia, X. (2014). Risk factors associated with impact severity of cyberbullying victimization: A qualitative study of adolescent online social networking. *Cyberpsychology, Behavior & Social Networking, 17*(5), 287–291.

7. Baumeister, R. F. (2005). *The cultural animal: Human nature, meaning, and social life.* New York: Oxford University Press; and Baumeister, R. F., Campbell, J. D., Krueger, J. I., & Vohs, K. D. Does high self-esteem cause better performance, interpersonal success, happiness, or healthier lifestyles? *Psychological Science in the Public Interest, 4*, 1–44.

8. Vohs, K. D., & Heatherton, T. F. (2004). Ego threats elicit different social comparison process among high and low self-esteem people: Implications for interpersonal perceptions. *Social Cognition, 22*, 168–191.

9. For more, see Kandler, C., Riemann, R., & Kämpfe, N. (2009). Genetic and environmental mediation between measures of personality and family environment in twins reared together. *Behavioral Genetics, 39*, 24–35; and Caspi, A., Harrington, H., Milne, B., Amell, J. W., Theodore, R. F., & Moffitt, T. E. (2003). Children's behavioral styles at age 3 are linked to their adult personality traits at age 26. *Journal of Personality, 71*, 495–514. doi: 10.1111/1467-6494.7104001.

10. Vukasović, T., & Bratko, D. (2015). Heritability of personality: A meta-analysis of behavior genetic studies. *Psychological Bulletin, 141*(4), 769–785.

11. See Gong, P., Zheng, A., Zhang, K., Lei, X., Li, F., Chen, D., … Zhang, F. (2010). Association analysis between 12 genetic variants of ten genes and personality traits in a young Chinese Han population. *Journal of Molecular Neuroscience, 42*, 120–126; and Heck, A., Lieb, R., Ellgas, A., Pfister, H., Lucae, S., Roeske, D.,… Ising, M. (2009). Investigation of 17 candidate genes for personality traits confirms effects of the HTR2A gene on novelty seeking. *Genes, Brain and Behavior, 8*, 464–472. doi: 10.1111/j.1601-183X.2009.00494.x.

12. Cole, J. G., & McCroskey, J. C. (2000). Temperament and sociocommunicative orientation. *Communication Research Reports, 17*, 105–114.

13. Dweck, C. (2008). Can personality be changed? The role of beliefs in personality and change. *Current Directions in Psychological Science, 6*, 391–394.

14. Begney, S. (2008, December 1). When DNA is not destiny. *Newsweek, 152*, 14.

15. National Coalition of Anti-Violence Programs. (2015). *Lesbian, gay, bisexual, transgender, queer, and HIV-affected hate violence in 2014* (2015 release edition). New York: New York City Anti-Violence Project. Retrieved from http://www.avp.org/storage/documents/Reports/2014_HV_Report-Final.pdf.

16. Zuro, A. (2014). *Measuring up: Social comparisons on Facebook and contributions to self-esteem and mental health* (master's thesis). Available from University of Michigan Deep Blue database (UMI No. 65059383), http://hdl.handle.net/2027.42/107346.

17. López-Guimerà, G., Levine, M. P., Sánchez-Carracedo, D., & Fauquet, J. (2010). Influence of mass media on body image and eating disordered attitudes and behaviors in females: A review of effects and processes. *Media Psychology, 13*, 387–416.

18. Slater, A., & Tiggeman, M. (2014, December). Media matters for boys too! The role of specific magazine types and television programs in the drive for thinness and muscularity in adolescent boys. *Eating Behaviors, 14*(4), 679–682.

19. Ata, R. N., Thompson, J. K., & Small, B. J. (2013). Effects of exposure to thin-ideal media images on body dissatisfaction: Testing the inclusion of a disclaimer versus warning label. *Body Image, 10*, 472–480.

20. Köber, C., Schmiedek, F., & Habermas, T. (2015). Characterizing lifespan development of three aspects of coherence in life narratives: A cohort-sequential study. *Developmental Psychology, 51*(2), 260–275.

21. Ask Matt: Why do trans people make me feel uncomfortable? (2013, February 25). *The Transadvocate*. Retrieved from http://transadvocate.com/ask-matt-why-do-trans-people-make-me-uncomfortable_n_8773.htm. Quote appears in para. 1 and 3.

22. Ask Matt (2013). Quote appears in third to last paragraph.

23. Sobo, E. J., & Loustaunau, M. O. (2010). *The cultural context of health, illness, and medicine.* Santa Barbara, CA: Praeger. Adapted from dialogue on p. 86.

24. Yamagishi, T., Hashimoto, H., Cook, K. S., Kiyonari, T., Shinada, M., Mifune, N., & . . . Li, Y. (2012). Modesty in self-presentation:

A comparison between the USA and Japan. *Asian Journal of Social Psychology, 15*(1), 60–68.

25. Brozovich, F. A., & Heimberg, R. G. (2013). Mental imagery and post-event processing in anticipation of a speech performance among socially anxious individuals. *Behavior Therapy, 44,* 701–716.

26. DiPaola, B. M., Roloff, M. E., & Peters, K. M. (2010). College students' expectations of conflict intensity: A self-fulfilling prophecy. *Communication Quarterly, 58*(1), 59–76.

27. Stinson, D. A., Cameron, J. J., Wood, J. V., Gaucher, D., & Holmes J. G. (2009). Deconstructing the "reign of error": Interpersonal warmth explains the self-fulfilling prophecy of anticipated acceptance. *Personality and Social Psychology, 35,* 1165–1178.

28. Dimberg, U., & Söderkvist, S. (2011). The voluntary facial action technique: A method to test the facial feedback hypothesis. *Journal of Nonverbal Behavior, 35,* 17–33.

29. Holmes, J. G. (2002). Interpersonal expectations as the building blocks of social cognition: An interdependence theory perspective. *Personal Relationships, 9,* 1–26.

30. Rosenthal, R., & Jacobson, L. (1968). *Pygmalion in the classroom.* New York: Holt, Rinehart and Winston.

31. For a detailed discussion of how self-fulfilling prophecies operate in relationships, see Watzlawick, P. (2005). Self-fulfilling prophecies. In J. O'Brien & P. Kollock (Eds.), *The production of reality* (3rd ed., pp. 382–394). Thousand Oaks, CA: Pine Forge Press.

32. Tan, T. (2016, June 21). What it's like to be friends with someone who has special needs. *Odyssey.* Retrieved from https://www .theodysseyonline.com/friends-someone-special-needs. Quote appears in para. 2.

33. James, W. (1920). *The letters of William James* (H. James, Ed.). Boston: Atlantic Monthly Press, p. 462.

34. Ruby, F. M., Smallwood, J., Engen, H., & Singer, T. (2013). How self-generated thought shapes mood—the relation between mind-wandering and mood depends on the socio-temporal content of thoughts. *PLOS ONE, 8*(10), e77554.

35. Knobloch, L. K., Miller, L. E., Bond, B. J., & Mannone, S. E. (2007). Relational uncertainty and message processing in marriage. *Communication Monographs, 74,* 154–180.

36. Heisler, J., & Crabill, S. (2006). Who are "stinkybug" and "packerfan4"? Email pseudonyms and participants' perceptions of demography, productivity, and personality. *Journal of Computer-Mediated Communication, 12,* article 6. Retrieved from http:// jcmc.indiana.edu/vol12/issue1/heisler.html.

37. Clark, A. (2000). *A theory of sentience.* New York: Oxford University Press.

38. Miró, E., Cano, M. C., Espinoza-Fernández, L., & Beula-Casal, G. (2003). Time estimation during prolonged sleep deprivation and its relation to activation measures. *Human Factors, 45,* 148–159.

39. Alaimo, K., Olson, C. M., & Frongillo, E. A. (2001). Food insufficiency and American school-aged children's cognitive, academic, and psychosocial development. *Pediatrics, 108,* 44–53.

40. Koukkari, W. L., & Sothern, R. B. (2006). *Introducing biological rhythms: A primer on the temporal organization of life, with implications for health, society, reproduction and the natural environment.* New York: Springer.

41. Hasler, B. P., & Troxel, W. M. (2010). Couples' nighttime sleep efficiency and concordance: Evidence of bidirectional associations with daytime relationship functioning. *Psychosomatic Medicine, 72,* 794–801.

42. Goldstein, S. (2008). Current literature in ADHD. *Journal of Attention Disorders, 11,* 614–616.

43. Von Briesen, P. D. (2007). Pragmatic language skills of adolescents with ADHD. *DAI, 68*(5-B), 3430.

44. Babinski, D. E., Pelham W. E., Jr., Molina, B. S. G., Gnagy, E. M., Waschbusch, D. A., Yu, J., & Karch, K. M. (2011). Late adolescent and young adult outcomes of girls diagnosed with ADHD in childhood: An exploratory investigation. *Journal of Attention Disorders, 15,* 204–214.

45. National Institute of Mental Health. (2008, April 3). *Attention deficit hyperactivity disorder.* Retrieved from http://www.nimh.nih .gov/health/publications/attention-deficit-hyperactivity-disorder/ index.shtml.

46. Sveinsdóttir, Á. K. (2011). The metaphysics of sex and gender. In C. Witt (Ed.), *Feminist metaphysics: Exploring the ontology of sex, gender and the self* (pp. 47–65). New York: Springer.

47. Bianchi, A. (2018). *Becoming an ally to the gender-expansive child: A guide for parents and carers.* London: Jessica Kingsley Publishers.

48. Francis, B., & Paechter, C. (2015). The problem of gender categorisation: Addressing dilemmas past and present in gender and education research. *Gender and Education, 7,* 776–790.

49. Kott, L. J. (2014, November 30). For these Millennials, gender norms have gone out of style. National Public Radio. https:// www.npr.org/2014/11/30/363345372/for-these-millennials-gender-norms-have-gone-out-of-style

50. Francis, B., & Paechter, C. (2015). The problem of gender categorisation: Addressing dilemmas past and present in gender and education research. *Gender and Education, 7,* 776–790.

51. 2018 study on sexual harassment and assault. (2018, February 21). StoptheHarrassment.org. Retrieved from http:// www.stopstreetharassment.org/resources/2018-national-sexual-abuse-report.

52. Ward, M. L. (2016, March 15). Media and sexualization: State of empirical research, 1995–2015. *Journal of Sex Research, 53*(4–5), 560–577.s

53. Spinner, L., Cameron, L., & Calogero, R. (2018, January). Peer toy play as a gateway to children's gender flexibility: The effect of (counter) stereotypic portrayals of peers in children's magazines. *Sex Roles,* online publication. Retrieved from https://link. springer.com/article/10.1007%2Fs11199-017-0883-3.

54. Schwabel, D. (2017, November 18). Are millennials putting an end to gender differences? *Newsweek.* Retrieved from https:// www.newsweek.com/are-millennials-putting-end-gender-differences-715922. Quote appears in para. 3.

55. Amodio, D. M., & Devine, P. G. (2006). Stereotyping and evaluation in implicit race bias: Evidence for independent constructs and unique effects on behavior. *Journal of Personality and Social Psychology, 4,* 652–661.

56. Block, C. J., Aumann, K., & Chelin, A. (2012). Assessing stereotypes of black and white managers: A diagnostic approach. *Journal of Applied Social Psychology.* Advance online publication retrieved from http://onlinelibrary.wiley.com/doi/10.1111/ j.1559-1816.2012.01014.x/abstract.

57. Samakov, J., & Noble, O. (2015a). 48 things women hear in a lifetime (that men just don't). (2015, December 9). *Huffington Post.* Retrieved from https://www.youtube.com/ watch?v=9yMFw_vWboE.

58. Samakov, J., & Noble, O. (2015b). 48 things men hear in a lifetime (that are bad for everyone). *Huffington Post.* Retrieved from https://www.huffingtonpost.com/entry/48-things-men-hear-in-a-lifetime_us_5671dd4de4b0dfd4bcc0705d.

59. Boscamp, E. (2015, December 11). 48 things women hear all the time (that men never do). *MindBodyGreen.* Retrieved from https://www.mindbodygreen.com/0-22931/48-things-women-hear-all-the-time-that-men-never-do.html. Quote appears in second to last paragraph.

60. Saulny, S. (2011, October 12). In strangers' glances at family, tensions linger. *New York Times*. Retrieved from http://www.nytimes.com/2011/10/13/us/for-mixed-family-old-racial-tensions-remain-part-of-life.html?pagewanted=1&_r=0&ref=raceremixed.

61. Swami, V., & Furnham, A. (2008). *The psychology of physical attraction*. New York: Routledge/Taylor & Francis.

62. Gonzaga, G. G., Haselton, M. G., Smurda J., Davies, M., & Poore, J. C. (2008). Love, desire, and the suppression of thoughts of romantic alternatives. *Evolution and Human Behavior, 29*, 119–126.

63. Shaw, C. L. M. (1997). Personal narrative: Revealing self and reflecting other. *Human Communication Research, 24*, 302–319.

64. Kellas, J. K., Baxter, L., LeClair-Underberg, C., Thatcher, M., Routsong, T., Normand, E. L., & Braithwaite, D. O. (2014). Telling the story of stepfamily beginnings: The relationship between young-adult stepchildren's stepfamily origin stories and their satisfaction with the stepfamily. *Journal of Family Communication, 14*(2), 149–166. Quotes appear on p. 160.

65. Martz, J. M., Verette, J., Arriaga, X. B., Slovik, L. F., Cox, C. L., & Rusbult, C. E. (1998). Positive illusion in close relationships. *Personal Relationships, 5*, 159–181.

66. Pearson, J. C. (2000). Positive distortion: "The most beautiful woman in the world." In K. M. Galvin & P. J. Cooper (Eds.), *Making connections: Readings in relational communication* (2nd ed., pp. 184–190). Los Angeles, CA: Roxbury.

67. Bad grammar is bad for business. (2014, September 19). Hallam. Retrieved from https://www.hallaminternet.com/2014/bad-grammar-bad-business.

68. Marek, C. I., Wanzer, M. B., & Knapp, J. L. (2004). An exploratory investigation of the relationship between roommates' first impressions and subsequent communication patterns. *Communication Research Reports, 21*, 210–220.

69. For a review of these perceptual biases, see Hamachek, D. (1992). *Encounters with the self* (3rd ed.). Fort Worth, TX: Harcourt Brace Jovanovich. See also Bradbury, T. N., & Fincham, F. D. (1990). Attributions in marriage: Review and critique. *Psychological Bulletin, 107*, 3–33. For information on the self-serving bias, see Shepperd, J., Malone, W., & Sweeny, K. (2008). Exploring causes of the self-serving bias. *Social and Personality Psychology Compass, 2/2*, 895–908.

70. Easton, S. S., & Bommelje, R. K. (2011). Interpersonal communication consequences of email non-response. *Florida Communication Journal, 39*(2), 45–63.

71. Kanouse, D. E., & Hanson, L. (1972). Negativity in evaluations. In E. E. Jones, D. E. Kanouse, S. Valins, H. H. Kelley, R. E. Nisbett, & B. Weiner (Eds.), *Attribution: Perceiving the causes of behavior*. Morristown, NJ: General Learning Press.

72. Rozin, P., & Royzman, E. B. (2001). Negativity bias, negativity dominance, and contagion. *Personality and Social Psychology Review, 5*(4), 296–320.

73. Williams, R. (2014, June 30). Are we hardwired to be positive or negative? *Psychology Today*. Retrieved from https://www.psychologytoday.com/blog/wired-success/201406/are-we-hardwired-be-positive-or-negative.

74. Black, S. L., & Johnson, A. F. (2012). Employers' use of social networking sites in the selection process. *Journal of Social Media in Society, 1*(1), 7–28.

75. Thorndike, E. L. (1920). A constant error in psychological ratings. *Journal of Applied Psychology, 4*(1), 25–29.

76. Talamas, S. N., Mavor, K. I., & Perrett, D. I. (2016). Blinded by beauty: Attractiveness bias and accurate perceptions of academic performance. *Plos ONE, 11*(2), 1–18.

77. Thorndike, E. L. (1920). A constant error in psychological ratings. *Journal of Applied Psychology, 4*(1), 25–29.

78. Bacev-Giles, C., & Haji, R. (2017). Person perception in social media profiles. *Computers in Human Behavior, 75*, 50–57.

79. Herrick, L. (2014, June 17). Respecting political differences: As told by a liberal living in a sea of conservatives. *Huffington Post*. Retrieved from https://www.huffingtonpost.com/lexi-herrick/respecting-political-diff_b_5500406.html. Quote appears in para. 2.

80. Stiff, J. B., Dillard, J. P., Somera, L., Kim, H., & Sleight, C. (1988). Empathy, communication, and prosocial behavior. *Communication Monographs, 55*, 198–213.

81. Walter, H. (2012). Social cognitive neuroscience of empathy: Concepts, circuits, and genes. *Emotion Review, 4*, 9–17.

82. Miklikowka, M., Duriez, M., & Soenens, B. (2011). Family roots of empathy-related characteristics: The role of perceived maternal and paternal needs support in adolescence. *Developmental Psychology, 47*, 1342–1352.

83. Spivak, A. L., & Farran, D. C. (2012). First-grade teacher behaviors and children's prosocial actions in classrooms. *Early Education & Development, 23*(5), 623.

84. Van der Graaff, J., Branje, S., De Wied, M., Hawk, S., Van Lier, P., & Meeus, W. (2014). Perspective taking and empathic concern in adolescence: Gender differences in developmental changes. *Developmental Psychology, 50*(3), 881–888.

85. Aradhye, C., Vonk, J., & Arida, D. (2015). Adults' responsiveness to children's facial expressions. *Journal of Experimental Child Psychology, 135*, 56–71.

86. Eyal, T., Steffal, M., & Epley, N. (2018). Perspective mistaking: Accurately understanding the mind of another requires getting perspective, not taking perspective. *Journal of Personality and Social Psychology, 114*, 547–571.

87. Goleman, D. (2006). *Emotional intelligence [why it can matter more than IQ]*. New York: Bantam Books.

88. Kozina, A., & Mleku, A. (2016). Intrinsic motivation as a key to school success: Predictive power of self-perceived autonomy, competence and relatedness on the achievement in international comparative studies. *Solsko Polje, 27*(1/2), 63–88.

89. Preece, J. (2004). Etiquette, empathy and trust in communities of practice: Stepping-stones to social capital. *Journal of Universal Computer Science, 10*(3), 294–302.

90. Ramanauskas, K. (2016). The impact of the manager's emotional intelligence on organisational performance. *Management Theory & Studies for Rural Business and Infrastructure Development, 38*(1), 58–69.

91. Goffman, E. (1971). *The presentation of self in everyday life*. Garden City, NY: Doubleday; and Goffman, E. (1971). *Relations in public*. New York: Basic Books.

92. Cupach, W. R., & Metts, S. (1994). *Facework*. Thousand Oaks, CA: Sage. See also Brown, P., & Levinson, S. C. (1987). *Politeness: Some universals in language usage*. Cambridge: Cambridge University Press.

93. Sharkey, W. F., Park, H. S., & Kim, R. K. (2004). Intentional self embarrassment. *Communication Studies, 55*, 379–399.

94. Vedantam, S. (2017, May 2). Why social media isn't always very social. *Hidden Brain*. Transcript retrieved from http://www.npr.org/2017/05/02/526514168/why-social-media-isnt-always-very-social. Quotes appear in para. 9.

95. Gonzales, A. L. (2014). Text-based communication influences self-esteem more than face-to-face or cellphone communication. *Computers in Human Behavior, 39*, 197–203.

96. Rogers, A. (2018, June 5). Draftee Rolison, Rockie's address "stupid" tweet. *Major League Baseball News*. Retrieved from https://www.mlb.com/news/ryan-rolison-addresses-president-obama-tweet/c-279941290. Quote appears in para. 2.

97. Rogers (2018, June 5). Quote appears in para. 4.

98. Pope, E. (2018, June 6). Twitter post retrieved from https://twitter.com/exavierpope/status/1004013947366846464?lang=en. Quote appears in June 6 tweet.

99. Groke, N. (2018, June 5). About the regrettable tweet from Rockies draft pick Ryan Rolison hoping someone would shoot Obama. *The Athletic.* Retrieved from https://theathletic.com/381610/2018/06/05/about-that-regrettable-tweet-from-rockies-draft-pick-ryan-rolison-hoping-someone-would-shoot-obama.

100. Mellor, R. (2018, June 6.) Rollins didn't know what "shoot him" means. Really? *We Know What's Up.* [blog]. Retrieved from https://weknowwhatsup.blogspot.com/2018/06/rolison-didnt-know-what-shoots-him.html.

101. Cornish, A. (2018, July 31). Baseball players issue apologies over old offensive tweets that resurfaced. National Public Radio. Retrieved from https://www.npr.org/2018/07/31/634369274/baseball-players-issue-apology-over-old-offensive-tweets-that-resurfaced. Quote appears in second to last statement by Walsh.

102. Siibak, A. (2009). Constructing the self through the photo selection: Visual impression management on social networking websites. *Cyberpsychology: Journal of Psychosocial Research on Cyberspace, 3,* article 1. Retrieved from http://www.cyberpsychology.eu/view.php?cisloclanku=2009061501&article=1.

103. Hancock, J. T., & Durham, P. J. (2001). Impression formation in computer-mediated communication revisited: An analysis of the breadth and intensity of impressions. *Communication Research, 28,* 325–347.

104. Yang, C.-c., Holden, S. M., & Carter, M. D. K. (2017). Emerging adults' social media self-presentation and identity development at college transition: Mindfulness as a moderator. *Journal of Applied Developmental Psychology, 52,* 212–221.

105. Desta, Y. (2015, March 13). Obsessing over the perfect social media post is ruining your life, study says. *Mashable.* Retrieved from http://mashable.com/2015/03/13/social-media-ruining-your-life/#qbjK5m5pKaq3.

106. Appiah, K. (August 10, 2018). Go ahead, speak for yourself. *New York Times.* Retrieved from https://www.nytimes.com/2018/08/10/opinion/sunday/speak-for-yourself.html.

107. Urciuoli, B. (2009). The political topography of Spanish and English: The view from a New York Puerto Rican neighborhood. *American Ethnologist, 10,* 295–310.

108. Benet-Martínez, V., Leu, J., Lee, F., & Morris, M. (2002). Negotiating biculturalism: Cultural frame switching in biculturals with oppositional versus compatible cultural identities. *Journal of Cross-Cultural Psychology, 33,* 492–516.

109. Toomey, A., Dorjee, T., & Ting-Toomey, S. (2013). Bicultural identity negotiation, conflicts, and intergroup communication strategies. *Journal of Intercultural Communication Research, 42*(2), 112–134. Quote appears on p. 120.

110. Leary, M. R., & Kowalski, R. M. (1990). Impression management: A literature review and two-component model. *Psychological Bulletin, 107,* 34–47.

111. Chovil, N. (1991). Social determinants of facial displays. *Journal of Nonverbal Behavior, 15,* 141–154.

112. Snyder, M. (1979). Self-monitoring processes. In L. Berkowitz (Ed.), *Advances in experimental social psychology.* New York: Academic Press; and Snyder, M. (1983, March). The many me's of the self-monitor. *Psychology Today,* p. 34f.

113. Hall, J. A., & Pennington, N. (2013). Self-monitoring, honesty, and cue use on Facebook: The relationship with user extraversion and conscientiousness. *Computers in Human Behavior, 29,* 1556–1564.

114. Fleming, P., & Sturdy, A. (2009). "Just be yourself!": Towards neo-normative control in organisations? *Employee Relations, 31,* 569–583.

115. Ragins, B. R. (2008). Disclosure disconnects: Antecedents and consequences of disclosing invisible stigmas across life domains. *Academy of Management Review, 33,* 194–215.

116. Ragins, B. R., Singh, R., & Cornwell, J. M. (2007). Making the invisible visible: Fear and disclosure of sexual orientation at work. *Journal of Applied Psychology, 92,* 1103–1118.

117. Pachankis, J. E. (2007). The psychological implications of concealing a stigma: A cognitive-affective-behavioral model. *Psychological Bulletin, 133,* 328–345.

118. Sezer, O., Gino, F., & Norton, M. I. (2015, April 15). *Humblebragging: A distinct—and ineffective—self-presentation strategy.* Harvard Busines School Marketing Unit Working Paper No. 15-080. Retrieved from http://papers.ssrn.com/sol3/papers.cfm?abstract_id=2597626.

119. Sezer, O., Gino, F., & Norton, M. I. (2015, April 15). *Humblebragging: A distinct—and ineffective—self-presentation strategy.* Harvard Busines School Marketing Unit Working Paper No. 15-080. Retrieved from http://papers.ssrn.com/sol3/papers.cfm?abstract_id=2597626. Quote appears on p. 16.

120. Toma, C., Hancock, J., & Ellison, N. (2008). Separating fact from fiction: An examination of deceptive self-presentation in online dating profiles. *Personality and Social Psychology Bulletin, 34,* 1023–1036.

121. Rubin, R. (2018, July 20). Bo Burnham wishes "Eighth Grade" wasn't rated R. *Variety.* Retrieved from https://variety.com/2018/film/features/bo-burnham-interview-eighth-grade-elsie-fisher-1202877506.

CHAPTER 4

1. Guo, W., and P. Vulchi. (2019). *Tell me who you are: Sharing our stories of race, culture, and Identity.* New York: Tarcher Perigee.

2. Tajfel, H., & Turner, J. C. (1986). The social identity theory of inter-group behavior. In S. Worchel & L. W. Austin (Eds.), *Psychology of intergroup relations.* Chicago: Nelson-Hall.

3. Smith, A., & Anderson, M. (2018, March 1). Social media use in 2018. Pew Research Center. Retrieved from http://www.pewinternet.org/2018/03/01/social-media-use-in-2018/.

4. Shenoy, R. (2016, January 6). First generation? Second? For immigrants and their children, a question without meaning. Public Radio International. Retrieved from https://www.pri.org/stories/2016-01-06/first-generation-second-immigrants-and-their-children-question-meaning. Quote appears in para. 3.

5. Brown, H. K., Ouellette-Kuntz, H., Lysaght, R., & Burge, P. (2011). Students' behavioural intentions towards peers with disability. *Journal of Applied Research in Intellectual Disabilities, 24,* 322–332.

6. Binder, J., Brown, R., Zagefka, H., Funke, F., Kessler, T., Mummendey, A., … Leyens, J.-F. (2009). Does contact reduce prejudice or does prejudice reduce contact? A longitudinal test of the contact hypothesis among majority and minority groups in three European countries. *Journal of Personality and Social Psychology, 96*(4), 843–856.

7. Caiola, S. (2017, March 29). New faces of autism in "Sesame Street," "Power Rangers" movie could help dispel stereotypes. North Bay Regional Center. Retrieved from https://nbrc.net/new-faces-of-autism-in-sesame-street-power-rangers-movie-could-help-dispel-stereotypes.

8. Oerta, S. (2009). *Affecting teen attitudes through positive media portrayals of teens with autism spectrum disorder* (unpublished doctoral dissertation). Walden University, Minneapolis, MN.

9. Tajfel, H., & Turner, J. C. (1986). The social identity theory of inter-group behavior. In S. Worchel & L. W. Austin (Eds.), *Psychology of intergroup relations* (pp. 7–24). Chicago: Nelson-Hall.

10. Samson, J. (2018, February 12). I'm an Asian American and I will never fit your stereotypes. *Odyssey.* Retrieved from https://www.theodysseyonline.com/asians-are-not-weak. Quote appears in paras. 1, 2, and 5.

11. van de Kemenade, D. (2013). Life lessons learnt from my intercultural relationship. [Personal blog.] Retrieved from http://www.daniellevandekemenade.com/life-lessons-learnt-from-my-intercultural-relationship. This quote appears in para. 5 and the next in para 12.

12. Merkin, R. (2015). The relationship between individualism/collectivism. *Journal of Intercultural Communication, 39,* 4.

13. Craig, T. J. (2000). *Japan pop: Inside the world of Japanese popular culture.* New York: Routledge.

14. Babe, A. (2017, December 18). How the South Korean language was designed to unify. *Travel.* British Broadcasting Corporation. Retrieved from http://www.bbc.com/travel/story/20171217-why-south-koreans-rarely-use-the-word-me. Quotes appear in paras. 1 and 16, respectively.

15. FitzGerald, C. (2015, May 11). 9 culture shocks Americans will have in Hungary. Matador Network. Retrieved from https://matadornetwork.com/abroad/9-culture-shocks-americans-will-hungary. Quote appears in item 6.

16. Cai, D. A., & Fink, E. L. (2002). Conflict style differences between individualists and collectivists. *Communication Monographs, 69,* 67–87.

17. 10 things to know about U.S. culture. (n.d.). InterExchange. Retrieved from https://www.interexchange.org/articles/career-training-abroad/10-things-to-know-about-u-s-culture. Quote appears in "Competition" section.

18. Merkin, The relationship between individualism/collectivism.

19. Wu, S., & Keysar, B. (2007). Cultural effects on perspective taking. *Psychological Science, 18,* 600–606.

20. Takano, Y., & Sogun, S. (2008). Are Japanese more collectivistic than Americans? Examining conformity in in-groups and the reference-group effect. *Journal of Cross-Cultural Psychology, 39*(3), 237–250.

21. Moghaddam, M. M. (2017). Politeness at the extremes: Iranian women's insincere responses to compliments. *Language & Dialogue, 7*(3), 413–431. Quotes appear on pages 422 and 421, respectively.

22. Merkin., R. S. (2009). Cross-cultural communication patterns—Korean and American communication. *Journal of Intercultural Communication, 20,* 5.

23. Hall, E. T. (1959). *Beyond culture.* New York: Doubleday.

24. Chlopicki, W. (2017). Communication styles—an overview. *Styles of Communication, 9*(2), 9–25.

25. Hahn, M., & Molinsky, A. (2016, March 25). Having a difficult conversation with someone from a different culture. *Harvard Business Review.* Retrieved from https://hbr.org/2016/03/having-a-difficult-conversation-with-someone-from-a-different-culture.

26. Hahn & Molinsky, Having a difficult conversation with someone from a different culture. Quote appears in para. 4.

27. Chen, Y.-S., Chen, C.-Y. D., & Chang, M.-H. (2011). American and Chinese complaints: Strategy use from a cross-cultural perspective. *Intercultural Pragmatics, 8,* 253–275.

28. Croucher, S., Bruno, A., McGrath, P., Adams, C., McGahan, C., Suits, A., & Huckins, A. (2012). Conflict styles and high-low context cultures: A cross-cultural extension. *Communication Research Reports, 29*(1), 64–73.

29. Hofstede, G. (2001). *Culture's consequences: Comparing values, behaviors, institutions, and organizations across nations* (2nd ed.). Thousand Oaks, CA: Sage.

30. Brassier-Rodrigues, C. (2015). How do American, Chinese and French students characterize their teachers' communication? *Journal of Intercultural Communication, 38,* unpaginated.

31. Brassier-Rodrigues, How do American, Chinese and French students characterize their teachers' communication?

32. Hofstede, G. (2001). *Culture's consequences: Comparing values, behaviors, institutions, and organizations across nations* (2nd ed.). London, UK: Sage.

33. Hofstede, *Culture's consequences.*

34. Spear, J., & Matusitz, J. (2015). Doctor-patient communication styles: A comparison between the United States and three Asian countries. *Journal of Human Behavior in the Social Environment, 25*(8), 871–884.

35. Sweetman, K. (2012, April 10). In Asia, power gets in the way. *Harvard Business Review.* Retrieved from https://hbr.org/2012/04/in-asia-power-gets-in-the-way. Quote appears in para. 1.

36. Dailey, R. M., Giles, H., & Jansma, L. L. (2005). Language attitudes in an Anglo-Hispanic context: The role of the linguistic landscape. *Language and Communication, 25*(1), 27–38.

37. Barker, G. G. (2016). Cross-cultural perspectives on intercultural communication competence. *Journal of Intercultural Communication Research, 45,* 13–30.

38. Basso, K. (2012). "To give up on words": Silence in Western Apache culture. In L. Monoghan, J. E. Goodman, & J. M. Robinson (Eds.), *A cultural approach to interpersonal communication: Essential readings* (2nd ed., pp. 73–83). Malden, MA: Blackwell, quote on p. 84.

39. Hofstede, *Culture's consequence.*

40. Hofstede, *Culture's consequences.*

41. Ten things everyone should know about race. (2003). *Race—The power of an illusion.* California Newsreel, Public Broadcasting System. Retrieved from http://www.pbs.org/race/000_About/002_04-background-01-x.htm.

42. Interview with Jonathan Marks. (2003). Background readings for *Race—The power of an illusion.* California Newsreel, Public Broadcasting System. Retrieved from http://www.pbs.org/race/000_About/002_04-background-01-08.htm.

43. Samovar, L. A., Porter, R. E., & McDaniel, E. R. (2013). *Communication between cultures* (8th ed.). Boston: Wadsworth.

44. Bowleg, L. (2008). When black + lesbian + woman ≠ black lesbian woman: The methodological challenges of qualitative and quantitative intersectionality research. *Sex Roles, 59*(5/6), 312–325. Quote appears on p. 312.

45. DeFrancisco, V. P., & Palczewski, C. H. (2014). *Gender in communication.* Thousand Oaks, CA: Sage.

46. Appiah, K. A. (2018, August 12). Go ahead, speak for yourself. *New York Times.* Retrieved from https://www.nytimes.com/2018/08/10/opinion/sunday/speak-for-yourself.html. Quote appears in para. 11.

47. Bailey, R. W. (2003). Ideologies, attitudes, and perceptions. *American Speech, 88,* 115–143.

48. The gender spectrum. (2013, Summer). Teaching Tolerance: A Project of the Southern Poverty Law Center. Retrieved from http://www.tolerance.org/gender-spectrum. Quote appears in para. 9.

49. Hancox, L. (n.d.). Top 8 tips for coming out as trans. *Ditch the Label.* Retrieved from http://www.ditchthelabel.org/8-tips-for-coming-out-as-trans. Quote appears in tip 1.

50. Russell, G. M., & Bohan, J. S. (2005, December). The gay generational gap: Communicating across the LGBT generational divide. *Institute for Gay and Lesbian Strategic Studies, 8*(1), 1–8. Quote appears on p. 3.

51. It Gets Better Project. (2013). About the It Gets Better Project. Retrieved from http://www.itgetsbetter.org/pages/about-it-gets-better-project.

52. Dan Savage: It gets better. (2013, January 14). *Take part.* Retrieved from http://www.itgetsbetter.org.

53. Potter, J. E. (2002). Do ask, do tell. *Annals of Internal Medicine, 137*(5), 341–343. Quote appears on p. 342.

54. 2016 hate crime statistics released. (2017, November 13). U.S. Federal Bureau of Investigation. Retrieved from https://www.fbi.gov/news/stories/2016-hate-crime-statistics.

55. Hussein, Y. (2015, December 3). Are you afraid to be Muslim in America? *Huffington Religion*. Retrieved from http://www.huffingtonpost.com/yasmin-hussein/are-you-afraid-to-be-muslim-in-america_b_8710826.html. Quote appears in para. 8.

56. Milevsky, A., Shifra Niman, D., Raab, A., & Gross, R. (2011). A phenomenological examination of dating attitudes in Ultra-Orthodox Jewish emerging adult women. *Mental Health, Religion and Culture, 14*, 311–322.

57. Colaner, C. (2009). Exploring the communication of evangelical families: The association between evangelical gender role ideology and family communication patterns. *Communication Studies, 60*, 97–113.

58. Pew Forum on Religion and Public Life (2008, June). *U.S. religious landscape survey. Religious beliefs and practices: Diverse and politically relevant*. Retrieved from http://religions.pewforum.org/pdf/report2-religious-landscape-study-full.pdf.

59. Miller, G. (2015, January 11). There's only one real difference between liberals and conservatives. *The Huffington Post*. Retrieved from http://www.huffingtonpost.com/galanty-miller/theres-only-one-real-diff_b_6135184.html. Quote appears in final paragraph.

60. Pew Research Internet Project. (2012). Social media and political engagement. Retrieved from http://www.pewinternet.org/2012/10/19/social-media-and-political-engagement.

61. Duggan, M., & Smith, A. (2016, October 25). Social media and political engagement. Pew Research Center: Internet and Technology. Retrieved from http://www.pewinternet.org/2016/10/25/political-engagement-and-social-media.

62. The future of free speech, trolls, anonymity and fake new online. (2017, March 29). Pew Research Center: Internet and Technology. Retrieved from http://www.pewinternet.org/2017/03/29/the-future-of-free-speech-trolls-anonymity-and-fake-news88/-online.

63. 12 easy tips for using social media responsibly. (2017, March 24). *Social Media Safety Blog*. Retrieved from https://safesmartsocial.com/using-social-media-responsibly.

64. Duyvis, C., & Whaley, K. (2016, July 8). Introduction to disability terminology. *Disability in Kidlit*. Retrieved from http://disabilityinkidlit.com/2016/07/08/introduction-to-disability-terminology.

65. Solomon, A. (2012). *Far from the tree: Parents, children, and the search for identity*. New York: Scribner, pp. 68–69.

66. Fitch, V. (1985). The psychological tasks of old age. *Naropa Institute Journal of Psychology, 3*, 90–106.

67. Gergen, K. J., & Gergen, M. M. (2000). The new aging: Self construction and social values. In K. W. Schae & J. Hendricks (Eds.), *The societal impact of the aging process* (pp. 281–306). New York: Springer.

68. The future of free speech (2017, March 29). Pew Research Center: Internet and Technology. Retrieved from http://www.pewinternet.org/2017/03/29/the-future-of-free-speech-trolls-anonymity-and-fake-news-online.

69. Editorial: Don't be a troll. (2017, April 9). *Iowa State Daily*. Retrieved from http://www.iowastatedaily.com/opinion/editorials/article_f2edd1e8-1d45-11e7-9ccf-ebde4f9f215c.html. Quote appears in para. 3.

70. Braithwaite, D. O., & Labrecque, D. (1994). Responding to the Americans with Disabilities Act: Contributions of interpersonal communication research and training. *Journal of Applied Communication Research, 22*, 285–294. See also Braithwaite, D. O. (1991). "Just how much did that wheelchair cost?": Management of privacy boundaries by persons with disabilities. *Western Journal of Speech Communication, 55*, 254–275; and Colvert, A. L., & Smith, J. (2000). What is reasonable? Workplace communication and people who are disabled. In D. O. Braithwaite & T. L. Thompson (Eds.), *Handbook of communication and people with disabilities: Research and application* (pp. 116–130). Mahwah, NJ: Erlbaum.

71. Braithwaite, D., & Braithwaite, C. (2012). "Which is my good leg?": Cultural communication of persons with disabilities. In L. A. Samovar, R. E. Porter, & E. R. McDaniel (Eds.), *Intercultural communication: A reader* (13th ed., pp. 241–254). Boston: Wadsworth.

72. Bailey, T. A. (2010). Ageism and media discourse: Newspaper framing of middle age. *Florida Communication Journal, 38*, 43–56.

73. Frijters, P., & Beatoon, T. (2012). The mystery of the U-shaped relationship between happiness and age. *Journal of Economic Behavior and Organization, 82*, 525–542.

74. Giles, H., Ballard, D., & McCann, R. M. (2002). Perceptions of intergenerational communication across cultures: An Italian case. *Perceptual and Motor Skills, 95*, 583–591.

75. Ryan, E. B., & Butler, R. N. (1996). Communication, aging, and health: Toward understanding health provider relationships with older clients. *Health Communication, 8*, 191–197.

76. Dukes, J. (2016, July 21). Self-imposed age discrimination: Being your own worst enemy. *Forbes*. Retrieved from https://www.forbes.com/sites/nextavenue/2015/07/21/self-imposed-age-discrimination-being-your-own-worst-enemy/#6b878e6e456c.

77. Harwood, J. (2007). *Understanding communication and aging: Developing knowledge and awareness*. Newbury Park, CA: Sage.

78. Kroger, J., Martinussen, M., & Marcia, J. E. (2010). Identity status change during adolescence and young adulthood: A meta-analysis. *Journal of Adolescence, 33*, 683–698.

79. Galanaki, E. P. (2012). The imaginary audience and the personal fable: A test of Elkind's theory of adolescent egocentrism. *Psychology, 3*, 457–466.

80. Myers, K. K., & Sadaghiani, K. (2010). Millennials in the workplace: A communication perspective on Millennials' organizational relationships and performance. *Journal of Business and Psychology, 25*(2), 225–238.

81. Lucas, K. (2011). The working class promise: A communicative account of mobility-based ambivalences. *Communication Monographs, 78*, 347–369.

82. Stuber, J. M. (2006). Talk of class. *Journal of Contemporary Ethnography, 35*, 285–318. Quote appears on p. 306.

83. Kim, Y. K., & Sax, L. J. (2009). Student–faculty interaction in research universities: Differences by student gender, race, social class, and first-generation status. *Research in Higher Education, 50*, 437–459..

84. Kaufman, P. (2003). Learning to not labor: How working-class individuals construct middle-class identities. *Sociological Quarterly, 44*, 481–504.

85. Orbe, M. P., & Groscurth, C. R. (2004). A co-cultural theoretical analysis of communicating on campus and at home: Exploring the negotiation strategies of first generation college (FGC) students. *Qualitative Research Reports in Communication, 5*, 41–47.

86. Hartnell, C. A., Ou, A., & Kinicki, A. (2011). Organizational culture and organizational effectiveness: A meta-analytic investigation of the competing values framework's theoretical suppositions. *Journal of Applied Psychology, 96*(4), 677–694.

87. Pettigrew, T. F., & Tropp, L. R. (2000). Does intergroup contact reduce prejudice? Recent meta-analytic findings. In S. Oskamp (Ed.), *Reducing prejudice and discrimination: Social psychological perspectives* (pp. 93–114). Mahwah, NJ: Erlbaum.

88. Paluck, E. L., Green, S. A., & Green, D. P. (2018, July 10). The contact hypothesis re-evaluated. *Behavioural Public Policy*. Online article made available by Cambridge University Press. Retrieved from https://osf.io/preprints/socarxiv/w2jkf.

89. Pettigrew, T. F., & Tropp, L. R. (2006, May). A meta-analytic test of intergroup contact theory. *Journal of Personality and Social Psychology, 90*, 751–783.

90. Broockman, D., & Kalla, J. (2016). Durably reducing transphobia: A field experiment on door-to-door canvassing. *Science, 352*(6282), 220–224.

91. Amichai-Hamburger, Y., & McKenna, K. Y. A. (2006). The contact hypothesis reconsidered: Interacting via the Internet. *Journal of Computer-Mediated Communication, 11*(3), 825–843. Retrieved from http://onlinelibrary.wiley.com/doi/10.1111/j.1083-6101.2006.00037.x/abstract.

92. Collins, L. (2016, October 2). I fell in love with a Frenchman—but didn't speak the language. *The Guardian*. Retrieved from https://www.theguardian.com/lifeandstyle/2016/oct/02/pardon-my-french-conscious-coupling-through-a-language-barrier. Quotes appear in paras. 14, 15, and 19, respectively.

93. Steves, R. (1996, May–September). Culture shock. *Europe Through the Back Door Newsletter, 50*, 9.

94. Madhani, A. (2018, May 23). Poll: Approval of same-sex marriage in U.S. reaches new high. *USA Today*. Retrieved from https://www.usatoday.com/story/news/nation/2018/05/23/same-sex-marriage-poll-americans/638587002.

95. Hu, E. (2018, February 5). In Seoul, a plastic surgery capital, residents frown on ads for cosmetic procedure. National Public Radio. Retrieved from https://www.npr.org/sections/parallels/2018/02/05/581765974/in-seoul-a-plastic-surgery-capital-residents-frown-on-ads-for-cosmetic-procedure.

96. Lah, K. (2011, March 23). Plastic surgery boom as Asians seek "Western" look. *CNN*. Retrieved from http://www.cnn.com/2011/WORLD/asiapcf/05/19/korea.beauty.

97. Kelly, N. (2013, July 30). Bad-luck numbers that scare off customers. *Harvard Business Review*. Retrieved from https://hbr.org/2013/07/the-bad-luck-numbers-that-scar.

98. A quick guide: Gift giving in Japan—dos and don'ts. Zooming Japan. (n.d.). Retrieved from http://zoomingjapan.com/culture/gift-giving-in-japan.

99. Working with sign language interpreters: The DOs and DON'Ts. (2014, September 29). Interpreting Services. Retrieved from http://www.signlanguagenyc.com/working-with-sign-language-interpreters-the-dos-and-donts.

100. How to eat in China—Chinese dining etiquette. (n.d.). *China Highlights*. Retrieved from https://www.chinahighlights.com/travelguide/chinese-food/dining-etiquette.htm.

101. DiMeo, D. F. (n.d.). Arabic greetings and good-byes. *Arabic for Dummies*. Retrieved from http://www.dummies.com/languages/arabic/arabic-greetings-and-good-byes.

102. Body language in Arab cultures. (n.d.). Word Press Culture Convo. Retrieved from https://tbell7.wordpress.com/2012/10/23/body-language-in-arab-cultures.

103. Traveling in a Muslim country. (n.d.). Embassy of the United Arab Emirates. Retrieved from https://www.uae-embassy.org/about-uae/travel-culture/traveling-muslim-country.

104. Chun, D. M. (2011). Developing intercultural communicative competence through online exchanges. *CALICO Journal, 28*(2), 392–419.

105. Kim, M. S., Hunter, J. E., Miyahara, A., Horvath, A. M., Bresnahan, M., & Yoon, H. (1996). Individual- vs. culture-level dimensions of individualism and collectivism: Effects on preferred conversational styles. *Communication Monographs, 63*, 28–49.

106. Gudykunst, W. B., & Nishida, T. (2001). Anxiety, uncertainty, and perceived effectiveness of communication across relationships and cultures. *Journal of Intercultural Relations, 25*, 55–71.

107. Berger, C. R. (1979). Beyond initial interactions: Uncertainty, understanding, and the development of interpersonal relationships. In H. Giles & R. St. Clair (Eds.), *Language and social psychology* (pp. 122–144). Oxford, UK: Blackwell.

108. Carrell, L. J. (1997). Diversity in the communication curriculum: Impact on student empathy. *Communication Education, 46*, 234–244.

109. Oberg, K. (1960). Cultural shock: Adjustment to new cultural environments. *Practical Anthropology, 7*, 177–182.

110. Oberg, Cultural shock.

111. Bruhwiler, B. (2012, November 12). Culture shock! [blog post]. Retrieved from http://www.joburgexpat.com/2012/11/culture-shock.html.

112. Chang, L. C.-N. (2011). My culture shock experience. *ETC: A Review of General Semantics, 68*(4), 403–405.

113. Kim, Y. Y. (2008). Intercultural personhood: Globalization and a way of being. *International Journal of Intercultural Relations, 32*, 359–368.

114. See Kim, Intercultural personhood.

115. Kim, Y. Y. (2005). Adapting to a new culture: An integrative communication theory. In W. B. Gudykunst (Ed.), *Theorizing about intercultural communication* (pp. 375–400). Thousand Oaks, CA: Sage.

CHAPTER 5

1. Masters, J. (2017, April 13). Asia Kate Dillon talks discovering the word non-binary: "I cried." HuffPost. Retrieved from https://www.huffingtonpost.com/entry/asia-kate-dillon-talks-discovering-the-word-non-binary_us_58ef1685e4b0156697224c7a. Quote appears in para. 4.

2. Nordstrom, L. (2017, May 16). Asia Kate Dillon is making a mark as "they." *Los Angeles Times*. Retrieved from http://www.latimes.com/fashion/la-ig-wwd-asia-kate-dillon-20170516-story.html. Quote appears in para. 1.

3. Smith, O., & Messer, L. (2018, January 15). "Billions" actor Asia Kate Dillon explains what it means to be non-binary. ABC News. Retrieved from https://abcnews.go.com/Entertainment/billions-actor-asia-kate-dillon-explains-means-binary/story?id=52317585. Quote appears in para. 4.

4. Wilson, R. (2013, December 2). What dialect do you speak? A map of American English. *Washington Post*. Retrieved from https://www.washingtonpost.com/blogs/govbeat/wp/2013/12/02/what-dialect-to-do-you-speak-a-map-of-american-english.

5. Sacks, O. (1989). *Seeing voices: A journey into the world of the deaf*. Berkeley: University of California Press, p. 17.

6. Adapted from O'Brien, J., & Kollock, P. (2001). *The production of reality* (3rd ed., p. 66). Thousand Oaks, CA: Pine Forge Press.

7. Ogden, C. K., & Richards, I. A. (1923). *The meaning of meaning*. New York: Harcourt Brace, p. 11.

8. Chomsky, N. (2007). "Approaching UG from Below." In Hans-Martin Gärtner; Uli Sauerland. *Interfaces + Recursion = Language? Chomsky's Minimalism and the View from Syntax-Semantics*. Studies in Generative Grammar. Berlin: Mouton de Gruyter.

9. Gaudin, S. (2011, March 25). OMG! Text shorthand makes the Oxford English Dictionary. *Computerworld*. Retrieved from http://www.computerworld.com/s/article/9215079/OMG_Text_shorthand_makes_the_dictionary.

10. Be careful—it's a gift! (2011). Languages. BBC. Retrieved from http://www.bbc.co.uk/languages/yoursay/false_friends/german/be_careful__its_a_gift_englishgerman.shtml.

11. People wash their clothing in Barf every day. (2009, June 17). *Adweek*. Retrieved from https://www.adweek.com/creativity/people-wash-their-clothing-barf-every-day-14028.

12. W. B. Pearce & V. Cronen. (1980). *Communication, action, and meaning*. New York: Praeger. See also J. K. Barge. (2004). Articulating CMM as a practical theory. *Human Systems: The Journal of Systemic Consultation and Management, 15*, 193–204, and E. M. Griffin. (2006). *A first look at communication theory* (6th ed.). New York: McGraw-Hill.

13. Croom, A. M. (2013). How to do things with slurs: Studies in the way of derogatory words. *Language and Communication, 33*(3), 177–204.

14. Braidwood, E. (2018, April 19). What does queer mean? PinkNews. Retrieved from https://www.pinknews.co.uk/2018/04/19/what-does-queer-mean. Quote appears in screen shot tweet by love, shauna.

15. Twenge, J. M., Van Landingham, H., & Campbell, W. K. (2017). The seven words you can never say on television: Increases in the use of swear words in American books, 1950–2008. *SAGE Open, 7*(3), 1–8.

16. New survey shows Americans believe civility is on the decline. (2016, April 15). Associated Press National Opinion Research Center (NORC) for Public Affairs Research. Retrieved from http://www.apnorc.org/PDFs/Rudeness/APNORC%20Rude%20Behavior%20Report%20%20PRESS%20RELEASE.pdf.

17. McLeod, L. (2011). *Swearing in the "tradie" environment as a tool for solidarity* (Vol. 4, pp. 1–10). Griffith Working Papers in Pragmatics and Intercultural Communication.

18. Scherer, C. R., & Sagarin, B. J. (2006). Indecent influence: The positive effects of obscenity on persuasion. Social Influence, 2, 138–146.

19. Sturt, D., & Nordstrom, T. (2015, March 19). Is swearing appropriate at your work? *Forbes*. Retrieved from https://www.forbes.com/sites/davidsturt/2015/03/19/is-the-f-bomb-appropriate-at-your-work/#575c5f8e48a2.

20. Pearce, W. B. & Cronen, V. E. (1980). *Communication, action and meaning: The creation of social realities*. New York: Praeger.

21. Dalton, D. (2015, March 18). 28 beautiful words the English language should steal. BuzzFeed. Retrieved from http://www.buzzfeed.com/danieldalton/ever-embasan#.abBQB66ep; DeMain, B., Sweetland Edwards, H., &Oltuski, R. 38 wonderful foreign words we could use in English.Retrieved from http://mentalfloss.com/article/50698/38-wonderful-foreign-words-we-could-use-english.

22. Granadillo, E. D., & Mendez, M. F.(2016). Pathological joking or witzelsucht revisited. *Journal of Neuropsychiatry & Clinical Neurosciences*. Advance online publication. Retrieved from http://www.ncbi.nlm.nih.gov/pubmed/26900737.

23. For a summary of scholarship supporting the notion of linguistic determinism, see Boroditsky, Lost in translation.

24. Whorf, B. (1956). The relation of habitual thought and behavior to language. In J. B. Carroll (Ed.), *Language, thought, and reality* (pp. 134–159). Cambridge, MA: MIT Press. See also Hoijer, H. (1994). The Sapir-Whorf hypothesis. In Larry A. Samovar & Richard E. Porter (Eds.), *Intercultural communication: A reader* (7th ed., pp. 194–200). Belmont, CA: Wadsworth.

25. Vedantham, S. [host]. (2018, January 29). Lost in translation: The power of language to shape how we view the world. *Hidden Brain*. Retrieved from https://www.npr.org/templates/transcript/transcript.php?storyId=581657754. Quote appears in para. 40 of transcript.

26. Vedantham, Lost in translation.

27. Boroditsky, L. (2012). How the languages we speak shape the way we think: The FAQs. In M. J. Spivey, K. McRae, & M. F. Joanisse (Eds.) *The Cambridge Handbook of Psycholinguistics* (pp. 615–631). New York: Cambridge University Press.

28. Davidoff, J., Goldstein, J., Tharp, I., Wakui, E., & Fagot, J. (2012). Perceptual and categorical judgements of colour similarity. *Journal of Cognitive Psychology, 24*(7), 871–892.

29. Pullum, G. K. (1991). *The great Eskimo vocabulary hoax and other irreverent essays on the study of language*. Chicago: University of Chicago Press.

30. Barrett, L. F. (2017). *How emotions are made: The secret life of the brain*. Boston: Houghton Mifflin Harcourt.

31. For a summary of scholarship supporting the notion of linguistic determinism, see Boroditsky, L. (2010, July 23). Lost in translation. *Wall Street Journal Online*. Retrieved from http://www.wsj.com/articles/SB10001424052748703467304575383131592767868.

32. Laham, S. M., Koval, P., & Alter, A. L. (2012). The name-pronunciation effect: Why people like Mr. Smith more than Mr. Colquhoun. *Journal of Experimental Social Psychology, 48*(3), 752–756.

33. Borget, J. (2012, November 9). Biracial names for biracial babies. *Mom Stories*. Retrieved from http://blogs.babycenter.com/mom_stories/biracial-baby-names-110912.

34. Edelman, B., Luca, M., & Svirsky, D. (2017.) Racial discrimination in the sharing economy: Evidence from a field experiment. *American Economic Journal: Applied Economics, 9*(2), 1–22.

35. Derous, E., Ryan, A. M., & Nguyen, H. D. (2012). Multiple categorization in resume screening: Examining effects on hiring discrimination against Arab applicants in field and lab settings. *Journal of Organizational Behavior, 33*(4), 544–570.

36. Bertrand, M., & Mullainathan, S. (2004). Are Emily and Greg more employable than Lakisha and Jamal? A field experiment on labor market discrimination. *The American Economic Review, 4*, 991–1013.

37. No names, no bias? (2015, October 31). *The Economist*. Retrieved from http://www.economist.com/news/business/21677214-anonymising-job-applications-eliminate-discrimination-not-easy-no-names-no-bias.

38. Derwing, T. M., & Munro, M. J. (2009). Putting accent in its place: Rethinking obstacles to communication. *Language Teaching, 42*(4), 476–490.

39. Agudo, R.R. (July 14, 2018). *New York Times*. Retrieved from https://www.nytimes.com/2018/07/14/opinion/sunday/everyone-has-an-accent.html.

40. Waxman, O. B. (2015, February 10). This the world's hottest accent. *Time*. Retrieved from http://time.com/3702961/worlds-hottest-accent.

41. Gluszek. A., & Dovidio, J. F. (2010). The way they speak: a social psychological perspective on the stigma of nonnative accents in communication. *Personality and Social Psychology Review, 14*(2), 214–237.

42. Hansen, K., & Dovidio, J. F. (2016). Social dominance orientation, nonnative accents, and hiring recommendations. *Cultural Diversity and Ethnic Minority Psychology, 22*(4), 544–551.

43. Cotton, J. L., O'Neill, B. S., & Griffin, A. (2008). The "name game": Affective and hiring reactions to first names. *Journal of Managerial Psychology, 23*, 18–39.

44. Brunning, J. L., Polinko, N. K., Zerbst, J. I., & Buckingham, J. T. (2000). The effect on expected job success of the connotative meanings of names and nicknames. *Journal of Social Psychology, 140*, 197–201.

45. Coffey, B., & McLaughlin, P. A. (2009). Do masculine names help female lawyers become judges? Evidence from South Carolina. *American Law and Economics Review, 11*, 112–133.

46. Karthikeyan, S., & Ficalora, S. (2017). Women's tendency to approach men speaking standard and non-standard accents varies with the nature of the help-seeking situation. *North American Journal of Psychology, 19*(1), 1–19.

47. Vervecken, D., & Hannover, B. (2015). Yes I can! Effects of gender fair job descriptions on children's perceptions of job status, job difficulty, and vocational self-efficacy. *Social Psychology, 46*(2), 76–92.

48. Lee, J. K. (2015). "Chairperson" or "chairman"?—A study of Chinese EFL teachers' gender inclusivity. *Australian Review of Applied Linguistics, 38*(1), 24–49.

49. Here in Finland, the language is completely gender-neutral—they don't have any gender specific pronouns like "he" or "she." (2015, May 2). *OMG Facts.* Retrieved from http://www.omgfacts.com/health/14822/Here-in-Finland-the-language-is-completely-gender-neutral-they-don-t-have-any-gender-specific-pronouns-like-he-or-she. Quotes appear in para. 2.

50. Prewitt-Freilino, J. L., Caswell, T. A., & Laakso, E. K. (2012). The gendering of language: A comparison of gender equality in countries with gendered, natural gender, and genderless languages. *Sex Roles, 66*(3/4), 268–281.

51. Erickson, B., Lind, E. A., Johnson, B. C., & O'Barr, W. M. (1978). Speech style and impression formation in a court setting: The effects of "powerful" and "powerless" speech. *Journal of Experimental Social Psychology, 14,* 266–279.

52. Parton, S., Siltanen, S. A., Hosman, L. A., & Langenderfer, J. (2002). Employment interview outcomes and speech style effects. *Journal of Language and Social Psychology, 21,* 144–161.

53. Reid, S. A., Keerie, N., & Palomares, N. A. (2003). Language, gender salience, and social influence. *Journal of Language and Social Psychology, 22,* 210–233.

54. Andrew. (2011). Manners in Spanish—The basics of being polite in Spanish-speaking cultures. How to learn Spanish online: Resources, tips, tricks, and techniques. Retrieved from http://howlearnspanish.com/2011/01/manners-in-spanish.

55. Guenzi, P., & Georges, L. (2010). Interpersonal trust in commercial relationships: Antecedents and consequences of customer trust in the salesperson. *European Journal of Marketing, 44,* 114–138.

56. See, for example, Bell, R. A., & Healey, J. G. (1992). Idiomatic communication and interpersonal solidarity in friends' relational cultures. *Human Communication Research, 18,* 307–335; and Bell, R. A., Buerkel-Rothfuss, N., & Gore, K. E. (1987). Did you bring the yarmulke for the Cabbage Patch Kid? The idiomatic communication of young lovers. *Human Communication Research, 14,* 47–67.

57. Cassell, J., & Tversky, D. (2005). The language of online intercultural community formation. *Journal of Computer-Mediated Communication, 10,* Article 2.

58. Blevins, J. (2016, March 25). Read this: The history of "jawn," Philadelphia's all-purpose word. *A.V. Club.* Retrieved from http://www.avclub.com/article/read-history-jawn-philadelphias-favorite-all-purpo-234368.

59. Maass, A., Salvi, D., Arcuri, L., & Semin, G. R. (1989). Language use in intergroup context. *Journal of Personality and Social Psychology, 57,* 981–993.

60. Kubanyu, E. S., Richard, D. C., Bower, G. B., & Muraoka, M. Y. (1992). Impact of assertive and accusatory communication of distress and anger: A verbal component analysis. *Aggressive Behavior, 18,* 337–347.

61. Motley M. T., & Reeder, H. M. (1995). Unwanted escalation of sexual intimacy: Male and female perceptions of connotations and relational consequences of resistance messages. *Communication Monographs, 62,* 356–382.

62. How many is a couple? A few? Several? [Forum post]. (2012, November 1). *The Escapist.* Retrieved from http://www.escapistmagazine.com/forums/read/18.392824-How-many-is-a-couple-A-few-Several.

63. Labov, T. (1992). Social and language boundaries among adolescents. *American Speech, 4,* 339–366.

64. Peters, M. (2017, January 27). The hidden dangers of euphemisms. British Broadcasting Corporation. Retrieved from http://www.bbc.com/capital/story/20170126-the-hidden-danger-of-euphemisms.

65. Hayakawa, S. I. (1964). *Language in thought and action.* New York: Harcourt Brace.

66. Alberts, J. K. (1988). An analysis of couples' conversational complaints. *Communication Monographs, 55,* 184–197.

67. Prath, C. (2016, December). The hidden toll of workplace incivility. *McKensey Quarterly.* Retrieved from http://www.mckinsey.com/business-functions/organization/our-insights/the-hidden-toll-of-workplace-incivility.

68. Poll finds Americans united in seeing an uncivil nation; divided about causes of civility of presidential candidates. (2016). Weber Shandwick. Retrieved from https://www.webershandwick.com/news/article/poll-finds-americans-united-in-seeing-an-uncivil-nation-divided-about-cause.

69. Culpepper, S. (2016). The lost art of civility or does being a jerk make you more authentic? *The Twelve.* Retrieved from https://blog.perspectivesjournal.org/2016/12/31/the-lost-art-of-civility-or-does-being-a-jerk-make-you-more-authentic/. Quote appears in the title.

70. Grant, A. (2013). *Give and take: Why helping others drivers our success.* New York: Penguin Books.

71. Heitler, S. (2012, October 2). The problem with over-emotional political rhetoric. *Psychology Today.* Retrieved from https://www.psychologytoday.com/blog/resolution-not-conflict/201210/the-problem-over-emotional-political-rhetoric. Quote appears in para. 2.

72. Andrews, T. M. (2017, June 1). Trump revives "crooked Hilary" nickname. Clinton fires back with "covfefe." *The Washington Post.* Retrieved from https://www.washingtonpost.com/news/morning-mix/wp/2017/06/01/trump-revives-insulting-crooked-hillary-nickname-on-twitter-clinton-fires-back-with-covfefe/?utm_term=.8b65c584118f.

73. Caldwell, L. A. (2016, January 20). Donald Trump sharpens attack against Bernie Sanders. *NBC News.* Retrieved from http://www.nbcnews.com/politics/2016-election/donald-trump-sharpens-attack-against-bernie-sanders-n500641.

74. Chavez, P., & Stracqualursi, V. (2016, May 11). From "crooked Hillary" to "Little Marco," Donald Trump's many nicknames. ABC News. http://abcnews.go.com/Politics/crooked-hillary-marco-donald-trumps-nicknames/story?id=39035114.

75. Saleem, H. M., Dillon, K. P., Benesch, S., & Ruths, D. (2016). A web of hate: Tackling hateful speech in online social spaces. Retrieved from https://dangerousspeech.org/a-web-of-hate-tackling-hateful-speech-in-online-social-spaces/. Workshop on Text Analytics for Cybersecutiy and Online Safety. Quote appears in title.

76. Hudson, P. (2015, January 16). "I don't understand women"—well read on the for the full explanation. *Mirror.* Retrieved from http://www.mirror.co.uk/lifestyle/dating/i-dont-understand-women—4993587.

77. Q&A: 8 things we don't understand about men. (n.d.). *Wewomen.* Retrieved from http://www.wewomen.com/understanding-men/what-women-dont-understand-about-men-questions-and-answers-d30896x64063.html.

78. McClure, L. (2017, January 12). How to tell fake news from real news. TEDEd. Retrieved from http://blog.ed.ted.com/2017/01/12/how-to-tell-fake-news-from-real-news.

79. Difference between facts and opinions. (n.d.). *Difference Between.* Retrieved from http://www.differencebetween.info/difference-between-facts-and-opinions.

80. Mehl, M. R., Vazire, S., Ramírez-Esparza, N., Slatcher, R. B., & Pennebaker, J. W. (2007, July). Are women really more talkative than men? *Science, 317,* 82.

81. Sehulster, J. R. (2006). Things we talk about, how frequently, and to whom: Frequency of topics in everyday conversation as a function of gender, age, and marital status. *The American Journal of Psychology, 119,* 407–432.

82. Sehulster, J. R. (2006). Things we talk about, how frequently, and to whom: Frequency of topics in everyday conversation as a function of gender, age, and marital status. *The American Journal of Psychology, 119,* 407–432.

83. Kapidzic, S., & Herring, S. C. (2011). Gender, communication, and self-presentation in teen chatrooms revisited: Have patterns changed? *Journal of Computer-Mediated Communication, 17*(1), 39–59.

84. Cohen, M. M. (2016). It's not you, it's me . . . no, actually it's you: Perceptions of what makes a first date successful or not. *Sexuality and Culture, 20*(1), 173–191.

85. Cohen, It's not you.

86. Wood, J. T. (2001). *Gendered lives: Communication, gender, and culture* (4th ed.). Belmont, CA: Wadsworth, p. 141.

87. Fox, A. B., Bukatko, D., Hallahan, M., & Crawford, M. (2007). The medium makes a difference: Gender similarities and differences in instant messaging. *Journal of Language and Social Psychology, 26,* 389–397.

88. Pfafman, T. M., & McEwan, B. (2014). Polite women at work: Negotiating professional identity through strategic assertiveness. *Women's Studies in Communication, 37*(2), 202–219.

89. Booth-Butterfield, M. M., Wanzer, M. B., Weil, N., & Krezmien, E. (2014). Communication of humor during bereavement: Intrapersonal and interpersonal emotion management strategies. *Communication Quarterly, 62*(4), 436–454.

90. Gesteland, R. R. (2012). *Cross-cultural business behavior: A guide for global management* (5th ed.). Copenhagen: Copenhagen Business School Press.

91. Menchhofer, T. O. (2015, April). Planting the seed of emotional literacy: Engaging men and boys in creating change. *The Vermont Connection, 24*(4), nonpaginated online version. Retrieved from http://scholarworks.uvm.edu/cgi/viewcontent.cgi?article=1197&context=tvc.

92. Schoenfeld, E. A., Bredow, C. A., & Huston, T. L. (2012). Do men and women show love differently in marriage? *Personality and Social Psychology Bulletin, 11,* 1396–1409.

93. Jones, A. C., & Josephs, R. A. (2006). Interspecies hormonal interactions between man and the domestic dog (*Canis familiaris*). *Hormones and Behavior, 50*(3), 393–400.

94. Pennebaker, J. W., Groom, C. J., Loew, D., & Dabbs, J. M. (2004). Testosterone as a social inhibitor: Two case studies of the effect of testosterone treatment on language. *Journal of Abnormal Psychology, 113*(1), 172.

95. Chen, C. P., Cheng, D. Z., Luo, Y.-J. (2011). Estrogen impacts on emotion: Psychological, neuroscience and endocrine studies. *Science China Life, 41*(11). Retrieved from http://www.eurekalert.org/pub_releases/2012-01/sicp-tio010912.php.

96. Premenstrual syndrome (PMS) fact sheet. How common is PMS? (2010). U.S. Department of Health and Human Services. Retrieved from http://womenshealth.gov/publications/our-publications/fact-sheet/premenstrual-syndrome.cfm#e.

97. Mulac, A., Giles, H., Bradac, J. J., & Palomares, N. A. (2013). The gender-linked language effect: An empirical test of a general process model. *Language Sciences, 38,* 22–31.

98. Hancock, A. B., & Rubin, B. A. (2015). Influence of communication partner's gender on language. *Journal of Language and Social Psychology, 34*(1), 46–64.

99. Fandrich, A. M., & Beck, S. J. (2012). Powerless language in health media: The influence of biological sex and magazine type on health language. *Communication Studies, 63*(1), 36–53.

CHAPTER 6

1. Goulston, M. (2009, September 9). "Just listen" . . . A story of the power of listening. YouTube video. Retrieved from https://www.youtube.com/watch?v=Rj7W-vucBKw. Quote is at 1:16 mins.

2. Goulston (2009). Quote is at 2 mins.

3. Goulston, M. (2010). *Just listen: Discover the secret to getting through to absolutely anyone.* New York: American Management Association.

4. Covey, S. (1989). *The 7 habits of highly effective people.* New York: Simon & Schuster.

5. Kalargyrou, V., & Woods, R. H. (2011). Wanted: Training competencies for the twenty-first century. *International Journal of Contemporary Hospitality Management, 23*(3), 361–376.

6. Suzuno, M. (2014, January 21). 5 things recruiters wish you knew about career fairs. *After College.* Retrieved from http://blog.aftercollege.com/5-things-recruiters-wish-knew-career-fairs. Quotes appear in tip 2 and in the title, respectively.

7. Kalargyrou, V., & Woods, R. H. (2011). Wanted: Training competencies for the twenty-first century. *International Journal of Contemporary Hospitality Management, 23*(3), 361–376.

8. Davis, J., Foley, A., Crigger, N., & Brannigan, M. C. (2008). Healthcare and listening: A relationship for caring. *International Journal of Listening, 22*(2), 168–175.

9. Pryor, S., Malshe, A., & Paradise, K. (2013). Salesperson listening in the extended sales relationship: an exploration of cognitive, affective, and temporal dimensions. *Journal of Personal Selling & Sales Management, 33*(2), 185–196.

10. Brockner, J., & Ames, D. (2010, December 1). Not just holding forth: The effect of listening on leadership effectiveness. *Social Science Electronic Publishing.* Retrieved from http://papers.ssrn.com/sol3/papers.cfm?abstract_id=1916263.

11. Ames, D., Maissen, L. B., & Brockner, J. (2012). The role of listening in interpersonal influence. *Journal of Research in Personality, 46,* 345–349.

12. Listening quotes. (n.d.) *Wise old sayings.* Retrieved from http://www.wiseoldsayings.com/listening-quotes. Quote appears sixth.

13. Gordon, P., James Allan, C., Nathaniel, B., Derek, J. K., & Jonathan, A. F. (2015). On the reception and detection of pseudo-profound bullshit. *Judgment and Decision Making, 10*(6), 549–563.

14. Brooks, A. W., Gino, F., & Schweitzer, M. E. (2015). Smart people ask for (my) advice: Seeking advice boosts perceptions of competence. *Management Science, 61*(6), 1421–1435. Quote appears on p. 1421.

15. Bodie, G. D., Vickery, A. J., & Gearhart, C. C. (2013). The nature of supportive listening, I: Exploring the relation between supportive listeners and supportive people. *International Journal of Listening, 27,* 39–49.

16. Fletcher, G. O., Kerr, P. G., Li, N. P., & Valentine, K. A. (2014). Predicting romantic interest and decisions in the very early stages of mate selection: Standards, accuracy, and sex differences. *Personality and Social Psychology Bulletin, 4,* 540–550.

17. What women want from men: A good listener. (2012, October 3.) Ingenio Advisor. Retrieved from http://www.ingenio.com/CommunityServer/UserBlogPosts/Advisor_Louise_PhD/What-Women-Want-from-Men—A-Good-Listener/630187.aspx. Quote appears in para. 3.

18. Bond, A. B. (2005, August 25). The power of listening: A true story. *Care2.* Retrieved from http://www.care2.com/greenliving/

power-of-listening-true-story.html. Quotes appear in para. 3. (Julio is a pseudonym for the person described.)

19. Ceraso, S. (2011, April 25). "I listen with my eyes": Deaf architecture and rhetorical space. *Humanities, Arts, Science, and Technology Alliance and Collaboratory.* Retrieved from http://www.hastac.org/blogs/stephceraso/i-listen-my-eyes-deaf-architecture-and-rhetorical-space.

20. Powers, W. G., & Witt, P. L. (2008). Expanding the theoretical framework of communication fidelity. *Communication Quarterly, 56,* 247–267; Fitch-Hauser, M., Powers, W. G., O'Brien, K., & Hanson, S. (2007). Extending the conceptualization of listening fidelity. *International Journal of Listening, 21,* 81–91; Powers, W. G., & Bodie, G. D. (2003). Listening fidelity: Seeking congruence between cognitions of the listener and the sender. *International Journal of Listening, 17,* 19–31.

21. Kim, Y. G. (2016). Direct and mediated effects of language and cognitive skills on comprehension of oral narrative texts (listening comprehension) for children. *Journal of Experimental Child Psychology, 141,* 101–120.

22. Fontana, P. C., Cohen, S. D., & Wolvin, A. D. (2015). Understanding listening competency: A systematic review of research scales. *International Journal of Listening, 29*(3), 148–176.

23. Thomas, T. L., & Levine, T. R. (1994). Disentangling listening and verbal recall: Related but separate constructs? *Human Communication Research, 21,* 103–127.

24. Humorous example of social media monitoring: Sydney University [blog post]. (2009, September 7). Retrieved from http://www.altimetergroup.com/2009/09/humorous-example-of-social-media-monitoring-sydney-university.html.

25. Caruso, R. (2011). A real example of effective social media monitoring and engagement. *Bundle Post.* Retrieved from http://bundlepost.wordpress.com/2011/11/07/a-real-example-of-effective-social-media-monitoring-and-engagement.

26. Petrocelli, T. (2012, December 20). One rule with social media and social networking: Don't be creepy [blog post]. Retrieved from http://www.esg-global.com/blogs/one-rule-with-social-media-and-social-networking-dont-be-creepy.

27. Cowan, N., & AuBuchon, A. M. (2008). Short-term memory loss over time without retroactive stimulus interference. *Psychonomic Bulletin and Review, 15,* 230–235.

28. Ames, D., Maissen, L. B., & Brockner, J. (2012, June). The role of listening in interpersonal influence. *Journal of Research in Personality, 46,* 345–349.

29. Accenture research finds listening more difficult in today's digital workplace. (2015, February 26). Accenture. Retrieved from https://newsroom.accenture.com/industries/global-media-industry-analyst-relations/accenture-research-finds-listening-more-difficult-in-todays-digital-workplace.htm.

30. rurounikenji. (2012, July 8). Girlfriend literally never listens to me [message board post]. *The Student Room.* Retrieved from http://www.thestudentroom.co.uk/showthread.php?t=2075372. Quote appears in para. 1.

31. Chapman, S. G. (2012). *The five keys to mindful communication: Using deep listening and mindful speech to strengthen relationships, heal conflicts, and accomplish your goals.* Boulder, CO: Shambhala Publications.

32. Ting-Toomey, S., & Chung, L. C. (2011). *Understanding intercultural communication* (2nd ed.). New York: Oxford University Press.

33. Robertson, R. R. (2016, October 12). Normani Kordei opens up about her struggle with cyberbullies and racist trolls. *Essence.* Retrieved from http://www.essence.com/celebrity/normani-kordei-cyberbullies-racist-trolls.

34. Accenture research finds listening more difficult in today's digital workplace. (2015, February 26). Accenture. Retrieved

from https://newsroom.accenture.com/industries/global-media-industry-analyst-relations/accenture-research-finds-listening-more-difficult-in-todays-digital-workplace.htm.

35. Hansen, J. (2007). *24/7: How cell phones and the Internet change the way we live, work, and play.* New York: Praeger. See also Turner, J. W., & Reinsch, N. L. (2007). The business communicator as presence allocator: Multicommunicating, equivocality, and status at work. *Journal of Business Communication, 44,* 36–58.

36. Hemp, P. (2009, December 4). Death by information overload. *Harvard Business Review* online. Retrieved from http://www.ocvets4pets.com/archive17/Death_by_Information_Overload_-_HBR.org.pdf.

37. Dean, M., & Street, J. L. (2014). Review: A 3-stage model of patient-centered communication for addressing cancer patients' emotional distress. *Patient Education and Counseling, 94,* 143–148.

38. Shafir, R. Z. (2003). *The Zen of listening: Mindful communication in the age of distraction.* Wheaton, IL: Quest Books.

39. Chapman, *The five keys to mindful communication.*

40. Info stupidity. (2005, April 30). *New Scientist, 186,* 6–7.

41. Lin, L. (2009, September 15). Breadth-biased versus focused cognitive control in media multitasking behaviors. *Proceedings of the National Academy of Sciences, 106,* 15521–15522. Retrieved from http://www.pnas.org/content/106/37/15521.full.pdf.

42. Ophir, E., Nass, C., & Wagner, A. (2009). Cognitive control in media multitaskers. *Proceedings of the National Academy of Sciences, 106,* 15583–15587.

43. Imhof, M. (2003). The social construction of the listener: Listening behaviors across situations, perceived listener status, and cultures. *Communication Research Reports, 20,* 357–366.

44. Zohoori, A. (2013). A cross-cultural comparison of the HURIER Listening Profile among Iranian and U.S. students. *International Journal of Listening, 27,* 50–60.

45. Imhof, The social construction of the listener.

46. Nelson, S. (2014, May 2). The problem: My friend doesn't ask me about life life! *Shasta's Friendship Blog.* Retrieved from http://www.girlfriendcircles.com/blog/index.php/2014/05/the-problem-my-friend-doesnt-ask-me-about-my-life/. Quotes appear in paras. 2 and 5, respectively.

47. Winnick, M. (2016, June 16). Putting a finger on our phone obsession. *Dscout.* Retrieved from https://blog.dscout.com/mobile-touches.

48. Turkle, S. (2012, April 21). The flight from conversation. *New York Times.* Retrieved from http://www.nytimes.com/2012/04/22/opinion/sunday/the-flight-from-conversation.html?pagewanted=all&_r=0.

49. Talking to your parents—or other adults. (2015, February). Nemours. Retrieved from https://kidshealth.org/en/teens/talk-to-parents.html?WT.ac=ctg. Quote appears in para. 7 of "Raising Difficult Topics" section.

50. Vangelisti, A. L., Knapp, M. L., & Daly, J. A. (1990). Conversational narcissism. *Communication Monographs,57,* 251–274.

51. Kline, N. (1999). *Time to think: Listening to ignite the human mind.* London: Ward Lock, p. 21.

52. Derber, C. (2000). *The pursuit of attention: Power and ego in everyday life* (2nd ed.). New York: Oxford University Press.

53. Poundstone, P. (2017). *The totally unscientific study of the search for human happiness.* Chapel Hill: NC. Algonquin Books. Quote appears on p. 189.

54. Wilson Mizner quotes. (n.d.). Brainy Quote. Retrieved from http://www.brainyquote.com/quotes/authors/w/wilson_mizner.html.

55. Halvorson, H. G. (2010, August 17). Stop being so defensive! A simple way to learn to take criticism gracefully [blog post].

Psychology Today. Retrieved from https://www.psychologytoday .com/blog/the-science-success/201008/stop-being-so-defensive.

56. Valdes, A. (2012, June 19). 8 tips to help you stop being defensive. *Mamiverse.* Retrieved from http://mamiverse.com/8-tips-to-help-you-stop-being-defensive-13577.

57. Pang, S. (2012). Be a good listener. Great Inspiring Stories. Retrieved from http://lifeaward.blogspot.com/2012/08/be-good-listener.html. Quote appears in para. 3.

58. Chia, H. L. (2009). Exploring facets of a social network to explicate the status of social support and its effects on stress. *Social Behavior and Personality: An International Journal, 37*(5), 701–710. See also Segrin, C., & Domschke, T. (2011). Social support, loneliness, recuperative processes, and their direct and indirect effects on health. *Health Communication, 26,* 221–232.

59. Bodie, G. D., Vickery, A. J., & Gearhart, C. C. (2013). The nature of supportive listening, I: Exploring the relation between supportive listeners and supportive people. *International Journal of Listening, 27,* 39–49.

60. Robinson, J. D., & Tian, Y. (2009). Cancer patients and the provision of informational social support. *Health Communication, 24,* 381–390.

61. Gearhart, C. G., & Bodie, G. D. (2011). Active-empathic listening as a general social skill: Evidence from bivariate and canonical correlations. *Communication Reports, 24,* 86–98.

62. Burleson, B. (2008). What counts as effective emotional support?" In M. T. Motley (Ed.), *Studies in Applied Interpersonal Communication* (pp. 207–227). Thousand Oaks, CA: Sage.

63. Guo, J., & Turan, B. (2016). Preferences for social support during social evaluation in men: The role of worry about a relationship partner's negative evaluation. *Journal of Social Psychology, 156*(1), 122–129.

64. Olson, R. (2014). A time-sovereignty approach to understanding carers of cancer patients' experiences and support preferences. *European Journal of Cancer Care, 23*(2), 239–248.

65. Young, R. W., & Cates, C. M. (2004). Emotional and directive listening in peer mentoring. *International Journal of Listening, 18,* 21–33.

66. Paraschos, S. (2013). Unconventional doctoring: A medical student's reflections on total suffering. *Journal of Palliative Medicine, 16,* 325.

67. Sarah Q. (n.d.). When I needed a friend. *TeenInk.* Retrieved from http://www.teenink.com/nonfiction/educator_of_the_year/ article/101090/When-I-Needed-a-Friend. Quotes appear in paras. 11 and 10, respectively.

68. Huerta-Wong, J. E., & Schoech, R. (2010). Experiential learning and learning environments: The case of active listening skills. *Journal of Social Work Education, 46,* 85–101.

69. Goldsmith, D. (2000). Soliciting advice: The role of sequential placement in mitigating face threat. *Communication Monographs, 67,* 1–19.

70. 10 things to say (and not to say) to someone with depression. (n.d.). *Health.* Retrieved from http://www.drbalternatives.com/ articles/cc2.html.

71. Stewart, M., Letourneau, N., Masuda, J. R., Anderson, S., Cicutto, L., McGhan, S., & Watt, S. (2012). Support needs and preferences of young adolescents with asthma and allergies: "Just no one really seems to understand." *Journal of Pediatric Nursing, 27*(5), 479–490.

72. Helping adults, children cope with grief. (2001, September 13). *Washington Post.*

73. Tannen, D. (2010). He said, she said. *Scientific American Mind, 21*(2), 55–59.

74. MacGeorge, E. L., Feng, B., & Thompson, E. R. (2008). "Good" and "bad" advice: How to advise more effectively. In M. T. Motley (Ed.), *Studies in applied interpersonal communication* (pp. 145–164). Thousand Oaks, CA: Sage.

75. Isaacson, W. (2016, December 15). How to fix the Internet. *The Atlantic.* Retrieved from https://www.theatlantic.com/ technology/archive/2016/12/how-to-fix-the-internet/510797. Quote appears in para. 6.

76. Desta, Y. (2014, May 1). 4 ways to reach out for mental health help anonymously. *Mashable.* Retrieved from https://mashable. com/2014/05/01/get-help-anonymously/#YOgOL_NnCmqQ. Quote appears in para. 1 and 2.

77. Isaacson, How to fix the Internet. Quote appears in para. 7.

78. Tannen, He said, she said.. Quote appears in para. 11.

79. Wood, Z. R. (2018, April). Why it's worth listening to people you disagree with. TED. Retrieved from https://www.ted.com/talks/ zachary_r_wood_why_it_s_worth_listening_to_people_we_ disagree_with. Quote appears in title.

80. DRB Alternatives, Inc. (n.d.). 10 things to say (and not to say) to someone with depression. Retrieved from http://www .drbalternatives.com/articles/cc2.html.

81. Weger, H., Jr., Castle, G. R., & Emmett, M. C. (2010). Active listening in peer interviews: The influence of message paraphrasing on perceptions of listening skill. *International Journal of Listening, 24,* 34–49.

82. Learn anything with the Feynman technique. (2016, December 17). *The Science Explorer.* Retrieved online from http:// thescienceexplorer.com/universe/learn-anything-feynman-technique.

83. Wood, Z. R. (2018, April). Why it's worth listening to people you disagree with. TED. Retrieved from https://www.ted.com/talks/ zachary_r_wood_why_it_s_worth_listening_to_people_we_ disagree_with. Quote appears in title.

84. Luedtke, K. (1987, January 7). What good is free speech if no one listens? *Los Angeles Times.* Retrieved from http://articles.latimes .com/1987-01-07/local/me-2347_1_free-speech.

85. Gearhart, C. C., Denham, J. P., & Bodie, G. D. (2013, November). *Listening is a goal-directed activity.* Paper presented at the annual meeting of the National Communication Association, Washington, DC.

86. Adapted from Infante, D. A. (1988). *Arguing constructively.* Prospect Heights, IL: Waveland, pp. 71–75.

CHAPTER 7

1. Amy J. C., C., Caroline A., W., Andy J., Y., & Dana R., C. (2015). Preparatory power posing affects nonverbal presence and job interview performance. *Journal of Applied Psychology, 4,* 1286–1295.

2. Cuddy, A. (2012, October). Your body language shapes who you are. TED.com. Retrieved from http://www.ted.com/talks/ amy_cuddy_your_body_language_shapes_who_you_are/trans cript?language=en#t-235180. All quotes in this profile are from this source.

3. Crede, M., & Phillips, L. A. (2010). Revisiting the power pose effect: How robust are the results reported by Carney, Cuddy, and Yap to data analytic decisions? *Social Psychological and Personality Science, 8*(5), 493–499.

4. Research supporting these claims is cited in Burgoon, J. K., & Hoobler, G. D. (2002). Nonverbal signals. In M. L. Knapp & J. A. Daly (Eds.)., *Handbook of interpersonal communication* (3rd ed., pp. 240–299). Thousand Oaks, CA: Sage.

5. Jones, S. E., & LeBaron, C. D. (2002). Research on the relationship between verbal and nonverbal communication: Emerging interactions. *Journal of Communication, 52,* 499–521.

6. For a survey of the issues surrounding the definition of nonverbal communication, see Knapp, M., & Hall, J. A. (2010). *Nonverbal communication in human interaction* (6th ed.). Belmont, CA: Wadsworth.

7. For a survey of the issues surrounding the definition of nonverbal communication, see Knapp, M., & Hall, J. A. (2010). *Nonverbal communication in human interaction* (6th ed.). Belmont, CA: Wadsworth.

8. Knapp, M., & Hall, J. A. (2010). *Nonverbal communication in human interaction* (6th ed.). Belmont, CA: Wadsworth.

9. Tracy, J. L., & Matsumoto, D. (2008, August 19). The spontaneous expression of pride and shame: Evidence for biologically innate nonverbal displays. *Proceedings from the National Academy of Science, 105*(33), 11655–11660.

10. Palmer, M. T., & Simmons, K. B. (1995). Communicating intentions through nonverbal behaviors: Conscious and nonconscious encoding of liking. *Human Communication Research, 22*, 128–160.

11. Dennis, A. R., Kinney, S. T., & Hung, Y. T. (1999). Gender differences in the effects of media richness. *Small Group Research, 30*, 405–437.

12. H. J. Miller, H. J., Kluver, D., Thebault-Spieker, J., Terveen, L. G., & Hecht, B. J. (2017). Understanding emoji ambiguity in context: the role of text in emoji-related miscommunication. In *Proceedings of the Eleventh International Conference on Web and Social Media*, ICWSM 2017, pp. 152–161.

13. Safeway clerks object to "service with a smile." (1998, September 2). *San Francisco Chronicle*.

14. Nowicki, S., & Duke, M. (2013). Accuracy in interpreting nonverbal cues. *Nonverbal Communication, 2*, 441.

15. Lieberman, M. D., & Rosenthal, R. (2001). Why introverts can't always tell who likes them: Multitasking and nonverbal decoding. *Journal of Personality and Social Psychology, 80*(2), 294.

16. Rosip, J. C., & Hall, J. A. (2004). Knowledge of nonverbal cues, gender, and nonverbal decoding accuracy. *Journal of Nonverbal Behavior, 28*(4), 267–286; Hall, J. A. (1979). Gender, gender roles, and nonverbal communication skills. In R. Rosenthal (Ed.), *Skill in nonverbal communication: Individual differences* (pp. 32–67). Cambridge, MA: Oelgeschlager, Gunn, and Hain.

17. Rourke, B. P. (1989). *Nonverbal learning disabilities: The syndrome and the model.* New York: Guilford Press.

18. Fudge, E. S. (n.d.). Nonverbal learning disorder syndrome? Retrieved from http://www.nldontheweb.org/fudge.htm.

19. Meyer, K. (2016, July 17). The four dimensions of tone of voice. Nielsen Norman Group. Retrieved from https://www.nngroup.com/articles/tone-of-voice-dimensions.

20. Carlson, E. N. (2013). Overcoming barriers to self-knowledge: Mindfulness as a path to seeing yourself as you really are. *Perspectives on Psychological Science, 8*, 173–186.

21. Burgoon, J. K., & Hale, J. L. (1988). Nonverbal expectancy violations: Model elaboration and application to immediacy behaviors. *Communication Monographs, 55*, 58–79.

22. Guerrero, L. K., & Bachman, G. F. (2008). Relational quality and relationships: An expectancy violations analysis. *Journal of Social and Personal Relationships, 23*(6), 943–963.

23. Miller-Ott, A., & Kelly, L. (2015). The presence of cell phones in romantic partner face-to-face interactions: An expectancy violation theory approach. *Southern Communication Journal, 4*, 253–270.

24. Cross, E. S., & Franz, E. A. (2003, March 30–April 1). Talking hands: Observation of bimanual gestures as a facilitative working memory mechanism. Paper presented at the 10th annual meeting of the Cognitive Neuroscience Society, New York.

25. Burnett, S. (2014, August 4). Have you ever wondered why Asians spontaneous make V-signs in photos? *Time*. Retrieved from http://time.com/2980357/asia-photos-peace-sign-v-janet-lynn-konica-jun-inoue.

26. Cosgrove, B. (2014, July 4). V for victory: Celebrating a gesture of solidarity and defiance. *Time*. Retrieved from http://time.com/3880345/v-for-victory-a-gesture-of-solidarity-and-defiance.

27. Shittu, H., & Query, C. (2006). *Absurdities, scandals and stupidities in politics.* Mexico: Genix Press.

28. Kleinke, C. R. (1977). Compliance to requests made by gazing and touching experimenters in field settings. *Journal of Experimental Social Psychology, 13*, 218–233.

29. Patel, S., & Scherer, K. (2013). Vocal behavior. In M. L. Knapp & J. A. Hall (Eds.), *Nonverbal communication* (pp. 167–204). Boston: De Gruyter Mouton.

30. Argyle, M. F., Alkema, F., & Gilmour, R. (1971). The communication of friendly and hostile attitudes: Verbal and nonverbal signals. *European Journal of Social Psychology, 1*, 385–402.

31. Vrij, A. (2006). Nonverbal communication and deception. In V. Manusov & M. L. Patterson (Eds.), *Sage handbook of nonverbal communication* (pp. 341–359). Thousand Oaks, CA: Sage.

32. Dunbar, N. E., Ramirez, A., Jr., & Burgoon, J. K. (2003). The effects of participation on the ability to judge deceit. *Communication Reports, 16*, 23–33.

33. DePaulo, B. M., Lindsay, J. J., Malone, B. E., Muhlenbruck, L., Charlton, K., & Cooper, H. (2003). Cues to deception. *Psychological Bulletin, 129*, 74–118; and Vrig, A., Edward, K., Roberts, K. P., & Bull, R. (2000). Detecting deceit via analysis of verbal and nonverbal behavior. *Journal of Nonverbal Behavior, 24*, 239–263.

34. Lock, C. (2004). Deception detection: Psychologists try to learn how to spot a liar. *Science News Online, 166*, 72.

35. Lee, K. (2016). Can you tell if a kid is lying? TED. Retrieved from https://www.ted.com/talks/kang_lee_can_you_really_tell_if_a_kid_is_lying/transcript.

36. Knapp, M. L. (2006). *Lying and deception in close relationships.* Cambridge, England: Cambridge University Press.

37. Morris, W. L., Sternglanz, R. W., Ansfield, M. E., Anderson, D. E., Snyder, J. H., & DePaulo, B. M. (2016). A longitudinal study of the development of emotional deception detection within new same-sex friendships. *Personality & Social Psychology Bulletin, 2*, 204–218.

38. Levine, T. R. (2014). Truth-default theory (TDT): A theory of human deception and deception detection. *Journal of Language & Social Psychology, 33*(4), 378–392.

39. Ein-Dor, T. T., Perry-Paldi, A., Zohar-Cohen, K., Efrati, Y., & Hirschberger, G. (2017). It takes an insecure liar to catch a liar: The link between attachment insecurity, deception, and detection of deception. *Personality & Individual Differences, 113*, 81–87.

40. Vrig, A., Akehurst, L., Soukara, S., & Bull, R. (2004). Detecting deceit via analyses of verbal and nonverbal behavior in children and adults. *Human Communication Research, 30*, 8–41.

41. Glass, L. (2012). *The body language advantage.* Beverly, MA: Fair Winds Press.

42. Waters, H. (2013, December 13). Fake it 'til you become it: Amy Cuddy's power poses, visualized. *TED Blog*. Retrieved from http://blog.ted.com/fake-it-til-you-become-it-amy-cuddys-power-poses-visualized. Quote appears in title.

43. Maurer, R. E., & Tindall, J. H. (1983). Effect of postural congruence on client's perception of counselor empathy. *Journal of Counseling Psychology, 30*, 158–163. See also Hustmyre, C., & Dixit, J. (2009, January 1). Marked for mayhem. *Psychology Today*. Retrieved from https://www.psychologytoday.com/articles/200901/marked-mayhem.

44. Ray, G., & Floyd, K. (2006). Nonverbal expressions of liking and disliking in initial interaction: Encoding and decoding perspectives. *Southern Communication Journal, 71*(1), 45–65.

45. Myers, M. B., Templer, D., & Brown, R. (1984). Coping ability of women who become victims of rape. *Journal of Consulting and Clinical Psychology, 52*, 73–78. See also Hustmyre & Dixit. Marked for mayhem.

46. Ekman, P. (1985). *Telling lies: Clues to deceit in the marketplace, politics, and marriage*. New York: Norton.

47. Musicus, A., Tal, A., & Wansink, B. (2014). Eyes in the aisles: Why is Cap'n Crunch looking down at my child? *Environment and Behavior, 47*(7), 715–733.

48. Murphy, K. (2014, May 18). Psst. Look over here . . . *New York Times*, p. SR6.

49. Farroni, T., Csibra, G., Simion, F., & Johnson, M. H. (2002). Eye contact detection in humans from birth. *Proceedings of the National Academy of Sciences of the United States of America, 99*(14), 9602–9605.

50. Akechi, H., Senju, A., Uibo, H., Kikuchi, Y., Hasegawa, T., & Hietanen, J. K. (2013). Attention to eye contact in the West and East: Autonomic responses and evaluative ratings. *Plos ONE, 8*(3), 1–10.

51. Ze, W., Huifang, M., Jessica, L., & Fan, L. (2016). The insidious effects of smiles on social judgments. *Advances in Consumer Research, 44*, 665–669.

52. Arapova, M. A. (2017). Cultural differences in Russian and Western smiling. *Russian Journal of Communication, 1*, 34–52.

53. Arapova, Cultural differences.

54. Starkweather, J. A. (1961). Vocal communication of personality and human feeling. *Journal of Communication, 11*(2), 63–72; and Scherer, K. R., Koiwunaki, J., & Rosenthal, R. (1972). Minimal cues in the vocal communication of affect: Judging emotions from content-masked speech. *Journal of Psycholinguistic Speech, 1*(3), 269–285. See also Cox, F. S., & Olney, C. (1985). *Vocalic communication of relational messages*. Paper presented at the annual meeting of the Speech Communication Association, Denver.

55. Burns, K. L., & Beier, E. G. (1973). Significance of vocal and visual channels for the decoding of emotional meaning. *Journal of Communication, 23*, 118–130. See also Hegstrom, T. G. (1979). Message impact: What percentage is nonverbal? *Western Journal of Speech Communication, 43*, 134–143; and McMahan, E. M. (1976). Nonverbal communication as a function of attribution in impression formation. *Communication Monographs, 43*, 287–294.

56. Kimble, C. E., & Seidel, S. D. (1991). Vocal signs of confidence. *Journal of Nonverbal Behavior, 15*, 99–105.

57. Tusing, K. J., & Dillard, J. P. (2000). The sounds of dominance: Vocal precursors of perceived dominance during interpersonal influence. *Human Communication Research, 26*, 148–171.

58. Hosoda, M., & Stone-Romero, E. (2010). The effects of foreign accents on employment-related decisions. *Journal of Managerial Psychology, 25*, 113–132.

59. Ritchel, M. (2015, June 27). The mouth is mightier than the pen. *New York Times*. Retrieved from https://www.nytimes.com/2015/06/28/business/the-mouth-is-mightier-than-the-pen.html. Quote appears in para. 4.

60. Schroeder, J., & Epley, N. (2015). The sound of intellect: Speech reveals a thoughtful mind, increasing a job candidate's appeal. *Psychological Science, 26*(6), 877–891.

61. Gupta, N. D., Etcoff, N. L., & Jaeger, M. M. (2015, June 14). Beauty in mind: The effects of physical attractiveness on psychological well-being and distress. *Journal of Happiness Studies, 17*(3), 1313–1325.

62. Milazzo, C., & Mattes, K. (2016). Looking good for election day: Does attractiveness predict electoral success in Britain? *British Journal of Politics and International Relations, 18*(1), 161–179.

63. Gunnell, J. J., & Ceci, S. J. (2010). When emotionality trumps reason: A study of individual processing style and juror bias. *Behavioral Sciences and the Law, 28*(6), 850–877.

64. Haas, A., & Gregory, S. W., Jr. (2005). The impact of physical attractiveness on women's social status and interactional power. *Sociological Forum, 20*(3), 449–471.

65. Bennett, J. (2010, July 19). The beauty advantage: How looks affect your work, your career, your life. *Newsweek*. Retrieved from http://www.newsweek.com/2010/07/19/the-beauty- advantage .html.

66. Behrend, T., Toaddy, S., Thompson, L. F., & Sharek, D. J. (2012.) The effects of avatar appearance on interviewer ratings in virtual employment interviews. *Computers in Human Behavior 28*(6), 2128–2133.

67. Mobius, M. M., & Rosenblat, T. S. (2005, June 24). Why beauty matters. *American Economic Review, 96*(1), 222–235.

68. Golle, J., Mast, F. W., & Lobmaier, J. S. (2014). Something to smile about: The interrelationship between attractiveness and emotional expression. *Cognition and Emotion, 28*(2), 298–310.

69. Abdala, K. F., Knapp, M. L., & Theune, K. E. (2002). Interaction appearance theory: Changing perceptions of physical attractiveness through social interaction. *Communication Theory, 12*, 8–40.

70. Agthe, M., Sporrle, M., & Maner, J. K. (2011). Does being attractive always help? Positive and negative effects of attractiveness on social decision making. *Personality and Social Psychology Bulletin, 37*, 1042–1054.

71. Frevert, T. K., & Walker, L. S. (2014). Physical attractiveness and social status. *Sociology Compass, 8*, 313–323.

72. Noh, M., Li, M., Martin, K., & Purpura, J. (2015). College men's fashion: Clothing preference, identity, and avoidance. *Fashion and Textiles, 2*(1), 1–12.

73. Noh, Li, Martin, & Purpura, College men's fashion.

74. Furnham, A. (2014, April 22). Lookism at work. *Psychology Today*; Gordon, R., Crosnoe, R., & Wang, X. (2013). Physical attractiveness and the accumulation of social and human capital in adolescence and young adulthood. *Monographs of the Society for Research in Child Development, 78*, 1–137.

75. McCall, T. (2013, August 1.) Why is it so difficult for the average American woman to shop for clothes? *Fashionista*. Retrieved from https://fashionista.com/2013/08/why-is-it-so-difficult-for-the-average-american-woman-to-shop-for-clothes.

76. Gurung, R. A. R., Brickner, M., Leet, M., & Punke, E. (2018). Dressing "in code": Clothing rules, propriety, and perceptions. *The Journal of Social Psychology, 158*(5), 553–557.

77. Bell, E. (2016, May 13). Wearing heals to work is a game women have been losing for decades. *The Conversation*. Retrieved from https://theconversation.com/wearing-heels-to-work-is-a-game-women-have-been-losing-for-decades-59337.

78. Reddy-Best, K. L., & Pedersen, E. L. (2015, October). Queer women's experiences purchasing clothing and looking for clothing styles. *Clothing and Textiles Research Journal, 33*(4), 265–279.

79. Roberts, S., Owen, R. C., & Havlicek, J. (2010). Distinguishing between perceiver and wearer effects in clothing color-associated attributions. *Evolutionary Psychology, 8*(3), 350–364.

80. Rehman S. U., Nietert P. J., Cope D. W., & Kilpatrick, A. O. (2005). What to wear today? Effect of doctor's attire on the trust and confidence of patients. *The American Journal of Medicine, 118*, 1279–1286.

81. Armstrong, M., & Fell, P. (2000). Body art: Regulatory issues and the NEHA Body Art Model Code. *Journal of Environmental Health, 62*(9), 25–30.

82. Dickson, L., Dukes, R. L., Smith, H., & Strapko, N. (2015). To ink or not to ink: The meaning of tattoos among college students. *College Student Journal, 49*(1), 106–120.

83. Musambira, G. W., Raymond, L., & Hastings, S. O. (2016). A comparison of college students' perceptions of older and younger tattooed women. *Journal of Women and Aging, 28*(1), 9–23.

84. Galbarczyk, A., & Ziomkiewicz, A. (2017). Tattooed men: Healthy bad boys and good-looking competitors. *Personality and Individual Differences, 106*, 122–125.

85. French, M. T., Maclean, J. C., Robins, P. K., Sayed, B., & Shiferaw, L. (2016). Tattoos, employment, and labor market earnings: Is there a link in the ink? *Southern Economic Journal, 82*(4), 1212–1246.

86. Hart, S., Field, T., Hernandez-Reif, M., & Lundy, B. (1998). Preschoolers' cognitive performance improves following massage. *Early Child Development and Care, 143*, 59–64. For more about the role of touch in relationships, see Keltner, D. (2009). *Born to be good: The science of a meaningful life.* New York: Norton, pp. 173–198.

87. Field, T. (2010). Touch for socioemotional and physical well-being: A review. *Developmental Review, 30*(4), 367–383.

88. Montagu, A. (1972). *Touching: The human significance of the skin.* New York: Harper & Row, p. 93.

89. Feldman, R. (2011). Maternal touch and the developing infant. In M. Hertenstein & S. Weiss (Eds.), *Handbook of touch* (pp. 373–407). New York: Springer.

90. Willis, F. N., & Hamm, H. K. (1980). The use of interpersonal touch in securing compliance. *Journal of Nonverbal Behavior, 5*, 49–55.

91. Jacob, C., & Guéguen, N. (2014). The effect of compliments on customers' compliance with a food server's suggestion. *International Journal of Hospitality Management, 40*, 59–61.

92. Gulledge, N., & Fischer-Lokou, J. (2003). Another evaluation of touch and helping behaviour. *Psychological Reports, 92*, 62–64.

93. Kraus, M. W., Huang, C., & Keltner, D. (2010). Tactile communication, cooperation, and performance: An ethological study of the NBA. *Emotion, 10*(5), 745–749.

94. Martin, B. S. (2012). A stranger's touch: Effects of accidental interpersonal touch on consumer evaluations and shopping time. *Journal of Consumer Research, 39*, 174–184.

95. Guéguen, N., & Vion, M. (2009). The effect of a practitioner's touch on a patient's medication compliance. *Psychology, Health and Medicine, 14*, 689–694.

96. Segrin, C. (1993). The effects of nonverbal behavior on outcomes of compliance gaining attempts. *Communication Studies, 11*, 169–187.

97. Hornik, J. (1992). Effects of physical contact on customers' shopping time and behavior. *Marketing Letters, 3*, 49–55.

98. Smith, D. E., Gier, J. A., & Willis, F. N. (1982). Interpersonal touch and compliance with a marketing request. *Basic and Applied Social Psychology, 3*, 35–38. See also Soars, B. (2009). Driving sales through shoppers' sense of sound, sight, smell and touch. *International Journal of Retail & Distribution Management, 37*(3), 286–298.

99. See Hall, E. (1969). *The hidden dimension.* Garden City, NY: Anchor Books, pp. 113–130.

100. LeFebvre, L., & Allen, M. (2014). Teacher immediacy and student learning: An examination of lecture/laboratory and self-contained course sections. *Journal of the Scholarship of Teaching and Learning, 14*(2), 29–45.

101. Wouda, J. C., & van de Wiel, H. B. (2013). Education in patient–physician communication: How to improve effectiveness? *Patient Education and Counseling, 90*(1), 46–53.

102. Mumm, J., & Mutlu, B. (2011, March). Human-robot proxemics: Physical and psychological distancing in human-robot interaction. Proceedings of the 6th International Conference on Human-Robot Interaction in Lausanne, Switzerland. Retrieved from http://www.cs.cmu.edu/~illah/CLASSDOCS/p331-mumm.pdf.

103. Barrett, P., Davies, F., Zhang, Y., & Barrett, L. (2017). The holistic impact of classroom spaces on learning in specific subjects. *Environment and Behavior, 49*(4), 425–451.

104. Horr, Y. A., Arif, M., Kaushik, A., Mazroei, A., Katafygiotou, M., & Elsarrag, E. (2016). Occupant productivity and office indoor environment quality: A review of the literature. *Building and Environment, 105*, 369–389.

105. Samani, S. A., Rasid, S. Z. A., Sofian, S. (2015, October). Perceived level of personal control over the work environment and employee satisfaction and work performance. *Performance Improvement, 54*(9), 28–35.

106. Jones, J. (2015, February 24). Innovative detail: The Superdesk at the Barbarian Group office. *Architect.* Retrieved from https://www.architectmagazine.com/technology/detail/innovative-detail-the-superdesk-at-the-barbarian-group-office_o.

107. Bruneau, T. J. (2012). Chronemics: Time-binding and the construction of personal time. *ETC: A Review of General Semantics, 69*(1), 72.

108. Ballard, D. I., & Seibold, D. R. (2000). Time orientation and temporal variation across work groups: Implications for group and organizational communication. *Western Journal of Communication, 64*, 218–242.

109. Levine, R. (1997). *A geography of time: The temporal misadventures of a social psychologist.* New York: Basic Books.

110. See, for example, Hill, O. W., Block, R. A., & Buggie, S. E. (2000). Culture and beliefs about time: Comparisons among black Americans, black Africans, and white Americans. *Journal of Psychology, 134*, 443–457.

111. Levine, R., & Wolff, E. (1985, March). Social time: The heartbeat of culture. *Psychology Today, 19*, 28–35. See also Levine, R. (1987, April). Waiting is a power game. *Psychology Today, 21*, 24–33.

112. Burgoon, J. K., Buller, D. B., & Woodall, W. G. (1996). *Nonverbal communication.* New York: McGraw-Hill, p. 148. See also White, L. T., Valk, R., & Dialmy, A. (2011). What is the meaning of "on time"? The sociocultural nature of punctuality. *Journal of Cross-Cultural Psychology, 42*(3), 482–493.

113. Matsumoto, D. (2006). Culture and nonverbal behavior. In V. Manusov & M. L. Patterson (Eds.), *Sage handbook of nonverbal communication* (pp. 219–235). Thousand Oaks, CA: Sage.

114. Birdwhistell, R. (1970). *Kinesics and context.* Philadelphia: University of Pennsylvania Press, Chapter 9.

115. Hall, E. (1969). *The hidden dimension.* Garden City, NY: Anchor Books.

116. Kelly, D. J., Liu, S., Rodger, H., Miellet, S., Ge, L., & Caldara, R. (2011). Developing cultural differences in face processing. *Developmental Science, 14*(5), 1176–1184.

117. Yuki, M., Maddux, W. W., & Masuda, T. (2007). Are the windows to the soul the same in the East and West? Cultural differences in using the eyes and mouth as cues to recognize emotions in Japan and the United States. *Journal of Experimental Social Psychology, 43*, 303–311.

118. Linneman, T. J. (2013). Gender in *Jeopardy!* Intonation variation on a television game show. *Gender & Society, 27*, 82–105; Wolk, L., Abdelli-Beruh, N. B., & Slavin, D. (2012). Habitual use of vocal fry in young adult female speakers. *Journal of Voice, 26*, 111–116.

119. Anderson, R. C., Klofstad, C. A., Mayew, W. J., & Venkatachalam, M. (2014). Vocal fry may undermine the success of young women in the labor market. *PLoS ONE, 9*, e97506.

120. Yuasa, I. P. (2010). Creaky voice: A new feminine voice quality for young urban-oriented upwardly mobile American women? *American Speech, 85*, 315–337.

121. Booth-Butterfield, M., & Jordan, F. (1988). *"Act like us": Communication adaptation among racially homogeneous and heterogeneous groups.* Paper presented at the Speech Communication Association meeting, New Orleans.

122. Warnecke, A. M., Masters, R. D., & Kempter, G. (1992). The roots of nationalism: Nonverbal behavior and xenophobia. *Ethnology and Sociobiology, 13*, 267–282.

123. Hall, J. A. (2006). Women and men's nonverbal communication. In V. Manusov & M. L. Patterson (Eds.), *The Sage handbook of nonverbal communication* (pp. 201–218). Thousand Oaks, CA: Sage.

124. Kirsch, A. C., & Murnen, S. K. (2015). "Hot" girls and "cool dudes": Examining the prevalence of the heterosexual script in American children's television media. *Psychology of Popular Media Culture, 4*(1), 18–30.

125. Kirsch & Murnen, "Hot" girls and "cool dudes." Quote appears in the title.

126. Stermer, S. P., & Burkley, M. (2015). SeX-Box: Exposure to sexist video games predicts benevolent sexism. *Psychology of Popular Media Culture, 4*(1), 47–55.

127. Matthes, J., Prieler, M, & Adam, K. (2016). Gender-role portrayals in television advertising across the globe. *Sex Roles, 75*(7/8), 314–327.

128. Townsend, M. (2017, June 1). GLAAD Studio Responsibility Index: Memorial Day update: Paramount's "Baywatch" opens summer blockbuster season with stale gay jokes. GLAAD (formerly the Gay & Lesbian Alliance Against Defamation). Retrieved from https://www.glaad.org/blog/glaad-finds-historic-low-percentage-hollywood-films-pass-vito-russo-test-fourth-annual-studio.

129. Bloomer, J., & Canfield, D. (2017, March 22). An exclusively gay breakdown of *Beauty and the Beast's* would-be queer moment. *Slate.* Retrieved from http://www.slate.com/blogs/browbeat/2017/03/22/beauty_and_the_beast_s_exclusively_gay_moment_dissected.html.

130. Stermer, S. P., & Burkley, M. (2015). SeX-Box: Exposure to sexist video games predicts benevolent sexism. *Psychology of Popular Media Culture, 4*(1), 47–55.

131. Döring, N., Reif, A., & Poeschl, S. (2016). How gender-stereotypical are selfies? A content analysis and comparison with magazine adverts. *Computers in Human Behavior, 55*(Part B), 955–962.

132. Hall, J. A., Carter, J. D., & Horgan, T. G. (2001). Status roles and recall of nonverbal cues. *Journal of Nonverbal Behavior, 25,* 79–100.

133. Mayo, C., & Henley, N. M. (Eds.). (2012). *Gender and nonverbal behavior.* Springer Science & Business Media.

134. Knöfler, T., & Imhof, M. (2007). Does sexual orientation have an impact on nonverbal behavior in interpersonal communication? *Journal of Nonverbal Behavior, 31,* 189–204.

CHAPTER 8

1. Eidell, L. (2018, November 2). *Bachelor* and *Bachelorette* couples: The complete list. *Glamour.* Retrieved from https://www.glamour.com/story/bachelor-bachelorette-couples-history.

2. Boardman, M. (2015, February 7). Sean Lowe, Catherine Giudici explain why so few *Bachelor* couples get married. *US Weekly.* Retrieved from http://www.usmagazine.com/entertainment/news/sean-lowe-catherine-giudici-explain-why-so-few-bachelor-couples-wed-201572. Quote appears in para. 6.

3. Luo, S., & Klohnen, E. (2005). Assortive mating and marital quality in newlyweds: A couple-centered approach. *Journal of Personality and Social Psychology, 88,* 304–326. See also Amodio, D. M., & Showers, C. J. (2005). Similarity breeds liking revisited: The moderating role of commitment. *Journal of Social and Personal Relationships, 22,* 817–836.

4. Mackinnon, S. P., Jordan, C., & Wilson, A. (2011). Birds of a feather sit together: Physical similarity predicts seating distance. *Personality and Social Psychology Bulletin, 37,* 879–892.

5. Nelson, P., Thorne, A., & Shapiro, L. (2001). I'm outgoing and she's reserved: The reciprocal dynamics of personality in close friendships in young adulthood. *Journal of Personality, 79,* 1113–1147.

6. Toma, C., Yzerbyt, V., & Corneille, O. (2012). Reports: Nice or smart? Task relevance of self-characteristics moderates inter-

personal projection. *Journal of Experimental Social Psychology, 48,* 335–340.

7. Dindia, K. (2002). Self-disclosure research: Knowledge through meta-analysis. In M. Allen & R. W. Preiss (Eds.), *Interpersonal communication research: Advances through meta-analysis* (pp. 169–185). Mahwah, NJ: Erlbaum.

8. Flora, C. (2004, January/February). Close quarters. *Psychology Today, 37,* 15–16.

9. Haythornthwaite, C., Kazmer, M. M., & Robbins, J. (2000). Community development among distance learners: Temporal and technological dimensions. *Journal of Computer-Mediated Communication, 6*(1), Article 2. Retrieved from http://jcmc.indiana.edu/vol6/issue1/haythornthwaite.html.

10. See, for example, Roloff, M. E. (1981). *Interpersonal communication: The social exchange approach.* Beverly Hills, CA: Sage.

11. Duck, S. W. (2011). Similarity and perceived similarity of personal constructs as influences on friendship choice. *British Journal of Clinical Psychology, 12,* 1–6.

12. Mackinnon, S. P., Jordan, C., & Wilson, A. (2011). Birds of a feather sit together: Physical similarity predicts seating distance. *Personality and Social Psychology Bulletin, 37,* 879–892.

13. Sias, P. M., Drzewiecka, J. A., Meares, M., Bent, R., Konomi, Y., Ortega, M., & White, C. (2008). Intercultural friendship development. *Communication Reports, 21,* 1–13.

14. Alley, T. R., & McCanless, E. R. (2002). *Body shape and muscularity preferences in short-term and long-term relationships.* Paper presented at the Biennial International Conference on Human Ethology, Vienna.

15. Hamachek, D. (1982). *Encounters with others: Interpersonal relationships and you.* New York: Holt, Rinehart and Winston.

16. Singj, R., & Tor, X. L. (2008). The relative effects of competence and likability on interpersonal attraction. *Journal of Social Psychology, 148*(2), 253–255. Quote appears on p. 253.

17. Finkel, E. J., Eastwick, P. W., Karney, B. R., Reis, H. T., & Sprecher, S. (2012). Online dating: A critical analysis from the perspective of psychological science. *Psychological Science in the Public Interest, 13*(1), 3–66.

18. Lim, T. S., & Bowers, J. W. (1991). Facework: Solidarity, approbation, and tact. *Human Communication Research,17,* 415–450.

19. Frei, J. R., & Shaver, P. R. (2002). Respect in close relationships: Prototype, definition, self-report assessment, and initial correlates. *Personal Relationships, 9,* 121–139.

20. Marano, H. E., (2014, January 1). Love and power. *Psychology Today.* Retrieved from https://www.psychologytoday.com/articles/201401/love-and-power.

21. See Rossiter, C. M., Jr. (1974). Instruction in metacommunication. *Central States Speech Journal, 25,* 36–42; and Wilmot, W. W. (1980). Metacommunication: A reexamination and extension. In D. Nimmo (Ed.), *Communication yearbook 4.* New Brunswick, NJ: Transaction.

22. Tamir, D. I., & Mitchell, J. P. (2012). Disclosing information about the self is intrinsically rewarding. *Proceedings of the National Academy of Science, 109*(21), 8038–8043.

23. Altman, I., & Taylor, D. A. (1973). *Social penetration: The development of interpersonal relationships.* New York: Holt, Rinehart and Winston.

24. Whitbourne, S. K. (2014, April 1). The secret to revealing your secrets. *Psychology Today.* Retrieved from https://www.psychologytoday.com/blog/fulfillment-any-age/201404/the-secret-revealing-your-secrets Quote appears in para. 2.

25. Luft, J. (1969). *Of human interaction.* Palo Alto, CA: National Press.

26. The friends I've never met. (2014, December 15). *Femsplain.* Retrieved from https://femsplain.com/the-friend-i-ve-never-met-

7ae521269047. Quotes appear in paras. 4 and 5. (Maya and Jad are pseudonyms. The post does not include names.)

27. Patton, B. R., & Giffin, K. (1974). *Interpersonal communication: Basic text and readings.* New York: Harper & Row.

28. Ledbetter, A. M. (2014). The past and future of technology in interpersonal communication theory and research. *Communication Studies, 65*(4), 456–459.

29. Lee, S. J. (2009). Online communication and adolescent social ties: Who benefits more from Internet use? *Journal of Computer-Mediated Communication, 14*, 509–531.

30. Jin, B., & Peña, J. F. (2010). Mobile communication in romantic relationships: Mobile phone use, relational uncertainty, love, commitment, and attachment styles. *Communication Reports, 23*, 39–51.

31. These are my "real" friends: Removing the stigma of online friendships. (2013, January 29). *Persephone.* Retrieved from http://persephonemagazine.com/2013/01/these-are-my-real-friends-removing-the-stigma-of-online-friendships. Quote appears in para. 4.

32. Durrotul, M. (2017). The use of social media in intercultural friendship development. *Profetik, 10*(1), 5–20.

33. Summarized in Pearson, J. (1989). *Communication in the family.* New York: Harper & Row, pp. 252–257.

34. Wright, K. B. (2012). Emotional support and perceived stress among college students using Facebook.com: An exploration of the relationship between source perceptions and emotional support. *Communication Research Reports, 29*, 175–184.

35. Hales, K. D. (2012). Multimedia use for relational maintenance in romantic couples. Presented at the annual meeting of the International Communication Association, Phoenix, AZ.

36. The phubbing truth. (2013, October 8). *Wordability.* Retrieved from http://wordability.net/2013/10/08/the-phubbing-truth.

37. Przybylski, A. K., & Weinstein, N. (2013). Can you connect with me now? How the presence of mobile communication technology influences face-to-face conversation quality. *Journal of Social and Personal Relationships, 30*, 237–246.

38. Yao, M. Z., & Zhong, Z. (2014). Loneliness, social contacts and Internet addiction: A cross-lagged panel study. *Computers in Human Behavior, 30*, 164–170.

39. Baiocco, R., Laghi, F., Schneider, B. H., Dalessio, M., Amichai-Hamburger, Y., Coplan, R. J. Koszycki, D. & Flament, M. (2011). Daily patterns of communication and contact between Italian early adolescents and their friends. *Cyberpsychology, Behavior, and Social Networking, 14*, 467–471.

40. Kerner, I. (2017, May 16). What counts as "cheating" in the digital age? CNN. Retrieved from https://www.cnn.com/2017/05/16/health/cheating-internet-sex-kerner/index.html.

41. Donnelly, M. (2017, May 3). Here's to the best friends who feel like family. *Thought Catalog.* Retrieved from https://thoughtcatalog.com/marisa-donnelly/2017/05/heres-to-the-best-friends-who-feel-like-family. Quote appears in para. 8.

42. Donnelly. Here's to the best friends who feel like family. Quote appears in para. 8.

43. Morrison, A. (2015, May 1). Straight teen asks gay best friend to prom. CNN. Retrieved from http://edition.cnn.com/2015/04/24/living/feat-straight-teen-asks-gay-best-friend-to-prom. Quote appears in para. 7.

44. A new kind of prom date. (2015, April 30). *The Ellen Show.* Retrieved from https://www.youtube.com/watch?v=znt3BPKUMCY.

45. Deci, E., La Guardia, J., Moller, A., Scheiner, M., & Ryan, R. (2006). On the benefits of giving as well as receiving autonomy support: Mutuality in close friendships. *Personality and Social Psychology Bulletin, 32*, 313–327.

46. Demir, M., & Özdemir, M. (2010). Friendship, need satisfaction and happiness. *Journal of Happiness Studies, 11*, 243–259.

47. Buote, V. M., Pancer, S., Pratt, M. W., Adams, G., Birnie-Lefcovitch, S., Polivy, J., & Wintre, M. (2007). The importance of friends: Friendship and adjustment among 1st-year university students. *Journal of Adolescent Research, 22*, 665–689.

48. Demir, M., Özdemir, M., & Marum, K. (2011). Perceived autonomy support, friendship maintenance, and happiness. *Journal of Psychology, 145*, 537–571.

49. Hall, J. A. (2019). How many hours does it take to make a friend? *Journal of Social and Personal Relationships, 36*, 1278–1296.

50. Becker, J. H., Johnson, A., Craig, E. A., Gilchrist, E. S., Haigh, M. M., & Lane, L. T. (2009). Friendships are flexible, not fragile: Turning points in geographically close and long-distance friendships. *Journal of Social and Personal Relationships, 26*, 347–369. Quote appears on p. 347.

51. Johnson, A., Haigh, M. M., Craig, E. A., & Becker, J. H. (2009). Relational closeness: Comparing undergraduate college students' geographically close and long-distance friendships. *Personal Relationships, 16*, 631–646.

52. Manago, A., Taylor, T., & Greenfield, P. (2012). Me and my 400 friends: The anatomy of college students' Facebook networks, their communication patterns, and well-being: Interactive media and human development. *Developmental Psychology, 48*, 369–380. Quote appears on p. 375.

53. Charleston, A. (2017, January 16). To those who overshare on social media: Don't. *Odyssey.* Retrieved from https://www.theodysseyonline.com/open-letter-those-overshare-social-media.

54. Migliaccio, T. (2009). Men's friendships: Performances of masculinity. *Journal of Men's Studies, 17*, 226–241.

55. Hall, J. A. (2011). Sex differences in friendship expectations: A meta-analysis. *Journal of Social and Personal Relationships, 28*, 723–747.

56. Bello, R. S., Brandau-Brown, F. E., Zhang, S., & Ragsdale, J. (2010). Verbal and nonverbal methods for expressing appreciation in friendships and romantic relationships: A cross-cultural comparison. *International Journal of Intercultural Relations, 34*, 294–302.

57. Quinn, B. (2011, May 8). Social network users have twice as many friends online as in real life. *The Guardian.* Retrieved from http://www.theguardian.com/media/2011/may/09/social-network-users-friends-online.

58. Whitty, M., & Joinson, A. (2009). *Truth, lies and trust on the Internet.* New York: Routledge.

59. Chan, D. K.-S., & Cheng, G. H.-L. (2004). A comparison of offline and online friendship qualities at different stages of relationship development. *Journal of Social and Personal Relationships, 21*, 305–320.

60. Walther, J. B. (1996). Computer-mediated communication: Impersonal, interpersonal, and hyperpersonal interaction. *Communication Research, 23*, 3–43; Okdie, B. M., Guadagno, R. E., Bernieri, F. J., Geers, A. L., & Mclarney-Vesotski, A. R. (2011). Getting to know you: Face-to-face versus online interactions. *Computers in Human Behavior, 27*, 153–159.

61. Chan, D. K.-S., & Cheng, G. H.-L. (2004). A comparison of offline and online friendship qualities at different stages of relationship development. *Journal of Social and Personal Relationships, 21*, 305–320.

62. Smith, S. (2016, February 9). An open letter to my guy best friend. *Odyssey.* Retrieved from https://www.theodysseyonline.com/letter-to-my-guy-bff. Quote appears in para. 4.

63. Hall, J. A. (2011). Sex differences in friendship expectations: A meta-analysis. *Journal of Social and Personal Relationships, 28*, 723–747.

64. Bleske-Rechek, A., Somers, E., Micke, C., Erickson, L., Matteson, L., Stocco, C., Schumacher, B., & Ritchie, L. (2012). Benefit or burden? Attraction in cross-sex friendship. *Journal of Social and Personal Relationships, 29,* 569–596.

65. Hall, Sex differences in friendship expectations.

66. Hall, Sex differences in friendship expectations.

67. Hall, Sex differences in friendship expectations.

68. Hall, Sex differences in friendship expectations.

69. Russell, E. M., DelPriore, D. J., Butterfield, M. E., & Hill, S. E. (2013). Friends with benefits, but without the sex: Straight women and gay men exchange trustworthy mating advice. *Evolutionary Psychology, 11*(1), 132–147.

70. Guerrero, L. K., Farinelli, L., & McEwan, B. (2009). Attachment and relational satisfaction: The mediating effect of emotional communication. *Communication Monographs, 76,* 487–514.

71. Bello, R. S., Brandau-Brown, F. E., Zhang, S., & Ragsdale, J. D. (2010). Verbal and nonverbal methods for expressing appreciation in friendships and romantic relationships: A cross-cultural comparison. *International Journal of Intercultural Relations, 34,* 294–302.

72. Tabak, B., McCullough, M., Luna, L., Bono, G., & Berry, J. (2012). Conciliatory gestures facilitate forgiveness and feelings of friendship by making transgressors appear more agreeable. *Journal of Personality, 80,* 503–536.

73. Davis, J. R., & Gold, G. J. (2011). An examination of emotional empathy, attributions of stability, and the link between perceived remorse and forgiveness. *Personality and Individual Differences, 50,* 392–397.

74. Rawlins, W. K., & Holl, M. (1987). The communicative achievement of friendship during adolescence: Predicaments of trust and violation. *Western Journal of Speech Communication, 51,* 345–363.

75. Deci, E., La Guardia, J., Moller, A., Scheiner, M., & Ryan, R. (2006). On the benefits of giving as well as receiving autonomy support: Mutuality in close friendships. *Personality and Social Psychology Bulletin, 32,* 313–327.

76. van der Horst, M., & Coffe, H. (2012). How friendship network characteristics influence subjective well-being. *Social Indicators Research, 107,* 509–529.

77. Allport, G. W. (1954). *The nature of prejudice.* Cambridge, MA: Perseus Books.

78. Baiocco, R., Laghi, F., Di Pomponio, I., & Nigito, C. S. (2012). Self-disclosure to the best friend: Friendship quality and internalized sexual stigma in Italian lesbian and gay adolescents. *Journal of Adolescence, 35,* 381–387.

79. Barbir, L. A., Vandevender, A. W., & Cohn, T. J. (2017). Friendship, attitudes, and behavioral intentions of cisgender heterosexuals toward transgender individuals. *Journal of Gay and Lesbian Mental Health, 21*(2), 154–170.

80. How can I support my friend who is transgender? (2014, November 11). *7 cups.* Retrieved from https://www.7cups .com/qa-lgbtq--17/how-can-i-support-my-friend-who-is-transgender-410. Quote appears in the post by MonBon.

81. Minow, M. (1998). Redefining families: Who's in and who's out? In K. V. Hansen & A. I. Garey (Eds.), *Families in the U.S.: Kinship and domestic policy* (pp. 7–19). Philadelphia: Temple University Press. (Originally published in the *University of Colorado Law Review,* 1991, 62, 269–285.)

82. Galvin, K. M. (2006). Diversity's impact of defining the family: Discourse-dependence and identity. In R. L. West & L. H. Turner (Eds.), *The family communication sourcebook* (pp. 3–20). Thousand Oaks, CA: Sage.

83. For background on this theory, see Baumrind, D. (1991). The influence of parenting styles on adolescent competence and substance use. *The Journal of Early Adolescence, 11,* 56–95.

84. Koerner, A. F., & Fitzpatrick, M. A. (2006). Family communication patterns theory: A social cognitive approach. In D. O. Braithwaite & L. A. Baxter (Eds.), *Engaging theories in family communication: Multiple perspectives* (pp. 50–65). Thousand Oaks, CA: Sage.

85. Koerner, A. F., & Fitzpatrick, M. A. (2002). Understanding family communication patterns and family functioning: The roles of conversation orientation and conformity orientation. *Communication Yearbook, 26,* 37–68.

86. Young, S. L. (2009). The function of parental communication patterns: Reflection-enhancing and reflection-discouraging approaches. *Communication Quarterly, 57,* 379–394.

87. Hamon, J. D., & Schrodt, P. (2012). Do parenting styles moderate the association between family conformity orientation and young adults' mental well-being? *Journal of Family Communication, 12,* 151–166.

88. Hamon & Schrodt, Do parenting styles moderate the association . . .? Quote appears on p. 162.

89. Edwards, R., Hadfield, L., Lucey, H., & Mauthner, M. (2006). *Sibling identity and relationships: Brothers and sisters.* New York: Routledge. Quote appears on p. 4.

90. Epstein, L. (2014, August 4). 16 things that only half-siblings understand. BuzzFeed. Retrieved from https://www.buzzfeed. com/leonoraepstein/things-only-half-siblings-understand?utm_ term=.gp1jQpMB1#.lmZROVGWY. Quote appears in item 7.

91. Stewart, R. B., Kozak, A. L., Tingley, L. M., Goddard, J. M., Blake, E. M., & Cassel, W. A. (2001). Adult sibling relationships: Validation of a typology. *Personal Relationships, 8,* 299–324.

92. Riggio, H. (2006). Structural features of sibling dyads and attitudes toward sibling relationships in young adulthood. *Journal of Family Issues, 27,* 1233–1254.

93. So . . . my brother just moved out. (2009, January 16). Grasscity. Retrieved from https://forum.grasscity.com/threads/so-my-brother-just-moved-out.322445. Quote appears in para. 7.

94. Scharf, M., Shulman, S., & Avigad-Spitz, L. (2005). Sibling relationships in emerging adulthood and in adolescence. *Journal of Adolescent Research, 20,* 64–90.

95. Riggio, Structural features of sibling dyads.

96. Duke, M. P. (2013, March 23). The stories that bind us: What are the twenty questions? *HuffPost: The Blog.* Retrieved from http://www.huffingtonpost.com/marshall-p-duke/the-stories-that-bind-us-_b_2918975.html.

97. Guerrero, L. K., Farinelli, L., & McEwan, B. (2009). Attachment and relational satisfaction: The mediating effect of emotional communication. *Communication Monographs, 76,* 487–514.

98. Young, S. L. (2009). The function of parental communication patterns: Reflection-enhancing and reflection-discouraging approaches. *Communication Quarterly, 57,* 379–394.

99. Information in this paragraph is from Petronio, S. (2010). Communication privacy management theory: What do we know about family privacy regulation? *Journal of Family Theory and Review, 2,* 175–196.

100. Baraldi, C., & Iervese, V. (2010). Dialogic mediation in conflict resolution education. *Conflict Resolution Quarterly, 27,* 423–445.

101. Strom, R. E., & Boster, F. J. (2011). Dropping out of high school: Assessing the relationship between supportive messages from family and educational attainment. *Communication Reports, 24,* 25–37.

102. Feiler, B. (2013). *The secrets of happy families: Improve your mornings, rethink family dinner, fight smarter, go out and play, and much more.* New York: HarperCollins.

103. Myers, S. A., & Goodboy, A. K. (2010). Relational maintenance behaviors and communication channel use among adult siblings. *North American Journal of Psychology, 12,* 103–116.

104. Ahmetoglu, G., Swami, V., & Chamorro-Premuzic, T. (2010). The relationship between dimensions of love, personality, and relationship length. *Archives of Sexual Behavior, 34,* 1181–1190.

105. Malouff, J. M., Schutte, N. S., & Thorsteinsson, E. B. (2013). Trait emotional intelligence and romantic relationship satisfaction: A meta-analysis. *American Journal of Family Therapy, 42,* 53–66.

106. Knapp, M. L., & Vangelisti, A. L. (2009). *Interpersonal communication and human relationships* (6th ed.). Boston: Allyn and Bacon.

107. Canary, D. J., & Stafford, L. (Eds.). (1994). *Communication and relational maintenance.* San Diego: Academic Press. See also Lee, J. (1998). Effective maintenance communication in superior-subordinate relationships. *Western Journal of Communication, 62,* 181–208.

108. Wilson, S. R., Kunkel, A. D., Robson, S. J., Olufowote, J. O., & Soliz, J. (2009). Identity implications of relationship (re)definition goals: An analysis of face threats and facework as young adults initiate, intensify, and disengage from romantic relationships. *Journal of Language and Social Psychology, 28,* 32–61.

109. Dunleavy, K., & Booth-Butterfield, M. (2009). Idiomatic communication in the stages of coming together and falling apart. *Communication Quarterly, 57,* 416–432.

110. Caughlin, J. P., & Sharabi, L. L. (2013). A communicative interdependence perspective of close relationships: The connections between mediated and unmediated interactions matter. *Journal of Communication, 63*(5), 873–893.

111. Flaa, J. (2013, October 29). I met my spouse online: 9 online dating lessons learned the hard way. *HuffPost: The Blog.* Retrieved from http://www.huffingtonpost.com/jennifer-flaa/9-online-dating-lessons_b_4174334.html. Quote appears in paragraph 13.

112. Aslay, J. (2012, November 3). You lost me at hello, how to get past the awkward first meeting. *Understand Men Now.* Retrieved from http://www.jonathonaslay.com/2012/11/03/you-lost-me-at-hello-how-to-get-past-the-awkward-first-meeting/. Quote appears in para. 11.

113. Flaa, J. (2013, October 29). I met my spouse online: 9 online dating lessons learned the hard way. *HuffPost: The Blog.* Retrieved from http://www.huffingtonpost.com/jennifer-flaa/9-online-dating-lessons_b_4174334.html. Quote appears in paragraph 13.

114. Meyers, S. (2014, December 23.) 5 ways to put your date at ease (and alleviate awkward tension). *Fox News Magazine.* Retrieved from http://magazine.foxnews.com/love/5-ways-put-your-date-at-ease-and-alleviate-awkward-tension. Quote appears in the last paragraph.

115. Knapp, M. L. (1984). *Interpersonal communication and human relationships.* Boston: Allyn & Bacon.

116. Brown, B. (2010). *The gifts of imperfection.* Center City, MN: Hazelden.

117. Chapman, G. (2010). *The five love languages: The secret to love that lasts.* Chicago: Northfield Publishing.

118. Frisby, B. N., & Booth-Butterfield, M. (2012). The "how" and "why" of flirtatious communication between marital partners. *Communication Quarterly, 60,* 465–480.

119. Merolla, A. J. (2010). Relational maintenance during military deployment: Perspectives of wives of deployed US soldiers. *Journal of Applied Communication Research, 38*(1), 4–26.

120. Haas, S. M., & Stafford, L. (2005). Maintenance behaviors in same-sex and marital relationships: A matched sample comparison. *Journal of Family Communication, 5,* 43–60.

121. Haas & Stafford, Maintenance behaviors in same-sex and marital relationships.

122. Soin, R. (2011). Romantic gift giving as chore or pleasure: The effects of attachment orientations on gift giving perceptions. *Journal of Business Research, 64,* 113–118.

123. Floyd, K., Boren, J. P., & Hannawa, A. F. (2009). Kissing in marital and cohabitating relationships: Effects of blood lipids, stress, and relationship satisfaction. *Western Journal of Communication, 73,* 113–133.

124. Egbert, N., & Polk, D. (2006). Speaking the language of relational maintenance: A validity test of Chapman's (1992) *Five Love Languages. Communication Research Reports, 23*(1), 19–26.

125. Bland, A. M., & McQueen, K. S. (2018). The distribution of Chapman's love languages in couples: An exploratory cluster analysis. *Couple and Family Psychology: Research and Practice, 7*(2), 103–126.

126. Chapman, G. (2010). *The five love languages: The secret to love that lasts.* Chicago: Northfield Publishing.

127. Moore, L. (2016, February 23). 15 things men don't understand about women. *Cosmopolitan.* Retrieved from https://www.cosmopolitan.com/sex-love/news/a54137/things-men-just-dont-understand-about-women. Quote is paraphrased from item 2.

128. 50 things men wish women knew. (2013, March 6). *Men's Health.* Retrieved from https://www.menshealth.com/sex-women/men-wish-women-knew. Quote appears in item 17.

129. Bond, B. J. (2009). He posted, she posted: Gender differences in self-disclosure on social network sites. *Rocky Mountain Communication Review, 6*(2), 29–37.

130. MacGeorge, E. L., Graves, A. R., Feng, B., Gillihan, S. J., & Burleson, B. R. (2004). The myth of gender cultures: Similarities outweigh differences in men's and women's provision of and responses to supportive communication. *Sex Roles, 50,* 143–175.

131. Hall, E., Travis, M., Anderson, S., & Henley, A. (2013). Complaining and Knapp's relationship stages: Gender differences in instrumental complaints. *Florida Communication Journal, 41,* 49–61.

132. Elliott, S., & Umberson, O. (2008). The performance of desire: Gender and sexual negotiation in long-term marriages. *Journal of Marriage and Family, 70,* 391–406.

133. Umberson, D., Thomeer, M. B., & Lodge, A. C. (2015). Intimacy and emotion work in lesbian, gay, and heterosexual relationships. *Journal of Marriage and Family, 77*(2), 542–556.

134. Mackey, R. A., Diemer, M. A., & O'Brien, B. A. (2000). Psychological intimacy in the lasting relationships of heterosexual and same-gender couples. *Sex Roles, 43*(3/4). 201–227.

135. See, for example, Baxter, L. A., & Montgomery, B. M. (1998). A guide to dialectical approaches to studying personal relationships. In B. M. Montgomery & L. A. Baxter (Eds.), *Dialectical approaches to studying personal relationships* (pp. 1–16). Mahwah, NJ: Erlbaum; and Ebert, L. A., & Duck, S. W. (1997). Rethinking satisfaction in personal relationships from a dialectical perspective. In R. J. Sternberg & M. Hojjatr (Eds.), *Satisfaction in close relationships* (pp. 190–216). New York: Guilford Press.

136. Summarized by Baxter, L. A. (1994). A dialogic approach to relationship maintenance. In D. J. Canary & L. Stafford (Eds.), *Communication and relational maintenance* (pp. 233–254). San Diego: Academic Press.

137. Baxter, A dialogic approach to relationship maintenance.

138. Morris, D. (1971). *Intimate behavior.* New York: Kodansha Globe, pp. 21–29.

139. Adapted from Baxter & Montgomery, *A guide to dialectical approaches,* pp. 1–16.

140. Serota, K. B., Levine, T. R., & Boster, F. J. (2010). The prevalence of lying in America: Three studies of self-reported lies. *Human Communication Research, 36*(1), 2–25.

141. Harrell, E. (2009, August 19). Why we lie so much. *Time.* Retrieved from http://www.time.com/time/health/article/0,8599,1917215,00.html.

142. McCornack, S. A., & Levine, T. R. (1990). When lies are uncovered: Emotional and relational outcomes of discovered deception. *Communication Monographs, 57,* 119–138.

143. Guthrie, J., & Kunkel, A. (2013). Tell me sweet (and not-so-sweet) little lies: Deception in romantic relationships. *Communication Studies, 64*(2), 141–157.

144. Kaplar, M. E., & Gordon, A. K. (2004). The enigma of altruistic lying: Perspective differences in what motivates and justifies lie telling within romantic relationships. *Personal Relationships, 11,* 489–507.

145. Bryant, E. (2008). Real lies, white lies and gray lies: Towards a typology of deception. *Kaleidoscope: A Graduate Journal of Qualitative Communication Research, 7,* 723–748.

146. Gunderson, P. R., & Ferrari, J. R. (2008). Forgiveness of sexual cheating in romantic relationships: Effects of discovery method, frequency of offense, and presence of apology. *North American Journal of Psychology, 10,* 1–14.

147. Lee, T. (2016, March 17). Jillian Harri's advice for new "Bachelorette" JoJo Fletcher: "It sounds cliché, but be yourself." *Hello!* Retrieved from http://us.hellomagazine.com/health-and-beauty/12016031712860/jillian-harris-bachelorette-advice-jojo-fletcher-motherhood/. Quote appears in headline.

CHAPTER 9

1. Brown, B. (2015). *Rising strong.* New York: Spiegel & Grau. Quotes appear on p. 31.

2. Brown, *Rising strong.* Quotes appear on p. 31.

3. Brown, *Rising strong.* Quote appears on p. 32.

4. Bernard Meltzer Quotes. (n.d.). *AZ Quotes.* Retrieved from https://www.azquotes.com/author/9957-Bernard Meltzer. Quote appears seventh.

5. Sillars, A. L. (2009). Interpersonal conflict. In C. Berger, M. Roloff, & D. R. Roskos-Ewoldsen (Eds.), *Handbook of communication science* (2nd ed., pp. 273–289). Thousand Oaks, CA: Sage.

6. Brown, B. (2015). *Rising strong.* New York: Spiegel & Grau. Quote appears on p. 33.

7. For a discussion of reactions to disconfirming responses, see Vangelisti, A. L., & Crumley, L. P. (1998). Reactions to messages that hurt: The influence of relational contexts. *Communication Monographs, 64,* 173–196. See also Cortina, L. M., Magley, V. J., Williams, J. H., & Langhout, R. D. (2001). Incivility in the workplace: Incidence and impact. *Journal of Occupational Health Psychology, 6,* 64–80.

8. Gottman, J. M., & Levenson, R. W. (2002). A two-factor model for predicting when a couple will divorce: Exploratory analyses using 14-year longitudinal data. *Family Process, 41*(1), 83–96; Gottman, J. M., Coan, J., Carrere, S., & Swanson, C. (1998). Predicting marital happiness and stability from newlywed interactions. *Journal of Marriage and the Family, 60*(1), 5–22. Retrieved from http://www.jstor.org/pss/353438; Carrere, S., Buehlman, K. T., Gottman, J. M., Coan, J. A., & Ruckstuhl, L. (2000). Predicting marital stability and divorce in newlywed couples. *Journal of Family Psychology, 14*(1), 42–58; Gottman, J. M. (1991). Predicting the longitudinal course of marriages. *Journal of Marital and Family Therapy, 17*(1), 3–7; Gottman, J. M., & Krokoff, L. J. (1989). The relationship between marital interaction and marital satisfaction: A longitudinal view. *Journal of Consulting and Clinical Psychology, 57,* 47–52; Carrere, S., & Gottman, J. M. (1999). Predicting divorce among newlyweds from the first three minutes of a marital conflict discussion. *Family Process, 38*(3), 293–301.

9. Gottman, J. (1994). *Why marriages succeed or fail: And how you can make yours last.* New York: Simon & Schuster.

10. Elium, D. (n.d.). What is the difference between a complaint and a criticism? Retrieved from http://www.donelium.com/complaints.html#.V46d0ZMrKCd.

11. Gottman, J. M. (2009). *The marriage clinic.* New York: Norton.

12. Cissna, K. N. L., & Seiburg, E. (1995). Patterns of interactional confirmation and disconfirmation. In M. V. Redmond (Ed.), *Interpersonal communication: Readings in theory and research* (pp. 301–317). Fort Worth, TX: Harcourt Brace.

13. Brown, B. (2015). *Rising strong.* New York: Spiegel & Grau. Quotes appear on p. 32.

14. Cissna & Seiburg, Patterns of interactional confirmation and disconfirmation.

15. Brown, *Rising strong.* Quote appears on p. 32.

16. Gibb, J. (1961). Defensive communication. *Journal of Communication,11,* 141–148. See also Eadie, W. F. (1982). Defensive communication revisited: A critical examination of Gibb's theory. *Southern Speech Communication Journal, 47,* 163–177.

17. De Vries, R. E., Bakker-Pieper, A., & Oostenveld, W. (2010). Leadership = communication? The relations of leaders' communication styles with leadership styles, knowledge sharing and leadership outcomes. *Journal of Business and Psychology, 25,* 367–380.

18. Singh, R., & Simons, J. J. P. (2010). Attitudes and attraction: Optimism and weight as explanations for the similarity-dissimilarity asymmetry. *Social and Personality Psychology Compass, 12,* 1206–1219.

19. Imai, T., & Vangelisti, A. L. (2011). *The influence of plans to marry in dating couples on relationship quality, confirmation, and desire for evaluation.* Presented at the annual meeting of the International Communication Association, Boston.

20. Burns, M. E., & Pearson, J. C. (2011). An exploration of family communication environment, everyday talk, and family satisfaction. *Communication Studies, 62*(2), 171–185.

21. Ellis, K. (2004). The impact of perceived teacher confirmation on receiver apprehension, motivation, and learning. *Communication Education, 53,* 1–20.

22. Turkle, S. (2007, May 7). Can you hear me now? *Forbes.* Retrieved from http://www.forbes.com/free_forbes/2007/0507/176.html.

23. See Wilmot, W. W. (1987). *Dyadic communication.* New York: Random House, pp. 149–158; and Andersson, L. M., & Pearson, C. M. (1999). Tit for tat? The spiraling effect of incivility in the workplace. *Academy of Management Review, 24,* 452–471. See also Olson, L. N., & Braithwaite, D. O. (2004). "If you hit me again, I'll hit you back": Conflict management strategies of individuals experiencing aggression during conflicts. *Communication Studies, 55,* 271–286.

24. Wilmot, W. W., & Hocker, J. L. (2007). *Interpersonal conflict* (7th ed., pp. 21–22). New York: McGraw-Hill.

25. Wilmot & Hocker, *Interpersonal conflict,* pp. 23–24.

26. Bates, C. E., & Samp, J. A. (2011). Examining the effects of planning and empathic accuracy on communication in relational and nonrelational conflict interactions. *Communication Studies, 62,* 207–223.

27. Brown, B. (2015). *Rising strong.* New York: Spiegel & Grau. Quote appears on p. 33.

28. Rahim, M. A. (1992). *Managing conflict in organizations* (2nd ed.). Westport, CT: Praeger.

29. Pikiewicz, K. (2015, Apri 24). How trying to make everyone happy can make you miserable. *Psychology Today.* Retrieved from https://www.psychologytoday.com/us/blog/meaningful-you/201504/how-trying-make-everyone-happy-can-make-you-miserable. Quote appears in title.

30. Riter, T., & Riter, S. (2005). Why can't women just come out and say what they mean? *FamilyLife.* Retrieved from https://www.familylife.com/articles/topics/marriage/staying-married/

husbands/why-cant-women-just-come-out-and-say-what-they-mean/. Quotes appear in paras. 6 and 8, respectively.

31. Bach, G. R., & Goldberg, H. (1974). *Creative aggression.* Garden City, NY: Doubleday.

32. For information on filing a formal complaint, see http://www.eeoc.gov/laws/types/sexual_harassment.cfm.

33. Information in this paragraph is from Rose, A. J., & Rudolph, K. D. (2006). A review of sex differences in peer relationship processes: Potential trade-offs for the emotional and behavioral development of girls and boys. *Psychological Bulletin, 132,* 98–131.

34. Meyer, J. R. (2004). Effect of verbal aggressiveness on the perceived importance of secondary goals in messages. *Communication Studies, 55,* 168–184; New Mexico Commission on the Status of Women. (2002). *Dealing with sexual harassment.* Retrieved from http://www.womenscommission.state.nm.us/Publications/sexhbrochre.pdf.

35. Brandon, J. (n.d.). 37 more quotes on handling workplace conflict. *Inc.* Retrieved from https://www.inc.com/john-brandon/37-more-quotes-on-handling-workplace-conflict.html. Quote appears in item 30. (It must be acknowledged that, although this quote is commonly attributed to Churchill, he may not have actually said it. See https://www.intellectualtakeout.org/blog/5-famous-things-churchill-didnt-actually-say for a rebuttal to that effect.)

36. The problem. What is battering? (2009). *National Coalition Against Domestic Violence.* Retrieved from https://www.ozarka.edu/blogs/dojgrant/index.cfm/2009/10/6/The-Problem—What-is-Battering.

37. Are you in a violent relationship? (2013). *The Center for Prevention of Abuse.* Retrieved from http://www.centerforpreventionofabuse.org/violent-relationship.php.

38. Zacchilli, T. L., Hendrick, C., & Hendrick, S. S. (2009). The romantic partner conflict scale: A new scale to measure relationship conflict. *Journal of Social and Personal Relationships, 26,* 1073–1096.

39. Shell, G. R. (2006). *Bargaining for advantage: Negotiation strategies for reasonable people.* New York: Penguin. Quote appears on p. 6.

40. Dickinson, A. (2017, May 18). How to negotiate with a bully. *Forbes.* Retrieved from https://www.forbes.com/sites/alexandradickinson/2017/05/18/how-to-negotiate-with-a-bully/#271b5e2e42bc. Quote appears in para. 1.

41. Dickinson, How to negotiate with a bully. Quote appears in para. 9.

42. Leonhardt, T. (2017, May 2). Five tips for winning business negotiations with bullies. Fast Company. Retrieved from https://www.fastcompany.com/40416298/five-tips-for-winning-business-negotiations-with-bullies. Quote appears in para. 7.

43. Fisher, R., & Ury, W. (1981). *Getting to yes: Negotiating agreement without giving in.* Boston: Houghton Mifflin.

44. Brown, B. (2015). *Rising strong.* New York: Spiegel & Grau. Quote appears on p. 56.

45. Brown, *Rising strong.* Quote appears on p. 34.

46. Brown, *Rising strong.* Quote appears on p. 35.

47. Brown, *Rising strong.* Quote appears on p. 37.

48. Turner, C. (2015, February 9). "Masculine" and "feminine" styles of handling conflict. *HuffPost.* Retrieved from https://www.huffingtonpost.com/caroline-turner/masculine-and-feminine-st_b_6633896.html.

49. Daly, M., & Wilson, M. (1983). *Sex, evolution, and behavior* (2nd ed.). Belmont, CA: Wadsworth.

50. Joseph, R. (2000). The evolution of sex differences in language, sexuality, and visual-spatial skills. *Archives of Sexual Behavior, 29,* 35–66.

51. Root, A., & Rubin, K. H. (2010). Gender and parents' reactions to children's emotion during the preschool years. *New Directions for Child & Adolescent Development, 128,* 51–64.

52. Niederle, M., & Versterlund, L. (2007). Do women shy away from competition? Do men complete too much? *Quarterly Journal of Economics, 122,* 1067–1101.

53. Holmstrom, A. J. (2009). Sex and gender similarities and differences in communication values in same-sex and cross-sex friendships. *Communication Quarterly, 57,* 224–238.

54. Benenson, J. F., & Wrangham, R. W. (2016). Cross-cultural sex differences in post-conflict affiliation following sports matches. *Current Biology, 26*(16), 2208–2212.

55. Baillargeon, R. H., Zoccolillo, M., Keenan, K., Côté, S., Pérusse, D., Wu, H.-X., . . . Tremblay, R. E. (2007). Gender differences in physical aggression: A prospective population-based survey of children before and after 2 years of age. *Developmental Psychology, 43,* 13–26.

56. The information in this paragraph is drawn from research summarized by Wood, J. T. (2005). *Gendered lives* (6th ed.). Belmont, CA: Wadsworth.

57. Wood, *Gendered Lives.*

58. Kapidzic, S., & Herring, S. C. (2011). Gender, communication, and self-presentation in teen chatrooms revisited: Have patterns changed? *Journal of Computer-Mediated Communication, 17,* 39–59.

59. Carothers, B. J., & Reis, H. T. (2013). Men and women are from Earth: Examining the latent structure of gender. *Journal of Personality and Social Psychology, 104,* 385–407.

60. Tan, R., Overall, N. C., & Taylor, J. K. (2012). Let's talk about us: Attachment, relationship-focused disclosure, and relationship quality. *Personal Relationships, 19,* 521–534.

61. Niemiec, E. (2010, September 27). Emotions and Italians. Retrieved from http://www.lifeinitaly.com/italian/emotions.

62. Kim-Jo, T., Benet-Martinez, V., & Ozer, D. J. (2010). Culture and interpersonal conflict resolution styles: Role of acculturation. *Journal of Cross-Cultural Psychology, 41,* 264–269.

63. Information in this paragraph is from Hammer, M. R. (2009)). Solving problems and resolving conflict using the Intercultural Style Model and Inventory. In M. A. Moodian (Ed.), *Contemporary leadership and intercultural competence* (pp. 219–232). Thousand Oaks, CA: Sage.

64. The following research is summarized in Tannen, D. (1990). *You just don't understand: Women and men in conversation.* New York: William Morrow, p. 160.

65. Modern Family Script VO, Season 4. (n.d.). HypnoseriesTV. Retrieved from https://www.hypnoseries.tv/modern-family/episodes/saison-4/404—the-butler-s-escape/script-vo-404.192.1220. Quote appears in third to last script block.

66. Hammer, M. R. (2009)). Solving problems and resolving conflict using the Intercultural Style Model and Inventory. In M. A. Moodian (Ed.), *Contemporary leadership and intercultural competence* (pp. 219–232). Thousand Oaks, CA: Sage.

67. Hammer, Solving problems and resolving conflict.

68. University of Colorado Conflict Research Consortium. (1998). *Shuttle diplomacy/mediated communication.* Boulder, CO: International Online Training Program on Intractable Conflict. Retrieved from http://www.colorado.edu/conflict/peace/treatment/shuttle.htm.

69. Brown, B. (2015). *Rising strong.* New York: Spiegel & Grau. Quote appears on p. 40.

CHAPTER 10

1. Baxter, C. (2012, October 27). She went for broke, and found a job. *New York Times.* Retrieved from http://www.nytimes.com/2012/10/28/jobs/taking-a-chance-and-finding-a-dream-job-in-new-york.html?_r=0. Quoted passages appear in para. 1 and 9, respectively.

2. Employers: Verbal communication most important candidate skill. (2016, February 24). National Association of Colleges and Employers Center for Career Development and Talent Acquisition. Retrieved from http://www.naceweb.org/career-readiness/competencies/employers-verbal-communication-most-important-candidate-skill.

3. Half, R. (2017, April 26). The value of teamwork in the workplace. *The Robert Half Blog.* Retrieved from https://www.roberthalf.com/blog/management-tips/the-value-of-teamwork-in-the-workplace. Quote is second subtitle.

4. Warrick, D. (2016). What leaders can learn about teamwork and developing high performance teams from organization development practitioners. *Performance Improvement, 3,* 13–21.

5. Employers want communication skills in new hires. (2014, August 7). The Graduate Management Admission Test [GMAT] blog. Retrieved at https://www.mba.com/us/the-gmat-blog-hub/the-official-gmat-blog/2014/aug/employers-want-communication-skills-in-new-hires.aspx.

6. Improved communication essential to enhance customer satisfaction with after-sale service. (2017, April 27). J. D. Power. Retrieved from http://india.jdpower.com/sites/default/files/2017042in.pdf.

7. Min, H., Lim, Y., & Magnini, V. P. (2015). Factors affecting customer satisfaction in responses to negative online hotel reviews: The impact of empathy, paraphrasing, and speed. *Cornell Hospitality Quarterly, 56*(2), 223–231.

8. Masoud, K., Mohamad Mehdi, M., & Alan, J. D. (2016). Cultural values and consumers' expectations and perceptions of service encounter quality. *International Journal of Pharmaceutical and Healthcare Marketing, 10*(1), 2–26.

9. Friez, D. (2015). 3 ways to measure the value of your social media marketing program. *Top Rank Marketing Blog.* Retrieved from http://www.toprankblog.com/2015/09/social-media-marketing-value.

10. Myatt, M. (2012, April 4). 10 communication secrets of great leaders. *Forbes.* Retrieved from https://www.forbes.com/sites/mikemyatt/2012/04/04/10-communication-secrets-of-great-leaders/#6cc9808a22fe. Quote appears in para. 3.

11. Mikkelson, A. C., York, J. A., & Arritola, J. (2015). Communication competence, leadership behaviors, and employee outcomes in supervisor-employee relationships. *Business and Professional Communication Quarterly, 78*(3), 336–354.

12. Dourado, P. (2014, June 5). A leader who inspired me: True story. Retrieved from http://phildourado.com/2014/06/leader-inspired-true-story. Quotes appear in paras. 13–14 and 16, respectively.

13. Profita, M. (2018, October 2). 8 tips for starting a college senior job search. The Balance Careers. Retrieved from https://www.thebalancecareers.com/starting-a-college-senior-job-search-2059879. Quote appears in para. 2.

14. Kellam, J. (2015, January 9). I highly recommend this student … *Saint Vincent College Faculty Blog.* Retrieved from http://info.stvincent.edu/faculty-blog/i-highly-recommend-this-student. Quote appears in third to last paragraph.

15. Landsbaum, C. (2015, June 9). "I got a job through social media": 5 Millennials share their stories. LEVO. Retrieved from https://www.levo.com/posts/i-got-a-job-through-social-media-5-millennials-share-their-stories. Quote appears in para. 2 of Rose McManus post.

16. Rosen, J. (2010, July 25). The end of forgetting. *New York Times Magazine,* pp. 30–35.

17. Lake, L. (2017, June 10). Tips on creating and growing your personal brand. *The Balance.* Retrieved from https://www.thebalance.com/creating-and-growing-personal-brand-2295814. Quote appears in tip 5.

18. St. John, A. (2017, August 1). Looking for a job? First, clean up your social media presence. *Consumer Reports.* Retrieved from https://www.consumerreports.org/employment-careers/clean-up-social-media-presence-when-looking-for-a-job.

19. Number of employers using social media to screen candidates at all-time high, find latest CareerBuilder study. (2017, June 15). CISION PR Newswire. Retrieved from http://www.prnewswire.com/news-releases/number-of-employers-using-social-media-to-screen-candidates-at-all-time-high-finds-latest-careerbuilder-study-300474228.html.

20. Number of employers (2017, June 15).

21. Knight, D. (2017, June 8). 5 social media mistakes most likely to cost you the job. *MyDomaine.* Retrieved from http://www.mydomaine.com/social-media-job-mistakes.

22. Poppick, S. (2014, September 5). 10 social media blunders that cost a millennial a job—or worse. *Money.* Retrieved from http://time.com/money/3019899/10-facebook-twitter-mistakes-lost-job-millennials-viral. Quote appears in para. 2.

23. Crispin, G., & Mehler, M. (2010). Impact of the Internet on source of hires. *CareerXRoads.* Retrieved from http://www.careerxroads.com/news/impactoftheinternet.doc.

24. Dodds, P. S., Muhamad, R., & Watts, D. J.(2003). An experimental study of search in global social networks. *Science, 301,* 827–829.

25. Brustein, D. (2014, July 22). 17 tips to survive your next networking event. *Forbes.* Retrieved from https://www.forbes.com/sites/yec/2014/07/22/17-tips-to-survive-your-next-networking-event/#59f1c18c7cd4. Quote appears in tip 2.

26. Kihn, S. (2013). Why it is so important to have a good resume. Career Miner. Retrieved from http://careerminer.infomine.com/why-it-is-so-important-to-have-a-good-resume. Quote appears in para. 1.

27. Tullier, M. (2002). The art and science of writing cover letters: The best way to make a first impression. *Monster.com.* Retrieved from http://resume.monster.com/coverletter/coverletters.

28. Graham, A. (2011, January 14). You won't land a job if you can't follow directions. *Forbes.* Retrieved from http://www.forbes.com/sites/work-in-progress/2011/01/14/you-wont-land-a-job-if-you-cant-follow-directions.

29. Moss, C. (2013, September 28). 14 weird, open-ended job interview questions asked at Apple, Amazon and Google. *Business Insider.* Retrieved from http://www.businessinsider.com/weird-interview-questions-from-apple-google-amazon-2013-9?op=1#ixzz37PdzyCTk. See also: Top 25 oddball interview questions for 2014. *Glassdoor.* Retrieved from http://www.glassdoor.com/Top-25-Oddball-Interview-Questions-LST_KQ0,34.htm.

30. Smith, A. (2012, April 20). 5 illegal interview questions and how to dodge them. *Forbes.* Retrieved from https://www.forbes.com/sites/dailymuse/2012/04/20/5-illegal-interview-questions-and-how-to-dodge-them/#6072984e191f. Quotes appear in para. 1.

31. Smith, 5 illegal interview questions. Quote appears in para. 8.

32. Rabin, M., & Schrag, J. L. (1999). First impressions matter: A model of confirmatory bias. *The Quarterly Journal of Economics, 14,* 37–82.

33. Mitchell, N. R. (n.d.). Top 10 interview tips from an etiquette professional. *Experience.* Retrieved from http://www.experience.com/entry-level-jobs/jobs-and-careers/interview-resources/top-10-interview-tips-from-an-etiquette-professional. Quote appears in para. 7.

34. 8 tips for success in a long distance interview. (2016, May 14). *The Everygirl.* Retrieved from http://theeverygirl.com/8-tips-for-success-in-a-long-distance-interview. Quote appears in tip 2.

35. Kaufman, C. Z. (n.d.). Job interview thank you: Is it better to send a letter or email? Monster.com. Retrieved from https://www.monster.com/career-advice/article/interview-thank-you-email-letter. Quote appears in para. 8.

36. Low, E. (2016, August 5). At Amazon's Zappos, CEO Tony Hsieh delivers on being different. *Investor's Business Daily*. Retrieved from https://www.investors.com/news/management/leaders-and-success/for-zappos-ceo-tony-hsieh-success-is-tied-to-company-culture.

37. Pascual, M. (2012, October 30). Zappos: 5 out-of-the-box ideas for keeping employees engaged. *US News & World Report*. Retrieved from https://money.usnews.com/money/blogs/outside-voices-careers/2012/10/30/zappos-5-out-of-the-box-ideas-for-keeping-employees-engaged.

38. Bergeron, J. (2018, July 3). At Zappos, every day is the dog days. Zappos.com. Retrieved from https://www.zappos.com/about/dog-days.

39. Ward, C. (1996). Acculturation. In D. Landis & R. S. Bhagat (Eds.), *Handbook of intercultural training* (pp. 124–147). Thousand Oaks, CA: Sage.

40. Global consumers are willing to put their money where their heart is when it comes to goods and services from companies committed to social responsibility. (2014, June 17). Nielsen. Retrieved from https://www.nielsen.com/us/en/press-room/2014/global-consumers-are-willing-to-put-their-money-where-their-heart-is.html.

41. Laforet, S. (2016). Effects of organisational culture on organisational innovation performance in family firms. *Journal of Small Business and Enterprise Development, 2*, 379–407.

42. Deloitte Millennial Survey 2018. (2018). Deloitte. Retrieved from https://www2.deloitte.com/global/en/pages/about-deloitte/articles/millennialsurvey.html.

43. Hoang, X. M. (n.d.). 7 most wanted work benefits to attract Millennials. The Undercover Recruiter. Retrieved from https://theundercoverrecruiter.com/benefits-attract-millennial-talent.

44. Rivers, T. B. (2015, November 16). 11 ways to make your workplace appealing to Millennials. iOffice. Retrieved from https://www.iofficecorp.com/blog/11-ways-to-make-your-workplace-appealing-to-millennials.

45. The following types of power are based on the categories developed by French, J. R., & Raven, B. (1968). The basis of social power. In D. Cartright & A. Zander (Eds.), *Group dynamics*. New York: Harper & Row, p. 565.

46. Rothwell, J. D. (2013). *In mixed company: Communicating in small groups* (8th ed.). Boston: Cengage Learning.

47. Rothwell, J. D. (2004). *In mixed company: Small group communication* (5th ed.). Belmont, CA: Wadsworth, pp. 247–282.

48. Foot-in-mouth stories … post your shame. (2008, August 14). The Straight Dope message board. https://boards.straightdope.com/sdmb/archive/index.php/t-479416.html. Quote appears in the post by Mister Rik.

49. Fleming, P., & Sturdy, A. (2009). "Just be yourself!": Towards neo-normative control in organisations? *Employee Relations, 31*, 569–583.

50. Ragins, B. R. (2008). Disclosure disconnects: Antecedents and consequences of disclosing invisible stigmas across life domains. *Academy of Management Review, 33*, 194–215. See also Ragins, B. R., Singh, R., & Cornwell, J. M. (2007). Making the invisible visible: Fear and disclosure of sexual orientation at work. *Journal of Applied Psychology, 92*, 1103–1118.

51. Rosh, L., & Offermann, L. (2013, October). Be yourself, but carefully. *Harvard Business Review*. Retrieved from http://hbr.org/2013/10/be-yourself-but-carefully/ar/1.

52. Parker, T. (2013, October 25). 30 non-Americans on the American norms they find weird. *Thought Catalogue*. Retrieved from http://thoughtcatalog.com/timmy-parker/2013/10/30-non-americans-on-the-weirdest-things-that-are-norms-to-americans.

53. Zaslow, J. (2010, January 6). Before you gossip, ask yourself this… . *Moving On*. Retrieved from http://online.wsj.com/article/SB10001424052748704160504574640011681307026.html.

54. Zaslow, Before you gossip, ask yourself this… .

55. "New hires—Stand out at work." (2010, May). OfficePro. Retrieved from http://web.ebscohost.com.libproxy.sbcc.edu:2048/bsi/detail?vid=4&hid=9&sid=079bf30df37c-4ad6897ec1626aa9bb49%40sessionmgr14&bdata=JnNpdGU9YnNpLWxpdmU%3d#db=buh&AN=50544049.

56. Nikravan, L. (2014, January 10). Employees behaving badly. *Talent Management*. Retrieved from http://talentmgt.com/articles/view/employees-behaving-badly/1.

57. Chandler, N. (n.d.). 10 tips for managing conflict in the workplace. *HowStuffWorks*. Retrieved from http://money.howstuffworks.com/business/starting-a-job/10-tips-for-managing-conflict-in-the-workplace1.htm#page=1.

58. Leaping lizards! OfficeTeam survey reveals managers' most embarrassing moment at work. (2011, January 18). *OfficeTeam*. Retrieved from http://rh-us.mediaroom.com/2011-01-18-LEAPING-LIZARDS.

59. Kelley, R. E. (1992). *The power of followership*. New York: Doubleday Business.

60. Kelley, R. E. (2008). Rethinking followership. In R. E. Riggio, I. Chaleff, & J. Lipman-Blumen (Eds.), *The art of followership: How great followers create great leaders and organizations* (pp. 5–16). San Francisco: Jossey-Bass, Quote appears on p. 8.

61. Agho, A. O. (2009). Perspectives of senior-level executives on effective followership and leadership. *Journal of Leadership and Organizational Studies, 16*, 159–166.

62. Kellerman, B. (2008). *Followership: How followers are creating change and changing leaders*. Boston: Harvard Business Press.

63. Kellerman, *Followership*. Quote appears on p. 179.

64. Radicati Group. (2015, March). Email statistics report. Palo Alto, CA: Author. Retrieved from https://www.radicati.com/wp/wp-content/uploads/2015/02/Email-Statistics-Report-2015-2019-Executive-Summary.pdf. Statistic appears on p. 4.

65. Myers, D. (2017, March 1). Research note—86 percent of businesses surveyed to use video conferencing as part of their UC [Unified Communication] Environment by 2018. Retrieved from https://technology.ihs.com/589990/research-note-86-percent-of-businesses-surveyed-to-use-video-conferencing-as-part-of-their-uc-environment-by-2018.

66. Capdeferro, N., & Romero, M. (2012). Are online learners frustrated with collaborative learning experiences? *International Review of Research in Open and Distance Learning, 13*, 26–44.

67. Boitnott, J. (2016, July 5). Bad email grammar ain't good for getting you a job or a date. *Entrepreneur*. Retrieved from https://www.entrepreneur.com/article/278526.

68. Lutz, J. (2016, April 12). The 9 worst—and funniest—reply-all email disasters. Remember to double-check, folks. *Someecards Life*. Retrieved from https://www.someecards.com/life/tech/reply-all-email-disasters.

69. Goleman, D. (2007, October 7). E-mail is easy to write (and to misread). *New York Times*. Retrieved from http://www.nytimes.com/2007/10/07/jobs/07pre.html?_r=0.

70. French, S. (2016, March 13). Ten rules of etiquette for videoconferencing. *Wall Street Journal*. Retrieved from https://www.wsj.com/articles/ten-rules-of-etiquette-for-videoconferencing-1457921535. Quote appears in para. 7.

71. Dempsey, B. (2010, February 10). The seven most universal job skills. *Forbes*. Retrieved from https://www.forbes.com/2010/02/18/most-important-job-skills-personal-finance-universal.html#78e44e375db2. Quote appears in Item 1.

CHAPTER 11

1. Stengel, R. (2009). *Mandela's way: Lessons on life, love, and courage.* New York: Random House.

2. Stengel, R. (2008, July 9). Mandela: His 8 lessons of leadership. *Time.* Retrieved from http://www.time.com/time/magazine/article/0,9171,1821659-1,00.html.

3. Nelson Mandela interview with Morgan Freeman. (2006). Audio transcripts. Retrieved from https://www.nelsonmandela.org/images/uploads/6.Audio_.pdf. Quote appears in first excerpt.

4. Information in this paragraph is from Marquard, L. (1969). *The people and policies of South Africa* (4th ed.). New York: Oxford University Press.

5. Limb, P. (2008). *Nelson Mandela: A biography.* Westport, CT: Greenwood Press.

6. Gallo, A. (2013, May 2). Act like a leader before you are one. *Harvard Business Review.* Retrieved from https://hbr.org/2013/05/act-like-a-leader-before-you-a. Quote appears in para. 1.

7. Prime, J., & Salbi, E. (2014, May 12). The best leaders are humble leaders. *Harvard Business Review Digital Articles,* pp. 2–5.

8. Collins, J. C. (2001). *Good to great: Why some companies make the leap—and others don't.* New York: HarperBusiness.

9. Van Wart, M. (2013). Lessons from leadership theory and the contemporary challenges of leaders. *Public Administration Review, 73*(4), 553–565.

10. Greenleaf, R. K., & Spears, L. C. (2002). *Servant leadership: A journey into the nature of legitimate power and greatness.* New York: Paulist Press.

11. Alonderiene, R., & Majauskaite, M. (2016). Leadership style and job satisfaction in higher education institutions. *International Journal of Educational Management, 30*(1), 140–164.

12. Bhatti, N., Maitlo, G. M., Shaikh, N., Hashmi, M. A., & Shaikh, F. M. (2012). The impact of autocratic and democratic leadership style on job satisfaction. *International Business Research, 5*(2), 192–201.

13. Skogstad, A., Hetland, J., Glasø, L., & Einarsen, S. (2014). Is avoidant leadership a root cause of subordinate stress? Longitudinal relationships between laissez-faire leadership and role ambiguity. *Work and Stress, 28*(4), 323–341.

14. Asplund, J., & Blacksmith, N. (2012, March 6). Strengths-based goal setting. *Gallup Business Journal.* Retrieved from http://www.gallup.com/businessjournal/152981/strengths-based-goal-setting.aspx.

15. Tischler, L., Giambatista, R., McKeage, R., & McCormick, D. (2016). Servant leadership and its relationships with core self-evaluation and job satisfaction. *The Journal of Values-Based Leadership, 9*(1), 1–20. Retrieved from http://scholar.valpo.edu/cgi/viewcontent.cgi?article=1148&context=jvbl.

16. Aristotle. (1958). *Politics, Book 7.* New York: Oxford University Press.

17. Sethuraman, K., & Suresh, J. (2014, August 25). Effective leadership styles. *International Business Research, 7*(9), 165–172.

18. Crwys-Williams, J. (Ed.). (2010). *In the words of Nelson Mandela.* New York: Walker. Quote appears on p. 62.

19. Fiedler, F. E. (1967). *A theory of leadership effectiveness.* New York: McGraw-Hill.

20. Hersey, P., & Blanchard, K. (2001). *Management of organizational behavior: Utilizing human resources* (8th ed.). Upper Saddle River, NJ: Prentice Hall.

21. Blake, R., & Mouton, J. (1964). *The Managerial Grid: The key to leadership excellence.* Houston: Gulf Publishing Co.

22. Blake & Mouton, *The Managerial Grid III.*

23. Kuhnert, K. W., & Lewis, P. (1987). Transactional and transformational leadership: A constructive/developmental analysis. *Academy of Management Review, 12,* 648–657.

24. Boies, K., Fiset, J., & Gill, H. (2015). Communication and trust are key: Unlocking the relationship between leadership and team performance and creativity. *Leadership Quarterly, 26,* 1080–1094.

25. Senge, P. M. (2006). *The fifth discipline: The art and practice of the learning organization.* New York: Doubleday/Currency.

26. Amin, M., Tatlah, I. A., & Khan, A. M. (2013). Which leadership style to use? An investigation of conducive and non-conducive leadership style(s) to faculty job satisfaction. *International Research Journal of Art and Humanities, 41,* 229–253.

27. Hersey, P., & Blanchard, K. (2001). *Management of organizational behavior: Utilizing human resources* (8th ed.). Upper Saddle River, NJ: Prentice-Hall.

28. Weinstein, B. (2007, August 17). How to handle an off-the-wall boss. CIO. Retrieved from http://www.cio.com/article/2438178/staff-management/how-to-deal-with-bully-bosses.html. Quotes appear in last two paragraphs on p. 1.

29. Costanzo, L. A., & Di Domenico, M. (2015). A multi-level dialectical-paradox lens for top management team strategic decision-making in a corporate venture. *British Journal of Management, 3,* 484–506.

30. Kuhnert, K. W., & Lewis, P. (1987). Transactional and transformational leadership: A constructive/developmental analysis. *Academy of Management Review, 12,* 648–657.

31. Bass, B. M. (1990). From transactional to transformational leadership: Learning to share the vision. *Organizational Dynamics, 3,* 19–31.

32. Pierro, A., Raven, B. H., Amato, C., & Bélanger, J. J. (2013). Bases of social power, leadership styles, and organizational commitment. *International Journal of Psychology, 48*(6), 1122–1134.

33. Van Wart, M. (2013). Lessons from leadership theory and the contemporary challenges of leaders. *Public Administration Review, 73*(4), 553–565.

34. Mandela, N. (1994). *The long walk to freedom.* London: Little, Brown. Quote appears on p. 617.

35. Claros Group. (2010). Leaving a job professionally: Wrapping up your current position before moving on. Retrieved from http://www.clarosgroup.com/leavingjob.pdf.

36. Mandela, N. (1990, February 11). Remarks by Nelson Mandela in Cape Town on 11 [sic] February 11, 1990 after his release from Victor Verster. Nelson Mandela Centre of Memory. Retrieved from https://www.nelsonmandela.org/omalley/index.php/site/q/03lv03445/04lv04015/05lv04154/06lv04191.htm. Quote appears in para. 3.

37. The long walk of Nelson Mandela. The prisoner. (n.d.). *Frontline.* PBS. Retrieved from http://www.pbs.org/wgbh/pages/frontline/shows/mandela/prison.

38. Freeland, G. (2018, June 1). Talent wins games, teamwork wins championships. *Forbes.* Retrieved from https://www.forbes.com/sites/grantfreeland/2018/06/01/talent-wins-games-teamwork-wins-championships/#18ad9ec04c8f. Quote appears in para. 4.

39. Robles, M. M. (2012). Executive perceptions of the top 10 soft skills needed in today's workplace. *Business Communication Quarterly, 75,* 453–465.

40. Lutgen-Sandvik, P., Riforgiate, S., & Fletcher, C. (2011). Work as a source of positive emotional experiences and the discourses informing positive assessment. *Western Journal of Communication, 75,* 2–27.

41. Marby, E. A. (1999). The systems metaphor in group communication. In L. R. Frey (Ed.), *Handbook of group communication theory and research* (pp. 71–91). Thousand Oaks, CA: Sage.

42. Rothwell, J. D. (2004). *In mixed company: Small group communication* (pp. 29–31, 5th ed.). Belmont, CA: Wadsworth.

43. Is your team too big? Too small? What's the right number? (2006, June 14). *Knowledge@Wharton.* Retrieved from http://knowledge.wharton.upenn.edu/article.cfm?articleid=1501.

44. Lowry, P., Roberts, T. L., Romano, N. C., Jr., Cheney, P. D., & Hightower, R. T. (2006). The impact of group size and social presence on small-group communication. *Small Group Research, 37,* 631–661.

45. Hackman, J. (1987). The design of work teams. In J. Lorsch (Ed.), *Handbook of organizational behavior* (pp. 315–342). Englewood Cliffs, NJ: Prentice-Hall.

46. LaFasto, F., & Carson, C. (2001). *When teams work best: 6,000 team members and leaders tell what it takes to succeed.* Thousand Oaks, CA: Sage; Larson, C. E., & LaFasto, F. M. J. (1989). *Teamwork: What must go right, what can go wrong.* Thousand Oaks, CA: Sage.

47. Kirschner, F., Paas, F., & Kirschner, P. A. (2010). Superiority of collaborative learning with complex tasks: A research note on alternative affective explanation. *Computers in Human Behavior, 27,* 53–57.

48. Simms, A., & Nichols, T. (2014). Social loafing: A review of the literature. *Journal of Management Policy and Practice, 15*(1), 58–67.

49. Wagner, R., & Harter, J. K. (2006). When there's a freeloader on your team. Excerpt from *The elements of great managing.* Washington, DC: Gallup Press. Retrieved from http://www.stybelpeabody.com/newsite/pdf/Freeloader_on_Your_Team.pdf.

50. Paknad, D. (n.d.). The 5 dynamics of low performing teams. Don't let freeloaders or fear undermine your team. *Workboard.* Retrieved from http://www.workboard.com/blog/dynamics-of-low-performing-teams.php.

51. Gouran, D. S., Hirokawa, R. Y., Julian, K. M., & Leatham, G. B. (1992). The evolution and current status of the functional perspective on communication in decision-making and problem-solving groups. In S. A. Deetz (Ed.), *Communication yearbook* 16 (pp. 573–600). Newbury Park, CA: Sage. See also Wittenbaum, G. M., Hollingshead, A. B., Paulus, P. B., Hirokawa, R. Y., Ancona, D. G., Peterson, R. S., Jehn, K. A., & Yoon, K. (2004). The functional perspective as a lens for understanding groups. *Small Group Research, 35,* 17–43.

52. Mayer, M. E. (1998). Behaviors leading to more effective decisions in small groups embedded in organizations. *Communication Reports, 11,* 123–132.

53. Motavalli, J. (2016, February 16). 5 inspiring companies that rely on teamwork to be successful. *Success.* Retrieved from https://www.success.com/5-inspiring-companies-that-rely-on-teamwork-to-be-successful.

54. Ryan, K. J. (2016, May 13). Casper co-founder on how to build a game-changing company. *Inc.* Retrieved from https://www.inc.com/kevin-j-ryan/casper-founder-on-how-to-build-a-game-changing-company.html.

55. Motavalli, 5 inspiring companies that rely on teamwork to be successful.

56. Clark, D. (2012, May 23). How to deal with difficult co-workers. *Forbes.* Retrieved from http://www.forbes.com/sites/dorieclark/2012/05/23/how-to-deal-with-difficult-co-workers/#6c21476a191d.

57. Dugan, D. (n.d.). Co-workers from hell: Dealing with difficult colleagues. Salary.com. Retrieved from http://www.salary.com/co-workers-from-hell-dealing-with-difficult-colleagues.

58. Hülsheger, U. R., Anderson, N., & Salgado, J. F. (2009). Team-level predictors of innovation at work: A comprehensive meta-analysis spanning three decades of research. *Journal of Applied Psychology, 94,* 1128–1145.

59. Waller, B. M., Hope, L., Burrowes, M., & Morrison, E. R. (2011). Twelve (not so) angry men: Managing conversational group size increases perceived contribution by decision makers. *Group Processes and Intergroup Relations, 14,* 835–843.

60. Degenaars, C. (2017, February 14). Why I don't talk in meetings. *Influence.* Retrieved from https://www.influencive.com/dont-talk-meetings. Quote appears in para. 10.

61. Bing, S. (2009, December 3). No one listens to me in meetings—and I'm the boss. *MoneyWatch.* Retrieved from https://www.cbsnews.com/news/no-one-listens-to-me-in-meetings-and-im-the-boss. Quote appears in para 1. (Maria is a pseudonym. The writer's name is not provided.)

62. Green, A. (2013, April 26). I talk too much in meetings! *Ask a Manager.* Retrieved from https://www.askamanager.org/2013/04/i-talk-too-much-in-meetings.html. Quote appears in para. 2. (Alex is a pseudonym. The writer's name is not provided.)

63. Rothwell, J. D. (2013). *In mixed company* (8th ed.). Boston: Wadsworth-Cengage.

64. Janis, I. (1982). *Groupthink: Psychological studies of policy decisions and fiascoes.* Boston: Houghton Mifflin. See also Baron, R. S. (2005). So right it's wrong: Groupthink and the ubiquitous nature of polarized group decision making. In M. P. Zanna (Ed.), *Advances in experimental social psychology* (Vol. 37, pp. 219–253). San Diego: Elsevier Academic Press.

65. Janis, I. L. (1972). *Victims of groupthink: A psychological study of foreign-policy decisions and fiascoes.* Boston: Houghton Mifflin.

66. *Challenger* explosion. (n.d.) *History.* Retrieved from https://www.history.com/topics/challenger-disaster.

67. Penn State scandal fast facts. (2018, March 28). CNN. Retrieved from https://www.cnn.com/2013/10/28/us/penn-state-scandal-fast-facts/index.html.

68. Decker, B. M. (2018, August 28). In a Catholic church where even the pope covers for sexual abuse, everywhere is as bad as Boston. *USA Today.* Retrieved from https://www.usatoday.com/story/opinion/2018/08/28/pope-francis-knew-cardinal-mccarrick-sexual-abuse-catholic-churchcolumn/1109251002.

69. Chavez, N. (2018, January 25). What others knew: Culture of denial protected Nassar for years. CNN. Retrieved from https://www.cnn.com/2018/01/23/us/nassar-sexual-abuse-who-knew/index.html.

70. Adapted from Rothwell, J. D. (2013). *In mixed company* (8th ed.). Boston: Wadsworth-Cengage, pp. 139–142; see note 49.

71. Rothwell, *In mixed company.*

72. Your brainstorming invitee list: Why diversity is the mother of innovation. SmartStorming. Retrieved from https://www.smart-storming.com/your-brainstorming-invitee-list-why-diversity-is-the-mother-of-innovation. Quote appears in the title.

73. Markman, A. (2017, May 18). Your team is brainstorming all wrong. *Harvard Business Review.* Retrieved from https://hbr.org/2017/05/your-team-is-brainstorming-all-wrong. Quotes appears in para. 3.

74. 10 brainstorming strategies that work. (2018, April 10). *Forbes.* Retrieved from https://www.forbes.com/sites/forbesagencycouncil/2018/04/10/10-brainstorming-strategies-that-work/#70db12aa5da7.

75. Athuraliya, A. (2018, June 1). The ultimate list of essential visual brainstorming techniques. *Creately.* Retrieved from https://creately.com/blog/diagrams/visual-brainstorming-techniques.

76. Bohm, D. (1996). *On dialogue* (L. Nichol, Ed.). London: Routledge & Kegan Paul.

77. Isaacs, W. (1999). *Dialogue: The art of thinking together.* New York: Currency.

78. Sampson, A. (1999). *Mandela: The authorized biography.* New York: Knopf.

79. Stengel, R. (2008, July 9). Mandela: His 8 lessons of leadership. *Time.* Retrieved from http://www.time.com/time/magazine/article/0,9171,1821659-1,00.html.

80. See, for example, Pavitt, C. (2003). Do interacting groups perform better than aggregates of individuals? *Human Communication Research, 29,* 592–599; Wittenbaum, G. M. (2004). Putting communication into the study of group memory. *Human Com-*

munication Research, 29, 616–623; and Frank, M. G., Feely, T. H., Paolantonio, N., & Servoss, T. J. (2004). Individual and small group accuracy in judging truthful and deceptive communication. Group Decision and Negotiation, 13, 45–54.

81. See, for example, Pavitt, Do interacting groups perform better than aggregates of individuals?; Wittenbaum, Putting communication into the study of group memory; Frank, Feely, Paolantonio, & Servoss, Individual and small group accuracy in judging truthful and deceptive communication.

82. Rae-Dupree, J. (2008, December 7). Innovation is a team sport. New York Times. Retrieved from http://www.nytimes.com/2008/12/07/business/worldbusiness/07iht-innovate.1.18456109.html?_r=0.

83. Aritz, J., & Walker, R. C. (2009). Group composition and communication styles: An analysis of multicultural teams in decision-making meetings. Journal of Intercultural Communication Research, 38(2), 99–114.

84. Stier, J., & Kjellin, M. (2010). Communicative challenges in multinational project work: Obstacles and tools for reaching common understandings. Journal of Intercultural Communication, 24, 1–12.

85. Information drawn from García, M., & Cañado, M. (2011). Multicultural teamwork as a source of experiential learning and intercultural development. Journal of English Studies, 9, 145–163; and van Knippenberg, D., van Ginkel, W. P., & Homan, A. C. (2013, July). Diversity mindsets and the performance of diverse teams. Organizational Behavior and Human Decision Processes, 121, 183–193.

86. van Knippenberg, van Ginkel, & Homan, Diversity mindsets and the performance of diverse teams.

87. van Knippenberg, van Ginkel, & Homan, Diversity mindsets and the performance of diverse teams.

88. Adler, R. B., & Elmhorst, J. M. (2010). Communicating at work: Principles and practices for business and the professions (10th ed.). New York: McGraw-Hill, pp. 278–279.

89. Fisher, B. A. (1970). Decision emergence: Phases in group decision making. Speech Monographs, 37, 53–66.

90. This terminology originated with Tuckman, B. W (1965). Developmental sequence in small groups. Psychological Bulletin, 63(6), 384–399.

91. Dewey, J. (1910). How we think. New York: Heath.

92. Poole, M. S. (1991). Procedures for managing meetings: Social and technological innovation. In R. A. Swanson & B. O. Knapp (Eds.), Innovative meeting management (pp. 53–109). Austin, TX: 3M Meeting Management Institute. See also Poole, M. S., & Holmes, M. E. (1995). Decision development in computer-assisted group decision making. Human Communication Research, 22, 90–127.

93. Lewin, K. (1951). Field theory in social science. New York: Harper & Row, pp. 30–59.

94. Dewey, How we think.

95. Hastle, R. (1983). Inside the jury. Cambridge, MA: Harvard University Press.

96. Automattic. (2018). Press. Retrieved from https://automattic.com/press.

97. Mullenweg, M. (2014, April). The CEO of Automattic on holding "auditions" to build a strong team. Harvard Business Review. Retrieved from https://hbr.org/2014/04/the-ceo-of-automattic-on-holding-auditions-to-build-a-strong-team. Quote appears in para. 2 of the "When 9 to 5 Fails" section.

98. Snow. S. (2014, September 11). How Matt's machine works. Fast Company. Retrieved from http://www.fastcompany.com/3035463/how-matts-machine-works. Quotes appear in para. 7 of "A Secret Sauce."

99. Berry, G. R. (2011). Enhancing effectiveness on virtual teams. Journal of Business Communication, 48, 186–206.

100. Altschuller, S., & Benbunan-Fich, R. (2010). Trust, performance, and the communication process in ad hoc decision-making virtual teams. Journal of Computer-Mediated Communication, 16, 27–47.

101. Kraut, R. E., Gergle, D., & Fussell, S. R. (2002). The use of visual information in shared visual spaces: Informing the development of virtual co-presence. Carnegie Mellon University Research Showcase.

102. Rico, R., Alcover, C., Sánchez-Manzanares, M., & Gil, F. (2009). The joint relationships of communication behaviors and task interdependence on trust building and change in virtual project teams. Social Science Information, 48, 229–255.

103. Johnson, G. M. (2006). Synchronous and asynchronous text-based CMC in educational contexts: A review of recent research. Tech Trends, 50, 46–53.

104. Baltes, B. B., Dickson, M. W., Sherman, M. P., Bauer, C. C., & LaGanke, J. S. (2002). Computer-mediated communication and group decision making: A meta-analysis. Organizational Behavior and Human Decision Processes, 156–179.

CHAPTER 12

1. Dwyer, K. K., & Davidson, M. M. (2012, April–June). Is public speaking really more feared than death? Communication Research Reports, 29, 99–107. This study found that public speaking was selected more often as a common fear than any other fear, including death. However, when students were asked to select a top fear, students selected death most often.

2. For an example of how demographics have been taken into consideration in great speeches, see Stephens, G. (1997, Fall). Frederick Douglass's multiracial abolitionism: "Antagonistic cooperation" and "redeemable ideals" in the July 5 speech. Communication Studies, 48, 175–194. On July 5, 1852, Douglass gave a speech titled "What to the Slave Is the 4th of July," attacking the hypocrisy of Independence Day in a slaveholding republic. It was one of the greatest antislavery speeches ever given, and part of its success stemmed from the way Douglass sought common ground with his multiracial audience.

3. See, for example, Jaschik, S. (2007, June 28). 2 kinds of part-time students. Inside Higher Education. Retrieved from https://www.insidehighered.com/news/2007/06/28/parttime.

4. Britton Ody, "Neglected Patients: Deliberate Indifference in State Prisons," Winning Orations, Interstate Oratorical Association, 2018. Britton was coached by Hope Willoughby and Matt Delzer.

5. See, for example, Rolfe-Redding, J., Maibach, E. W., Feldman, L., & Leiserowitz, A. (2011, November 7). Republicans and climate change: An audience analysis of predictors for belief and policy preferences (SSRN Scholarly Paper 2026002). Retrieved from http://papers.ssrn.com/sol3/papers.cfm?abstract_id=2026002.

6. Reproduced by permission of Robert Reich. Prof. Reich posts his videos on Facebook's Inequality Media page and on YouTube. A small portion of this transcript was edited to reduce potential political reaction. That excision is as follows: "Trump has intentionally cleaved America into two warring camps: pro-Trump or anti-Trump. Now most Americans aren't passionate conservatives or liberals, Republicans or Democrats. But they have become impassioned for or against Trump. As a result … "

7. Different analyses might use these terms in slightly different ways. See, for example, Mohan Kumar, "The Relationship Between Beliefs, Values, Attitudes and Behaviours," Owlcation.com, Updated on July 23, 2018. Available at https://owlcation.com/social-sciences/Teaching-and-Assessing-Attitudes.

8. Stutman, R. K., & Newell, S. E. (1984, Fall). Beliefs versus values: Silent beliefs in designing a persuasive message. *Western Journal of Speech Communication, 48*(4), 364.

9. Mahoney, B. (2015). They're not a burden: The inhumanity of anti-homeless legislation. *Winning Orations, 2015* (pp. 20–22). Mankato, MN: Interstate Oratorical Association. Brianna was coached by Kellie Roberts.

10. Mahoney (2015).

11. In information science parlance, these are referred to as Boolean terms.

12. Allison McKibban, "Criminalizing Survivors," *Winning Orations,* Interstate Oratorical Association, 2017, p. 163.

13. Some recent works specifically refer to public speaking anxiety, or PSA. See, for example, Bodie, G. D. (2010, January). A racing heart, rattling knees, and ruminative thoughts: Defining, explaining, and treating public speaking anxiety. *Communication Education, 59*(1), 70–105.

14. See, for example, Alicia H. Clark, "How to Harness Your Anxiety," *New York Times,* October 16, 2018, online, available at https://www.nytimes.com/2018/10/16/well/mind/how-to-harness-your-anxiety.html.

15. See, for example, Borhis, J., & Allen, M. (1992, January). Meta-analysis of the relationship between communication apprehension and cognitive performance. *Communication Education, 41*(1), 68–76.

16. Daly, J. A., Vangelisti, A. L., & Weber, D. J. (1995, December). Speech anxiety affects how people prepare speeches: A protocol analysis of the preparation process of speakers. *Communication Monographs, 62,* 123–134.

17. Researchers generally agree that communication apprehension has three causes: genetics, social learning, and inadequate skills acquisition. See, for example, Finn, A. N. (2009). Public speaking: What causes some to panic? *Communication Currents, 4*(4), 1–2.

18. See, for example, Sawyer, C. R., & Behnke, R. R. (1997, Summer). Communication apprehension and implicit memories of public speaking state anxiety. *Communication Quarterly, 45*(3), 211–222.

19. Adapted from Ellis, A. (1977). *A new guide to rational living.* North Hollywood, CA: Wilshire Books. G. M. Philips listed a different set of beliefs that he believes contributes to reticence. The beliefs are as follows: (1) an exaggerated sense of self-importance (reticent people tend to see themselves as more important to others than others see them); (2) effective speakers are born, not made; (3) skillful speaking is manipulative; (4) speaking is not that important; (5) I can speak whenever I want to; I just choose not to; (6) it is better to be quiet and let people think you are a fool than prove it by talking (they assume they will be evaluated negatively); and (7) what is wrong with me requires a (quick) cure. See Keaten, J. A., Kelly, L., & Finch, C. (2000). Effectiveness of the Penn State Program in changing beliefs associated with reticence. *Communication Education, 49*(2), 134–145.

20. Behnke, R. R., Sawyer, C. R., & King, P. E. (1987, April). The communication of public speaking anxiety. *Communication Education, 36,* 138–141.

21. Honeycutt, J. M., Choi, C. W., & DeBerry, J. R. (2009, July). Communication apprehension and imagined interactions. *Communication Research Reports, 26*(2), 228–236.

22. Behnke, R. R., & Sawyer, C. R. (1999, April). Milestones of anticipatory public speaking anxiety. *Communication Education, 48*(2), 165.

23. See, for example, Olivia Goldhill, "Rhetoric scholars pinpoint why Trump's inarticulate speaking style is so persuasive," Quartz, April 22, 2017. Available at https://qz.com/965004/ rhetoric-scholars-pinpoint-why-trumps-inarticulate-speaking-style-is-so-persuasive/.

24. Hinton, J. S., & Kramer, M. W. (1998, April). The impact of self-directed videotape feedback on students' self-reported levels of communication competence and apprehension. *Communication Education, 47*(2), 151–161. Significant increases in competency and decreases in apprehension were found using this method.

25. Peter Economy, "These Three Body Language Mistakes Make Millennials Look Really Unprofessional," Inc., May 11, 2017. Available at https://www.inc.com/peter-economy/these-3-body-language-mistakes-make-millennials-look-really-unprofessional.html.

26. See, for example, Rosenfeld, L. R., & Civikly, J. M. (1976). *With words unspoken.* New York: Holt, Rinehart and Winston, p. 62. Also see Chaiken, S. (1979). Communicator physical attractiveness and persuasion. *Journal of Personality and Social Psychology, 37,* 1387–1397.

27. Research has confirmed that speeches practiced in front of other people tend to be more successful. See, for example, Smith, T. E., & Frymier, A. B. (2006, February). Get "real": Does practicing speeches before an audience improve performance? *Communication Quarterly, 54,* 111–125.

28. A study demonstrating this stereotype is Street, R. L., Jr., & Brady, R. M. (1982, December). Speech rate acceptance ranges as a function of evaluative domain, listener speech rate, and communication context. *Speech Monographs, 49,* 290–308.

29. See, for example, Mulac, A., & Rudd, M. J. (1977). Effects of selected American regional dialects upon regional audience members. *Communication Monographs, 44,* 184–195. Some research, however, suggests that nonstandard dialects do not have the detrimental effects on listeners that were once believed. See, for example, Johnson, F. L., & Buttny, R. (1982, March). White listeners' responses to "sounding black" and "sounding white": The effect of message content on judgments about language. *Communication Monographs, 49,* 33–39. See also Mairym Lloréns Monteserín and Jason Zevin, "Investigating the impact of dialect prestige on lexical decision," Interspeech Sept. 8, 2016 2214–2218. Available at https://www.isca-speech.org/archive/ Interspeech_2016/pdfs/1549.PDF.

30. Smith, V., Siltanen, S. A., & Hosman, L. A. (1998, Fall). The effects of powerful and powerless speech styles and speaker expertise on impression formation and attitude change. *Communication Research Reports, 15*(1), 27–35. In this study, a powerful speech style was defined as one without hedges and hesitations such as *uh* and *anda.*

CHAPTER 13

1. See, for example, Stern, L. (1985). *The structures and strategies of human memory.* Homewood, IL: Dorsey Press. See also Turner, C. (1987, June 15). Organizing information: Principles and practices. *Library Journal, 112*(11), 58.

2. Booth, W. C., Colomb, G. C., & Williams, J. M. (2003). *The craft of research.* Chicago: University of Chicago Press.

3. Koehler, C. (1998, June 15). Mending the body by lending an ear: The healing power of listening. *Vital Speeches of the Day,* 543.

4. Hallum, E. (1998). Untitled. In L. Schnoor (Ed.), *Winning orations, 1998* (pp. 4–6). Mankato, MN: Interstate Oratorical Association, p. 4. Hallum was coached by Clark Olson.

5. Monroe, A. (1935). *Principles and types of speech.* Glenview, IL: Scott, Foresman.

6. Adapted from http://vaughnkohler.com/wp-content/uploads/ 2013/01/Monroe-Motivated-Sequence-Outline-Handout1.pdf, accessed May 23, 2013.

7. Graham, K. (1976, April 15). The press and its responsibilities. *Vital Speeches of the Day*, 42.

8. Tucker, A.C. (2018). Bad medicine: $765 billion of medical waste. In L. Schnoor (Ed.), *Winning Orations, 2018* (pp. 15–16). Mankato, MN: Interstate Oratorical Association, p. 15. Tucker was coached by Hope Willoughby and Matt Delzer.

9. Tucker, Bad medicine.

10. Anderson, G. (2009). Don't reject my homoglobin. In L. Schnoor (Ed.), *Winning Orations, 2009* (pp. 33–35). Mankato, MN: Interstate Oratorical Association, p. 33. Anderson was coached by Leah White.

11. Chakrabarty, P. (2018). Four billion years of evolution in six minutes. TED Conference, Vancouver BC, April 10, 2018. Available at https://www.ted.com/talks/prosanta_chakrabarty_four_billion_years_of_evolution_in_six_minutes.

12. Robbins, A. (2019). A frat boy and a gentleman. *New York Times*, January 26, 2019. Available at https://www.nytimes.com/2019/01/26/opinion/sunday/fraternity-sexual-assault-college.html.

13. Eggleston, T. S. (n.d.). The key steps to an effective presentation. Retrieved April 4, 2016 from http://seggleston.com/1/writing-and-communications/key-steps.

14. Cook, E. (n.d.). Making business presentations work. Retrieved May 19, 2010 from www.businessknowhow.com/manage/presentation101.htm.

15. Kotelnikov, V. (n.d.). Effective presentations. Retrieved May 19, 2010 from http://www.1000ventures.com/business_guide/crosscuttings/presentations_main.html.

16. Toastmasters International, Inc.'s Communication and Leadership Program. Retrieved May 19, 2010 from www.toastmasters.org.

17. Mehta, A. (2018) Where do your online returns go? TED@UPS, Atlanta, GA. July 19, 2018. Available at https://www.ted.com/talks/aparna_mehta_where_do_your_online_returns_go.

18. Hill, N. (2018). Shut up or get out: How nuisance ordinances fail our communities. In L. Schnoor (Ed.), *Winning orations, 2018* (pp. 82–84). Mankato, MN: Interstate Oratorical Association, p. 82. Hill was coached by Mark Rittenour.

19. Williams, R. (2017). Asking for it: The lies of the rape culture. In L. Schnoor (Ed.), *Winning orations, 2017* (pp. 22–26). Mankato, MN: Interstate Oratorical Association. Williams was coached by Michael Gray and Baker Weilert.

20. See Williams, Asking for it.

21. Wideman, S. (2006). Planning for peak oil: Legislation and conservation. In L. Schnoor (Ed.), *Winning orations, 2006* (pp. 7–9). Mankato, MN: Interstate Oratorical Association, p. 7. Wideman was coached by Brendan Kelly.

22. Teresa Fishman, director of the Center for Academic Integrity at Clemson University, quoted in Gabriel, T. (2010, August 1). Plagiarism lines blur for students in digital age. *New York Times*. Retrieved from http://www.nytimes.com/2010/08/02/education/02cheat.html?pagewanted-all.

23. Herbert, B. (2007, April 26). Hooked on violence. *New York Times*. Retrieved from http://www.nytimes.com/2007/04/26/opinion/26herbert.html.

24. Sherriff, D. (1998, April 1). Bill Gates too rich [Online forum comment]. Retrieved from CRTNET discussion group.

25. Elsesser, K. (2010, March 4). And the gender-neutral Oscar goes to . . . *New York Times*. Retrieved from http://www.nytimes.com/2010/03/04/opinion/04elsesser.html.

26. Gieseck, A. (2015). Problem of police brutality with regard to the deaf community. In L. Schnoor (Ed.), *Winning orations, 2015* (pp. 82–84). Mankato, MN: Interstate Oratorical Association, p. 83.

27. Boyle, J. (2015). Spyware stalking. In L. Schnoor (Ed.), *Winning orations, 2015* (pp. 32–34). Mankato, MN: Interstate Oratorical Association, p. 32. Julia was coached by Judy Santacaterina and Lisa Roth.

28. Notebaert, R. C. (1998, November 1). Leveraging diversity: Adding value to the bottom line. *Vital Speeches of the Day*, 47.

29. McCarley, E. (2009). On the importance of drug courts. In L. Schnoor (Ed.), *Winning orations, 2009* (pp. 36–38). Mankato, MN: Interstate Oratorical Association, p. 36.

30. See McCarley, On the importance of drug courts.

31. Barclay, H. Personal correspondence with George Rodman, April 2019.

CHAPTER 14

1. Sigler, M. G. (2010, August 4). Eric Schmidt: Every 2 days we create as much information as we did up to 2003. *TechCrunch*. Retrieved from http://techcrunch.com/2010/08/04/schmidt-data.

2. See, for example, Wurman, R. S. (2000). *Information anxiety 2*. Indianapolis: Que.

3. Buffett, W. E. (2012, November 25). A minimum tax for the wealthy. *New York Times*. Retrieved from http://www.nytimes.com/2012/11/26/opinion/buffett-a-minimum-tax-for-the-wealthy.html.

4. See Fransden, K. D., & Clement, D. A. (1984). The functions of human communication in informing: Communicating and processing information. In C. C. Arnold & J. W. Bowers (Eds.), *Handbook of rhetorical and communication theory* (pp. 338–399). Boston: Allyn and Bacon.

5. Allocca, K. (2001, November). *Why videos go viral*. TEDYouth talk, New York City. Retrieved from http://www.ted.com/talks/kevin_allocca_why_videos_go_viral.html.

6. Davis, B (20180. The importance of black male teachers, in L. Schnoor (Ed.), *Winning Orations, 2018* (pp. 9–11). Mankato, MN: Interstate Oratorical Association, (p. 10).

7. Cabalan, S. (2018). Eleven and engaged: On America's unseen child marriage crisis, in L. Schnoor (Ed.), *Winning Orations, 2018* (pp. 31–33). Mankato, MN: Interstate Oratorical Association, (p. 31).

8. Cacioppo, J. T., & Petty, R. E. (1979). Effects of message repetition and position on cognitive response, recall, and persuasion. *Journal of Personality and Social Psychology, 37*, 97–109.

9. Schulz, K. (2011, March). *On being wrong*. Speech presented at the TED Conference. Retrieved from http://www.ted.com/talks/kathryn_schulz_on_being_wrong.html.

10. Boroditsky, L. (2009). How does our language shape the way we think? *Edge*, June 11, 2009. Available at https://www.edge.org/conversation/how-does-our-language-shape-the-way-we-think.

11. Parker, I. (2001, May 28). Absolute PowerPoint. *New Yorker*, p. 78.

12. Steven Pinker, a psychology professor at MIT, quoted in Zuckerman, L. (1999, April 17). Words go right to the brain, but can they stir the heart? *New York Times*, p. 9.

13. Tufte, E. (2003, September). PowerPoint is evil: Power corrupts. PowerPoint corrupts absolutely. *Wired*. Retrieved from http://www.wired.com/wired/archive/11.09/ppt2.html.

14. Simons, T. (2004, March). Does PowerPoint make you stupid? *Presentations*, p. 25.

15. Tufte, E. R. (2003). *The cognitive style of PowerPoint*. Cheshire, CT: Graphics Press.

16. See Parker, Absolute PowerPoint.

17. Lera Boroditsky, "How Language Shapes the Way We Think," TEDWomen 2017, New Orleans, November 1, 2017. Available at https://www.ted.com/talks/lera_boroditsky_how_language_shapes_the_way_we_think.

CHAPTER 15

1. Cabalan, S. (2019). Personal correspondence with George Rodman, April 2019.
2. Cabalan, Personal correspondence with George Rodman.
3. For an explanation of social judgment theory, see Griffin, E. (2012). *A first look at communication theory* (8th ed.). New York: McGraw-Hill.
4. See, for example, Jaska, J. A., & Pritchard, M. S. (1994). *Communication ethics: Methods of analysis* (2nd ed.). Boston: Wadsworth.
5. Some research suggests that audiences may perceive a direct strategy as a threat to their freedom to form their own opinions. This perception hampers persuasion. See Brehm, J. W. (1966). *A theory of psychological reactance.* New York: Academic Press.
6. Golden, J. L., Berquist, G. F., Coleman, W. E., Golden, R., & Sproule, J. M., eds. (2011). *The Rhetoric of Western thought: From the Mediterranean world to the global setting,* 10th ed. Dubuque, IA: Kendall/Hunt, p. 72.
7. Heath, D., & Heath, C. (2010) *Switch: How to change things when change is hard.* New York: Currency/Random House, p. 117.
8. Haidt, J. (2012). *The righteous mind.* New York: Pantheon, p. 50.
9. Haidt, *The righteous mind,* p. 68.
10. Yocum, R. (2015). A deadly miscalculation. In L. Schnoor (Ed.), *Winning orations 2015* (pp. 97–99). Mankato, MN: Interstate Oratorical Association. Rebeka was coached by Mark Hickman.
11. For an excellent review of the effects of evidence, see Reinard, J. C. (1988, Fall). The empirical study of persuasive effects of evidence: The status after fifty years of research. *Human Communication Research, 15*(1), 3–59.
12. There are, of course, other classifications of logical fallacies than those presented here. See, for example, Warnick, B., & Inch, E. (1994). *Critical thinking and communication: The use of reason in argument* (2nd ed.). New York: Macmillan, pp. 137–161.
13. Sprague, J., & Stuart, D. (1992). *The speaker's handbook* (3rd ed.). Fort Worth, TX: Harcourt Brace Jovanovich, p. 172.
14. Myers, L. J. (1981). The nature of pluralism and the African-American case. *Theory into Practice, 20,* 3–4. Cited in Samovar, L. A., & Porter, R. E. (1995). *Communication between cultures* (2nd ed.). New York: Wadsworth, p. 251.
15. Samovar & Porter, *Communication between cultures,* pp. 154–155.
16. For an example of how one politician failed to adapt to his audience's attitudes, see Hostetler, M. J. (1998, Winter). Gov. Al Smith confronts the Catholic question: The rhetorical legacy of the 1928 campaign. *Communication Quarterly, 46*(1), 12–24. Smith was reluctant to discuss religion, attributed bigotry to anyone who brought it up, and was impatient with the whole issue. He lost the election. Many years later, John F. Kennedy dealt with "the Catholic question" more reasonably and won.
17. Mount, A. (1973). Speech before the Southern Baptist Convention. In W. A. Linkugel, R. R. Allen, & R. Johannessen (Eds.). *Contemporary American speeches* (3rd ed., pp. 203–205). Belmont, CA: Wadsworth, p. 204.
18. Rudolf, H. J. (1983, Summer). Robert F. Kennedy at Stellenbosch University. *Communication Quarterly, 31,* 205–211.
19. DeVito, J. A. (1986). *The communication handbook: A dictionary.* New York: Harper & Row, pp. 84–86.
20. Rodriguez, G. (2010). How to develop a winning sales plan. Retrieved from www.powerhomebiz.com/062006/salesplan.htm.
21. Sanfilippo, B. (2010). Winning sales strategies of top performers. Retrieved from www.selfgrowth.com/articles/Sanfilippo2.html.
22. Cabalan, S. (2019). Personal correspondence with George Rodman, April 2019.

Glossary

abstract language Language that lacks specificity or does not refer to observable behavior or other sensory data.

abstraction ladder A range of more to less abstract terms describing a person, object, or event.

accent Pronunciation perceived as different from the locally accepted speech style.

acculturation The process of adapting to a culture

activists Followers who are energetically and passionately engaged.

actuate To move members of an audience toward a specific behavior.

ad hominem fallacy A problematic strategy of attacking a person's character rather than debating the issues at hand.

addition The articulation error that involves adding extra parts to words.

advising response Helping response in which the receiver offers suggestions about how the speaker should deal with a problem.

affect blend The combination of two or more expressions, each showing a different emotion.

affect displays Facial expressions, body movements, and vocal traits that reveal emotional states.

affinity The degree to which people like or appreciate one another, whether or not they display that outwardly.

all-channel network A communication network pattern in which group members are frequently together and share a great deal of information with one another.

altruistic lies Deception intended to be unmalicious, or even helpful, to the person to whom it is told.

analogies An extended comparison that can be used as supporting material in a speech.

analytical listening A response style in which the primary goal is to understand a message.

anecdote A brief, personal story used to illustrate or support a point in a speech.

argumentum ad populum fallacy Fallacious reasoning based on the dubious notion that because many people favor an idea, you should, too.

argumentum ad verecundiam fallacy Fallacious reasoning that tries to support a belief by relying on the testimony of someone who is not an authority on the issue being argued.

Aristotle's rhetorical triad According to the ancient Greek philosopher Aristotle, there are three elements of influencing others: *ethos* (speaker credibility), *pathos* (emotional appeals), and *logos* (logical reasoning).

articulation The process of pronouncing all the necessary parts of a word.

assertive communication A style of communicating that directly expresses the sender's needs, thoughts, or feelings, delivered in a way that does not attack the receiver.

asynchronous communication Communication that occurs when there's a lag between the creation and reception of a message.

attending The process of focusing on certain stimuli in the environment.

attitude The predisposition to respond to an idea, person, or thing favorably or unfavorably.

attribution The process of attaching meaning.

audience analysis A consideration of characteristics, including the type, goals, demographics, beliefs, attitudes, and values of listeners.

audience involvement The level of commitment and attention that listeners devote to a speech.

audience participation Listener activity during a speech; a technique to increase audience involvement.

authoritarian leadership A style in which a designated leader uses coercive and reward power to dictate the group's actions.

avoidance spiral Occurs when relational partners reduce their dependence on one another, withdraw, and become less invested in the relationship.

bar chart A visual aid that compares two or more values by showing them as elongated horizontal rectangles.

basic speech structure The division of a speech into introduction, body, and conclusion.

behavioral description An account that refers only to observable phenomena.

belief An underlying conviction about the truth of an idea, often based on cultural training.

brainstorming A group activity in which members share as many ideas about a topic as possible without stopping to judge them or rule anything out.

breakout groups A strategy used when the number of members is too large for effective discussion. Subgroups simultaneously address an issue and then report back to the group at large.

bystanders Followers who are aware of what's going on around them but tend to hang back and watch rather than play an active role.

cause-effect pattern An organizing plan for a speech that demonstrates how one or more events result in another event or events.

chain network A communication network in which information passes sequentially from one member to another.

channel The medium through which a message passes from sender to receiver.

chronemics The study of how humans use and structure time.

citation A brief statement of supporting material in a speech.

climax pattern An organizing plan for a speech that builds ideas to the point of maximum interest or tension.

coculture A group that is part of an encompassing culture.

coercive power The ability to influence others by threatening or imposing unpleasant consequences.

cognitive complexity The ability to understand issues from a variety of perspectives.

cohesiveness The degree to which members feel part of a group and want to remain in it.

collectivistic culture A culture in which members focus more on the welfare of the group as a whole than on individual identity.

column chart A visual aid that compares two or more values by showing them as elongated vertical rectangles.

comforting A response style in which a listener reassures, supports, encourages, or distracts the person seeking help.

communication climate The emotional tone of a relationship as it is expressed in the messages that the partners send and receive.

communication competence The ability to achieve one's goals through communication and, ideally, maintain healthy relationships.

communication The process of creating meaning through symbolic interaction.

compromise An agreement that gives both parties at least some of what they wanted, although both sacrifice part of their goals.

conclusion (of a speech) The final part of a speech, in which the main points are reviewed and final remarks are made to motivate the audience or help listeners remember key ideas.

confirmation bias The emotional tendency to interpret new information as confirmation of one's existing beliefs.

confirming messages Actions and words that express respect for another person.

conflict (storming) stage When group members openly defend their positions and question those of others.

conflict An expressed struggle between at least two interdependent parties who perceive incompatible goals, scarce rewards, and/or interference from the other party in achieving their goals.

connection power The ability to influence others based on having relationships that might help the group reach its goal.

connotative meanings Informal, implied interpretations for words and phrases that reflect the people, culture, emotions, and situations involved.

contact hypothesis A proposition based on evidence that prejudice tends to diminish when people have personal contact with those they might otherwise stereotype.

contempt Verbal and nonverbal messages that ridicule or belittle the other person.

content message A message that communicates information about the subject being discussed.

control The amount of influence one has over others.

convergence Accommodating one's speaking style to another person, usually a person who is desirable or has higher status.

conversational narcissists People who focus on themselves and their interests instead of listening to and encouraging others.

convincing A speech goal that aims at changing audience members' beliefs, values, or attitudes.

coordinated management of meaning (CMM) The notion that people co-create meaning in the process of communicating with one another.

counterfeit question A question that is not truly a request for new information.

credibility The believability of a speaker or other source of information.

critical listening Listening in which the goal is to evaluate the quality or accuracy of the speaker's remarks.

criticism A message that is personal, all-encompassing, and accusatory.

culture The language, values, beliefs, traditions, and customs people share and learn.

cyberbullying A malicious act in which one or more parties aggressively harass a victim online, often in public forums.

cyberstalking Ongoing monitoring of the social presence of a person.

database A computerized collection of information that can be searched in a variety of ways to locate information that the user is seeking.

debilitative communication apprehension An intense level of anxiety about speaking before an audience, resulting in poor performance.

deception bias The tendency to assume that others are lying.

decode To attach meaning to a message.

defensive listening A response style in which the receiver perceives a speaker's comments as an attack.

defensiveness Protecting oneself by counterattacking the other person.

deletion An articulation error that involves leaving off parts of words.

democratic leadership A style in which a leader invites the group's participation in decision making.

demographics Audience characteristics that can be analyzed statistically, such as age, gender, education, and group membership.

denotative meanings Formally recognized definitions for words, as in those found in a dictionary.

description A type of speech that uses details to create a "word picture" of something's essential factors.

developmental model (of relational maintenance) A theoretical framework based on the idea that communication patterns are different in various stages of interpersonal relationships.

diagram A line drawing that shows the most important components of an object.

dialect A version of the same language that includes substantially different words and meanings.

dialogue A process in which people let go of the notion that their ideas are more correct or superior to those of others and instead seek to understand an issue from many different perspectives.

diehards Followers who will, sometimes literally, sacrifice themselves for a cause they believe in.

direct persuasion Persuasion that does not try to hide or disguise the speaker's persuasive purpose.

directly aggressive message A message that openly attacks the position and perhaps the dignity of the receiver.

disconfirming messages Actions and words that imply a lack of agreement or respect for another person.

disfluencies Vocal interruptions such as stammering and use of "*uh*," "*um*," and "*er*."

disinhibition The tendency to transmit messages without considering their consequences.

divergence A linguistic strategy in which speakers emphasize differences between their communicative style and that of others to create distance.

dyadic communication Two-person communication.

either–or fallacy Fallacious reasoning that sets up false alternatives, suggesting that if the inferior one must be rejected, then the other must be accepted.

emblems Deliberate gestures with precise meanings, known to virtually all members of a cultural group.

emergence (norming) stage When a group moves from conflict toward a single solution.

emergent leader A member who assumes leadership without being appointed by higher-ups.

emotional intelligence (EI) The ability to understand and manage one's own emotions and to deal effectively with the emotions of others.

emotive language Language that conveys an attitude rather than simply offering an objective description.

empathy The ability to imagine another person's point of view.

encode To put thoughts into symbols, most commonly words.

environment Both the physical setting in which communication occurs and the personal perspectives of the parties involved.

equivocal words Words that have more than one dictionary definition.

equivocation A deliberately vague statement that can be interpreted in more than one way.

escalatory conflict spiral A reciprocal pattern of communication in which messages, either confirming or disconfirming, between two or more communicators reinforce one another.

ethical persuasion Persuasion in an audience's best interest that does not depend on false or misleading information to induce change in that audience.

ethnicity A social construct that refers to the degree to which a person identifies with a particular group, usually on the basis of nationality, culture, religion, or some other unifying perspective.

ethnocentrism The attitude that one's own culture is superior to other cultures.

ethos Appeals based on the credibility of the speaker.

euphemism A mild or indirect term or expression used in place of a more direct but less pleasant one.

evasion The act of making a deliberately vague statement.

expectancy violation theory The proposition that nonverbal cues cause physical and/or emotional arousal, especially if they deviate from what is considered normal. People who consider a particular nonverbal violation to be positive may compensate (respond similarly), but they are likely to compensate (respond with opposite or distancing behavior) if they find it unpleasant.

expert power The ability to influence others by virtue of one's perceived expertise on the subject in question.

explanations Speeches or presentations that clarify ideas and concepts already known but not understood by an audience.

extemporaneous speech A speech that is planned in advance but presented in a direct, conversational manner.

face The socially approved identity that a communicator tries to present.

facework Verbal and nonverbal behavior designed to create and maintain a communicator's face and the face of others.

facilitative communication apprehension A moderate level of anxiety about speaking before an audience that helps improve the speaker's performance.

factual example A true, specific case that is used to demonstrate a general idea.

factual statement A statement that can be verified as being true or false.

fallacy An error in logic.

fallacy of approval The irrational belief that it is vital to win the approval of virtually every person a communicator deals with.

fallacy of catastrophic failure The irrational belief that the worst possible outcome will probably occur.

fallacy of overgeneralization Irrational beliefs in which (1) conclusions (usually negative) are based on limited evidence or (2) communicators exaggerate their shortcomings.

fallacy of perfection The irrational belief that a worthwhile communicator should be able to handle every situation with complete confidence and skill.

family People who share affection and resources as a family and who think of themselves and present themselves as a family, regardless of genetics.

feedback A receiver's response to a sender's message.

Feynman technique A process proposed by physicist Richard Feynman that involves depicting a complex concept as best one can on paper, describing the concept as if teaching it to a child, considering what aspects of the idea are still unclear, and then reviewing information further to achieve even deeper understanding of it.

flow chart A diagram that depicts the steps in a process with shapes and arrows.

force field analysis A method of problem analysis that identifies the forces that work in favor of a desired outcome and those that will make it difficult to achieve.

formal outline A consistent format and set of symbols used to identify the structure of ideas.

formal role A role assigned to a person by group members or an organization, usually to establish order.

frame switching Adopting the perspectives of different cultures.

gatekeeper (media) In mass media, a professional who controls the content of public messages.

gatekeeper (small-group interaction) The person in a small group through whom communication among other members flows.

gender matrix A construct that recognizes gender as a multidimensional collection of qualities.

gender Socially constructed roles, behaviors, activities, and attributes that a society considers appropriate.

general purpose One of three basic ways a speaker seeks to affect an audience: to entertain, inform, or persuade.

groupthink The tendency in some groups to support ideas without challenging them or providing alternatives.

halo/horns effect A form of bias that overgeneralizes positive or negative traits.

haptics The study of touch.

hearing The process wherein sound waves strike the eardrum and cause vibrations that are transmitted to the brain.

hegemony The dominance of one culture over another.

high self-monitors People who pay close attention to their own behavior and to others' reactions, adjusting their communication to create the desired impression.

high-context culture A culture that relies heavily on subtle, often nonverbal cues to maintain social harmony.

hyperpersonal communication The phenomenon in which digital interaction creates deeper relationships than arise through face-to-face communication.

hypothetical example An example that asks an audience to imagine an object or event.

identity management Strategies used by communicators to influence the way others view them.

illustrators Nonverbal behaviors that accompany and support verbal messages.

imaginary audience A heightened self-consciousness that makes it seem as if people are always observing and judging you.

immediacy Expression of interest and attraction communicated verbally and/or nonverbally.

implicit bias Unconsciously held associations about a social group.

impromptu speech A speech given "off the top of one's head," without preparation.

indirect communication Hinting at a message instead of expressing thoughts and feelings directly.

indirect persuasion Persuasion that disguises or deemphasizes the speaker's persuasive goal.

individualistic culture A culture in which members focus on the value and welfare of individual members more than on the group as a whole.

inferential statement A conclusion arrived at from an interpretation of evidence.

informal role A role usually not explicitly recognized by a group that describes functions of group members rather than their positions.

information anxiety The psychological stress of dealing with too much information.

information hunger Audience desire, created by a speaker, to learn information.

informational interview A structured meeting in which a person seeks answers from someone whose knowledge can help them succeed.

informative purpose statement A complete statement of the objective of a speech, worded to stress audience knowledge and/or ability.

in-groups Groups with which one identifies.

insensitive listening The failure to recognize the thoughts or feelings that are not directly expressed by a speaker, and instead accepting the speaker's words at face value.

instructions Remarks that teach something to an audience in a logical, step-by-step manner.

instructions Remarks that teach something to an audience in a logical, step-by-step manner.

insulated listening A style in which the receiver ignores undesirable information.

intergroup communication Interaction between members of different cultures or cocultures.

interpersonal communication Two-way interactions between people who are part of a close and irreplaceable relationship in which they treat each other as unique individuals.

interpretation The perceptual process of attaching meaning to stimuli that have previously been selected and organized.

intersectionality theory The idea that people are influenced in unique ways by the complex overlap and interactions of multiple identities and social factors.

intimate distance A distance between people that ranges from touching to about 18 inches; usually used by those who are emotionally close.

intrapersonal communication Communication that occurs within a single person.

intro A brief explanation or comment before a visual aid is used.

introduction (of a speech) The first structural unit of a speech, in which the speaker captures the audience's attention and previews the main points to be covered.

irrational thinking Beliefs that have no basis in reality or logic; one source of debilitative communication apprehension.

isolates Followers who are indifferent to the overall goals of the organization and communicate very little with people outside their immediate environment.

jargon Specialized vocabulary used as a kind of shorthand by people with common backgrounds and experience.

Johari Window A model that describes the relationship between self-disclosure and self-awareness.

judging response Feedback that indicates a listener is evaluating the sender's thoughts or behaviors.

kinesics The study of body movement, facial expression, gesture, and posture.

laissez-faire leadership A style in which a leader takes a hands-off approach, imposing few rules on a team.

language A collection of symbols governed by rules and used to convey messages between individuals.

latitude anchor The position supported by audience members before a persuasion attempt.

latitude of acceptance In social judgment theory, statements that a receiver would not reject.

latitude of noncommitment In social judgment theory, statements that a receiver would not care strongly about one way or another.

latitude of rejection In social judgment theory, statements that a receiver could not possibly accept.

leanness The lack of nonverbal cues to clarify a message.

legitimate power The ability to influence group members based on one's official position.

line chart A visual aid that maps out the direction of a moving point; it is ideally suited for showing changes over time.

linear communication model A characterization of communication as a one-way event in which a message flows from sender to receiver.

linguistic intergroup bias The tendency to label people and behaviors in terms that reflect their in-group or out-group status.

linguistic relativism The notion that language influences the way we experience the world.

listening fidelity The degree of congruence between what a listener understands and what the message sender was attempting to communicate.

listening The process wherein the brain recognizes sounds and gives them meaning.

logos Appeals based on logical reasoning.

lose–lose problem solving An approach to conflict resolution in which neither party achieves its goals.

low self-monitors People who express what they are thinking and feeling without much attention to the impression their behavior creates.

low-context culture A culture that uses language primarily to express thoughts, feelings, and ideas as directly as possible.

manipulators Movements in which a person grooms, massages, rubs, holds, pinches, picks, or otherwise manipulates an object or body part.

manuscript speech A speech that is read word for word from a prepared text.

mass communication The transmission of messages to large, usually widespread audiences via TV, Internet, movies, magazines, and other forms of mass media.

masspersonal communication The overlap between personal and public communication.

media Communication mechanisms such as phones and computers used to convey messages between people.

memorized speech A speech learned and delivered by rote without a written text.

message A sender's planned and unplanned words and nonverbal behaviors.

metacommunication Messages that refer to other messages; communication about communication.

mindful listening Being fully present with people—paying close attention to their gestures, manner, and silences, as well as to what they say.

model (in speeches and presentations) A replica of an object being discussed. It is usually used when it would be difficult or impossible to use the actual object.

monochronic The use of time that emphasizes punctuality, schedules, and completing one task at a time.

narration The presentation of speech supporting material as a story with a beginning, middle, and end.

narratives The stories people create and use to make sense of their personal worlds.

negativity bias The perceptual tendency to focus more on negative indicators than on positive ones.

negotiation An interactive process meant to help people reach agreement when one person wants something from another.

networking The strategic process of meeting people and maintaining contacts to gain information and advice.

noise External, physiological, and psychological distractions that interfere with the accurate transmission and reception of a message.

nominal group technique A method for including the ideas of all group members in a problem-solving session, alternating between individual contributions and group discussion.

nonassertion The inability or unwillingness to express one's thoughts or feelings.

nonverbal communication Messages expressed without words, as through body movements, facial expressions, eye contact, tone of voice, and so on.

norms Typically unspoken shared values, beliefs, behaviors, and processes that influence group members.

number chart A visual aid that lists numbers in tabular form to clarify information.

online surveillance Discreet monitoring of the social media presence of unknowing targets.

opinion statement A statement based on the speaker's beliefs.

organization The perceptual process of organizing stimuli into patterns.

organizational communication Interaction among members of a relatively large, permanent structure (such as a nonprofit agency or business) in order to pursue shared goals.

organizational culture A relatively stable, shared set of rules about how to behave and a set of values about what is important.

orientation (forming) stage When group members become familiar with one another's positions and tentatively volunteer their own.

out-groups Groups one views as different from oneself.

outro A brief summary or conclusion after a visual aid has been used.

paralanguage Nonlinguistic means of vocal expression: rate, pitch, tone, and so on.

paraphrasing Feedback in which the receiver rewords the speaker's thoughts and feelings.

participants Followers who are moderately involved and attempt to have an impact either by supporting leaders' efforts or working in opposition to them.

participative decision making A process in which people contribute to the decisions that will affect them.

passive aggression An indirect expression of aggression, delivered in a way that allows the sender to maintain a facade of innocence.

passive narcissists People who are so wrapped up in themselves that they fail to be supportive or encouraging of listeners.

pathos Appeals based on emotion.

perceived self The person we believe ourselves to be in moments of candor. It may be identical to or different from the presenting and ideal selves.

perception A process in which people use sensory date to reach conclusions about others and the world around them.

perception checking A three-part method for verifying the accuracy of interpretations, including an objective description of the behavior, two possible interpretations, and a request for more information.

personal distance A distance between people that ranges from 18 inches to 4 feet; common among most relational partners.

personal fable A sense common in adolescence that one is different from everybody else.

personality The set of enduring characteristics that define a person's temperament, thought processes, and social behavior.

persuasion The act of motivating a listener, through communication, to change a particular belief, attitude, value, or behavior.

phonological rules Linguistic rules governing how sounds are combined to form words.

phubbing A mixture of the words *phoning* and *snubbing,* used to describe episodes in which people pay attention to their devices rather than they do to the people around them.

pictogram A visual aid that conveys its meaning through an image of an actual object.

pie chart A visual aid that divides a circle into wedges, representing percentages of the whole.

pitch The highness or lowness of one's voice.

polychronic The use of time that emphasizes flexible schedules in which multiple tasks are pursued at the same time.

polymediation The range of communication channel options available to communicators.

positive spiral Occurs when one person's confirming message leads to a similar or even more confirming response from the other person.

post hoc **fallacy** Fallacious reasoning that mistakenly assumes that one event causes another because they occur sequentially.

power distance The degree to which members of a group are willing to accept a difference in power and status.

power The ability to influence others' thoughts and/or actions.

pragmatic rules Rules that govern how people use language in everyday interaction.

prejudice An unfairly biased and intolerant attitude toward others who belong to an out-group.

presenting self The image a person presents to others. It may be identical to or different from the perceived and ideal selves.

problem census A technique in which members write their ideas on cards, which are then posted and grouped by a leader to generate key ideas the group can then discuss.

problem-solution pattern An organizing pattern for a speech that describes an unsatisfactory state of affairs and then proposes a plan to remedy the problem.

prompting Using silence and brief statements to encourage a speaker to continue talking.

proposition of fact A claim bearing on issue in which there are two or more sides of conflicting factual evidence.

proposition of policy A claim bearing on an issue that involves adopting or rejecting a specific course of action.

proposition of value A claim bearing on an issue involving the worth of some idea, person, or object.

proxemics The study of how people and animals use space.

pseudolistening An imitation of true listening.

public communication Communication that occurs when a group is too large for everyone to contribute. It is characterized by an unequal amount of speaking and by limited verbal feedback.

public distance A distance between people that exceeds 12 feet; common in public speaking.

purpose statement A complete sentence that describes precisely what a speaker wants to accomplish.

race A social construct to describe a group of people who share physical and cultural traits and potentially a common ancestry.

rate The speed at which a speaker utters words.

reappropriation The process by which members of a marginalized group reframe the meaning of a term that has historically been used in a derogatory way.

receiver One who notices and attends to a message.

reductio ad absurdum **fallacy** Fallacious reasoning that unfairly attacks an argument by extending it to such extreme lengths that it looks ridiculous.

referent power The ability to influence others by virtue of being liked or respected.

reflected appraisal The influence of others on one's self-concept.

reflecting Listening that helps the person speaking hear and think about the words they have just spoken.

reflective thinking method A systematic approach to solving problems that involves identifying a problem, analyzing it, developing and evaluating creative solutions, implementing a plan, and following up on the solution. Also known as Dewey's reflective thinking method.

reinforcement (performing) stage When group members endorse the decision they have made.

relational dialectics The perspective that partners in interpersonal relationships must deal with simultaneous and opposing forces of connection versus autonomy, predictability versus novelty, and openness versus privacy.

relational listening A listening style that is driven primarily by the desire to build emotional closeness with the speaker.

relational message A message that expresses the social relationship between two or more people.

relational spiral A reciprocal communication pattern in which each person's message reinforces the other's.

relative words Terms that gain their meaning by comparison.

remembering The act of recalling previously introduced information. The amount of recall drops off in two phases: short term and long term.

residual message The part of a message a receiver can recall after short- and long-term memory loss.

respect The degree to which a person holds another in esteem, whether or not they like them.

responding Providing observable feedback to another person's behavior or speech.

reward power The ability to influence others by granting or promising desirable consequences.

richness The degree to which nonverbal cues can clarify a verbal message.

role pattern of behavior enacted by particular members

round robin Session in which members each address the group for specified amount of time.

rules Explicit, official guidelines that govern group functions and member behavior.

salience How much weight people attach to a particular phenomenon or characteristic.

script Habitual, reflexive way of behaving.

selection The perceptual act of attending to some stimuli in the environment and ignoring others.

selection interview A formal meeting (in person or via communication technology) to evaluate a candidate for a job or other opportunity.

selective listening A listening style in which the receiver responds only to messages that interest them.

self-concept A set of largely stable perceptions about oneself.

self-disclosure The process of deliberately revealing information about oneself that is significant and that would not normally be known by others.

self-esteem The part of the self-concept that involves evaluations of self-worth.

self-fulfilling prophecy A prediction or expectation of an event that makes the outcome more likely to occur than would otherwise have been the case.

self-monitoring Paying close attention to one's own behavior and using these observations to make effective choices.

self-serving bias The tendency to judge others harshly but to cast oneself in a favorable light.

semantic rules Rules that govern the meaning of language as opposed to its structure.

sender The originator of a message.

servant leadership A style in which a leader recruits outstanding team members and provides the support they need to do a good job.

sex A biological category such as male, female, or intersex.

signpost A phrase that emphasizes the importance of upcoming material in a speech.

situational leadership A theory that argues that the most effective leadership style varies according to the circumstances involved.

slang Language used by a group of people whose members belong to a similar coculture or other group.

slurring The articulation error that involves overlapping the end of one word with the beginning of the next.

small group A limited number of interdependent people who interact with one another over time to pursue shared goals.

small-group communication Communication within a group of a size such that every member can participate actively with the other members.

social comparison Evaluating oneself in comparison to others.

social distance A distance between people that ranges from 4 to 12 feet; commonly used by people who work together or interact during sales transactions.

social exchange theory The idea that relationships seem worthwhile if the rewards are greater than or equal to the costs involved.

social judgment theory The theory that opinions will change only in small increments and only when the target opinions lie within the receiver's latitudes of acceptance and noncommitment.

social loafing The tendency of some people to do less work as a group member than they would as an individual.

social media Dynamic websites and applications that enable individual users to create and share content or to participate in personal networking.

social media bots Automated systems that generate and distribute social media posts.

social media trolls Individuals whose principal goal is to disrupt public discourse by posting false claims and prejudiced remarks, usually anonymously.

social penetration model A theory that describes how intimacy can be achieved via the breadth and depth of self-disclosure.

sociogram A graphic representation of the interaction patterns in a group.

sound bite A brief recorded excerpt from a longer statement.

space pattern An organizing plan in a speech that arranges points according to their physical location.

specific purpose The precise effect that the speaker wants to have on an audience. It is expressed in the form of a purpose statement.

speech act The purpose or intention of a communication episode (e.g., greeting, breaking up, kidding around).

stage hogs People who are overly invested in being the center of attention.

statistic Numbers arranged or organized to show how a fact or principle is true for a large percentage of cases.

stereotype A widely held but oversimplified or inaccurate idea tied to social categorization.

stonewalling Refusing to engage with the other person.

substitution The articulation error that involves replacing part of a word with an incorrect sound.

supportive listening The reception approach to use when others seek help for personal dilemmas.

survey research Information gathering in which the responses of a population sample are collected to disclose information about the larger group.

symbol An arbitrary sign used to represent a thing, person, idea, or event in ways that make communication possible.

sympathy Compassion for another's situation.

synchronous communication Communication that occurs in real time.

syntactic rules Rules that govern how symbols can be arranged (as opposed to the meanings of those symbols).

target audience That part of an audience that must be influenced to achieve a persuasive goal.

task-oriented listening A response style in which the goal is to secure information necessary to get a job done.

teams Groups of people who are highly successful at reaching clear, inspiring, and lofty goals.

territoriality The tendency to claim spaces or things as one's own, at least temporarily.

testimony Supporting material that proves or illustrates a point by citing an authoritative source.

thesis statement A complete sentence describing the central idea of a speech.

time pattern An organizing plan for a speech based on chronology.

topic pattern An organizing plan for a speech that arranges points according to logical types or categories.

trait theories of leadership A school of thought based on the belief that some people are born to be leaders and others are not.

transactional communication model A characterization of communication as the simultaneous sending and receiving of messages in an ongoing process that involves feedback and includes unintentional (often ambiguous) messages.

transformational leadership Defined by a leader's devotion to help a team fulfill an important mission.

transition A phrase that connects ideas in a speech by showing how one relates to the other.

trolling Attacking others via online channels.

truth bias The tendency to assume that others are being honest.

uncertainty avoidance The cultural tendency to seek stability and to honor tradition instead of welcoming risk, uncertainty, and change.

understanding The act of interpreting a message by following syntactic, semantic, and pragmatic rules.

value A deeply rooted belief about a concept's inherent worth.

visual aids Graphic devices used in a speech to illustrate or support ideas.

visualization A technique for rehearsal using a mental visualization of the successful completion of a speech.

vocal citation A concise, spoken statement of the source of your evidence.

wheel network A communication network in which a gatekeeper regulates the flow of information from all other members.

win–lose problem solving An approach to conflict resolution in which one party reaches their goal at the expense of the other.

win–win problem solving An approach to conflict resolution in which the parties work together to satisfy all their goals.

word chart A visual aid that lists words or terms in tabular form in order to clarify information.

working outline A constantly changing organizational aid used in planning a speech.

wraparound A brief introduction before a visual aid is presented, accompanied by a brief conclusion afterward.

Credits

CARTOONS

Page 10 Warren Miller/The New Yorker Collection/The Cartoon Bank.

17 CALVIN and HOBBES © 1994 Watterson. Distributed by UNIVERSAL PRESS SYNDICATE. Reprinted with permission. All rights reserved.

19 Mike Twohy/The New Yorker Collection/The Cartoon Bank.

36 Rina Piccolo Cartoon used with the permission of Rina Piccolo and the Cartoonist Group. All rights reserved.

41 Ron Barrett.

47 Mick Stevens/The New Yorker Collection/The Cartoon Bank.

56 Edward Frascino/The New Yorker Collection/The Cartoon Bank.

59 Edward Koren/The New Yorker Collection/The Cartoon Bank.

64 William Steig/The New Yorker Collection/The Cartoon Bank.

68 Peter Steiner/The New Yorker Collection/The Cartoon Bank.

88 Paul Noth The New Yorker Collection/The Cartoon Bank.

95 Malcolm Evans/Cagle Cartoons.

108 Drew Dernavich/The New Yorker Collection/The Cartoon Bank.

115 Leo Cullum/Cartoonstock.

130 Zits © 2003 Zits Partnership distributed by King Features Syndicat Inc.

134 DILBERT © 2009 Scott Adams. Used By permission of ANDREWS MCMEEL SYNDICATION. All rights reserved.

135 CALVIN AND HOBBES © 1995 Watterson. Reprinted with permission of ANDREWS MCMEEL SYNDICATION. All rights reserved.

145 Courtesy of Ted Goff.

165 DILBERT © 2006 Scott Adams. Used By permission of ANDREWS MCMEEL SYNDICATION. All rights reserved.

168 Alex Gregory/Cartoon Collections.

186 Leo Cullum The New Yorker Collection/The Cartoon Bank.

187 Leo Cullum The New Yorker Collection/The Cartoon Bank.

204 *Zits © 2006 Zits Partnership distributed by King Features Syndicate, Inc.*

215 *Crock © 2019 North American Syndicate, World rights Reserved.*

225 Leo Cullum The New Yorker Collection/The Cartoon Bank.

275 Sizemore, Jim/Cartoonstock.

281 Universal/Kobal/Shutterstock.

282 Dragon Images/Shutterstock.

338 DILBERT © 2006 Scott Adams. Used By permission of ANDREWS MCMEEL SYNDICATION. All rights reserved.

358 *By permission of Leigh Rubin and Creators Syndicate, Inc.*

TABLES AND FIGURES

Page 83 *Sources*: Cai, D. A., & Fink, E. L. (2002). Conflict style differences between individualists and collectivists. *Communication Monographs, 69*, 67–87. Croucher, S. M., Galy-Badenas, F., Jäntti, P., Carlson, E., & Cheng, Z. (2016). A test of the relationship between argumentativeness and individualism/collectivism in the United States and Finland. *Communication Research Reports, 33*(2), 128–136. Merkin, R. (2015). The relationship between individualism/collectivism. *Journal of Intercultural Communication, 39*, 4.

92 *Source*: J. Harwood. (2007). *Understanding communication and aging: Developing knowledge and awareness.* Newbury Park, CA: Sage, p. 76.

157 *Source*: Adapted from Stewart, J., & D'Angelo, G. (1980). *Together: Communicating interpersonally* (2nd ed.). Reading, MA: Addison-Wesley, p. 22. Copyright © 1993 by McGraw-Hill. Reprinted/adapted by permission.

165 *Source*: Based on material from Ekman, P. (1981). Mistakes when deceiving. In T. A. Sebok & R. Rosenthal (Eds.), *The Clever Hans phenomenon: Communication with horses, whales, apes and people* (pp. 269–278). New York: New York Academy of Sciences. See also Samhita, L., & Gross, H. J. (2013). The "Clever Hans phenomenon" revisited. *Communicative and Integrative Biology, 6*(6), e27122. doi:10.4161/cib.27122.

205 *Source*: Adapted from categories originally presented in Camden, C., Motley, M. T., & Wilson, A. (1984, Fall). White lies in interpersonal communication: A taxonomy and preliminary investigation of social motivations. *Western Journal of Speech Communication, 48*, 315.

217 *Source*: Adapted from Hess, J. A. (2002). Distance regulation in personal relationships: The development of a conceptual model and a test of representational validity. *Journal of Social and Personal Relationships, 19*, 663–683.

256 *Source*: Adapted from Rothwell, J. D. (1998). *In mixed company: Small group communication* (3rd ed.). Fort Worth, TX: Harcourt Brace, pp. 252–272. Reprinted with permission of Wadsworth, an imprint of the Wadsworth Group, a division of Thomson Learning. Fax 800-730-2215.

285 *Source*: Adapted from J. Brilhart & G. Galanes. (2001). Adapting problem-solving methods. In *Effective group discussion* (10th ed., p. 291). Copyright © 2001. Reprinted by permission of McGraw-Hill Companies, Inc.

323 Source: Adapted from "Outlining the Speech," Speech Program, University of Colorado at Colorado Springs. Available at https://www.uccs.edu/Documents/eberhardt/Outlining%20a%20Speech.doc.

360 choosemyplate.gov.

362 *Source:* Kim Goodwin Twitter post July 19, 2018 at https://twitter.com/kimgoodwin/status/102002957226643865 7?lang=en.

375 *Source*: Adapted from Andersen, M. K. (1979). *An analysis of the treatment of ethos in selected speech communication textbooks* (Unpublished dissertation). University of Michigan, Ann Arbor, pp. 244–247.

Index

Ability, coculture based on, 90
Abstraction ladder, 116
Abstract language, 116–18
Abusive partners, 223
Academic Search Premier, 302
Accenting behavior, 163
Accents, 87, 88, 109, 110, 168
Acceptance, 285, 373–74
Accommodating behavior, 160
Accommodation, 220, 221, 233
Accomplishments, 56, 251
Acculturation, 253
Accuracy, of group problem solving, 281
Acknowledgment, 143, 217
Acronyms, 115
Action zone, 309
Active strategies for learning about
    other cultures, 97
Activists (follower type), 259
Actors, in behavioral descriptions, 117, 118
Acts of service, 201
Actuating, 377
Addition, 311
ADHD (attention deficit hyperactivity
    disorder), 60
Ad hominem fallacy, 120, 382–83
Adichie, Chimamanda Ngozi, 62
Adjustment shock, 97–98
Admiration, of relational partners, 184
Adoption, 377
Advice, asking for, 129
Advising responses, 142, 143
Advocacy, with social media, 35–36
Affect, 160
Affect blends, 167
Affect displays, 160, 175
Affective communication style,
    121–22, 124
Affiliation, 112–13
Affinity, 186
Affirming language, 201
Afia, Nura, 60

Age:
    of audience members, 298
    cocultures based on, 90–92
    external influence on self-concept
        and, 54
    listening habits and, 134
    mediated communication use
        and, 41–42
    paralanguage and, 175
Aggression, 222–23, 233
Agreement, 142, 217
"Ain't I a Woman?" (Truth), 329
AirBnB, 109, 174
Alibaba, 266
All-channel networks, 254
Allocca, Kevin, 354–55
Altman, Irwin, 187
Altruistic lies, 205–6
Ambiguity:
    with laissez-faire leadership, 266
    in mediated messages, 32, 45
    of nonverbal communication, 158–60
    in relational messages, 185
    tolerance for, 94
American Sign Language, 157, 158
Analogies, 338
Analysis, with metacommunication, 187
Analytical listening, 141, 147–48
Anchor (in persuasion), 373
Anderson, Grant, 332–33
Anecdotes, 330, 338
Anonymity, online, 144
Anticlimactic organization, 327
Anxiety, 189–90, 249
Apathetic siblings, 196
Apollo 13 (film), 281
Appearance:
    nonverbal communication related to,
        168–71
    for selection interviews, 249–50
    during speech delivery, 308
    for video conferences, 260

Appiah, Kwame Anthony, 87
Application materials, 244–47
Appraisal, reflected, 54
Appreciation, 118, 184, 244
Appropriateness, 33, 45, 363
Approval, 14, 305
Areas of communication competence, 16
Argumentum ad populum fallacy, 384–85
Argumentum ad verecundiam fallacy, 384
Aristotle, 15, 378, 389
Arousal, 160, 232
Articulation, 311
Assertions, as questions, 145
Assertive communication, 222–27
    culture and, 234
    defined, 222
    process of, 223–25
    self-assessment, 226–27
Assumptions, 58, 145
Asterisks, 164
Asynchronous communication, 32,
    190, 287, 288
Attack-and-defend communication
    patterns, 215
Attending, 130, 138
Attention, 43, 217, 257, 260, 330
Attention deficit hyperactivity disorder
    (ADHD), 60
Attitudes:
    of audience members, 298–300
    defined, 299
    language as a reflection of, 111–13
    language in shaping of, 107–11
    positive, 306
Attractiveness:
    physical, 168, 169
    of visual aids, 362
Attribution, 62–65
Atypical Familia (blog), 27
Audience(s), 297–300
    action by, 381
    capturing attention of, 330

Audience(s) (*Cont.*)
  demographics of, 297–98
  desired response from, 381
  distasteful, 310
  effect of problem on, 380–81
  expectations of, 301
  imaginary, 91
  outlines for, 327
  for persuasive speeches, 385–86
  purpose of, 297
  referring to, 330
  for social media, 27
  target, 385
Audience analysis, 297–300, 388
Audience involvement, 356–58
Audience participation, 357–58
Audio files (for speeches), 359
Authenticity, 242
Authoritarian leadership, 265, 266
Authoritarian parents, 195, 196
Authoritative parents, 195, 196
Authority, appeal to, 384
Authority–obedience leaders, 267, 268
Authority rule, 286
Automattic, 287
Autonomy-connection dialectic, 203–4
Avoidance, 217, 220, 221
Avoidance spirals, 218
Avoiding stage of relationships, 199
Awareness, of nonverbal
      communication, 158

Babe, Ann, 81
Baby Boomers, 91
Baby-ish terms, 92
Bach, George, 222
Bachelor, The (television program),
      182, 189
Bachelorette, The (television program),
      182, 189, 207
Bandwagon appeals, 384–85
Barbarian Group, 173
Bar charts, 360
Barclay, Hunter, 320, 340–44
Basic speech structure, 320
Baxter, Courtney, 240, 243
Beauty and the Beast (film), 176
Beauty premium, 169
Begging the question, 384
Behavior:
  attribution of, 62–65
  in behavioral descriptions, 117, 118
  interpretation of, 59, 224
  referencing, in perception checking, 65
  self-concept and, 55–57
Behavioral descriptions, 117, 118, 224
Being-oriented friendships, 192
Being Wrong (Schulz) (book and
      TED talk), 357

Beliefs:
  of audience members, 298–300
  defined, 299
  language as a reflection of, 111–13
  language in shaping of, 107–11
  and self-concept, 53
  and self-fulfilling prophecy, 56–57
Bell, Kristen, 107
Bell, W. Kamau, 69–70
Biases:
  confirmation, 389
  deception, 165
  halo and horn effects, 63
  implicit, 61
  linguistic intergroup, 113
  negativity, 63
  self-serving, 63
  truth, 165
Bibliography (speech), 315, 322,
      341, 393–94
Bieber, Justin, 174
Billions (television program), 102
Biological cycles, perception and, 59
Bipolar disorder, 60
Biracial individuals, 61
Black Lives Matter movement, 35
Blake, Robert R., 266, 268
Blended families, 62, 196
Blind area (Johari window), 188
Blindspotting (film), 82
Blogs, 34–35
Boasting, 82
Body (speech), 355
Body art, 170–71
Body movements, 166–67, 309
Bona fide occupational qualifications, 248
Bonding stage of relationships, 198–99
Booth, Wayne, 322
Boroditsky, Lera, 108, 348, 355, 357–58,
      363–67
Bosses, difficult, 269 (*See also* Leaders)
Boyle, Julia, 339
Brainstorming, 230–31, 278, 279, 282
Breadth, of self-disclosure, 188
Breakout groups, 278, 279
Brevity, of workplace messages, 259
#BringBackOurGirls campaign, 35–36
Brown, Brené, 212–14, 216, 217, 220,
      229–30, 235
Bruhwiler, Barbara, 97
Buffett, Warren, 265, 354
Bullies and bullying, 39–40, 53, 227
Burnham, Bo, 74
Business at hand, focusing on, 83
"But" statements, 113
Bystanders, 45, 259

Cabalan, Shayla, 372, 378, 389–94
Cadman, Joseph, 242

Camaraderie, 259
Canons of Rhetoric (Aristotle), 15
Capitalization, in online
      communication, 164
Career success:
  and communication skills, 12
  LinkedIn use and, 42
  name and, 110
  setting stage for, 241–44
  vocal cues and, 169
CARE mnemonic, 328
Carson, Ben, 120
Caruso, Robert, 131
Casal, Rafael, 82
Casper (company), 275
Catastrophic failure, fallacy of, 305
Categorization, 62–63
Catfish (film and television program), 38
Cause-effect patterns, 328, 329
Cavoli riscaldati, 108
Cell phones, 14, 38, 44, 47
Chain networks, 254
Chakrabarty, Prosanta, 334
Chalkboards, 361, 362
Challenger explosion, 278
Challenges, interview questions about,
      250, 251
Chang, Lynn Chih-Ning, 97–98
Channels, 9, 30, 32, 41
Chapman, Gary, 201
Character, 388–89
Charisma, 389
Charts, 359–62
Chomsky, Noam, 104
Chronemics, 173–74
Circumscribing stage of relationships, 199
Circumstances, in behavioral
      descriptions, 117, 118
Cisgender individuals, 88
Citations, 339, 355–56
Civility, 43–45, 95, 117, 118
Clarification, supporting material for, 334
Clarity, 353–55, 379
Client satisfaction, 240
Climax patterns, 327
Clinton, Hillary, 117, 120
Clothing, 169–71, 249–50
CMM (coordinated management of
      meaning), 107
Cocultures:
  defined, 78
  nonverbal communication in, 176
  types, 86–92
  understanding, 78–81
Coercion, 373
Coercive power, 255, 256
Cognitive complexity, 16, 94
Cohesiveness, 276
Collaboration, 70, 281, 282

Collectivistic cultures, 56, 81–83, 234
Collins, Lauren, 94
"Color blindness," 7
Color of clothing, 170
Color words, describing people with, 7
Column charts, 360, 361
Comforting, 142–44
Comments, encouraging, 139
Commitment, 18–19, 281
Common ground, 45, 57, 385–86
Communication:
   in a changing world, 13–15
   characteristics of, 5–10
   defined, 5–8
   formats for team problem solving, 282
   misconceptions about, 19–21
   models of, 8–10
   reasons for, 121
   and relational partner selection, 185
   types of, 10–13
   unethical behaviors, 375
Communication and Mass Media
   Complete, 302
Communication apprehension, 304–7
   culture and, 82–83
   overcoming, 306
   sources of, 304–5
Communication climates, 214–22
   confirming and disconfirming
      messages in, 215–17
   defined, 214
   development of, 217–19
Communication competence, 15–19
   defined, 15
   with dissimilar individuals, 64
   elements of, 16–19
   and identity construction/
      management, 70, 73
   intercultural, 93–98
   with social media, 43–47
Communications, defined, 5, 6
Communication skills, 5, 12, 17
Communication technology:
   appropriate use of, 44, 45
   changes in, 13–15
   in conflicts, 220
   connection and, 218
   ending romantic relationships via, 200
   job interviews using, 252
   and nonverbal communication, 164
   offensive posts from the past, 69
   pace of, 29
   plagiarism on, 337
   social media listening, 131
   syntactic rules for, 104–5
   taking breaks from social media, 14
   trust development with, 288
   See also Mediated communication;
      specific types

Communicators, 9
Comparison-contrast statements, 338
Compensating behavior, 160
Competence, 149, 387, 388 (See also
      Communication competence)
Competition, 233, 276
Competitive cultures, 81–82, 86
Competitive siblings, 196
Complementary differences, 184
Complementing behavior, 161, 163
Comprehension checking, 358
Compromise, 228
Concepts, speeches about, 350
Concern, 64, 94
Conclusion (speech), 332–33, 355
Conferred power, 255
Confidence, 166, 168, 169
Confirmation bias, 389
Confirming messages, 216–17
Conflict, 210–37
   altruistic lies to avoid, 205
   and communication climate, 214–19
   cultural approaches to, 234–35
   defined, 213
   gender and, 232–34
   negotiation strategies, 225–31
   over meaning of symbols, 7, 8
   and self in culture, 83
   styles of expressing, 220–25
   understanding, 213–14
Conflict communication styles, 220–25
   assertiveness, 223–25
   direct aggression, 222, 223
   indirect communication, 221–22
   nonassertiveness, 220, 221
   passive aggression, 222
Conflict management, 212–13
   cultural influences on, 81, 84
   rules for "fighting fair," 216
Conflict resolution, 228–31
Conflict (storming) stage of groups, 283
Conformity, 195, 196, 278–79
Confusion, over abstract language, 116–17
Connection:
   eye contact and, 166
   mediated communication and, 189, 218
   social media and, 34, 37
Connection-autonomy dialectic, 203–4
Connection power, 255
Connotative meanings, 103
Conscious identity management, 70–71
Consensus, 286
Consequence statements, 224
Constructive criticism, 142
Contact frequency, in friendships, 192
Contact hypothesis, 94, 195
Contempt, 215
Content:
   of gender-based language, 121

informative speaking by, 350
   user-generated, 27
Content messages, 185–86
Contradicting, 163
Contrasting stimuli, 58
Control, 158, 186, 257
Convergence, 113
Conversational narcissism, 136, 137
Convincing, 377
Cook, Ethel, 335
Cooper, Anderson, 88
Cooperative cultures, 81–82, 86
Coordinated management of
      meaning (CMM), 107
Core values, 253
Counterfeit questions, 145
Country club leaders, 267–68
Cover letters, 244, 246, 247
Covey, Stephen, 128–29
Cox, Laverne, 54
Craft of Research, The (Booth et al.), 322
Credibility:
   critical listening and, 149
   defined, 386
   ethos-based appeals, 378
   persuasive speaking and, 386–89
   public communication and, 332
   of websites, 302
Criminals, postural clues used by, 166
Critical listening, 148–49
Criticism, 142, 215
Cronen, Vernon, 107
Cross-cultural friendships, 189–90
Cruz, Ted, 120
Cuddy, Amy, 156, 177
Culpepper, Scott, 118
Culture, 77–99
   of audience members, 297
   communication competence
      and, 93–98
   conflict styles and, 234–35
   defined, 78
   emblems in, 161
   eye contact and, 167
   facial expression and, 167
   frame switching and, 71
   friendships and, 189–90, 192
   information and, 352
   interpretation and, 106
   listening and, 134
   multicultural teams, 281–82
   names as indicators of, 109
   nonverbal communication and, 174–76
   persuasion and, 383
   powerful language use by, 112
   respecting differences in, 257
   self-concept and, 53, 55–56
   time and, 174
   understanding, 78–81

Culture (*Cont.*)
  values and norms in, 80–86, 233–34
  See also Cocultures; Diversity
Culture shock, 97–98
Currency, of evidence, 382
Cyberbullying, 39–40, 53
Cyber relationships, 190–91
Cyberstalking, 39–40

Dailey, Spike, 81
Damage control, online, 242
"Danger of a Single Story, The"
    (Adichie), 62
Danson, Ted, 107
Databases, 302
Davis, Brian, 356
Debilitative communication apprehension,
    304–7
Deception:
  as distancing tactic, 217
  and ethical persuasion, 375
  with lies and evasions, 205–7
  nonverbal communication and, 163–65
  via social media, 38
  See also Honesty
Deception bias, 165
Decoding, 8
Defensive listening, 135, 136
Defensiveness, 216, 217, 358
Definitions, 336
Degrees of communication
    competence, 16
Degrees of power, 255
DelhiGangRape movement, 35
De Klerk, F. W., 270–71
Delegation, 267, 268
Deletion, 311
Delivery (speech), 308–12
  auditory aspects of, 310–12
  types of, 307–8
  visual aspects of, 308–10
Democratic leadership, 266
Demographics, audience, 297–98
Demonstrations, 381–82
Denial, of truth, 39
Denotative meanings, 103
Depth, of self-disclosure, 188
Derived credibility, 386
Deschanel, Zooey, 175
Descriptions, 350, 380, 381
Desire to work for company, questions
    about, 251
Detachment, 217
Developmental model (of relationship
    maintenance), 197–200
Developmental stage, 59
Devil's advocate, 279
Dewey, John, 283
Diagrams, 359

Dialectical model and tensions, 203–5
Dialects, 102, 311
Dialogue, 279–80
Dickinson, Alexandra, 227
Dictionary.com, 353
Diehards (follower type), 259
Differentiating stage of relationships, 199
Difficult tasks, group problem
    solving for, 285
Diggs, Daveed, 82
Digital infidelity, 190
Digital natives, 41
Dillon, Asia Kate, 102
Directly aggressive message, 222, 223
Direct persuasion, 377
Direct rewards, 381–82
Direct speech, 81
Disability(-ies):
  cocultures based on, 79, 90
  communicating with people with, 91
  International Symbol for Access, 7, 8
Disagreement, respectful, 278–79
Disclaimers, 112
Disclosures, 143, 192 (*See also*
    Self-disclosure)
Disconfirming messages, 215–17
Discontinuance, 377
Discounting, 217
Discrimination, name-based, 109
Discussione, 235
Discussions, problem solving, 277,
    279–80, 285
Disengagement, 216
Disfluencies, 168
Disinhibition, 44
Disrespect, 217
Disruptive language, 117–20
Distance, 85, 171–73
Distortions, 39
Distractions, 134, 135, 190
Divergence, 113
Diversion, 143
Diversity:
  of American dialects, 311
  of audience members, 297
  communicating with people who have
    disabilities, 91
  cultural differences in persuasion, 383
  in emotional expressiveness, 235
  friendship and gender, 195
  group, 281–82
  information and culture, 352
  language and worldview, 108
  language for describing people, 7
  of multicultural teams, 282
  nontraditional organization
    patterns, 329
  in online communication networks, 189
  in other-sex friendships, 193, 195

sexist assumptions in everyday
    language, 60
  see also Culture
Division, rule of, 325
Divorce, 215
Doing-oriented friendships, 192
Donnelly, Marisa, 191
Dourado, Phil, 241
Dress rehearsals, for interviews, 252
Driving, cell phone use while, 47
Dunbar, Robin, 36
Dunbar's number, 36
Dyadic communication, 11
Dynamism, 389
Dysfunctional roles, 274, 275

Education:
  cocultures based on level of, 92
  communication for, 5
  distance in, 172–73
  and problem solving by multicultural
    teams, 282
Effort, communication competence
    and, 17
Eggleston, T. Stephen, 335
Ego, 14
EI (emotional intelligence), 65–67, 156
Eighth Grade (film), 52, 74
Either–or fallacy, 225, 383
Electronic media, 26
"Eleven and Engaged" (Cabalan), 389–94
Ellen DeGeneres Show, The (television
    program), 191
Ellis, Albert, 305
Email:
  in all-channel networks, 254
  ending romantic relationships over, 200
  for group problem solving, 287
  richness of, 158
  as verbal communication, 157
  in workplace, 260
Emblems, 161
Emergence (norming) stage
    of groups, 283
Emergencies, planning for, 286
Emergent leader, 265
Emoji, 159, 164
Emoticons, 164, 175
Emotional appeals, 149, 378
Emotional expressiveness, 121–22,
    201, 235
Emotional intelligence (EI), 65–67, 156
Emotions:
  and analytical listening, 147
  describing, 224
  emotional tone of relationship
    (*see* Communication climate)
  empathy and, 64
  granularity of, 108–9

group problem solving and, 285
nonverbal communication and, 159–61, 167, 175
over involvement with others', 137
perception and, 57–59
and task-oriented listening, 145
unexpressed, 138–39
Emotive language, 119–20
Empathy, 16
  defined, 64
  dimensions of, 64–65
  emotional intelligence and, 65–67
  intercultural communication and, 94
  perception and, 64–67
Empowerment, 269, 271
Encoding, 8
Encouragement, 264
Endearing terms, 92
Endorsement, 217
Entertainment, social media for, 28
Environments, 9, 106, 173
Epley, Nicholas, 169
Equivocal words, 114
Equivocation, 114, 206
Escalatory conflict spirals, 218
Estrogen, 122
Ethical persuasion, 374–75, 385
Ethics:
  adapting to speaking situations, 299
  appearance and impression management, 170
  civility when values clash, 95
  communicating vs. not communicating, 20
  conflict management, 221
  digital infidelity, 190
  distasteful audiences for speeches, 310
  freedom of speech and hate speech, 111
  group participation inequality, 277
  honesty on social media, 39
  illegal interview questions, 249
  indirect speech, 116
  negotiating with bullies, 227
  online anonymity, 144
  stereotyping, 55
  unethical communication behaviors, 375
Ethnicity, 7, 61, 87
Ethnocentrism, 95, 352
Ethos, 378
Euphemisms, 115
Evasions, 206
Events, speeches about, 350
Evidence, 149, 382
Exaggerations, 39
Examples, 336
Expectancy violation theory, 160
Expectations:
  of audience, 301

behavior and, 57, 58
gendered, 60–61
listening and, 134
organizing speech based on, 386
and self in culture, 83
team problem solving and, 282
Experimenting stage of relationships, 197
Expert opinion, 286
Expert power, 254–56
Explanations, 248, 350, 381
Expletives, 106–7
Expressed struggle, 213
Extemporaneous speeches, 307
External noise (physical noise), 8–9, 133
Eye contact, 167, 175, 252, 309–10

Fabrications, 39, 206
Face, 68, 71, 221
Facebook, 15, 33, 34, 36, 41, 45, 71, 164
FaceTime, 35
Face-to-face communication, 10
  face management in, 71
  as interpersonal communication, 189
  mediated and, 29–33, 190
  with people of different backgrounds, 94
  self-esteem and, 68, 69
  time spent in, 46
Facework, 68, 71
Facial expressions, 166, 167, 309
Facilitative communication apprehension, 304
Fact(s):
  citing startling, 330
  inferences vs., 118–19
  opinions presented as, 119–20
  propositions of, 376
Factiva, 302
Factual examples, 336
fake news, 90
Fallacies:
  with communication apprehension, 305
  and emotional appeals, 149
  and personal attacks, 120
  in persuasive speaking, 382–85
  and win–lose problem solving, 225
Fallacy of approval, 305
Fallacy of catastrophic failure, 305
Fallacy of overgeneralization, 305
Fallacy of perfection, 305
False cause fallacy, 383, 384
False friends, 105
Familiarity:
  capturing attention with familiar references, 330
  gravitating to the familiar, 63–64
  in informative speeches, 353, 354
  and interpretation of behavior, 58
  and organization of speech, 327
Family, 195

Family relationships, 195–97
Faulty analogies, 384
Feedback, 9–10, 56, 91, 130–31
Feelings (See Emotions)
Feminine, 60, 88
Feynman, Richard, 147
Feynman Technique, 147–48
Fiber-optic communication technology, 14
Fidgeting, 166
Fifth Harmony, 132
Figure–ground principle, 58
First impressions, 62, 185
Fisher, Aubrey, 283
Fisher, Elsie, 52, 74
Fisher, Roger, 228
Flip pads, 362
Flirting, 201
Flow charts, 360, 362
Focus, 14, 175
Foggy mirror concept, 74
Follower (role), 257–59
Follow up:
  after interviews, 251, 252
  in group problem solving, 287
  on solution to conflict, 231
Force field analysis, 284, 285
Forgotten, right to be, 33
Formality, 250
Formal outlines, 322
Formal roles, 273–74
Forming stage of groups, 283
48 Things Men Hear in a Lifetime (That Are Bad for Everyone) (video), 60
48 Things Women Hear in a Lifetime (That Men Just Don't) (video), 60
Forwarding emails, 260
Four Horsemen of the Apocalypse, 215–16
Frame switching, 71
Freedom of speech, 111
Friendship, 191–95
  communication and, 16
  culture and, 93
  development of, 191–93
  gender and, 193, 195
  listening and, 129
  mediated communication and, 30, 34, 36
  qualities of, 191
Functional roles, 274
Fundamental attribution error, 63
Fundraising, 36

Galileo Galilei, 384–85
Gallo, Amy, 265
Game of Thrones (television program), 232
Gandhi, Mahatma, 265
Gatekeepers, 26–27, 254
Gates, Bill, 265, 271, 337
Gates, Melinda, 271

Gay individuals (*See* LGBTQ individuals)
Gender:
   of audience members, 297–98
   conflict styles and, 232–34
   defined, 60, 61
   empathy development and, 64–65
   friendship and, 193, 195
   intimacy styles and, 201–3
   language and, 108, 120–24
   mediated communication and, 40–41
   nonverbal communication and, 170–71,
     176–77
   powerful language use and, 112
   references to, in language, 110–11
   stereotypes related to, 60–61
   supportive listening and, 143–44
Gender binary individuals, 88
Gendered expectations, 60–61
Gender identity, 61, 88
Gender matrix, 60–61
Gender nonbinary individuals, 54, 88, 102
Gender roles, 86, 121
Generalizations, 384
General purpose, 295
Generation:
   cocultures based on, 90–92
   listening habits and, 134
   mediated communication use and,
     41–42
   paralanguage use and, 175
Generation Z, 79
Gergen, Kenneth, 6
Gieseck, Alyssa, 338
Gifts, 201
Goals:
   behavioral descriptions of, 118
   communication to accomplish, 5
   group, 272–73
   individual, 272–73
   interview questions about, 251
   online identity management and, 242
   perceived incompatibility of, 213–14
   perception and, 58
   personal, 58, 72, 73
   relational, 266–69
   shared, 276
   task, 266–69
Goffman, Erving, 68
Gofundme, 36
Golden rule, 257
Goldilocks principle, 188
Goleman, Daniel, 65, 67
Good, inherent, of communication, 19
Good Doctor, The (television program),
   79, 80
Good Place, The (television program), 107
Google, 242, 302, 337
Google Scholar, 302
Gossip, 257

Gottman, John, 215–16
Goulston, Mark, 128
Graham, Katharine, 331, 333
Grammar, 92, 259
Greatest weakness question, 248
Great man or woman approach to
   leadership, 266
Greenwood, Heather, 61
Griffith, Eric, 14
Group(s), 271–88
   audience membership in, 298
   followership in, 257–59
   individual roles in, 273–75
   leadership in (*see* Leadership)
   methods for acquiring power in, 256
   motivational factors in, 272–73
   reaching decisions in, 286
   rules in, 273
   sociograms for analyzing, 253–54
   as teams, 271–72
Group goals, 272–73
Group problem solving, 280–88
   advantages of, 281–83
   decision-making methods in, 286
   developmental stages in, 283
   in special circumstances, 285
   structured approach to, 283–87
   in virtual groups, 287–88
Group productivity, 275–80
   cohesiveness and, 276
   discussion format use, 279–80
   in meetings, 276–79
Groupthink, 278
Grumstrip, Christen, 14
Guiltmakers, 222
Guo, Winona, 78

Hahn, Melissa, 83
Haidt, Jonathan, 379
Half, Robert, 240
Hall, Edward T., 171, 172, 175
Hallum, Elizabeth, 328, 329
Halo effect, 63
Hamilton (musical), 112
Hamilton, Josh, 52
Hancox, Lewis, 88
Handouts, 362
Haptics, 171, 172
Harassment, 39–40, 223, 294
Harris, Jillian, 207
Harvard Negotiation Project, 228
Hasty generalizations, 384
Hate speech, 111
Health, perception and, 59
Health and helping professions, 172, 173
Hearing, 130–32
Hedges, 112
Hegemony, 95
Heitler, Susan, 120

Help, offering, 142
Helping style, 137–38
Hesitations, 112
Hidden agendas, 145
Hidden area (Johari window), 188
High-context cultures:
   communication in, 83–84
   conflict communication in, 234, 235
   friendships in, 192
   listening in, 134
High-disclosure friends, 192
Highmore, Freddie, 80
High-obligation friends, 192
High self-monitors, 71
Hill, Angela, 260
Hill, Nathan, 336
Hinting, 221–22
Hip-hop, 112
Hiring, 129, 243
Honesty, 38–39, 73–74, 164–65, 388
   (*See also* Deception; Lies)
Hormones, communication differences
   and, 122
Horns effect, 63
Hostile siblings, 197
Hostility, neutralizing, 386
"How Language Shapes the Way We Think"
   (Boroditsky), 363–67
Humblebragging, 73
Humility, 82
Humor:
   capturing attention, 330
   insensitive, 256
   personalizing speeches with, 357
   self-deprecating, 68
   in speeches, 300
Humoring, 217
Hussein, Yasmin, 89
Hyperpersonal communication, 30
Hypothetical examples, 336

"Ideal" communication, 16
Identity(-ies):
   interview questions about, 251
   multiple, 69–70
   name as part of, 109
   professional, 242
Identity-first language, 90
Identity management, 67–74
   characteristics of, 69–72
   communication for, 4
   defined, 67
   honesty and, 73–74
   in interviews, 252
   with mediated communication, 30, 32,
     241–43
   nonverbal communication and, 160–61
   public and private selves in, 67–68
   reasons for, 72–73

on social media, 28, 68–69
in the workplace, 72
Identity needs, 353
"I Forgot My Phone" (video), 38
"I Have a Dream" (King), 329
Illegal interview questions, 248–49
Illustrators, 163
Image management, 205
Imaginary audience, 91
Immediacy, 186
Immigrant rights, 294, 313–15
Impartiality, 149, 388
Impersonality, 183, 217
Implicit bias, 61
Impoverished leaders, 266–67
Impression management, 73–74, 170
Impromptu speeches, 308
Inappropriate content, posts with, 33, 45
Inattention, 217
Incremental persuasion, 373–74
Indirect communication, 221–22, 233
Indirect persuasion, 377–78
Indirect speech, 81, 116
Individual goals, 272–73
Individualistic cultures, 56, 81–83, 234
Individual responsibilities, 286
Individual roles, in groups, 273–75
Inferential statements, 118–19
Infidelity, digital, 190
Informal roles, 274
Information:
    analytical listening for, 147–48
    culture and method of conveying,
        83, 352
    gathering, for speech, 301–3
    for group problem solving, 284
    sharing, 68
    social media use for, 28
    value of, 148
Informational interviews, 243, 244
Information anxiety, 348, 353
Information hunger, 353
Information overload, 278, 348, 353
Information power, 256
Information underload, 278
Informative purpose statement, 351–52
Informative speaking, 347–69
    indirect persuasion with, 378
    persuasive speaking vs., 351
    sample speech, 363–67
    supporting material for, 355–63
    techniques of, 351–55
    types of, 349–50
In-groups, 80, 113
Initial credibility, 386
Initiating stage of relationships, 197
In-person friendships, 192–93
Input, asking for, 277
Insensitive humor, 256

Insensitive listening, 136, 137
Inside Out (film), 167
Inspiration, 241
Instant intimacy, 257
Instant messaging, 200, 287
Instructions, 247, 350
Instrumental communication style,
    122, 124
Insulated listening, 136, 137
Integrating stage of relationships, 198
Integrity, 257, 299, 388–89
Intensifiers, 112
Intensifying stage of relationships,
    197, 198
Intensity of stimuli, 58
Intention statements, 225
Interaction:
    altruistic lies to continue, 205
    persuasion and, 374
    shortening of, 217
    in small groups, 272
    on social media, 27
    workplace patterns of, 253–54
Intercultural communication, 9, 93–98
Interdependence, 213, 272, 276
Interest(s):
    focusing on, in negotiations, 228
    supporting material to add, 335
Intergroup communication, 79
Internal conflict, 213
Internal motivation, 67
International Symbol of Access, 7, 8
Internet:
    evaluation of websites, 302
    impact of, 14, 15
    plagiarism and, 337
    social isolation with overuse of, 37
    speech research on, 302
    See also Communication technology;
        Mediated communication
Interpersonal communication, 181–209
    characteristics of, 182–90
    choosing relational partners for, 183–85
    communication climates in
        (see Communication climates)
    conflict in (see Conflict)
    defined, 11, 183
    in family relationships, 195–96
    in friendships, 191–95
    lies and evasions in, 205–7
    and masspersonal communication,
        28, 29
    online, 189–90
    and relational dialectics, 203–5
    in romantic relationships, 197–203
    self-disclosure in (see Self-disclosure)
Interpretation, 59, 224
Intersectionality theory, 87
Interviews:

gathering information for
    speeches with, 302–3
informational, 243, 244
job (selection), 56, 73, 247–52
Intimacy, 30, 37, 197, 201–3, 257
Intimate distance, 172
Intrapersonal communication, 10–11
Introduction (speech), 330–32, 354–55
Intros, 359
Inverted pyramid format, 335
Invisible stigmas, 72
Irrational thinking, 305, 306
Irreplaceable relationships, 183
Isaacs, William, 280
Isaacson, Walter, 144
Isolates (follower type), 258
Isolation, 37
"I" statements, 113
It Gets Better Project, 89
"It" statements, 113

Jackson, Jesse, Sr., 339
James, William, 57
Jane the Virgin (television program), 79
Jargon, 113, 115
Je Suis Charlie movement, 35
JetBlue, 257
Job applicants, 244–52
    application material preparation,
        244–45
    participation in interviews, 249–52
    planning for interviews, 247–49
Job application materials, 244–45
Job interviews (selection interviews):
    culture and communication in, 56
    humblebragging, 73
    participating in, 249–52
    planning for, 247–49
Jobs, Steve, 33
Johari Window, 188–89
Jokers, 222
Jokes (See Humor)
Jongintaba, 280–81
Jordan, Michael, 271
Judging response, 141–42
Judgments, 142, 224
Just Listen (Goulston), 128

Kailey, Matt, 55
Kardashian, Kim, 175
Kellam, Jim, 241
Kellerman, Barbara, 258, 259
Kelley, Robert, 258
Kennedy, Robert, 386
Key ideas, 145
Keynote (software), 360, 361
Khan, Sadiq, 87
Kihn, Susan, 244
Kim, Young Yum, 98

Kinesics, 166
King, Martin Luther, Jr., 329
Knapp, Mark, 197, 198
Knowledge, 58, 97
Koehler, Carol, 328
Koi no yokan, 108
Kordei, Normani, 132
Kotelnikov, Vadim, 335

La Guardia, Fiorello, 174–75
Laissez-faire leadership, 266, 267
Language, 101–25
    defined, 102
    for describing people, 7
    disruptive, 117–20
    gender and, 120–24
    in informative speaking, 353–54
    of love, 201, 202
    nature of, 102–7
    power of, 107–13
    troublesome, 114–17
Larson, Doug, 129
Late arrival, for interviews, 249
Latitudes of acceptance, 373–74
Latitudes of noncommitment, 373–74
Latitudes of rejection, 373–74
Laughter, 157
Lawrence, Jennifer, 304
Leaders:
    authority–obedience, 267, 268
    country club, 267–68
    effective, 265
    emergent, 265
    impoverished, 266–67
    middle-of-the-road, 267, 268
    team, 267, 269
Leadership, 265–71
    approaches to, 265–66, 270
    characteristics of effective leaders, 265
    as encouragement, 264
    by good communicators, 240
    listening and, 129
    potential for, 265
    situational, 266–69
    trait theories of, 266
    transformational, 269–71
Leading questions, 145
Leanness, message, 30
Learning mindset, 45
Leave taking, altruistic lies for, 205
Legitimate power, 254–56
Leonard, Rachel, 68
Lesbians (*See* LGBTQ individuals)
"Lessons of Cultural Intimacy"
    (Guo and Vulchi), 78
Letters of recommendation, 241
Levine, Robert, 173
LexisNexis, 302
LGBTQ individuals:

clothing options for queer women, 170
cocultures of, 87–89
gender differences in nonverbal
    communication for, 177
gender-diverse friendships for, 193, 195
nonverbal communication in
    media by, 176
Library research, 302
Lies, 205–7 (*See also* Deception; Honesty)
Life satisfaction, attractiveness and,
    168, 169
Lincoln, Abraham, 305
Linear communication model, 8–9
Line charts, 360, 361
Lines That Make Us, The (Vass), 4
Linguistic intergroup bias, 113
Linguistic relativism, 108, 109
LinkedIn, 42, 243
Listening, 127–53
    as acknowledgment, 217
    analytical, 147–48
    critical, 148–49
    defined, 130
    faulty habits in, 135–36
    gender and, 143–44
    in high- vs. low-context cultures, 83
    to informative speeches, 353
    misconceptions about, 129–32
    as natural process, 132
    overcoming challenges to effective,
        133–34
    relational, 137–43
    supportive, 137–44
    task-oriented, 144–47
    types of, 137–49
    value of, 128–29
Listening fidelity, 130
Literacy rates, 14
Location, interview, 249
Logos (logical reasoning), 378–79
Loneliness, 34
Long-distance romantic
    relationships, 35
Longing siblings, 196
Long-term friends, 192
Lookism, 170
Lose-lose problem solving, 228
Love languages, 201, 202
Low-context cultures, 83–84, 192, 234
Low-disclosure friends, 192
Lowe, Sean, 182
Low-obligation friends, 192
Low self-monitors, 71
Luo, Bin ("Robin"), 93

Ma, Jack, 266
Mahoney, Brianna, 300
Main points of speech, 324–26, 330–31
Maintenance roles, 274

Majority control, 286
Managerial grid, 266–69
Mandela, Nelson, 264–66, 269–71,
    280–81, 288
Manipulators, 166
Manners, 72, 250
Manuscript speeches, 308
Marital satisfaction, 34, 37
Mars-Venus metaphor of gender
    differences, 120–21
Martin, Courtney E., 240
Masculine, 60, 88
Mass communication, 12, 26–29
Mass media, 26–29
Masspersonal communication, 28, 29
McCarley, Evan, 339
McKibban, Allison, 303
Meaning(s):
    in communication, 20
    coordinated management of, 107
    in language, 103–4
    of nonverbal communication, 158–59
    of symbols, 7, 8
Media, 26–29
    culture shaped by, 79, 95
    defined, 9
    listening challenges caused by, 133
    nonverbal communication in, 176–77
    uses and gratifications theory, 27–28
Mediated communication:
    attribution in, 62
    in conflicts, 220, 233–34
    convergence in, 113
    digital infidelity, 190
    ending relationships via, 200
    face-to-face vs., 29–33
    friendship quality and, 192–93
    influences on, 40–42
    interpersonal relationships in, 189–90
    meeting a date in person, 199
    metacommunication in, 187
    nonverbal expressiveness for, 164
    oversharing on, 256–57
    with people of different backgrounds, 94
    perception in, 59
    preference for, 37
    in professional environments, 259–60
    self-esteem and, 68, 69
    time spent on, 46
    for virtual groups, 287–88
    See also Communication technology;
        specific types
Meetings, productivity in, 276–79
Mehta, Aparna, 335
Meltzer, Bernard, 213
Memorable points, supporting materials
    for, 336
Memorized speeches, 308
Mental health, social media use and, 37

Message(s):
    confirming, 216–17
    content, 185–86
    differences in reception of, 132
    directly aggressive, 222–23
    disconfirming, 215–17
    evaluating, 149
    in linear model, 8
    nonverbal communication of, 158
    for persuasive speaking, 379–85
    relational, 185–86
    residual, 132
    richness of, 29–32
    separating the speaker from, 148
    unexpressed, 138–39
Message overload, 135
Metacommunication, 186–87
Metaphors, 338
MeToo movement, 35
Microaggressions, 61
Micromanagement, 268
Middle-of-the-road leaders, 267, 268
Military personnel, space for, 173
Millennials, 79, 91, 253
Mindful listening, 132, 133
Mindful nonverbal communication, 159
Mindless listening, 132
Minions, 104
Minnow, Martha, 195
Minor conflicts, 221
Mirroring, 166
Misinformation, on social media, 45–46
Misrepresentation, 38–39, 73–74
Mission, 269
Mistaken identity, online, 242
Mistakes in workplace communication,
    256–57
Misunderstandings, 104, 114–17
Mizner, Wilson, 136
Mnemonics, 328
Models, as visual aids, 359
Moderation, 190
Molinsky, Andy, 83
Momenian, Donya, 7
Money, as scarce resource, 214
Monochronic cultures, 174
Monroe, Alan, 329
Mood, 159, 331–32
Morale, 266
Moroney, Francesca, 144
Morris, Desmond, 204
Motivated Sequence, 329–30
Motivation, 67, 269, 272–73
Mouton, Jane S., 266, 268
Mudita, 108
Mullen, Chris, 14
Mullenweg, Matt, 287
Multicultural teams, 281–82
Multitasking, 134, 260

Murphy's Law, 286
Muslims, 89
Mutually satisfying answers, 214
Myatt, Matt, 240
My Fair Lady (musical), 109

Naddaff, Tricia, 275
Name(s):
    culture and, 81
    online presence of people
        with same, 242
    shaping of beliefs and attitudes by,
        109, 110
    and symbols in language, 103
Name calling, 120
Namesakes, 109
Narration, 339
Narratives, 61–62
Needs, 229–30, 353
Negativity bias, 63
Negotiation strategies, 225–31
    compromise, 228
    lose-lose problem solving, 228
    for use with bullies, 227
    win-lose problem solving, 225–27
    win-win problem solving, 228–31
Nervous energy, 306
Networking, 242–44
Networks, communication, 189, 254
Neurological conditions, 60
NeverAgain movement, 35
Niemiec, Ewa, 235
Nivea, 10
Noise, 8–9, 133–34
Nominal group technique, 284, 285
Nominal leader, 254
Nonassertion, 220, 221
Noncommitment, latitude of, 373–74
Nondefensive listening, 136
Nonimmediacy, 217
Nonsexist language, 111
Nontraditional interview questions,
    247–48
Nontraditional organization patterns, 329
Nonverbal communication, 10, 155–79
    characteristics of, 157–61
    and deception via social media, 38
    defined, 156–57
    demonstrating charisma with, 389
    of direct aggression, 222, 223
    for expressing contempt, 215
    functions of, 161–65
    influences on, 174–77
    in interviews, 250
    and message richness, 29–33
    and relational messages, 186
    in supportive listening, 139
    trust building with, 288
    types of, 166–74

Nonverbal cues, 94
Nonverbal learning disorder (NVLD), 159
Nordstrom, 93
Norming stage of groups, 283
Norms:
    cultural, 80–86, 233–34
    group, 273
    social, 67–68, 72, 273
Notes for speaking, 324
Note-taking, 146–47
#NotInMyName movement, 89
Novel solutions, from multicultural
    teams, 282
Novelty-predictability dialectic, 204
Number charts, 359–60
Nutrition, perception and, 59
NVLD (nonverbal learning disorder), 159
Nye, Bill, 45

Obama, Barack, 69
Objections, to illegal questions, 248
Objects, 350, 359
Occasion, for speech, 301, 330
Ody, Britton, 298
Offensive language, 106
Ogden, C. K., 103
Older adults, 90–92
Omissions, 39, 206
Online communication (See Mediated
    communication)
Online dating, 34, 199
Online networking, 243
Online research, 302
Online support groups, 35
Online surveillance, 39
Open area (Johari window), 188
Open-mindedness, 94–95, 97
Openness, 184, 204
Opinion(s), 119–20, 286, 330
Orange Is the New Black (television
    program), 54
Organization:
    of informative speeches, 354
    of outlines, 326–30
    in perception process, 58–59
    of persuasive speech, 386
Organizational communication, 11–12,
    252–56
Organizational culture, 72, 92, 93, 253
Orientation (forming) stage of groups, 283
OscarsSoWhite movement, 35
Other-sex friendships, 193, 195
Outcome categorization of persuasion, 377
Outlines, 324–30
    formal, 322
    organization of, 326–30
    principles of, 324–26
    samples, 321, 322, 340–41, 390

Outlines (*Cont.*)
  standard format for, 325
  standard symbols used in, 324–25
  working, 321
Outros, 359
Overcommunicating, 20
Overgeneralization, 63, 305
Oversharing, 256–57

Pang, Samarnh, 137
Paralanguage, 168, 175
Parallel wording, rule of, 325
Paraphrasing, 339
  of audience questions, 358
  in conflict resolution, 230
  in supportive listening, 138–39
  in task-oriented listening, 145–46
Parents and parenting relationships, 61,
  195–96
Parks, Rosa, 265
Parroting, 146
Participants (follower type), 259
Participation, meeting, 277
Participative decision making, 281
Passive aggression, 222
Passive narcissists, 136, 137
Passive observation, 97
Pathos, 378–79
Patience, 97–98
Pearce, Barnett, 107
Pease, Jonathan, 131
Perceived self, 67
Perception, 57–67
  common tendencies in, 62–64
  defined, 57
  empathy and, 64–67
  factors affecting, 58
  influences on, 59–62
  language and, 108
  listening and, 132
  narratives and, 61–62
  in self-concept, 52
  steps in, 57–59
Perception checking, 45, 62, 65, 119, 216
Perfection, 185, 305
Performing stage of groups, 283
Permanence, of mediated communication,
  32–33
Permissive parents, 195, 196
Perry, Katy, 176
Perseverance, 97–98
Personal attacks, 120
Personal distance, 172
Personal experience, 58, 387
Personal fables, 91
Personal factors, in listening, 137–38
Personal goals, 58, 72, 73
Personal identity (*See* Identity
  management)

Personality, 53
Personalization, 357
Personal relationships (*See*
  Relationship(s))
Person-centered language (person-first
  language), 90, 103–4
Perspective taking, 64
Persuasion, 5, 171, 373–75
Persuasive appeals, types of, 378–79
Persuasive speaking, 371–96
  adapting, to audience, 385–86
  categorizing types of, 376–79
  characteristics of, 373–75
  creating the message, 379–85
  credibility in, 386–89
  informative speaking vs., 351
  sample speech, 389–94
Petrocelli, Tom, 131
Phone interviews, 252
Phonological rules, 104
Photos (visual aids), 359
Phubbing, 190
Physical attractiveness, 168, 169
Physical needs, 353
Physical noise, 8–9, 133
Physical qualities, in self-concept, 53
Physical touch (*See* Touch)
"Physician Suicide" (Barclay), 340–44
Physiological influences:
  on conflict style, 232
  on perception, 59–60
Physiological noise, 9, 134
Pictograms, 360
Pie charts, 360
Pitch, voice, 92, 311
Place, for speech, 301
Placebo effect, 57
Plagiarism, 337
Plan implementation, 286
Platinum rule, 45
Platonic other-sex friendships, 193, 195
Policy, propositions of, 376–77
Polite forms, 112
Political viewpoints, 89–90
Polychronic cultures, 174
Polymediation, 32
Polymer marking surfaces, 361, 362
Pope, Alexander, 339
Porter, Richard, 352, 383
Portfolio, work, 249
Position power, 254–56
Positive spirals, 218
Poster boards, 362
Post hoc fallacy, 383, 384
Posture, 166, 309
Potter, Jennifer, 89
Poundstone, Paula, 136
Power:
  defined, 254

distribution of, 255
of language, 107–13
nonverbal communication of, 156
win-lose conflicts and, 225, 226
in workplace, 254–56
Power distance, 85
Powerful language, 111–12, 122
Powerless language, 111–12, 122
PowerPoint, 360, 361, 363
Power Rangers (television program), 79
Practical needs, 353
Practice, for speech, 302, 308
Pragmatic rules, 105–6
Praise, 142, 217, 241
Predictability-novelty dialectic, 204
Preference for online social interaction, 37
Prejudice, 95
Preparation, for speech, 306, 387, 388
Presence, 14, 143
Presentation (speech), 307–8
Presentations, organizing, 335
Presentation software, 360–61, 363
Presenting self, 67–68
Prezi, 360, 361
Print media, 26
Priorities, 83, 185
Privacy, 45, 173, 188, 204
Problem(s):
  analyzing, 284–85
  communication limits in solving, 19
  demonstrating, 380–81
  describing, 229–30, 380
  identifying, 229, 284
Problem census, 278, 279
Problem-solution patterns, 328
Problem solving:
  behavioral descriptions for, 118
  metacommunication for, 187
  negotiation strategies for, 225–31
  and supportive listening by men, 143
  See also Group problem solving
Procedural norms, 273
Processes, 6, 350
Productivity, multitasking and, 134
  (*See also* Group productivity)
Profanity, 106–7
Professional environments, 256–60
  communication mistakes in, 256–57
  follower communication in, 257–59
  online communication in, 259–60
Professional identity, 242
Professional network, 243–44
Profita, Mike, 241
Progress, celebrating, 276
Promotions, 129
Prompting, 139
Pronouns, 81, 88, 102, 110
Proof, 334
Propaganda, 15

Propositions, 376–77
Propositions of fact, 376
Propositions of policy, 376–77
Propositions of value, 376
ProQuest, 302
Prosocial behaviors, 233
Proxemics, 171–73
Proximity, for relational partners, 184
Pseudoaccommodators, 222
Pseudolistening, 135
Psychological noise, 9, 133
Public awareness, 240
Public communication, 293–317, 319–45
    analyzing the situation, 297–301
    choosing the topic, 295
    communication apprehension, 304–7
    conclusion, 332–33
    defined, 12
    defining the purpose, 295–96
    delivery guidelines, 308–12
    gathering information for, 301–3
    introduction, 330–32
    outlining, 324–30
    presenting, 307–8
    sample speeches, 312–15, 340–44,
        363–67, 389–94
    stating the thesis, 296
    structuring, 320–24
    supporting material for, 334–39
    transitions in, 333–34
    See also Informative speaking; Persuasive
        speaking
Public distance, 172
Punctuation, 41–42, 259
Purpose:
    audience, 297
    of informative speaking, 350–52
    of persuasive speaking, 351
    of public communication, 295–96
Purpose statement, 295–96, 351–52, 379

Qualifications, 248, 251
Quality time, 201–3
Queer, reappropriation of, 106
Question(s):
    in affective communication style, 122
    asking, in job interviews, 251
    for audience, 330
    in conflict resolution, 230
    counterfeit, 145
    in group problem solving, 284
    leading, 145
    perception checking by asking, 65
    selection interview, 247–51
    statements vs., 113
    in supportive listening, 138, 139
    tag, 112, 311
    in task-oriented listening, 145
Question-and-answer period, 358

Quiet team members, in meetings, 277
Quinones-Fontanez, Lisa, 27
Quinones-Fontanez, Norrin, 27
Quotations, 330, 338–39

Race, 7, 61, 87
Radio, 14, 15
Ranking, of solutions, 285
Rate, speech, 92, 310–11
Realistic purpose statements, 296
Reappropriation, 106
Reasoning, examining, 149
Reassurance, 143
Receivers, 8, 9, 306
Receiving, 130
Recognition, 216, 217
Recommendation, letters of, 241
Recordkeeping, for job application
        process, 247
Reddit, 144
Red herrings, 384
Redirection, 248
Reductio ad absurdum fallacy, 383
Referent power, 255, 256
Referrals, seeking, 243, 244
Reflected appraisal, 54
Reflecting, 140–41
Reflective thinking method, 283–87
Reframing, 249
Refusal, to answer illegal questions, 248
Regional differences, 87
Regionalisms, 115
Regulating, 163
Reich, Robert, 299
Reinforcement (performing) stage of
        groups, 283
Rejection, latitude of, 373–74
Relational dialectics, 203–5
Relational goals, 266–69
Relational intimacy, 30, 37
Relational listening, 137–39
Relational maintenance, 197, 198
Relational messages, 185–86
Relational nature:
    of communication, 6, 15
    of communication competence, 16
    of nonverbal communication,
        160–61
Relational partners, choosing, 183–85
Relational roles, 61
Relational satisfaction, 58, 62, 171
Relational spirals, 218–19
Relationship(s):
    cohesiveness and building of, 276
    communication for management of, 5
    first impressions in, 62
    interpersonal communication in, 183
    interpretation based on, 106
    listening and, 129

mediated communication
        competence and, 43–45
    metacommunication for
        reinforcing, 187
    nonassertion in new or sensitive, 221
    references to, 330
    social media use and, 28, 34–38
    touch in, 171
Relative words, 114–15
Reliability, of visual aids, 363
Religion, 89
Remembering, 131, 132
Repetition, 92, 161, 356
Reply to All emails, 260
Reputation, 38, 241, 242, 257
Reputationdefender.com, 242
Requests, 118
Research, 247, 302–3
Reserve, 217
Residual message, 132
Resilience, 53
Resistance messages, 114
Resources, 205, 214, 281, 286
Respect, 43, 45, 186, 257, 279
Responding:
    advising responses, 142, 143
    describing desired response, 381
    by giving comfort, 142–43
    to illegal questions, 248, 249
    to interview questions, 251
    judging responses, 141–42
    in listening process, 130–31
    perception checking before, 65
Responsibility, 113
Restatement, 145–46
Restraining forces, 285
Restraint, 217
Restriction of topics, 217
Result orientation, 295
Résumés, 244, 245, 249
Reward power, 255, 256
Rewards, 184–85, 381–82
Rhetoric, 15
Rhetoric (Aristotle), 15, 378
Rhetorical Triad, 378–79
Richards, I. A., 103
Richness, message, 29–32, 158
Right to be forgotten, 33
Riley, Connor, 12
Risk level, nonassertion and, 221
Robots, proxemics for, 173
Rodriguez, George, 388
Rodriguez, Gina, 79
Role(s):
    in groups, 273–75
    relational, 61
    social, 53, 274
Role fixation, 274
Rolison, Ryan, 69

Romantic relationships, 121, 197–207
  ending, 200
  gender and intimacy styles, 201–3
  lies and evasions in, 205–7
  listening in, 129
  love languages, 201
  stages of, 197–200
  touch in, 171
Roque, Rocky, 294, 295, 312–14
Round robins, 278, 279, 285
Rule(s):
  group, 273
  of language, 104–7
Rule of seven, 362
Rumors, 254

Safety, on social media, 46–47
Salary, interview questions about, 251
Sales and marketing, 172
Sales plan development, 388
Salience, 79–81
Same-sex couples, 95, 203 (*See also*
    LGBTQ individuals)
Samovar, Larry A., 352, 383
Samson, Jaclyn, 80, 86
Sanders, Bernie, 120
Sanfilippo, Barbara, 388
Sapir–Whorf hypothesis, 103
Sarcasm, 168
Schulman, Nev, 38
Schulz, Kathryn, 357, 358
Schwabel, Dan, 61
Scripts, 71
Search engines, 305
Selection, in perception process, 58
Selection interviews (*See*
    Job interviews)
Selective listening, 135
Self-acceptance, 69
Self-assessment:
  Are You Overloaded?, 349
  How Assertive Are You?, 226–27
  How Do You Use Language?, 123
  How Emotionally Intelligent
    Are You?, 66
  How Good a Follower Are You?, 258
  How Much Do You Know About Other
    Cultures?, 96
  How Worldly Are Your Nonverbal
    Communication Skills?, 162
  Main Points and Subpoints, 326
  Persuasive Speech, 387
  Speech Anxiety Symptoms, 307
  What Are Your Listening Strengths?, 150
  What Is Your love Language?, 202
  What Kind of Friendship
    Do You Have?, 194
  What's the Forecast for Your
    Communication Climate?, 219

What's Your Leadership Style?, 270
What Type of Communicator
  Are You?, 18
What Type of Social Media
  Communicator Are You?, 31
Self-awareness, 65
Self-concept, 52–57
  biology, personality and, 53–54
  culture and, 55–56, 83
  defined, 52–53
  external influences on, 54–55
  and interpretation, 106
  and perceived self, 67
  and self-fulfilling prophecies, 56–57
Self-deprecating humor, 68
Self-disclosure:
  defined, 187
  in friendships, 192
  and identity management, 72
  intercultural competence and, 97
  models of, 187–89
  questions to ask before, 189
  to relational partners, 184
  and self-esteem, 68
  via mediated communication, 30, 46
Self-esteem, 37, 53, 68, 69
Self-fulfilling prophecy, 56–57, 306
Selfies, 177
Self-image, 177
Selflessness, 221
Self-monitoring, 17, 94
Self-promotion, 82
Self-protection, on social media, 45–47
Self-regulation, 67
Self-serving bias, 63
Self-serving lies, 206–7
Self-talk, 10–11
Semantics and semantic rules, 105, 130
Senders, 8, 9
Sensitive information, 260
Servant leadership, 266
Service, acts of, 201
Sesame Street (television program), 79
7 Habits of Highly Effective People, The
  (Covey), 128–29
Sex (act), 114, 203
Sex (biological), 60, 61
Sexism, 60
Sexting, 45–46
Sexual harassment, 223, 294
Sexual orientation, 88–89
Shared goals, 276
Shenoy, Rupa, 79
Shoneye, Talani, 7
Shortening of interaction, 217
Short-lived conflicts, 221
Short-term friends, 192
Shyness, 34, 54
Siblings, 196–97

Significance, of self-disclosure, 188
Sign language, 103, 157, 158
Signposts, 356
Silence, 20, 86
Similarity, 64, 184
Similes, 338
Simplicity, 20, 353–54, 362
Sincerity, 161, 310
Situational factors:
  in communication competence, 16
  in relational and supportive listening,
    137–38
Situational leadership, 266–69
Six degrees of separation hypothesis, 243
Size:
  of audience, 27, 298
  of groups, 271, 272, 276
  of visual aids, 362
Skype, 35
Slackers (*See* Social loafing)
Slang, 115
Slurring, 311, 312
Small groups, 11, 271–72, 276 (*See also*
    Group(s))
Small talk, 197
Smartphones (*See* Cell phones)
Smiling, 159, 167, 169
Smith, Angela, 249
Snapchat, 27, 33
Snap judgments, 62
Social comparison, 54–55
Social connection, 4–5
Social desirability, 205
Social distance, 172
Social exchange theory, 184–85
Social isolation, 37
Socialization, 122, 232–33, 253
Social judgment theory, 373–74
Social loafing, 273
Social media, 25–49
  benefits of, 33–36
  characteristics of, 27
  communication competence with,
    43–47
  communication via, 12–13
  courteous and wise use of, 43
  defined, 27
  drawbacks of, 36–40
  face-to-face communication vs., 29–33
  identity management on, 68–69,
    241–42
  influences on, 40–42
  interpersonal communication via, 190
  limiting distractions related to, 134, 135
  limiting time on, 14, 46
  and listening, 131
  personal attacks on, 120
  political discussions on, 89–90
  role of mass media and, 26–29

social comparison on, 54
types of content, 28
uses of, 27–28
venting on, 257
See also Mediated communication
Social media bots, 90
Social media listening, 131
Social media trolls, 90
Social needs, 353
Social norms, 67–68, 72, 273
Social penetration model, 187–88
Social roles, 53, 274
Social skills, 37, 67
Social status, 109
Social support, 35
Socioeconomic status, 92
Sociograms, 253–54
Software, presentation, 360–61, 363
Solution(s):
    advantages of, 381
    demonstrating, 381–82
    describing, 381
    developing, 285
    evaluating, 286
    following up on, 287
    problems with many possible, 285
Sound bites, 359
Space, 171–75
Space patterns, 327
Specific purpose, 295–96
Speech acts, 105
Speech anxiety (*See* Communication apprehension)
Speeches (*See* Public communication)
Speech rubric, 323
Spelling, 259
Stage fright (*See* Communication apprehension)
Stage hogs, 136, 137
Stagnating stage of relationships, 199
Stalking, via social media, 39–40
"Start over" button, for social media, 33
Star Wars: The Last Jedi (film), 13
Statistics, 337, 359
Statue of Liberty, 376
Status, 173–74, 177, 278
Stereotypes:
    abstract language and, 116
    avoiding, 63
    and conflict resolution, 234
    contact to break down, 94
    culture-based, 79, 94
    defined, 60
    gender-related, 40, 120–22, 176–77
    and perception, 55, 60–61
    and self-concept, 54
Steves, Rick, 95
Stimuli, intensity of, 58
Stonewalling, 216, 217

Storming stage of groups, 283
Straw man arguments, 384
Strengths, personal, 242, 251
Structure (speech), 320–24
    basic, 320
    of informative speeches, 354
    of persuasive speeches, 380–82
Stuart-Ulin, Chloe Rose, 26
Subpoints of speech, 324–26
Substituting, 161
Substitution, 311
Success criteria, 284
Succinct answers, to interview questions, 250
Superficial relationships, 36
Supporting forces, 285
Supporting material, 334–39
    emphasizing important points with, 356
    functions of, 334–36
    generating audience involvement in, 356–58
    for informative speeches, 355–63
    styles of, 339
    types of, 336–39
Supportive listening, 137–44
    advising responses in, 142, 143
    analyzing in, 142–43
    asking questions in, 138
    comfort and, 142–43
    encouraging further comments, 139
    gender differences in, 143–44
    judging responses in, 141–42
    and passive narcissists, 136
    reflecting speaker's thoughts, 140–41
    situational factors, 137–38
    time for, 138
    unexpressed thoughts and feelings, 138–39
Supportive siblings, 196
Surveillance, online, 39
Survey research, 303
Sweetman, Kate, 85
Symbols and symbolism, 6–7, 103
Sympathy, 64
Synchronous communication, 32
Syntax and syntactic rules, 104–5, 130

Tag questions, 112, 311
Talkative team members, in meetings, 277
Talking, 86, 121, 136
Tan, Tiffany, 57
Tannen, Deborah, 143, 144
Target audience, 385
Task assignments, in meetings, 277
Task goals, 266–69, 286
Task norms, 273
Task-oriented listening, 144–47
Task roles, 274
Tattoos, 170–71

Taylor, Dalmas, 187
Team leaders, 267, 269
Team members, difficult, 275
Teams, 271–72, 283 (*See also* Group(s))
Teamwork, 240, 271–72
Technical quality, problems with high degree of, 285
Technology (*See* Communication technology)
TED Talks, 331
Telegraph, 14
Telephone communication, 14
Television, 14, 15
Tell Me Who You Are (Guo and Vulchi), 78
Terminal credibility, 386
Terminating stage of relationships, 199, 200
Territoriality, 173
Testimony, 339
Testosterone, 122
Text messaging, 30, 32, 36, 104–5, 159, 200
Thank-you notes, after interviews, 251, 252
Thesis statement, 296, 352
They/them, as singular pronouns, 88
Thoughts, unexpressed, 138–39
Tian, Siew, 85
Time:
    factors influencing use of, 173–74
    group formation over, 272
    listening and, 138
    mediated vs. face-to-face communication and, 46, 190
    for multicultural team interactions, 282
    quality, 201–3
    as scarce resource, 214
    for speech, 301
Time patterns, 326–27
Timing, of discussions about conflict, 229
Tinder, 168
Ting, 132
Toastmasters International, 336
Tone, of speech, 159, 331–32
Topic, 217, 277, 295, 332, 351
Topic patterns, 327–28
Touch, 171, 172, 201
Training, on mediated communication, 259
Trait theories of leadership, 266
Tran, Kelly, 13
Transactional communication model, 9–10, 131
Transactional model of communication, 374
Transformational leaders, 269–71
Transgender individuals, 54, 55, 88, 195 (*See also* LGBTQ individuals)
Transitions, 333–34
Transmediation, 32

Transparency, 269
Triangle of meaning, 103
Triple delivery, 363
Trivial tyrannizers, 222
Trolling, 44
Trophy hunting, 69
Trump, Donald, 117, 119, 120
Trust, 267, 288
Truth, Sojourner, 329
Truth bias, 165
Tucker, Anna Claire, 332
Tufte, Edward R., 363
Turkle, Sherry, 36, 218
Turner, Sophie, 232
Turn-taking cues, 163
Twain, Mark, 338
Twitch, 28, 29
Twitter, 14, 29, 69, 106

Uncensored (Wood), 147
Uncertainty avoidance, 84
Unconscious identity management, 70–71
Understanding, 19, 45, 129, 130, 230
Undivided attention, 43
Unexpressed thoughts and feelings, 138–39
Universal Grammar, 104
Unknown area (Johari window), 188
Unmet needs, 229–30
Uptalk, 175
Ury, William, 228
User-generated content, 27
Uses and gratifications theory, 27–28

Validation, 190
Values:
    of audience members, 298–300
    cultural, 80–86
    defined, 299
    differences in, 95
    language as a reflection of, 111–13
    language in shaping of, 107–11
    organizational, 253
    propositions of, 376
    and self in culture, 83

Vass, Nathan, 4, 7, 10, 18–19, 21
Venn diagram, 359
Verbal communication:
    accenting, 163
    complementing, 161, 163
    contradicting, 163
    nonverbal communication
        compared with, 157
    regulating, 163
    repeating, 161
    substituting nonverbal
        communication for, 161
Video conferencing, 14, 252, 260, 287, 288
Videos, as visual aids, 359
Video sharing services, 28
View from Nathan's Bus, The (blog), 4
Virtual communities or groups, 34, 287–88
Visual aids, 359–63
Visualization, 306
Vocabulary, simplified, 92
Vocal citations, 355–56
Vocal fry, 175, 176
Voice, 163, 168, 169
Volume, voice, 92, 310
Volunteers, 358
Voorpret, 108
Vulchi, Priya, 78
Vulnerability, 187, 200, 229

Walk, Hunter, 33
Walsh, Lauren, 69
Walther, Joseph, 30
Wansink, Brian, 167
Warning signs, groupthink, 278
Wave pattern, 329
Websites, evaluating, 302
Webster, Daniel, 310
Weinstein, Bob, 268
Wheel networks, 254
Whiteboards, 361, 362
"White Is Purity" campaign, 10
"Why Videos Go Viral" (Allocca), 354–55
Wideman, Stephanie, 336
Wikipedia, 302, 337
Williams, Reagan, 336

Win-lose problem solving, 225–27
Win-win problem solving, 228–31
Withdrawal, after illegal questions, 248
Withholders, 222
Witzelsucht, 108
Wood, Zachary R., 147
Word charts, 359–60
Word choice, 121
Work environment, use of space in, 173, 174
Working outlines, 321
"Working to Document the Common Neighbor" (Roque), 312–15
Workplace communication, 239–61
    communication skills and success at, 12
    constructive dialogues, 280
    for delegation, 268
    first impressions in, 62
    humblebragging in job interviews, 73
    identity management in, 72
    for job applicants, 244–52
    LinkedIn use, 42
    multitasking and, 134
    names and, 110
    organizational communication factors, 252–56
    organizational culture and, 93
    organizing business presentations, 335
    persuasion skills in sales, 388
    power distance and culture in, 85
    presentation software for, 363
    in professional environment, 256–60
    setting stage for career success, 241–44
    sexual harassment, 223
    touch in, 172
    vocal cues and, 169
Worldview, 108
Wraparounds, 359

Yocum, Rebecca, 380
"You" statements, 113
YouTube, 27, 28, 354

Zappos, 253
"Zero in on," 358